ATOMIC PHYSICS 16

ATOMIC PHYSICS 16

Sixteenth International Conference on Atomic Physics

Windsor, Ontario, Canada   August 1998

EDITORS
William E. Baylis
Gordon W. F. Drake
*University of Windsor,
Windsor, Ontario, Canada*

American Institute of Physics

AIP CONFERENCE
PROCEEDINGS 477

Woodbury, New York

Editors:

William E. Baylis and Gordon W. F. Drake
Department of Physics
University of Windsor
Windsor, Ontario N9B 3P4
CANADA

E-mail: baylis@uwindsor.ca
A36@uwindsor.ca

The articles on pp. 29–41, 87–99, and 144–153 were authored by U. S. Government employees and are not covered by the below mentioned copyright.

Authorization to photocopy items for internal or personal use, beyond the free copying permitted under the 1978 U.S. Copyright Law (see statement below), is granted by the American Institute of Physics for users registered with the Copyright Clearance Center (CCC) Transactional Reporting Service, provided that the base fee of $15.00 per copy is paid directly to CCC, 222 Rosewood Drive, Danvers, MA 01923. For those organizations that have been granted a photocopy license by CCC, a separate system of payment has been arranged. The fee code for users of the Transactional Reporting Service is: 1-56396-752-9/99/$15.00.

© 1999 American Institute of Physics

Individual readers of this volume and nonprofit libraries, acting for them, are permitted to make fair use of the material in it, such as copying an article for use in teaching or research. Permission is granted to quote from this volume in scientific work with the customary acknowledgment of the source. To reprint a figure, table, or other excerpt requires the consent of one of the original authors and notification to AIP. Republication or systematic or multiple reproduction of any material in this volume is permitted only under license from AIP. Address inquiries to Office of Rights and Permissions, 500 Sunnyside Boulevard, Woodbury, NY 11797-2999; phone: 516-576-2268; fax: 516-576-2499; e-mail: rights@aip.org.

L.C. Catalog Card No. 99-62377
ISBN 1-56396-752-9
ISSN 0094-243X
DOE CONF- 980844

Printed in the United States of America

Contents

Preface vii

Recent Improvements in Measurement of Parity Violation in Atoms 1
    C. E. Wieman

Nuclear Anapole Moment and Tests of the Standard Model 14
    V. V. Flambaum

High-Resolution, High-Accuracy Spectroscopy of Trapped Ions 29
    D. J. Berkeland, J. D. Miller, F. C. Cruz, B. C. Young, R. J. Rafac,
    X.-P. Huang, W. M. Itano, J. C. Bergquist, and D. J. Wineland

Precision Spectroscopy of Helium 42
    P. Cancio, M. Artoni, G. Giusfredi, F. Minardi, F. S. Pavone,
    and M. Inguscio

Collective Collapse of a Bose-Einstein Condensate with Attractive Interactions 58
    C. A. Sackett, J. M. Gerton, M. Welling, and R. G. Hulet

Interactions in Trapped Bose-Einstein Condensates 74
    D. A. W. Hutchinson, R. J. Dodd, N. P. Proukakis, S. A. Morgan,
    S. Choi, M. Rusch, and K. Burnett

Atomic Ion Crystals in Non-Neutral Plasmas 87
    J. J. Bollinger, T. B. Mitchell, X.-P. Huang, W. M. Itano, J. N. Tan,
    B. M. Jelenković, and D. J. Wineland

Novel Dipole-Force Atom Traps 100
    I. Manek, U. Moslener, Yu. B. Ovchinnikov, P. Rosenbusch,
    A. I. Sidorov, G. Wasik, M. Zielonkowski, and R. Grimm

Single Atoms in a MOT 118
    D. Meschede, B. Ueberholz, V. Gomer, S. Knappe, U. Reiter,
    H. Schadwinkel, and F. Strauch

Atomic Collisions at Sub-Microkelvin Temperatures 132
    D. J. Heinzen

Ultracold Collisions: Exploring the Quantum Threshold Regime 144
    P. S. Julienne, E. Tiesinga, P. Leo, and C. J. Williams

BEC: The Alkali Gases from the Perspective of Research on Liquid Helium 154
    A. J. Leggett

Quantum Communication and Computation 170
    H. Briegel, W. Dür, S. Van Enk, J. I. Cirac, and P. Zoller

Laser Manipulation and Cavity QED with Trapped Ions 179
    H. Walther

Dynamics of Autoionizing Rydberg Atoms in an Electric Field 197
    L. D. Noordam and F. Robicheaux

Atoms and Cavities: Explorations of Quantum Entanglement 209
    J. M. Raimond, E. Hagley, X. Maître, G. Nogues, C. Wunderlich,
    M. Brune, and S. Haroche

*Italicized name indicates author who presented the paper.

Atom Interferometers and Atom Holography 223
    F. Shimizu, J. Fujita, M. Morinaga, T. Kishimoto, and S. Mitake

Alignment and Orientation: Opening Remarks 234
    J. H. Macek

Propensity Rules for Orientation and Alignment in Atomic Collisions 237
    N. Andersen

Polarization, Alignment, and Orientation in Electron-Atom Collisions: Benchmarks for Atomic Collision Theory 254
    K. Bartschat

Correlation Studies of Inelastic Electron Scattering from Simple Atoms: Recent Advances 278
    A. Crowe

Measured Correlated Motion in the Continuum of Three Coulomb-Interacting Particles, H^+, H^-, H^+ 296
    D. H. Jaecks, L. M. Wiese, and O. Yenen

Precision Measurements of Atomic Polarizabilities 305
    W. A. van Wijngaarden

Mass Measurements Far From Stability: Modern Approaches 322
    G. Bollen

The Muon $g-2$ Experiment at Brookhaven 334
    F. J. M. Farley, H. N. Brown, G. Bunce, R. M. Carey, P. Cushman,
    G. T. Danby, P. T. Debevec, H. Deng, W. Deninger, S. K. Dhawan,
    V. P. Druzhinin, L. Duong, W. Earle, E. Efstathiadis, G. V. Fedotovich,
    S. Giron, F. Gray, M. Grosse Perdekamp, A. Grossmann, U. Haeberlen,
    M. Hare, E. S. Hazen, D. W. Hertzog, *V. W. Hughes*, M. Iwasaki,
    K. Jungmann, D. Kawall, M. Kawamura, B. I. Khazin, J. Kindem,
    F. Krienen, I. Kronkvist, R. Larsen, Y. Y. Lee, W. Liu, I. Logashenko,
    R. McNabb, W. Meng, J.-L. Mi, J. P. Miller, W. M. Morse,
    C. J. G. Onderwater, Y. Orlov, C. Pai, C. Polly, J. Pretz, R. Prigl,
    G. zu Putlitz, S. I. Redin, O. Rind, B. L. Roberts, N. Ryskulov, R. Sanders,
    S. Sedykh, Y. K. Semertzidis S. Serednyakov, Yu. M. Shatunov, E. Solodov,
    M. Sossong, A. Steinmetz, L. R. Sulak, C. Timmermans, A. Trofimov,
    D. Urner, D. Warburton, D. Winn, Q. Xu, A. Yamamoto, and D. Zimmerman

Rhenium-187 and the Age of the Galaxy 344
    F. Bosch

X-Ray Emission from Comets 361
    K. Dennerl

Ultrafast Structural Studies on Biological Molecules by X-Rays 377
    J. Hajdu, R. Neutze, R. Wouts, and D. van der Spoel

Atomic Electron Correlations in Intense Laser Fields 386
    L. F. DiMauro, B. Sheehy, B. Walker, P. A. Agostini, and K. C. Kulander

Recollisions and High-Harmonic Generation 400
    P. L. Knight, A. Patel, M. Protopapas, N. J. Kylstra, D. G. Lappas,
    K. Burnett, A. Sanpera, S. Shaw, and J. Watson

Author Index 411

*Italicized name indicates author who presented the paper.

PREFACE

The Sixteenth International Conference on Atomic Physics (ICAP XVI) was held at the University of Windsor in Windsor, Canada, August 3 through August 7, 1998. This biennial meeting has stood for many years as the single most important conference world wide for the presentation of fundamental advances in atomic physics, high precision measurements of critical importance, and key applications to other fields of science. The field of atomic physics deals broadly with the structure and properties of atoms, and their interactions with electromagnetic fields. Its origins in nineteenth century spectroscopy led directly to the development of quantum mechanics in the 1920s, the Dirac equation, and early tests of the special theory of relativity. The measurement of the Lamb shift in atomic hydrogen in 1947 stimulated the development of modern quantum electrodynamics, and atomic physics measurements continue to be of crucial importance in testing this most successful of theories.

As the field has matured over the past 70 years, it has left in its wake a remarkable trail of technological advances such as magnetic resonance imaging (from nuclear magnetic resonance), atomic beam techniques, lasers and their many varied applications, standards of time and frequency, and measurements of the fundamental constants of nature. It also provides essential input data for the modeling of plasmas and the earth's atmosphere, as well as astrophysical applications. Atomic physics provides the fundamental testing ground for theories of many-body systems, including their sometimes chaotic behavior at the quantum/classical interface. In recent years, the field has been constantly revitalized by the relentless advance in the power and versatility of lasers for high precision measurements, and for the study of nonlinear interactions between matter and high intensity electromagnetic fields. The use of lasers for the cooling and trapping of atoms, leading eventually to Bose-Einstein condensation, has had a particularly profound impact, as recognized by the award of the 1997 Nobel Prize in Physics to Steve Chu, Claude Cohen-Tannoudji, and Bill Phillips. Ongoing studies of the unique quantum properties of Bose-Einstein condensates will undoubtedly lead to further dramatic advances, such as the creation of an atom laser.

The conference program was organized into plenary sessions for the 40 invited talks (including five in a *Hot Topics* session), each authored by a leading expert in the field. We are particularly grateful to Steve Chu, Claude Cohen-Tannoudji, and Bill Phillips for their participation as speakers in a special Nobel Prize session. A unique feature of the 1998 conference was a special session on the applications of atomic physics to other areas of science and technology. This volume presents the written texts submitted by most of the speakers. These proceedings provide a valuable resource of authoritative review articles covering the areas of atomic physics where the most significant advances are being made, and their likely directions for future development.

The chair of the Local Organizing Committee for the 1998 conference was Gordon W. F. Drake of the University of Windsor. His work would not have been possible without the able assistance of Tim Rausch to handle finances, Americo Buzzeo to oversee technical arrangements, and the conference secretaries Petrona Parungo and Nancy Sadler. The program of 12 regular plenary sessions for the invited talks was determined by the International Program Committee: K. Burnett, J. Dalibard, G. Drake, T. F. Gallagher, M. Gavrila, T. W. Hänsch, J.-C. Kieffer, D. Kleppner, J. Kluge, J. W. McConkey (Chair), J. Walraven, and D. J. Wineland. In addition, there were 317 contributed papers assembled and organized into three poster sessions by W. E. Baylis, who also edited the book of Abstracts of Contributed Papers. Policy guidance and numerous suggestions relating to the scientific program were provided by the International Organizing Committee: E. Arimondo, V. I. Balykin, S. Chu, C. Cohen-Tannoudji, G. Drake, N. Fortson, T. W. Hänsch, S. Haroche, V. W. Hughes, M. Inguscio, D. Kleppner, K. Kulander, R. R. Lewis, I. Martinson, H. Narumi, E. Otten, P. G. H. Sandars, F. Shimizu, J. Walraven, H. Walther, C. E. Wieman, and P. Zoller.

There were approximately 380 registered participants coming from 32 countries. The two largest representations were 171 from the U.S. and 42 from Canada. The University of Windsor Conference Services provided housing and meals for many of the participants. Other conference activities were arranged by the members of the Local Organizing Committee: J. B. Atkinson (local transportation and conference outings), W. E. Baylis (contributed papers), M. Czajkowski (poster sessions), E. N. Glass (publicity), R. K. B. Helbing (registration and records), R. D. Kent (exhibitors), L. Krause (banquet and outings), R. Maev (printing and publicity), J. W. McConkey (program), M. Schlesinger (finance), and A. van Wijngaarden (printing and publicity). Special thanks are due to Mary Louise Drake and her committee members, Anne Atkinson, Hanni Hedgecock, Elena Maeva, and Maureen McConkey, for making the companions program a great success.

A long tradition of ICAP meetings has been to maintain a plenary structure for the conference with talks that are broadly accessible to the audience. Since this provides a valuable learning environment for graduate students, we particularly encouraged their participation. A total of twenty students received travel assistance to attend the meeting, thanks to financial support from the U.S. Department of Energy and the National Science Foundation. We are grateful to the Oak Ridge Institute of Science and Education, and to R. R. Lewis of the University of Michigan, for their help in applying for and administering these funds.

We are grateful for the sponsorship of the International Union of Pure and Applied Physics, the National Institute of Standards and Technology, the U.S. Department of Energy and the National Science Foundation. We are also grateful to the University of Windsor for the provision of conference facilities, and to the Pelee Island and Colio Wineries for their support of the conference receptions and banquet. Special thanks are due to the authors for the fine collection of reports contained herein. Finally we want to thank Jens Zorn of the University of Michigan for a magnificent composite photograph of a poster session in action.

W. E. Baylis and Gordon W. F. Drake
Windsor, Ont., Canada, January 1999

Recent Improvements in Measurement of Parity Violation in Atoms

Carl E. Wieman

*Department of Physics, University of Colorado, and
JILA, University of Colorado and National Institute of Standards and Technology
Boulder, CO 80309-0440*

Abstract. A review of recent measurements of parity violation in atoms is given. In the past few years several experiments that measure atomic parity violation by optical rotation have achieved fractional accuracies between one and a few percent. A recent measurement in atomic cesium using the Stark interference technique has achieved 0.35%. This experiment also provided the first measurement of a nuclear spin dependence to atomic parity violation. This dependence provides the first measurement of a nuclear anapole moment.

INTRODUCTION

The measurement of parity violation in atoms has been actively pursued for the past 25 years, and there has been a steady improvement in accuracy. In the past several years (which is "recent" in this field), a few experiments have achieved accuracies near, and in one case well below, the 1% level. These experiments are providing important information for the understanding of both elementary particle and nuclear physics.

Two interests that are connected with the source of the parity violation (PV) motivate this work. First, these experiments provide a precision test of the Standard Model of elementary particle physics that unifies the weak and electromagnetic interactions. Many extensions to this model have been proposed, and the atomic physics tests are uniquely sensitive to the new physics introduced by a number of these, particularly those involving extra Z bosons, leptoquarks, or compositeness to the fermions [1]. Second, there is a small component to atomic parity violation that depends on the nuclear spin. The measurement of this component provides a unique and valuable probe of parity violation in purely hadronic nuclear interactions, for which there are few if any good probes. As discussed in the paper by Flambaum [2], this component is called the "anapole moment" contribution.

This paper will briefly review the history and basic concepts utilized in PV measurements in atoms, and will then discuss recent experiments, with emphasis on the measurement in cesium. It will conclude with a brief examination of the future directions. A discussion of the implications of these results for elementary particle and nuclear physics will be left to the paper by Flambaum.

History of Parity Violation in Atoms

The earliest relevant work on parity violation in atoms was done by Michels in the 1960s [3]. He proposed this as a means to look for the existence of hypothetical PV neutral current interactions, pointed out how these interactions would lead to parity mixed states in atoms, and offered a few proposals for possible experiments. Unfortunately, at that time no one thought neutral currents existed and so his work was largely ignored, and led to no experiments.

The field really began with the development of the electroweak unification theory of Weinberg, Salaam, and Glashow that predicted PV neutral currents, and the subsequent papers of M.-A. and C. Bouchiat that applied that theory to atoms [4]. These seminal papers discussed in detail how this interaction would be manifested in atoms, pointed out that the effects would scale as the cube of the atomic number, Z^3, and suggested that the effects should be observable for large Z. They also discussed a number of possible experimental approaches. This led to a number of experiments. The late '70s saw several results on the PV optical rotation in bismuth vapor that did not agree, and several calculations for the expected size of the effect that showed similar variations. During the early to mid '80s, optical rotation experiments of greater consistency [5] and measurements using Stark interference in thallium [6] and cesium [7] were obtained. Parity violation effects were convincingly detected at the few, and then many sigma levels. By the end of the '80s, the best measurement had obtained a 2% fractional uncertainty [8], and the best atomic structure calculations needed to interpret the results were good to about 1% [9]. At this level, atomic PV became an important test of the Standard Model, and provided the tightest constraints on many possible extensions [10]. The '90s have now seen three measurements at the 1–2% level, and one very recent result of 0.35%.

Basic Concepts of Parity Violation in Atoms

Although the details are complicated, the basic concepts behind PV in atoms, and its measurement, are quite simple [3,4]. The neutral current interaction arising from the exchange of a Z boson between the electrons and the nucleons in an atom leads to a mixing of the S and P states in the atoms. The amount of mixing scales roughly as Z^3, but even for large Z it is only about 1 part in 10^{11}. This mixing is detected by observing the electric dipole transition amplitude between two S (or two P) states in an atom. Such amplitude can only exist if parity is violated. Because the amplitude is so small it can only be observed by using interference

New Optical Rotation Results - Bi, Pb, Tl

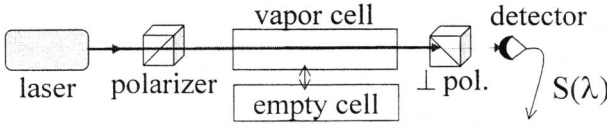

FIGURE 1. Schematic of optical rotation apparatus.

techniques. In such a technique the nS to $n'S$ (or P) transition rate is given by

$$R = |A_0 \pm A_{\rm pv}|^2 = A_0^2 \pm 2A_0 A_{\rm pv} + A_{\rm pv}^2, \qquad (1)$$

where $A_{\rm pv}$ is the desired PV amplitude and A_0 is a larger parity-conserving amplitude. Because the interference term is linear in $A_{\rm pv}$ it can be large enough to measure; however, it must be distinguished from the large background due to the A_0^2 contribution. For there to be a nonzero interference term, the experiment must have a "handedness," and if the handedness is reversed, the interference term will change sign, and can thereby be distinguished.

There have been two choices used for A_0. The first is an allowed M1 transition between P states. All experiments that have used this amplitude have chosen to detect the PV by observing optical rotation of polarized light in an atomic vapor. The second choice for A_0 is a "Stark induced" E1 amplitude due to an applied DC electric field. This field mixes S and P states giving rise to an E1 amplitude that can interfere with $A_{\rm pv}$ under the proper conditions. This approach has been used in the first thallium experiments [6], and all the experiments in cesium [7,8,11]. As a crude generalization, the first approach involves a considerably simpler experimental apparatus, but has fewer reversals and hence tends to be more limited by systematic errors.

Recent Optical Rotation Experiments

Optical rotation experiments have now begun to approach 1% accuracy in Bi, Pb, and Tl. The basic experiment for all these cases is shown schematically in Fig. 1. A beam of linearly polarized laser light is sent through a vapor of the desired atom, and then the light is sent through a nearly crossed linear polarizer followed by a detector. When the laser is tuned to the P-P transition of interest, the PV interference causes the plane of polarization to rotate. This results in an increase in the amount of light reaching the detector. To distinguish the signal of interest from spurious effects in the optics, the vapor cell is interchanged with a

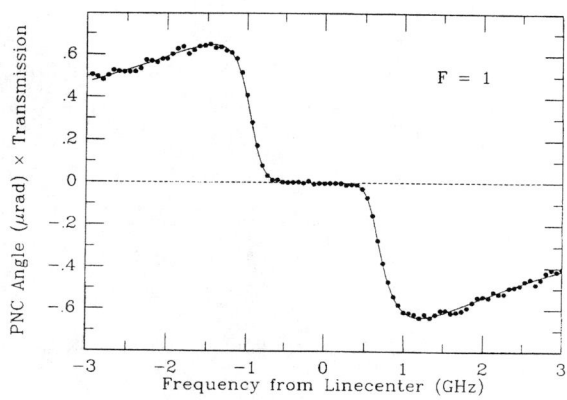

FIGURE 2. PNC data, from Vetter et al. (see Ref. 12).

TABLE 1. Results from recent optical rotation experiments.

Atom	Im($E1_{\text{PNC}}/M1$) $\times 10^8$	Reference
Bi	10.12(20)	Oxford '91 (Ref. 13)
Pb	9.86(12)	Seattle '93 (Ref. 14)
	(Bi, Pb theory $\sim$ 10%)	
Tl	15.68(45)	Oxford '95 (Ref. 15)
	14.68(17)	Seattle '95 (Ref. 12)
	(Tl theory 3%)	(Ref. 2)

"dummy" cell that has identical optics but no atomic vapor. Also, the wavelength of the laser is scanned over the transition, and the resulting line shape is fit to the known PNC (parity non-conserving) line shape. An example of the data obtained in this fashion is shown in Fig. 2, taken from Ref. [12].

It can be seen that the agreement between predicted and observed signals is excellent. Similar results are obtained with Pb and Bi. In Table 1 is a tabulation of recent precise measurements in these three atoms. Although the experimental accuracy ranges from 1.2 to 3%, in all cases the interpretation of the results in terms of the PV couplings of the Standard Model is limited by the uncertainties in the necessary atomic physics (see Flambaum paper [2] for further discussion). This uncertainty is about 10% for Pb and Bi, and 3% for Tl. In all these experiments researchers have searched for nuclear spin dependent contributions to the atomic PV by comparing the amount of rotation on different hyperfine transitions. No such contributions have been seen, with limits at the 1–2% level, depending on the experiment.

Although the basic experimental approach is the same as has been used since the mid '70s, there are several reasons for the improved accuracy. The primary one is having better lasers that provide good tuning over the wavelengths of interest. The second reason is improved detectors, and the third, which is highly dependent on the first two, is the use of atomic transitions that are more favorable, for reasons such as less molecular background. A last, and very important, reason is that the experimental groups are older and wiser. These same groups have now carried out several generations of experiments, and in the process have learned a great deal about possible sources of problems, and how to avoid them.

NEW PARITY VIOLATION EXPERIMENT AT JILA

Now I will switch topics to discuss the 1997 Cs PV measurement [16]. This work was carried out at JILA over a period of nearly 14 years by Wood, Bennett, Roberts, Cho, Masterson, Tanner, and myself. In common with all the previous measurements of PV in cesium carried out at ENS [7,17] and JILA [11,16], the Stark interference technique is used to detect the PV E1 amplitude on the 6S to 7S transition. In the JILA approach, this transition is excited in an atomic beam by a laser that intersects the beam at right angles. The directions of three perpendicular vectors, those of an applied electric field, an applied DC magnetic field, and the angular momentum of the laser photon, define the handedness of the apparatus. The PV interference term then changes sign, causing a modulation in the excitation rate as this coordinate system is reversed back and forth between right and left handed. There are four such reversals: changing sign of each of the E and B fields, reversing the laser light from left to right circular polarization, and reversing the sign of the m level that is being excited. This use of multiple redundant reversals greatly suppresses possible systematic errors.

It is hard to convey how difficult this and similar PV experiments are to those who have not worked on such projects. In attempt to provide some perspective on this, I would like to compare the efforts of the members of my group working on this PV experiment with the most similar "control group," namely, the other half of my research group. There were similar levels of effort involved in terms of the number and quality of people and how hard they worked, as well as the level of guidance (or misguidance as the case may be), but half of the people were working on the cesium PV experiment, while the other half worked on more typical sorts of atomic physics experiments involving laser cooling and trapping. Both had their start in 1983 when we began to develop diode laser technology to use for optical pumping and detection in the PV experiment. After a couple of years of work and initial results on optical pumping had been obtained, a parallel effort in using diode lasers for laser cooling was begun. While that effort was demonstrating nice results in low cost laser cooling, the PV work was leading to improvements in the laser technology that were necessary to achieve the required noise levels for optical pumping and detection. During the mid to late '80s, the PV experiment worked on low noise optical pumping, fluorescence detection, and the development of laser power buildup cavities. In the same period, the improved diode lasers were allowing the other side of the lab to explore developments in laser trapping and cooling, precision spectroscopy in ultracold atoms, pioneer experiments in ultracold collisional loss from traps, and discover the radiation repulsion limits on density. By 1990 the PV apparatus was giving initial data on the parity violation and starting serious study of PV systematic errors. Meanwhile, "the other side" had demonstrated the combination of laser cooling and trapping and magnetic trapping, and had started a focused effort to achieve Bose-Einstein condensation (BEC) using that approach. (There was also the notable addition of Eric Cornell to the group.) Over the subsequent five years the cooling side developed several new types of trapping and cooling techniques, measured a number of relevant low temperature cross sections, and finally achieved BEC. At that point the PV experiment was commencing its fifth year of studying and eliminating possible systematic errors! That painful task, and with it the PV experiment, was completed in late 1996, about the same time as the second generation BEC apparatus was completed. The one-half of the group had produced some 18 Physical Review Letter-level publications plus many others over this period from 1985–1997, while the other half, with a similar level of effort, was carrying out a single measurement of parity violation. This example provides some idea of how truly difficult and time consuming it is to carry out such an experiment.

The primary features of the improvements of this PV experiment over its predecessors are shown in Fig. 3. The 6S and 7S levels in cesium are each split into two hyperfine levels.

We drive transitions from a single m state of either the $F = 3$ ground state up to a single m state of the $F = 4$ excited state, or the ground 4 to excited 3 transition. We have improved the signal over the earlier experiments by optically pumping all of the atoms into the m state that we are exciting, rather than using the 1/16 of the atoms that normally populates a single m state. This optical pumping also allows us to more efficiently detect the 7S excitation using a technique similar to the "atom shelving" used in ion trapping. About half of the time the 7S atoms decay back down into the empty 6S F level. We can excite atoms in the "empty" F level on a cycling transition to the 6P state, and therefore scatter and detect many (200) 6P6S photons for each 6S–7S excitation. We have also substantially increased the laser power used to drive the 6S–7S transition over previous experiments, by improvements in the optical power buildup cavity. There are many other less obvious changes in the apparatus that primarily serve to reduce the systematic errors.

The apparatus is shown schematically in Fig. 4. An intense beam of atomic cesium passes through beams from two diode lasers that optically pump it into a single F, m state. It then continues on to intersect an intense green standing wave laser field between electric field plates. The standing wave is produced by a Fabry-Perot power buildup cavity that increases the incident laser power by a factor of 30,000 when the laser frequency is locked in resonance with the cavity. After

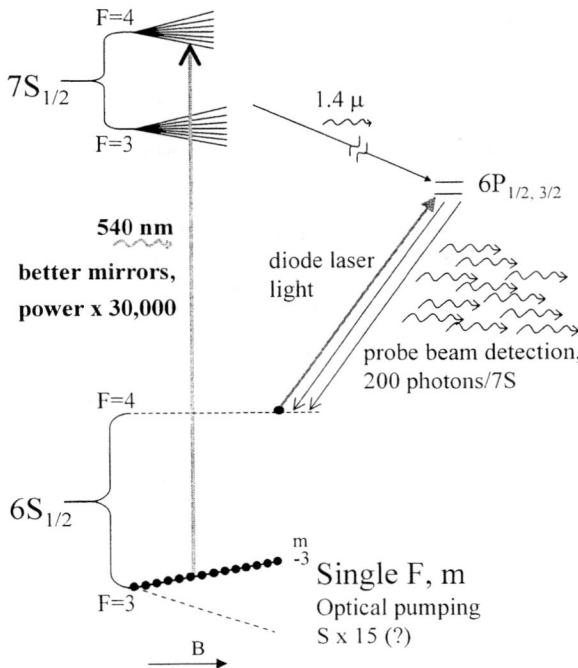

FIGURE 3. Cs energy level diagram illustrating transitions used in PV experiment.

the interaction region, the beam goes into the detection region where it intersects another diode laser beam that drives a cycling transition for the atoms in the previously emptied F level. The resulting fluorescence is detected on a photodiode. Not shown are the 23 coils required to produce the appropriate magnetic fields throughout the apparatus. There are a number of aspects of this apparatus that are rather challenging. It requires five lasers (four diode, one dye) that must be stabilized to nearly state-of-the-art performance. This means frequency stabilities on the order of 1 part in $10^{14}/\mathrm{s}^{1/2}$ and intensity stabilities of 1 part in $10^6/\mathrm{s}^{1/2}$. The power buildup cavity also involves some challenges, since the spacing between the mirrors must be held constant to about 1/100 of an atomic diameter per $\mathrm{s}^{1/2}$. Meeting any one of these requirements is not easy, but by far the most difficult aspect of the technology is that all these things must work at the same time, and they must do that for hours, days, and months in order to acquire the necessary data.

After achieving this necessary level of performance, we then look at the signal as we perform five parity reversals (E, B, polarization, and the two ways to reverse m in the optical pumping process). The quantity of primary interest is the fraction

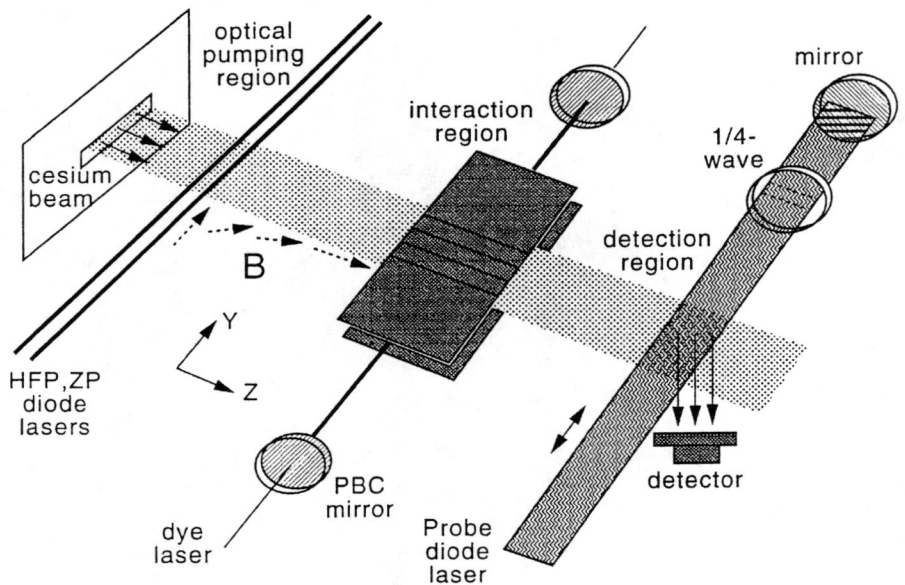

FIGURE 4. Schematic of JILA PV experiment.

(about 6 parts in 10^6) of the signal that modulates with all five reversals. However, there are many other modulation channels that we monitor that measure the modulation with different subsets of these five reversals. These channels provide a great deal information about the experiment, for example, the size of many stray (nonreversing) and misaligned field components. This information is essential for dealing with possible systematic errors.

The bulk of the time and effort spent on this experiment was devoted to the study and elimination of such systematic errors. This required about 7000 hours of data, or roughly 95% of data that was acquired. A full discussion of the systematic errors is too long and tedious to give here, so I will just outline the general approach that was used. First, we calculate all the possible ways that combinations of stray (nonreversing) and misaligned electric and magnetic fields (either oscillating or DC EM fields) can mimic the PV signal. This involves looking at all field components. One then finds a way to measure and control all the relevant field components to the necessary level. One of the painful aspects of this experiment was that it was necessary to consider all the possible combinations of gradients of these fields as well. The measurement and control of all these fields is done by appropriate construction of the apparatus, monitoring 32 modulation channels as mentioned earlier, and using 31 servosystems to stabilize everything about the experiment. Many of the modulation channels needed come directly out of the same data as used for the parity measurement, but there are several others that come from a set of auxiliary experiments that are interspersed with the acquisition of the PV data.

The second part of eliminating systematic errors is based on extensive statistical tests. These are chi squared tests of various sorts. These tests examine whether there are any variations in the results that can not be explained by the noise that is observed on the shortest possible time scales. The noise on these time scales is dominated by simple shot noise in the excitation rate. The data used in these tests encompass many different conditions for operation of the experiment, including varying the eccentricity of the polarization ellipse of the green light between 1 and 2, the electric field between 400 and 900 V/cm, rotating the buildup cavity mirrors, changing the laser power, redoing the optical alignments, etc. We have set the unusually stringent requirement that it is unacceptable to have any variation that is not consistent with the short time noise. Another way to say this is that the noise per $t^{1/2}$ that is obtained from 350 hours of data that is taken under all these different conditions is the same as the noise per root time obtained in looking at the detector output for a fraction of a second. These statistical tests that indicate consistency in the PV number under widely different conditions provides confidence that our analysis has not overlooked any systematic errors. We believe that the systematic errors are all below 0.1% of the PV signal, with the largest error being due to a product of a stray electric field gradient and a gradient of a misaligned component of the B field.

The final result for the two different hyperfine lines is:

$$\text{Im}(E1_\text{pv}/\beta) = \begin{cases} 1.6349(80)\,\text{mV/cm} & F = 4 \text{ to } F' = 3 \\ 1.5576(77)\,\text{mV/cm} & F = 3 \text{ to } F' = 4 \end{cases} \quad (2)$$

Since we are detecting the PV as a fractional modulation in an electric field induced rate, it is given in units of the equivalent electric field required to give the same mixing of S and P states and hence the same modulation in the signal.

In Fig. 5 we show a summary of all the measurements of PV in cesium. It can be seen that all the experiments are consistent, and that the first hint of a hyperfine dependence seen in our 1988 result is confirmed by the current work. In fact, the current results provide a measurement of this difference, 0.077 mV/cm, which has a fractional uncertainty of 14%. The primary difference between these two hyperfine lines is that the nuclear spin is reversed relative to the electron spin. Thus the observed difference provides a direct determination of the nuclear spin dependence of atomic PV. As discussed in the paper by Flambaum [2], this difference is primarily due to the nuclear anapole moment. This is the first measurement of an anapole moment, some forty years after Zeldovich first proposed its existence.

If we take an appropriately weighted average of the two lines, all of the nuclear dependent parts cancel out, and we are left with a quantity, 1.5935(56) mV/cm, that can be used to compare with the predictions of the standard model. As discussed by Flambaum, this comparison is limited by the atomic theory at the 1% level. At that level it provides the best current constraints on a variety of possible new physics such as extra Z bosons, leptoquarks, and composite fermions.

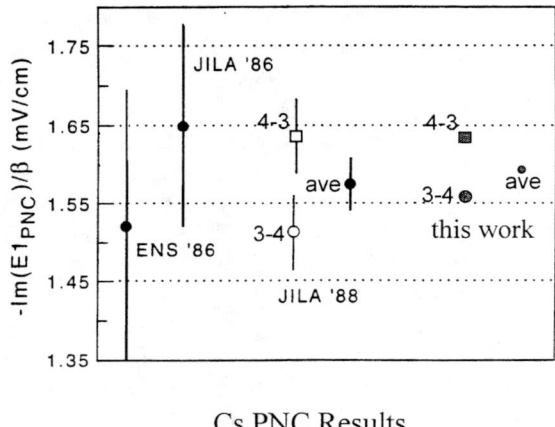

Cs PNC Results

FIGURE 5. Results of different measurements of PV in cesium. (The Paris '82 and '86 results have been combined in order to fit on the scale.)

CONCLUSIONS

Now let me turn to the ongoing and future work in atomic parity violation. There is still some room for improvement. The weak charge can be predicted by the Standard Model to about 1 part in 10^3 before the interpretation becomes clouded by small poorly known corrections. That level is nearly a factor of 10 better than the current determination. Measurements of the anapole moment would be worthwhile at almost any level, assuming they were for nuclei with not too messy structure. The first issue in improving the test of the Standard Model is the atomic theory. It can definitely be improved by difficult but straightforward calculations in Cs, and quite possibly also in Tl. When this will be done, and how large an improvement can be made, remain open questions. One key issue is that the best way to evaluate the accuracy of such calculations is to see how accurately they can calculate measured atomic properties, particularly lifetimes and oscillator strengths. Preliminary results from a recent effort at JILA to carry out such a program suggest that the actual uncertainty is probably about half of the 1% that has been used. However, the relevant level of interest for any further improvement in the theory is 0.1%, and it is very difficult to measure any atomic quantity except frequencies to such levels of precision. The obtaining of such precise atomic data is a clear challenge for experimental atomic physicists, and in many respects matches the challenge this problem poses for theorists.

On the experimental side, although there is unlikely to be any more PV measurements at JILA, there are several other groups around the world pursuing further improvements. A new experiment in cesium has been underway for many years in Paris [16]. This uses stimulated emission detection of the PV, and expects to reach a precision of 1%. Modest improvements in the Tl optical rotation experiment in Seattle are being sought using the technique of electromagnetically induced transparency [19]. The principal goal of this work is to reach a fraction of a percent where it is expected that the anapole moment contribution can be seen. A more ambitious project of the Seattle group is the measurement of PV in a single barium ion [20]. The atomic theory for this case should be comparable to the cesium case, and thus relatively tractable and accurate. However this experiment was started fairly recently, and judging from the history of this field, that means we should probably check back in about 2008 to see how it is progressing.

Measurements of PV in rare earth elements have been pursued by at least two groups. The appeal of these systems is that there should be nearly degenerate S and P states, offering the hope that the PV signals would be enormous. There is no hope for doing precise atomic theory for these atoms, but the hope is that the large signals will allow very precise measurements and hence precise determinations of isotope ratios that would reduce or eliminate the dependence on the atomic theory. Also anapole moments might be investigated. Unfortunately, up to now the expectations for seeing large PV signals have not been fulfilled. Samarium has been studied at Oxford [21] and dysprosium at Berkeley [22], and in both cases PV has not been detected at a level far below what was anticipated. Currently the Berkeley group is studying ytterbium [23] with the continued hope that it will still show large signals. On a somewhat longer time scale, there are plans to measure atomic PV in francium at Stonybrook [24] and TRIUMF and possibly also radioactive cesium isotopes at TRIUMF. Because all the atoms in question are relatively short lived radioactive isotopes, it is necessary to use optical trapping technology to collect the atoms and hold them for the experiment. The necessary trapping technology has been demonstrated [25,26]. The principle remaining hurdle is to develop the technology for carrying out PV experiments in traps. It is likely that this will be done, but it has not been demonstrated yet. One can anticipate further improvements in atomic PV measurements and their interpretation, but it is likely that these improvements will neither be rapid nor easy.

ACKNOWLEDGMENTS

I am happy to acknowledge the work of C. Wood, S. Bennett, J. Roberts, D. Cho, B. Masterson, and C. Tanner who carried out the JILA PV experiment. This work was supported by the National Science Foundation.

REFERENCES

1. *Precision Tests of the Standard Electroweak Model*, ed. P. Lanacker, World Scientific, Singapore, 1995.
2. Flambaum, V., this volume.
3. Curtis-Michel, F., *Phys. Rev.* **138B**, 408 (1965).
4. Bouchiat, M. A., and Bouchiat, C., *Phys. Lett.* **48B**, 111 (1974a); ibid., *J. Physique* **35**, 899 (1974b); and ibid. *J. Physique* **36**, 493 (1975).
5. Birich, G. N., Bogdanov, Yu V., Kanorskii, S. I., Sobel'man, I. I. Sorokin, V. N., Stuck I. I., and, Yukov, E. A., *Sov. Phys.-JETP* **60**, 442 (1984); Hollister, J. H., Apperson, G. R., Lewis, L. L., Emmons, T. P., Vold, T. G., and Fortson, E. N., *Phys. Rev. Lett.* **46**, 642 (1981).
6. Conti, R., Bucksbaum, P., Chu, S., Commins, E., and Hunter, L., *Phys Rev. Lett.* **42**, 343 (1979); Commins, E., Bucksbaum, P., and Hunter, L., *Phys. Rev. Lett.* **46**, 640 (1981); Bucksbaum, P. H., Commins, E. D., and Hunter, L. R., *Phys. Rev.* **D24**, 1134 (1981); Drell, P. S., and Commins, E. D., *Phys. Rev. Lett* **53**, 968 (1984).
7. Bouchiat, M. A. Guéna, J., Hunter, L., and Pottier, L., *Phys. Lett.* **117B**, 358 (1982).
8. Noecker, M. C., Masterson, B. P., and Wieman, C. E., *Phys. Rev. Lett.* **61**, 310 (1988).
9. Blundell, S. A., Johnson, W. R., and Sapirstein, J., *Phys. Rev. Lett.* **65**, 1411 (1990); Dzuba, V. A., Flambaum, V. V., and Sushkov, O. P., *Phys. Lett.* **141A**, 147 (1989).
10. Langacker, P., Luo, M.-X., and Mann, A., *Rev. Mod. Phys.* **64**, 87 (1992).
11. Gilbert, S. L., Noeker, M. C., Watts, R. N., and Wieman, C. E., *Phys. Rev. Lett.* **55**, 2680 (1985).
12. Vetter, P., Meekhof, D. M., Majumder, P. K., Lamoreauxs, S. K., and Fortson, E. N., *Phys. Rev. Lett.* **74**, 2658 (1995).
13. Macpherson, M. J. D., Zetie, K. P., Warrington, R. B., Stacey, D. N., and Hoare, J. P., *Phys. Rev. Lett.* **67**, 2784 (1991).
14. Meekhof, D. M., Vetter, P., Majumder, P. K., Lamoreaux, S. K., and Fortson, E. N., *Phys. Rev. Lett.* **71**, 3442 (1993).
15. Edwards, J. N. H., Phipp, S. J., Baird, P. E. G., and Nakayama, S., *Phys. Rev. Lett.* **74**, 2654 (1995).
16. Wood, C. S., Bennett, S. C., Cho, D. Masterson, B. P., Roberto, J. L., Tanner, C. E., and Wieman, C. E., *Science* **275**, 1759 (1997).
17. Bouchiat, M. A. Guéna, J., Hunter, L., and Pottier, L., *Phys. Lett* **134B**, 463 (1984); ibid., *J. Physique* **47**, 1709 (1986).
18. Bouchiat, M. A., Guéna, J., Jacquier, P., Lintz, M., and Pottier, L., *Opt. Commun.* **77**, 374 (1990); Bouchiat, J., Jacquier, P., Lintz, M., and Pottier, L., *Opt. Commun.* **56**, 100 (1985b).
19. Cronin, A. D., Warrington, R. B., Lamoreaux, S. K., and Fortson, E. N., *Phys. Rev. Lett.* **80**, 3719 (1998).
20. Fortson, E. N., *Phys. Rev. Lett.* **70**, 2383 (1993).
21. Lucas, D. M., Warrington, R. B., and Stacey, D. N., "A search for parity violation in atomic samarium," *Phys. Rev. A*, submitted.
22. Budker, D., DeMille, D., Commins, E. D., and Zolotorev, M., *Phys. Lett.* **50A**, 132

(1994); DeMille, D., Budker, D., and Commins, E. D., *Phys. Rev. A* **50**, 4657 (1994). Also see contributed papers, this conference.
23. DeMille, D., *Phys. Rev. Lett.* **74**, 4165 (1995). Also see contributed papers, this conference.
24. L. Orozco, private communication.
25. Simsarian, J. E., Ghosh, A., Gwinner, G., Orozco, L. A., Sprouse, G. D., and Voytas, P. A., *Phys. Rev. Lett.* **76**, 3522 (1996).
26. Stephens, M., and Wieman, C., *Phys. Rev. Lett.* **72**, 3787 (1994); Lu, Z.-T., Corwin, K. L., Vogel, K. R., Wieman, C. E., Dinneen, T. P., Maddi, J., and Gould, H., *Phys. Rev. Lett.* **79**, 994 (1997).

Nuclear Anapole Moment and Tests of the Standard Model

V. V. Flambaum

School of Physics, University of New South Wales, Sydney, 2052, Australia

Abstract. There are two sources of parity nonconservation (PNC) in atoms: the electron-nucleus weak interaction and the magnetic interaction of electrons with the nuclear anapole moment. A nuclear anapole moment has recently been observed. This is the first discovery of an electromagnetic moment violating fundamental symmetries—the anapole moment violates parity and charge-conjugation invariance.

We describe the anapole moment and how it can be produced. The anapole moment creates a circular magnetic field inside the nucleus. The interesting point is that measurements of the anapole allow one to study parity violation inside the *nucleus* through *atomic* experiments. We use the experimental result for the nuclear anapole moment of ^{133}Cs to find the strengths of the parity violating proton-nucleus and meson-nucleon forces.

Measurements of the weak charge characterizing the strength of the electron-nucleon weak interaction provide tests of the Standard Model and a way of searching for new physics beyond the Standard Model. Atomic experiments give limits on the extra Z-boson, leptoquarks, composite fermions, and radiative corrections produced by particles that are predicted by new theories.

The weak charge and nuclear anapole moment can be measured in the same experiment. The weak charge gives the mean value of the PNC effect while the anapole gives the difference of the PNC effects for the different hyperfine components of an electromagnetic transition. The interaction between atomic electrons and the nuclear anapole moment may be called the "PNC hyperfine interaction."

DESCRIPTION OF THE ANAPOLE MOMENT

The notion of the anapole moment was introduced by Zel'dovich [1] just after the discovery of parity violation. He noted that a particle may have a parity violating electromagnetic form factor, in addition to the usual electric and magnetic form factors. The first realistic example, the anapole moment of the nucleus, was considered in Ref. [2] and calculated in Ref. [3]. In these works it was also demonstrated that atomic and molecular experiments could detect anapole moments. Subsequently, a number of experiments were performed in Paris, Boulder, Oxford, and Seattle [4–6] and some limits on the magnitude of the anapole moment were established.

However, it was not until recently that a nuclear anapole moment was unambiguously detected — in 1997 a group in Boulder detected a nuclear anapole moment in ^{133}Cs (using atomic experiments) to an accuracy of 14% [7].

Multipole moments arise from expansions of the electrostatic and vector potentials as a series in R^{-1}, where R is the distance from the centre of the charge or current distribution. Examples of multipole moments which obey the three discrete symmetries of charge conjugation invariance, parity conservation and time reversal invariance (C, P and T) are the electric monopole (i.e., charge), magnetic dipole, electric quadrupole and magnetic octupole moments. There are also multipole moments that violate both parity conservation and time reversal invariance, i.e., they are both P- and T-odd. These moments are the magnetic monopole, electric dipole, magnetic quadrupole, electric octupole, and so on.

There are also other electromagnetic multipole moments, which are not usually dealt with in multipole moment expansions as they give rise to contact, rather than long-range, potentials. The anapole moment is such a moment. It obeys time reversal invariance but violates parity conservation and charge conjugation invariance (i.e., T-even and P- and C-odd). We will show how the anapole moment arises out of an expansion of the vector potential — the following is based on Ref. [8]. Consider the usual expression for the vector potential in terms of the current distribution, $\mathbf{j}(\mathbf{r})$:

$$\mathbf{A}(\mathbf{R}) = \int \frac{\mathbf{j}(\mathbf{r})}{|\mathbf{R}-\mathbf{r}|} \, d^3r. \tag{1}$$

When we expand this as a series in R^{-1} the first term that we get is the vector potential of the "normal" (i.e., T,P-even) magnetic dipole moment. The second-order term is

$$\mathbf{A}^{(2)}(\mathbf{R}) = \frac{1}{2} \int \mathbf{j}(\mathbf{r}) r_m r_n \, d^3r \, \partial_m \partial_n \frac{1}{R}. \tag{2}$$

Since $j_i r_m r_n$ is a reducible tensor (its trace is not zero) the vector potential contains two irreducible contributions, that of the T,P-odd magnetic quadrupole moment and that of the T-even, P-odd anapole moment. The vector potential due to the anapole moment is

$$\mathbf{A}^a(\mathbf{r}) = \mathbf{a}\delta(\mathbf{r}), \tag{3}$$

where

$$\mathbf{a} = -\pi \int r^2 \mathbf{j}(\mathbf{r}) \, d^3r \tag{4}$$

can be taken as the definition of the anapole moment. The $\delta(\mathbf{r})$ is a result of $\partial_m \partial_m (1/R) = -4\pi\delta(\mathbf{r})$. Note that even though the anapole moment vector arises in the second order of the vector-potential expansion, it is a rank-one tensor. Notice the contact form of the potential — this is true for any T-even, P-odd moment; see, e.g., [2,9] for a proof and references.

Now to see what an anapole moment actually looks like, we have to find what kind of a current distribution can give a nonzero value in Eq. (4). Let's begin by considering a current distribution comprising a current going around a single circular loop. A current circle whose centre is the origin would give a zero value in Eq. (4) since the value of r is constant for this circle and so all the currents from different parts of the circle will cancel each other out. It turns out that for a current circle to give a nonzero value of the anapole moment it must be displaced from the origin, with the normal to the plane of the circle being at right angles to its displacement (see Fig. 1). The current distribution in Fig. 1 will give an upward pointing anapole moment, as the side of the circle with the downward going current is further from the origin (and so has larger values of r^2) and so it dominates in Eq. (4); note the negative sign in the equation. If we take many of these current circles and connect them together so that the current flows through them all continuously we get a current distribution as in Fig. 2, i.e., a toroidal solenoid. (This is the same current distribution as in a tokamak.) The anapole moment corresponding to this current distribution points in the upward direction. A magnetic field is produced inside the current distribution, as shown in the figure.

The expression for the anapole moment (4) contains the current vector $\mathbf{j}$, which changes sign under reflection of coordinates (when $\mathbf{r} \rightarrow -\mathbf{r}$, $\mathbf{j} \propto e d\mathbf{r}/dt \rightarrow -\mathbf{j}$). According to the Wigner-Eckart theorem, the anapole moment vector must be directed along the nuclear spin $\mathbf{I}$: $\langle \hat{\mathbf{a}} \rangle = -\pi \langle r^2 \mathbf{j} \rangle = a\mathbf{I}/I$. However, the spin $\mathbf{I}$ does not change its sign under coordinate reflection (similar to the orbital angular momentum $\mathbf{l} = \mathbf{r} \times \mathbf{p}$). The different behaviour of the right and left hand sides of the relation $\langle r^2 \mathbf{j} \rangle \propto \mathbf{I}$ under reflection of coordinates means that the existence of the anapole moment violates parity, i.e., symmetry under the reflection of coordinates

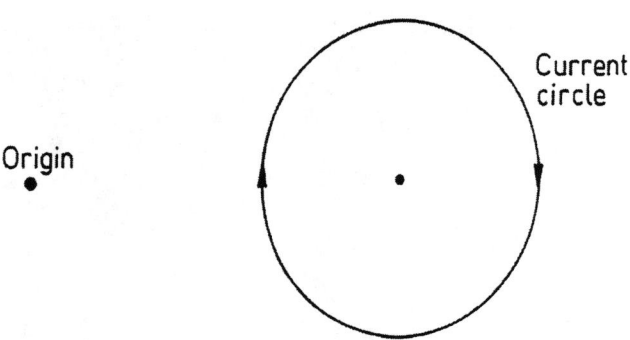

FIGURE 1. A current circle that produces an anapole moment.

(but it does not violate time-reversal invariance).

HOW AN ANAPOLE MOMENT CAN BE PRODUCED

Now that we have seen what an anapole moment is, we turn to the question of how such a thing can be produced. A T-even, P-odd object like the anapole can only arise if there is some kind of P-odd force present; for this the weak interaction is needed. The potential for the P-odd weak interaction between a valence (i.e., unpaired) nucleon and the nuclear core can be written

$$\hat{W} = \frac{G}{2\sqrt{2}m} g[\boldsymbol{\sigma} \cdot \mathbf{p}\rho(r) + \rho(r)\boldsymbol{\sigma} \cdot \mathbf{p}], \tag{5}$$

where $G = 1.0 \times 10^{-5}/m_p{}^2$ is the Fermi constant of the weak interaction, m, $\mathbf{p}$ and $\boldsymbol{\sigma}$ are the mass, momentum and twice the spin of the unpaired nucleon, $\rho(r)$ is the number density of core nucleons, and g is a dimensionless strength constant. For the ^{133}Cs atom, the unpaired nucleon is a proton and so $g \equiv g_p$.

This interaction perturbs the wave function of the unpaired nucleon, resulting in the mixing of opposite parity states: $\psi = \psi_0 + \delta\psi$, where $\psi_0 \equiv |0\rangle$ is the unperturbed

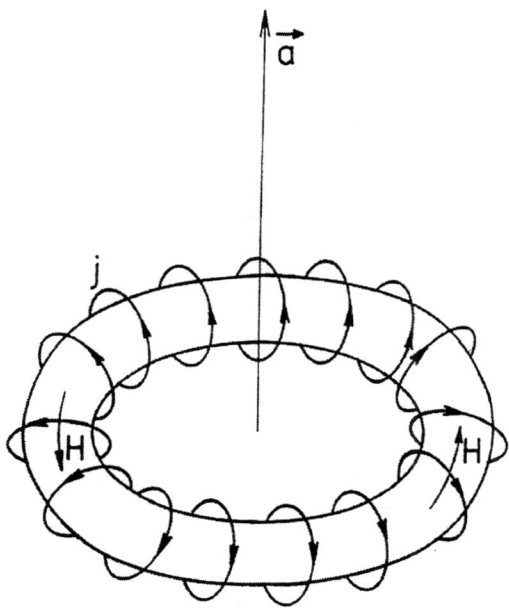

FIGURE 2. The anapole moment, **a**, the toroidal current that produces it, **j**, and the magnetic field that the current creates, **H**.

wave function and $\delta\psi = \sum_n |n\rangle\langle n|\hat{W}|0\rangle(E_0 - E_n)^{-1}$. An approximate analytical solution for the perturbed Schrödinger equation $(\hat{H}_0 + \hat{W})\psi = E\psi$ (which assumes that the nuclear density is constant) gives [10,3].

$$\psi = e^{i\theta\boldsymbol{\sigma}\cdot\mathbf{r}}\psi_0, \tag{6}$$

where $\theta = -gG\rho/\sqrt{2}$. What this means is that the spin ($\mathbf{s} = \frac{1}{2}\boldsymbol{\sigma}$) of the unperturbed wave function will be rotated around the vector $\mathbf{r}$ by an angle of $2\theta r$. If, for example, the unperturbed wave function is in a spin-up state, the spin at different points for the perturbed wave function will be as shown in Fig. 3. Thus we have a spin helix, with a definite chirality, i.e., right- or left-handedness. This means that the parity symmetry has been broken [9]. Let's see what the current and magnetic field produced by such a spin helix are. The electromagnetic current of the unpaired nucleon can be written

$$\mathbf{j} = -\frac{ie}{2m}q[\psi^*\boldsymbol{\nabla}\psi - (\boldsymbol{\nabla}\psi^*)\psi] + \frac{e\mu}{2m}\boldsymbol{\nabla}\times(\psi^*\boldsymbol{\sigma}\psi), \tag{7}$$

where $q = 0$ (1) for a neutron (proton) and μ is the nucleon magnetic moment in nuclear magnetons. The first term comes from the orbital motion of the nucleon, while the second term is a magnetic moment current term, which produces the dominating contribution. The current distribution and the magnetic field produced by the wave function ψ of Eq. (6) have been presented, e.g., in Ref. [11]. Cross-sections of these are shown in Figs. 4 and 5. Note their toroidal shapes; this means that they will produce an anapole moment.

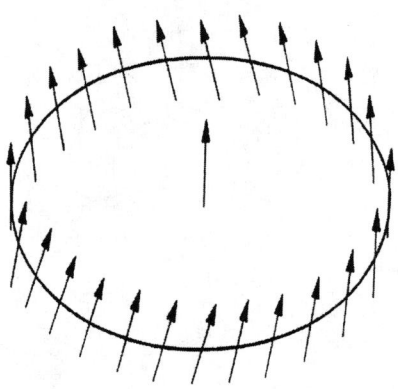

FIGURE 3. The spin helix that occurs due to the parity violating nucleon-nucleus interaction. The degree of spin rotation is proportional to the distance from the origin and the strength of the weak interaction.

Using the expression for the electromagnetic current (7) in Eq. (4), the operator of the anapole moment, $\hat{\mathbf{a}}$ ($\mathbf{a} = \langle \psi | \hat{\mathbf{a}} | \psi \rangle$) can be written as

$$\hat{\mathbf{a}} = \frac{\pi e}{m} \left[\mu (\mathbf{r} \times \boldsymbol{\sigma}) - \frac{q}{2}(\mathbf{p}r^2 + r^2 \mathbf{p}) \right], \tag{8}$$

where $\mathbf{r}$ and $\mathbf{p}$ are the position and momentum operators of the nucleon. The dominant contribution to the nuclear anapole comes from the first, spin term, and thus we can express the anapole moment operator in terms of the magnetic dipole moment operator $\hat{\mathbf{M}} = \boldsymbol{\sigma}(e\mu)/(2m)$ as $\hat{\mathbf{a}} \approx 2\pi (\mathbf{r} \times \hat{\mathbf{M}})$.

The anapole moment is usually described by a dimensionless parameter, κ_a, defined by the following equation:

$$\mathbf{a} = \frac{1}{e} \frac{G}{\sqrt{2}} \frac{K \mathbf{I}}{I(I+1)} \kappa_a, \tag{9}$$

where $K = (I + \frac{1}{2})(-1)^{I+1/2-l}$ (l is the orbital angular momentum of the external nucleon) and e is the electric charge of the proton. Note that the anapole moment is directed along the nuclear spin $\mathbf{I}$.

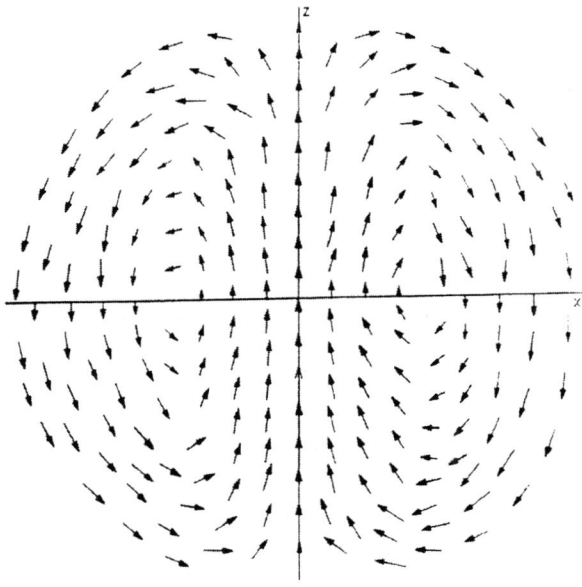

FIGURE 4. A cross-section (in the x-z plane) of the current distribution due to the spin helix (the anapole moment points along the z direction).

Calculations of κ_a in terms of g_p have been made in Refs. [3,12–16]. In Ref. [3] an approximate analytical formula was obtained by using the wave function (6) to calculate the mean value of the anapole moment operator (8). The result is

$$\kappa_a = \frac{9}{10}\frac{\alpha\mu}{mr_0}A^{2/3}g_p = 0.08 g_p \tag{10}$$

(for the mass number, $A = 133$), where $\alpha = 1/137$ and $r_0 = 1.2$ fm is the internucleon distance. More accurate numerical calculations for ^{133}Cs give a close result:

$$\kappa_a = 0.06 g_p. \tag{11}$$

THE DETECTION OF THE NUCLEAR ANAPOLE IN ATOMIC EXPERIMENTS

The nuclear anapole moment interacts with an atom's electrons due to its magnetic field. The interaction is [using Eqs. (3) and (9)]

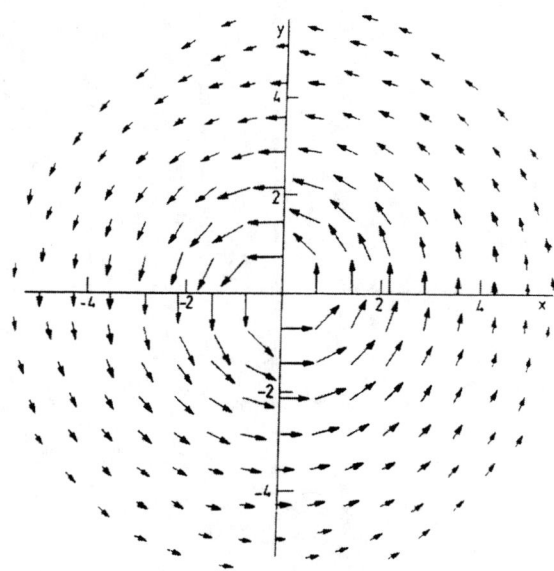

FIGURE 5. A cross-section (in the x-y plane) of the magnetic field due to the spin helix (the anapole moment points along the z direction).

$$V_a = e\boldsymbol{\alpha} \cdot \mathbf{A} = e\boldsymbol{\alpha} \cdot \mathbf{a}\delta(r) = \frac{G}{\sqrt{2}} \frac{K\mathbf{I} \cdot \boldsymbol{\alpha}}{I(I+1)} \kappa_a \delta(r), \qquad (12)$$

where $\mathbf{A}$ is the anapole vector-potential and $\boldsymbol{\alpha}$ is the relativistic velocity operator (Dirac matrices).

The anapole moment can be detected by observing transitions between atomic levels that violate parity. The Boulder experiment [7] used an electric-dipole (E1) transition between the $6S$ and $7S$ states of the cesium atom. The problem is that the anapole moment is just one small P-odd effect and there are other parity violating effects present, such as that caused by the weak interaction between the atomic electron and the weak charge of the nucleus (due to Z-boson exchange). To solve this problem we can use the fact that the nuclear anapole moment's interaction with an atomic electron depends on the nuclear spin $\mathbf{I}$ [see Eq. (12)], unlike the other effects (in fact, there are P-odd effects that do depend on the nuclear spin, but they are small compared to the anapole moment, and can easily be taken into account). The nuclear spin has different relative orientations in the different hyperfine states $F = 3$ and $F = 4$, where F is the total angular momentum of the atom ($\mathbf{F} = \mathbf{j}_e + \mathbf{I}$, where $\mathbf{j}_e$ is the electron's angular momentum). Therefore, if instead of observing just a $6S \to 7S$ transition, we resolve the hyperfine structure of the atomic levels and observe the two different hyperfine transitions $6S_{F=4} \to 7S_{F=3}$ and $6S_{F=3} \to 7S_{F=4}$, we should be able to see the effect of the anapole moment. This is what was done in the Boulder experiment [7], with the following result:

$$-\mathrm{Im}(E1_{\mathrm{PNC}})/\beta = \begin{cases} 1.6349(80) \text{ mV/cm}, & 6S_{F=4} \to 7S_{F=3} \\ 1.5576(77) \text{ mV/cm}, & 6S_{F=3} \to 7S_{F=4}, \end{cases} \qquad (13)$$

where $E1_{\mathrm{PNC}}$ is the amplitude of the transition and β is the vector transition polarizability. The difference between the two values, $0.077(11)$ mV/cm, is the signal of the anapole moment.

We will use these experimental results to derive an estimate of the anapole moment constant. To reduce the theoretical error we calculate the ratio of the nuclear-spin dependent parity nonconserving (PNC) amplitude to the main spin-independent PNC amplitude. By using accurate atomic calculations of the nuclear-spin dependent PNC amplitude produced by the interaction (12) [17] (see also [18,19,13,20]) and the main (i.e., spin-independent) PNC amplitude [21], together with the experimental results (13) we obtain $\kappa = 0.442(63)$. As mentioned earlier, there are other mechanisms that produce small effects similar to the anapole moment (the part of the weak electron-nucleus interaction that depends on the nuclear spin and the combined action of the "usual" electron-nucleus weak interaction and the hyperfine interaction). After taking these into account we obtain the following result for the dimensionless anapole moment constant:

$$\kappa_a = 0.364(62). \qquad (14)$$

One can find the complete list of the references and details of this calculation in Ref. [22].

STRENGTH OF PARITY-VIOLATING NUCLEAR FORCES DERIVED FROM THE ANAPOLE MEASUREMENT

Using the estimate of κ_a and Eq. (11) we can obtain an estimate for g_p, the dimensionless strength constant for the parity violating interaction of an unpaired proton with the nuclear core:

$$g_p = 6 \pm 1 (\text{exp.}). \qquad (15)$$

There is also an error which is produced by the uncertainty in the theoretical value of the anapole moment. The smallest published error is that of Ref. [16]: 20 %. A new detailed calculation by W. C. Haxton and M. J. Musolf is in progress. The proton-nucleus constant g_p can be expressed in terms of the meson-nucleon parity nonconserving interaction constants as follows [3,23] (we use the notation of Ref. [24]):

$$g_p = 8.0 \times 10^4 \left[70.4 f_\pi - 19.5 h_\rho^0 - 4.7 h_\rho^1 + 1.3 h_\rho^2 - 11.3 (h_\omega^0 + h_\omega^1) \right], \qquad (16)$$

where $f_\pi \equiv h_\pi^1$ and the h's are meson-nucleon coupling constants — the subscript denotes the type of meson involved and the superscript indicates whether it is an isoscalar, isovector or isotensor interaction (0, 1 or 2). As stated in the recent review [25], the ρ and ω constants (i.e., the h_ρ's and h_ω's) are known well and can be taken to be those values listed as the "best" values in [24]. However there is uncertainty about the value of f_π. Using our estimate of g_p, the above equation, and the "best" values of the h_ρ's and h_ω's from [24], we can obtain an estimate of f_π:

$$f_\pi \equiv h_\pi^1 = [7 \pm 2 \text{ (exp.)}] \times 10^{-7}. \qquad (17)$$

Now we will compare this estimate of f_π with other estimates in the literature. This seems to be in contradiction with the limit on f_π derived from a ^{18}F PNC measurement: $|f_\pi| < 1.3 \times 10^{-7}$ (see, e.g., the review [26]). However, the above result (17) does agree with QCD calculations of f_π, which give $f_\pi \equiv h_\pi^1 = 5\text{--}6 \times 10^{-7}$ [27,28]. It is also in agreement with the "best" value of f_π in Ref. [24]: $f_\pi = 4.6 \times 10^{-7}$.

A detailed discussion of this question can be found in Ref. [22]. Analyses of the nucleon weak interactions, based on the experiment [7], have also been done in the work [29].

TESTS OF THE STANDARD MODEL AND "NEW PHYSICS"

The experiments that were suggested in [30] for measuring parity nonconservation (PNC) in heavy atoms have provided an important confirmation [31,6,32,7] of the

Standard Model of elementary particles. Now the problem is different. There is a general belief that the Standard Model is only an intermediate step in the process of the unification of all interactions and particles into a new theory. Also, the foundation of the Standard Model is incomplete. A very important element of this model, the Higgs boson, has not been found. It seems that "new physics" beyond the Standard Model is inevitable. New particles would contribute to PNC in atoms as intermediate particles in the electron-nucleus PNC interaction or as a contribution to the radiative corrections to the weak charge. The Standard Model radiative corrections to the weak charge are about 2.5%. This gives an estimate of the accuracy that is needed for atomic experiments to be sensitive to new physics. Of course, we need high accuracy in both the atomic experiments and the atomic calculations that allow one to extract the value of the weak charge from the atomic experiments. Currently, the most accurate results are for the Cs atom: the accuracy of the measurements is 0.35% [7], the accuracy of the atomic theory is 1% [33,34,20]. This gives one the chance to study new physics beyond the Standard Model.

The measured nuclear spin-independent part of the PNC effect in Cs is of the form

$$-\frac{\text{Im}(E_{\text{PNC}})}{\beta} = 1.5939(56)\frac{\text{mV}}{\text{cm}}. \qquad (18)$$

The PNC E1 amplitude of the 6s-7s transition E_{PNC} is due to the weak electron-nucleus interaction which admixes p-states with s-states. The vector polarizability β of the transition comes from the Stark amplitude of the transition, which appears due to an admixture of p-states and s-states by an external electric field. The measured PNC effect in Cs is due to the interference of the PNC E1 amplitude and the Stark amplitude. Therefore, the relative magnitude of the PNC effect is proportional to the ratio E_{PNC}/β.

The most accurate theoretical values of E_{PNC} are as follows:

$$E_{\text{PNC}} = -i|e|a_0 10^{-11}\left(-\frac{Q_W}{N}\right)\begin{cases} 0.908(10) & \text{Ref. [33]} \\ 0.905(9) & \text{Ref. [34,20]} \end{cases}. \qquad (19)$$

Here Q_W is the weak charge of the cesium nucleus and N is the number of neutrons.

The method for *ab initio* calculations of E_{PNC} that we used in [33] was based on an all-orders summation of the dominant diagrams of the many-body perturbation theory in the residual Coulomb interaction and used a relativistic Hartree-Fock basis set and Green's functions. This technique has been described in [33,35].

We took into account direct and exchange polarization of the atomic core by the external electric field and the weak nuclear potential using the time-dependent Hartree-Fock method (summation of the "RPA with exchange" chain of diagrams) and we calculated second-order correlation corrections and three series of dominating higher-order diagrams:

1. Screening of the electron-electron interaction. This is a collective phenomenon and so the corresponding chain of diagrams is enhanced by a factor approximately equal to the number of electrons in the external closed subshell (the

5p electrons in Cs). We stress that our approach takes into account screening diagrams with double, triple and higher core-electron excitations [36], in contrast to popular pair equations (coupled cluster) method, where only double excitations are considered.

2. Hole-particle interaction. This effect is enhanced by the large zero-multipolarity diagonal matrix elements of the Coulomb interaction.

3. Iterations of the self-energy operator ("correlation potential"). This chain of diagrams describes the nonlinear effects of the correlation potential and is enhanced by the small denominator, which is the energy for the excitation of an external electron (in comparison with the excitation energy of a core electron).

The error in the theoretical value was tested in many different ways: by estimating the contribution of the unaccounted higher-order diagrams, by comparing the calculated and measured values of the energy levels, the fine and hyperfine structure intervals, the probabilities of electromagnetic transitions, etc. (see Ref. [33]). The result for the PNC amplitude hardly changed when we introduced factors into the correlation potential to fit the energy levels (in imitation of the unaccounted higher-order diagrams). Important tests of our method include predictions of the spectrum [37] and electromagnetic transition amplitudes for the Fr atom [38], which is an analogue of Cs. Recently the positions of many energy levels [39] and some transition rates [40] of Fr were measured and found to be in excellent agreement with our predictions.

Our calculations of PNC for atoms with electron structures more complex than those of the alkali atoms were proved to be accurate as well. In a series of works done about ten years ago we claimed an accuracy of 3% for Tl [21], 8% for Pb and 11% for Bi [41]. All these PNC effects were recently measured to an accuracy of about 1% [6,32] and found to be in good agreement with our predictions. This means that our estimates for the theoretical accuracy were correct and probably even too pessimistic. For example, in our first calculation of the Fr energy levels [37] we claimed the accuracy of our predictions to be about 0.5% while the actual agreement with latter measurements was found to be 0.1%. The situation was similar for the electromagnetic transitions 6s-6p$_{1/2}$ and 6s-6p$_{3/2}$ in Cs (see below). These numerous tests give us firm ground to believe that the theoretical error in E_{PNC} (19) indeed does not exceed 1%.

Very careful calculations of the PNC effect have been done by a different method in Refs. [34,20] and the calculations were compared with numerous experimental results. The presentation in Ref. [20] is very detailed so I refer the reader to this excellent work. It is important to note that the difference between the theoretical results [33] and [34,20] is only 0.3%. There are also accurate semiempirical calculations in Refs. [42,43] which have an accuracy of a few per cent and are in agreement with the many-body calculations [33,34,20].

As can be seen from (18) an accurate value of the vector transition polarizability

β is also required for the interpretation of the PNC measurements. The values of β are the following:

$$\beta = a_0^3 \begin{cases} 27.30(40) & \text{Ref. [5]}, \\ 27.17(35) & \text{Ref. [44]}, \\ 27.20(40) & \text{Ref. [42]}, \\ 27.00(20) & \text{Ref. [20]}, \\ 27.15(13) & \text{Ref. [45]}. \end{cases} \qquad (20)$$

The last calculation of β used the measured ratio of the vector and scalar polarizabilities of the 6s-7s transition and the most accurate values of the $E1$ electromagnetic amplitudes derived from both the calculations and measurements. We would like to stress that accurate measurements of the $E1$ amplitudes are very desirable for an improvement of the interpretation of the PNC measurements in Cs. The most important improvement would be a more accurate value of the 6s-7p amplitude. An improvement for the 7s-7p amplitude is also very important because of some disagreement between theory and existing data.

Using the last value of β (which is also very close to the mean value $\beta = 27.16$), the measurement (18), the mean value of the theoretical amplitudes (19), and $|e|/a_0^2 = 5.1422 \times 10^{12}\text{mV/cm}$, we obtain

$$Q_W(\text{exper}) = -72.41(25)_{\text{exper}}(80)_{\text{theor}}. \qquad (21)$$

Comparing this result for Q_W with the theoretical value [46]

$$Q_W(\text{theor}) = -73.20(13) - 0.8S - 0.005T, \qquad (22)$$

we can find the Peskin-Takeuchi parameter S characterizing new physics beyond the Standard Model (the parameter S stands for weak isospin conserving radiative corrections produced by new particles).

$$S + 0.006T = -1.0(0.3)_{\text{exper}}(1.0)_{\text{theor}}. \qquad (23)$$

This result is already important for high energy physics. For example, the prediction of the Technicolor Model is $S \approx 2$ [46] therefore the atomic result above seems to rule out this model. We can also use the calculation of the extra Z_x-boson contribution in the $SO(10)$ model [46]

$$\Delta Q_W = 0.4(2N + Z)(\frac{M_W}{M_{Z_x}})^2 = 84.4(\frac{M_W}{M_{Z_x}})^2 \qquad (24)$$

to find the limit for the mass of this boson

$$M_{Z_x} > 550 \text{ GeV}. \qquad (25)$$

In conclusion, I present a table from the talk of D. Budker at the WEIN-98 conference, comparing the limits on new physics obtained from the single atomic Cs PNC experiment and all High-Energy Physics (HEP) data (see Table 1).

TABLE 1. Limits on new physics beyond the standard model currently obtained from atomic PNC and directly from high-energy physics (HEP).

New Physics	Parameter	Constraint from atomic PNC	Direct constraints from HEP
Oblique radiative corrections	$S + 0.006T$	-1.0 ± 1.2	$S = -0.28 \pm 0.19$, $T = -0.20 \pm 0.26$
Z_x-boson in SO(10) model	$M(Z_x)$	> 550 GeV	> 425 GeV
Leptoquarks	M_S	> 0.7 TeV	> 0.28 TeV
Composite Fermions	L	> 14 TeV	> 6 TeV

ACKNOWLEDGMENTS

This work was supported by the Australian Research Council.

REFERENCES

1. Zel'dovich, Ya. B., *Zh. Éksp. Teor. Fiz.* **33**, 1531 (1957) [*Sov. Phys. JETP* **6**, 1184 (1958)]. (This reference also contains a mention of analogous results found by V. G. Vaks.)
2. Flambaum, V. V. and Khriplovich, I. B., *Zh. Éksp. Teor. Fiz.* **79**, 1656 (1980) [*Sov. Phys. JETP* **52**, 835 (1980)].
3. Flambaum, V. V., Khriplovich, I. B. and Sushkov, O. P., *Phys. Lett. B* **146**, 367 (1984).
4. Bouchiat, M. A., Guéna, J., Pottier, L. and Hunter, L., *Phys. Lett. B* **134**, 463 (1984); Noecker, M. C., Masterson, B. P. and Wieman, C. E., *Phys. Rev. Lett.* **61**, 310 (1988).
5. Gilbert, S. L. and Wieman, C. E., *Phys. Rev. A* **34**, 792 (1986).
6. Edwards, N. H., Phipp, S. J., Baird, P. E. G. and Nakayama, S., *Phys. Rev. Lett.* **74**, 2654 (1995); Vetter, P. A., Meekhof, D. M., Majumder, P. K., Lamoreaux, S. K. and Fortson, E. N., *Phys. Rev. Lett.* **74**, 2658 (1995).
7. Wood, C. S., Bennett, S. C., Cho, D., Masterson, B. P., Roberts, J. L., Tanner, C. E. and Wieman, C. E., *Science* **275**, 1759 (1997).
8. Sushkov, O. P., Flambaum, V. V. and Khriplovich, I. B., *Zh. Éksp. Teor. Fiz.* **87**, 1521 (1984) [*Sov. Phys. JETP* **60**, 873 (1984)]. Flambaum, V. V., in *"Modern Developments in Nuclear Physics"*, ed. Sushkov, O. P., Singapore: World Scientific, 1987, p.556.
9. Khriplovich, I. B., *Parity Nonconservation in Atomic Phenomena*, Philadelphia: Gordon and Breach, 1991.
10. Michel, Curtis F., *Phys. Rev.* **133**, B329 (1964).
11. Flambaum, V. V. and Hanhart, C., *Phys. Rev. C* **48**, 1329 (1993).
12. Haxton, W. C., Henley, E. M. and Musolf, M. J., *Phys. Rev. Lett.* **63**, 949 (1989).
13. Bouchiat, C. and Piketty, C. A., *Z. Phys. C* **49**, 91 (1991).
14. Bouchiat, C. and Piketty, C. A., *Phys. Lett. B* **269**, 195 (1991); erratum **274**, 526 (1992).

15. Dmitriev, V. F., Khriplovich, I. B. and Telitsin, V. B., *Nucl. Phys. A* **577**, 691 (1994).
16. Dmitriev, V. F. and Telitsin, V. B., *Nucl. Phys. A* **613**, 237 (1997).
17. Kraftmakher, A. Ya., *Phys. Lett. A* **132**, 167 (1988).
18. Novikov, V. N., Sushkov, O. P., Flambaum, V. V. and Khriplovich, I. B., *Zh. Éksp. Teor. Fiz.* **73**, 802 (1977) [*Sov. Phys. JETP* **46**, 420 (1977)].
19. Frantsuzov, P. A. and Khriplovich, I. B., *Z. Phys. D* **7**, 297 (1988).
20. Blundell, S. A., Sapirstein, J. and Johnson, W. R., *Phys. Rev. D* **45**, 1602 (1992).
21. Dzuba, V. A., Flambaum, V. V., Silvestrov, P. G. and Sushkov, O. P., *J. Phys. B* **20**, 3297 (1987).
22. Flambaum, V. V. and Murray, D. W., *Phys. Rev. C* **56**, 1641 (1997).
23. Flambaum, V. V., *Physica Scripta* **T46**, 198 (1993).
24. Desplanques, B., Donoghue, J. F. and Holstein, B. R., *Annals of Physics* **124**, 449 (1980).
25. Brown, B. A., in *Parity and Time Reversal Violation in Compound Nuclear States and Related Topics*, edited by Auerbach, N. and Bowman, J. D., Singapore: World Scientific, 1996.
26. Desplanques, B. in *Parity and Time Reversal Violation in Compound Nuclear States and Related Topics*, edited by Auerbach, N. and Bowman, J. D., Singapore: World Scientific, 1996.
27. Khatsimovskii, V. M., *Yad. Fiz.* **42**, 1236 (1985) [*Sov. J. Nucl. Phys.* **42**, 781 (1985)].
28. Kaplan, D. B. and Savage, M. J., *Nucl. Phys. A* **556**, 653 (1993).
29. Haxton, W. C., *Science* **275**, 1753 (1997).
30. Bouchiat, M. A. and Bouchiat, C., *Phys. Lett. B* **48**, 111 (1974); Khriplovich, I. B., *Pis'ma v ZhETF* **20**, 686 (1974) [*Sov. Phys. JETP Lett.* **20**, 315 (1974)]; Sandars, P. G. H. in *Atomic Physics*, edited by Putlitz, G. zu, **4**, 71, New York: Plenum, 1975; Sorede, D. S. and Fortson, E. N., *Bull. Am. Phys. Soc.* **20**, 491 (1975).
31. Barkov, L. M. and Zolotorev, M. S., *Pis'ma v ZhETF* **27**, 379 and **28**, 544 (1978) [*Sov. Phys. JETP Lett.* **27**, 357 and **28**, 503 (1978)]; Drell, P. S. and Commins, E. D., *Phys. Rev. A* **32**, 2196 (1985); DeMille, D., Budker, D. and Commins, E. D., *Phys. Rev. A* **50**, 4657 (1994); Birich, G. N., Bogdanov, Yu. V., Kanorskii, S. I., Sobel'man, I. I., Sorokin, V. N., Struk, I. I. and Yukov, E. A., *Zhur. Eksp. Teor. Fiz.* **87**, 776 (1984) [*Sov. Phys. JETP* **60**, 442 (1984)]; Bouchiat, M. A., Guena, J., Pottier, L. and Hunter, L., *J. Phys. (Paris)* **47**, 1709 (1986).
32. Meekhof, D. M., Vetter, P. A., Majumder, P. K., Lamoreaux, S. K., Fortson, E. N., *Phys. Rev. A.* **52**, 1895 (1995); Phipp, S. J., Edwards, N. H., Baird, P. E. G., Nakayama, S., *J. Phys. B.* **29**, 1861 (1996); Macpherson, M. J. D., Zetie, K. P., Warrington, R. B., Stacey, D. N. and Hoare, J. P., *Phys. Rev. Lett.* **67**, 2784 (1991).
33. Dzuba, V. A., Flambaum, V. V. and Sushkov, O. P., *Phys. Lett.* **A 141**, 147 (1989).
34. Blundell, S. A., Johnson, W. R. and Sapirstein, J., *Phys. Rev. Lett.* **65**, 1411 (1990).
35. Dzuba, V. A., Flambaum, V. V., Silvestrov, P. G. and Sushkov, O. P., *J. Phys. B* **20**, 1399 (1987); *Phys. Lett. A* **131**, 461 (1988); Dzuba, V. A., Flambaum, V. V. and Sushkov, O. P., *Phys. Lett. A* **140**, 493 (1989); Dzuba, V. A., Flambaum, V. V., Kraftmakher, A. Ya., Sushkov, O. P., *Phys. Lett. A* **142**, 373 (1989).
36. The point is that we exploit the Feynman diagram technique which contains all

possible "time ordering" of the loops and therefore screening diagrams contain any number of excited electrons.
37. Dzuba, V. A., Flambaum, V. V. and Sushkov, O. P., *Phys. Lett.* **95A**, 230 (1983).
38. Dzuba, V. A., Flambaum, V. V. and Sushkov, O. P., *Phys. Rev. A* **51**, 3454 (1995).
39. Liberman, S. *et al.*, *C. R. Acad. Sci.* Paris **286B**, 253 (1978); Andreev, S. V., Letokhov, V. S. and Mishin, V. I., *JETP. Lett.* **43**, 736 (1986); *Phys. Rev. Lett.* **59**, 1274 (1987); *J. Opt. Soc. Am.* **B5**, 2190 (1988); Bauche, J. *et al.*, *J. Phys.* **B 19**, L593 (1986); Duong, H. T. *et al.*, *Europhys. Lett.* **3**, 175 (1987); Arnold, E., Borchers, W., Carré, M. *et al.*, *J. Phys.* **B 22**, L391 (1989); Arnold, E. *et al.*, *J. Phys.* **B 25**, 3511 (1990); Simsarian, J. E., Shi, W., Orozco, L. A., Sprouse, G. D., Zhao, W. Z., *Optics Lett.* **21**, 1939 (1996).
40. Zhao, W. Z., Simsarian, J. E., Orozco, L. A., Shi, W. and Sprouse, G. D., *Phys. Rev. Lett.* **78**, 4169 (1997).
41. Dzuba, V. A., Flambaum, V. V., Silvestrov, P. G. and Sushkov O. P., *Europhys. Lett.* **7**, 413 (1988).
42. Bouchiat, C. and Piketty, C. A., *Europhys. Lett.* **2**, 511 (1986).
43. Hartley, A. C. and Sandars, P. G. H., *J. Phys. B: At. Mol. Opt. Phys.* **23**, 2649 (1990).
44. Bouchiat, M. A. and Guena, J., *J. Phys. (Paris)* **49**, 2037 (1988).
45. Dzuba, V. A., Flambaum, V. V., Sushkov, O. P., *Phys. Rev. A* **141**, R4357 (1997).
46. Marciano, W. J. and Rosner, J. L., *Phys. Rev. Lett.* **65**, 2963 (1990).

High-Resolution, High-Accuracy Spectroscopy of Trapped Ions[*]

D. J. Berkeland[*], J. D. Miller[†], F. C. Cruz[‡], B. C. Young,
R. J. Rafac, X.-P. Huang[§], W. M. Itano, J. C. Bergquist and
D. J. Wineland

National Institute of Standards and Technology (NIST), 325 Broadway, Boulder Colorado 80303

Abstract. Microwave spectroscopy using trapped and cooled ions can achieve precision and accuracy comparable to the best cesium frequency standards. We discuss standards based on ^{199}Hg$^+$ ions trapped in linear Paul traps: the Jet Propulsion Laboratory (JPL) standard, which uses up to 10^7 atoms confined near the trap axis, and the recently evaluated NIST standard, which uses approximately ten ions laser cooled and crystalized on the trap axis. We consider future directions in trapped ion frequency standard work, including the use of entangled states for achieving higher precision, and progress on trapped ion optical frequency standards. Finally, we discuss scientific and technical applications of extremely stable frequency standards.

INTRODUCTION

Precise and accurate atomic spectroscopy can rigorously test theories of atomic structure, quantum electrodynamics [1] and other fundamental physics [2,3], determine fundamental constants [4], and provide time and frequency standards [3]. Precise and accurate spectroscopy has two basic requirements. First, the measurement must reach the desired precision in a reasonable averaging time. This requires a good signal to noise ratio and a narrow transition. Second, systematic frequency shifts and broadening mechanisms of the atomic transition being studied must be either very small, or stable and very well measured.

As seen at this conference, many experiments with cooled and trapped ions and with laser-cooled neutral atoms can satisfy these requirements. Here we limit

---

[*]) Work performed at NIST is supported by the Office for Naval Research and the Army Research Office. Work of U.S. Government; not subject to U.S. copyright.
[*]) Los Alamos National Laboratory, P23, MS H803, Los Alamos, NM
[†]) KLA-Tencor Corp., 1701 Directors Blvd., Suite 1000, Austin, TX 78744
[‡]) Universidade Estadual de Campinas, UNICAMP-IFGW-DEQ, CP.6165 Campinas, SP, 13083-970, Brazil
[§]) Colorado MED-Tech-RELA, Inc., 6175 Longbow Dr., Boulder, CO 80301

our discussion to precision experiments using trapped and cooled ions [5]. The systematic shifts of atomic transition frequencies in these systems can be small and well-characterized. For example, Stark shifts are small since $<\mathbf{E}> = 0$ and $<E^2>$ is small when the ions are cold. The magnetic field can be small and easy to characterize because the ions occupy a small volume. The corresponding Zeeman shift is typically small because usually transitions with only a second-order field dependence are used. Because background gas pressure can be negligible, collisional shifts and broadening can be small. The ions can be cooled using either buffer-gas cooling or laser cooling [6], reducing Doppler shifts. The statistical precision can be high if large numbers of ions are stored, or if the shot noise limit is reached. Finally, free precession times of several minutes have been reported [7,8], giving extremely narrow transition linewidths.

The statistical precision of a frequency standard can be predicted quantitatively for various cases. For example, if an atomic transition is probed using the Ramsey technique [9] and the measurement precision is limited only by quantum fluctuations in the atomic state populations [10], the fractional precision of the frequency measurement is given by [11]

$$\frac{\Delta \omega_{measured}}{\omega_0} \equiv \sigma_y(\tau) = \frac{1}{\omega_0 \sqrt{NT_R}} \tau^{-1/2}. \qquad (1)$$

Here, ω_0 is the transition angular frequency, $\Delta\omega_{measured}$ is the precision with which the frequency is measured, and the Allan deviation $\sigma_y(\tau)$ is related to the fractional frequency instability. N is the number of atoms used, T_R is the free precession time between the two $\pi/2$ Rabi pulses, and τ ($> T_R$) is the averaging time of the measurement. Ideally, N, T_R, and τ are large, although various experimental constraints may limit these values.

Table 1 compares these parameters for two types of microwave frequency standards—those based on ions confined in an rf Paul trap and those based on a pulsed fountain of cesium atoms [12]. The number N of atoms used in the trapped ion standards is limited in part by the second-order Doppler shift due to micromotion. This motion is driven by the trap's rf electromagnetic fields, and becomes greater as the Coulomb repulsion between the ions forces them further from the field nodal point or line [13]. Thus there is a trade-off between using very large N but with a substantial Doppler shift, or a negligible Doppler shift but smaller N. In the Cs fountain standard, N is limited in part by collisional shifts. For laser-cooled trapped ion standards, T_R is limited by the time the ions remain cold in the absence of cooling radiation. In a fountain standard it is limited to about 1 s by the maximum practical height the atoms can be tossed (about one meter). Finally, the averaging time τ is not fundamentally limited in either type of standard.

Ions in Penning traps have been used previously to realize the first laser-cooled frequency standard [7,16]. An important limitation to accuracy in those experiments was the uncertainty in the second-order Doppler shift due to the overall rotational motion of the ion cloud. This motion can now be precisely controlled

TABLE 1. Example parameters for trapped ion and cesium fountain microwave frequency standards.

Parameter	Ions	Cs
$\omega_0/(2\pi)$	40.5 GHz (^{199}Hg$^+$)	9.2 GHz
N	$\approx$ 10 [14] to 10^7 [15]	10^4 to 10^6
T_R	Up to 600 s [7,8]	$\sim$ 1 s

[17], and it should therefore be possible to realize an rf or microwave frequency standard with accuracy comparable to what is possible with Paul traps [18]. To date, the highest accuracies and stabilities have been obtained in linear Paul traps; therefore, we highlight this work below.

Figure 1 shows a schematic diagram of a linear Paul trap [19–21]. In this trap, the ions are confined axially by the two cylindrical sections (endcaps) held at static potential U_0. The shape of these endcaps is not critical, and axial confinement can be produced by thin conducting rods located on the trap axis at both ends of the trap [15], by small rings [20] or by segmented electrodes [21]. Two of the long, thin rods of Fig. 1 are held at ground potential, while the other two are held at an rf potential. This gives an oscillating electric potential that traps the ions in a radial quadratic pseudopotential [22]. The advantage of the linear trap is that many ions can be confined near or on the electric field nodal line, where Doppler shifts from micromotion are minimized.

MICROWAVE SPECTROSCOPY

Microwave spectroscopy with clouds of trapped ions has been used for many years in atomic structure measurements [23]. Some recent experiments report measurements of g factors [24] and hyperfine constants [25–29]. Because ion traps can use small samples, they are well-suited for measurements of hyperfine anomalies using

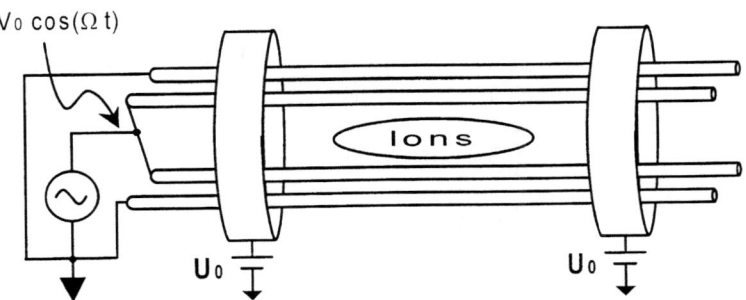

FIGURE 1. Schematic diagram of a linear Paul trap.

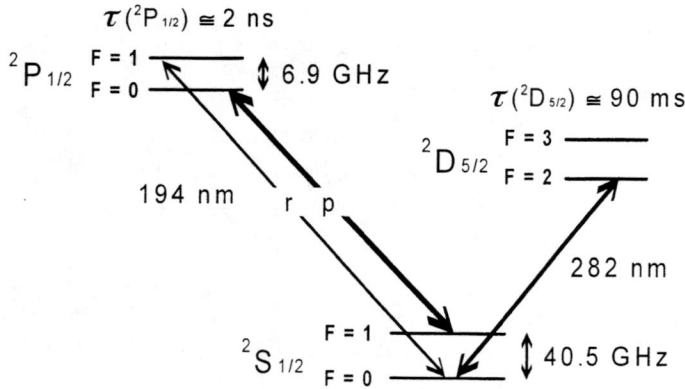

FIGURE 2. Partial energy level diagram of ^{199}Hg$^+$.

different isotopes, some of which are radioactive [25,26,29].

Taking advantage of the precision and accuracy offered by the ground state hyperfine transition of trapped ions, several groups are working on atomic frequency standards based on ^{113}Cd$^+$ ($\omega_0 = 2\pi \times 15.2$ GHz) [27,30] and ^{171}Yb$^+$ ($\omega_0 = 2\pi \times 12.6$ GHz) [20,31–33]. Here we describe microwave frequency standards based on trapped ^{199}Hg$^+$ ions [14,15,34]. Figure 2 shows a partial energy level diagram of this atom. The electric dipole transitions at 194 nm are used for state preparation and detection, and can be used for laser cooling. The ground state hyperfine splitting frequency is $2\pi \times 40.5$ GHz, the highest of routinely trapped ions. The $\Delta m_F = 0$ hyperfine transition depends only quadratically on the magnetic field, when the field is near zero.

At JPL, much work has been done with clouds of ^{199}Hg$^+$ ions in a linear Paul trap [15,34]. The methods and performance of the Commonwealth Scientific and Industrial Research Organization (CSIRO) experiments on ^{171}Yb$^+$ are similar [20]. The four trap rods of the JPL linear ion trap standard (LITS) in reference [15] are evenly spaced on a circle of 1 cm radius, with two endcap rods on the trap axis 7.5 cm apart. This trap confines clouds of up to 10^7 ions, which are cooled with a helium buffer gas. States are prepared and detected with a ^{202}Hg lamp, which emits broad-line radiation that partially overlaps the resonances of the two $^2S_{1/2}, F = 1 \to {}^2P_{1/2}, F = 0, 1$ transitions. Because the $^2P_{1/2}, F = 1$ state can decay to the $^2S_{1/2}, F = 0$ state, illuminating the ions with the lamp radiation pumps them into the $^2S_{1/2} F = 0$ state in preparation for the Ramsey interrogation. At the end of the Ramsey interrogation, the lamp radiation is returned to the ions, and the detected fluorescence indicates how many ions are in the $^2S_{1/2} F = 1$ state. Because the ions scatter only a few photons before optically pumping out of the $^2S_{1/2} F = 1$ state, and because only a small fraction of the scattered photons are detected, the state detection efficiency is much less than unity, so the state measurements are not quantum noise limited and Eq. (1) is not applicable.

However, because N is large, this device is very stable. With a Ramsey time T_R of 8 s, LITS has recently demonstrated an instability of $\sigma_y(\tau) = 3 \times 10^{-14} \tau^{-1/2}$ [35]. At CSIRO, a similar standard using clouds of laser-cooled ^{171}Yb$^+$ ions in a linear trap has $\sigma_y(\tau) = 4.7 \times 10^{-14} \tau^{-1/2}$ [20].

Although these standards are very stable, the second-order Doppler shift from the ion motion is substantial (about 9×10^{-13}). At NIST, the goal is to develop a frequency standard that is both stable and accurate. We use a linear trap such as that shown in Fig. 1, with the endcaps approximately 4 mm apart and the 0.2 mm radius rods on a 0.64 mm radius [36]. Approximately ten ions are used, and they are laser cooled so that they crystallize along the nodal line of the rf electric field, near the trap axis. Using a small number of ions sacrifices precision according to Eq. (1), but greatly reduces Doppler shifts. Groups at CSIRO [37] and the Communications Research Laboratory (CRL) [38] have also crystalized laser-cooled ions in linear Paul traps in order to improve the accuracy of their frequency standards.

The NIST standard has other advantages. The trap is enclosed in a copper container that forms the bottom of a liquid helium reservoir [36]. Because of the cryogenic environment, the pressure of background neutral mercury atoms is negligible. This is critical because the background mercury pressure leads to ion loss, presumably due to dimer formation. This ion loss limits the storage time of trapped ions in a room temperature trap to about ten minutes. At 4 K, the ions can be trapped for days at a time without loss. Also, in the 4 K environment the pressure of all other background gases is negligible, with the possible exception of helium. This greatly reduces collisional shifts and heating from collisions. Additionally, the black body shift, which is already over two orders of magnitude smaller than in cesium at room temperature [39], is dramatically reduced. To ensure that the ions are on the nodal line, we minimize the micromotion observed in three non-coplanar directions [13]. Because the rf electric field at the site of the ions in minimized, the rf heating while the cooling lasers are off during the free precession time T_R is minimized. Finally, laser cooling (instead of buffer gas cooling) significantly reduces Doppler shifts; we have measured the second-order Doppler shift to be less than 3×10^{-17} [14].

The steps in operating the frequency standard are as follows. We Doppler cool the ions by using radiation from a primary laser whose frequency is slightly red-detuned from transition p (see Fig. 2). Although this is a cycling transition, the laser weakly couples the $S_{1/2}$, $F = 1$ state to the $P_{1/2}$, $F = 1$ state, which decays to the $S_{1/2}$, $F = 0$ state. To optically pump the ions out of this state, we overlap a less intense repumping laser beam with the primary laser beam to drive the r transition. After a Doppler cooling period of 200 to 300 ms, we pump the ions into the $S_{1/2}$, $F = 0$ state with high efficiency by applying only the primary laser for 10 ms. Then we drive the microwave transition using the Ramsey method, with a free precession time T_R of 2 to 100 s. After this, to determine the ensemble average of the $S_{1/2}$, $F = 1$ state populations, we again apply only the primary laser for about 10 ms. If an ion is found to be in the $F = 1$ state, it will scatter

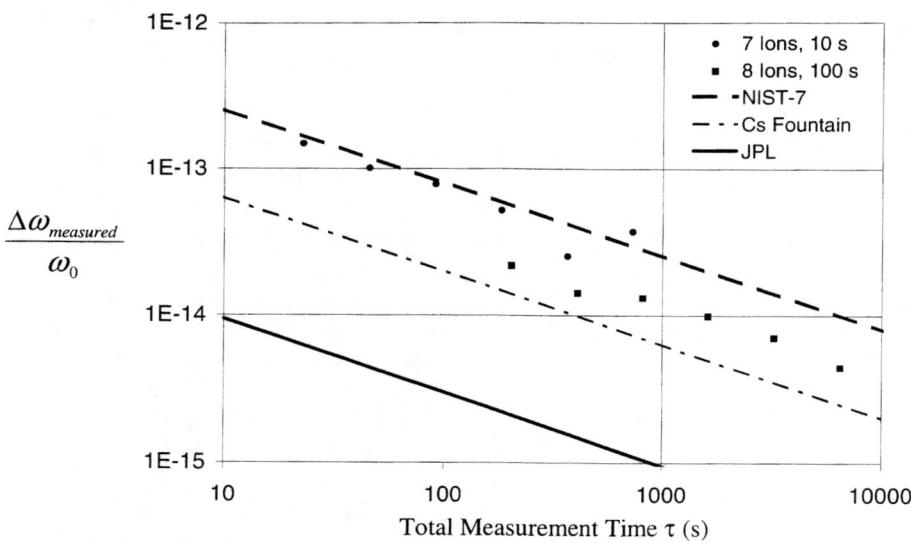

FIGURE 3. Stabilities of several standards.

approximately 10,000 photons before being optically pumped into the $S_{1/2}$, $F = 0$ state. Otherwise, the atoms scatter only a few photons in the same time interval, due to the off-resonant $S_{1/2}$, $F = 0$ to $P_{1/2}$, $F = 1$ transition. Our detection signal is the observed fluorescence from all of the ions. Repeating this measurement process and scanning the microwave frequency produces a Ramsey fringe pattern. We lock the frequency to the central Ramsey fringe by stepping the frequency from one side of the fringe to the other, while a digital servo works to keep the detection signal constant at each step.

Figure 3 shows the instability $\sigma_y(\tau)$ of the microwave oscillator when it is locked to the central Ramsey fringe with $T_R = 10$ s and $T_R = 100$ s. For $T_R = 100$ s, the instability is $\sigma_y(\tau) = 3.3 \times 10^{-13} \tau^{-1/2}$. The measured instability of the standard is consistently twice that expected from Eq. (1), due to fluctuations of the laser intensity at the site of the ions. The figure also shows the stability of the cesium beam standard NIST-7 (8×10^{-13} $\tau^{-1/2}$) [40] and the Paris cesium fountain standard (2×10^{-13} $\tau^{-1/2}$) [12]. The instabilities of these three standards are comparable, while that of the JPL ^{199}Hg$^+$ standard is significantly better ($\sigma_y(\tau) = 3 \times 10^{-14} \tau^{-1/2}$).

The accuracy of the JPL standard is limited by the second-order Doppler shift to around 10^{-13} [41]. The other standards in Fig. 3 have significantly greater accuracy, and the accuracies of these standards are comparable. The fractional accuracy of NIST-7 is 5×10^{-15}, limited by the distributed cavity phase shift [42]. This high accuracy corresponds to splitting the central Ramsey fringe to about a part in 10^6. A natural next step toward a more accurate frequency standard is to reduce the linewidth. The cesium fountain frequency standard does this by increasing the Ramsey free precession time T_R. In this standard, the accuracy is currently limited by black body and collisional shifts to 1.4×10^{-15} [43]. The accuracy of the NIST ^{199}Hg$^+$ standard is 3.4×10^{-15}, and is limited by uncertainties in the Zeeman shift [14]. The dominant shift is caused by asymmetries in currents running through the trap electrodes, which cause a net rf magnetic field at the site of the ions.

The trapped ion and cesium fountain standard are both emerging technologies that promise even higher accuracy. At NIST we have constructed a new, smaller trap, which we will run with lower trap drive frequency Ω. Less rf potential is needed for the same radial confinement in the trap, reducing the currents in the trap rods and, from scaling arguments, the associated magnetic field at the site of the ions. Also, we have improved our magnetic shielding to reduce fluctuations in the ambient magnetic field. We expect that the next generation of this standard will be accurate to approximately 10^{-16}.

FUTURE DIRECTIONS

One way to reduce the instability $\sigma_y(\tau)$ of a standard is to increase the transition frequency ω_0. Therefore, many groups are working on optical frequency standards. The laser-cooled stored ion efforts include trapped ^{172}Yb$^+$ at 411 nm [44] and 3.43 μm [45], ^{171}Yb$^+$ at 435 nm [46], ^{88}Sr$^+$ at 674 nm [47,48], ^{138}Ba$^+$ at 12.5 μm [49], ^{40}Ca$^+$ at 729 nm [50–52], ^{115}In$^+$ at 237 nm [53,54], and ^{199}Hg$^+$ at 282 nm [55].

Earlier work at NIST using ^{199}Hg$^+$ in a room temperature trap produced a narrow transition with structure due to Rabi oscillations [55]. The width of the central feature was about 40 Hz at 563 nm. Our pursuit of higher resolution and a study of systematic effects was hampered by the limited lifetime of the ion in the room temperature trap. However, we have now built a second cryogenic system that will provide long ion lifetimes and more detailed investigations of this transition. Also, we have recently made substantial improvements in the laser system; the laser linewidth in about a one minute averaging time is now less than one hertz at 563 nm [56].

For the absolute frequency of a transition to be determined, ω_0 must be compared to an accepted frequency standard. We are now comparing the frequency of the 282 nm transition to a that of the narrow $^1S_0 - ^3P_1$ intercombination line of Ca at 657 nm [57,58] by mixing a CO overtone line with the fundamental 563 nm light [59]. The NIST Ca standard has been compared to the Physikalisch-Technische Bundesanstalt (PTB) Ca standard, which has in turn been compared to the 9.2 GHz line in Cs [60]. In other laboratories, the absolute frequency of the ^{88}Sr$^+$ S to D transition at 674 nm has been measured with a frequency chain [48]. Interferometric measurements have also been made on this transition [61] and on the S to D transition at 411 nm in ^{172}Yb$^+$ [44], with frequency chain measurements in progress [62].

Another way to reduce $\sigma_y(\tau)$ is by using entangled states [63,64]. Consider an atomic system with two states labeled ↑ and ↓, separated by frequency ω_0. We use the spin 1/2 analog for these two states [65], so the total angular momentum for N ions is given by $\mathbf{J} = \sum_{i=1}^{N} \mathbf{S_i}$, where $\mathbf{S_i}$ is the spin of the ith atom ($S_i = 1/2$). For uncorrelated atoms, the Ramsey spectrum is given by $<J_z> = N\cos((\omega_0 - \omega) T_R)$. The best possible uncertainty in the measured value of ω_0 is the shot noise limit of $\Delta\omega_0 = 1/\sqrt{NT_R\tau}$ [10]. This uncertainty can be reduced if we use an entangled state in the following way. Suppose that the N-atom state at the beginning of the free precession time is given by

$$(e^{\frac{-iN\omega_0 t}{2}} |\uparrow_1 \uparrow_2 ... \uparrow_N> + e^{\frac{+iN\omega_0 t}{2}} |\downarrow_1 \downarrow_2 ... \downarrow_N>)/\sqrt{2}. \qquad (2)$$

After the free precession time, a $\pi/2$ Rabi pulse is applied as in usual Ramsey spectroscopy. The signal is obtained by measuring an operator that is the product of the z-components of the Pauli spin matrices, $O = \prod_{i=1}^{N}(\sigma_z)_i$. This gives a signal $\langle O \rangle = (-1)^N \cos(N(\omega_0 - \omega) T_R)$, where ω is the frequency of the applied radiation. The uncertainty in ω_0 is now given by the exact Heisenberg limit $\Delta\omega_0 = 1/N\sqrt{T\tau}$ [63]. This method results in a factor of N decrease in the averaging time τ required to achieve a given precision. Although similar precision gains can be made by increasing N, the optimal value of N in high accuracy ion trap standards might be limited by other experimental constraints, thereby making entangled state spectroscopy advantageous.

APPLICATIONS

Improved frequency standards benefit communications and navigation [66,67] and help in determining some fundamental physical constants [68]. Another possible application of improved frequency standards is in detecting the stochastic background of gravitational radiation [69,70]. This background is similar to the observed cosmic microwave background at 4 K, but has not been detected. Estimates for the density ρ of these gravitational waves vary by many orders of magnitude and are extremely model dependent. A limit on ρ may be determined by timing the bursts of light emitted by pulsars, which are more stable for long times than the best clocks on earth [71]. If gravitational radiation present at the pulsar location is not correlated with the gravitational radiation present at earth, clocks at the two locations will become decorrelated after some time τ_d, and the measured instability $\sigma_y(\tau)$ of the pulsar bursts will begin to increase with averaging time $\tau > \tau_d$. For example, using the estimates from [72], timing measurements from the pulsar J1713+0747 should destabilize at $\tau = \tau_d \sim 2$ y and $\sigma_y(\tau_d) \sim 10^{-14}$ [73]. However, because the current global time scale is stable at only the 10^{-14} level on a time scale of about one year, any observed destabilization could also be from the long-term drifts in the terrestrial time scale. Better clocks are thus required to verify or reject these theories.

Another scientific application of better frequency standards is the laboratory measurement of possible changes in fundamental constants, such as the fine structure constant α. Nonlaboratory measurements include estimates from Sm isotope distributions at the Oklo mine, giving -6.7×10^{-17}/year $< \dot{\alpha}/\alpha < 5.0 \times 10^{-17}$/year [74]. However, this measurement assumes a linear change in α over time and is model dependent. Laboratory measurements can in principle detect nonlinear changes in α over a relatively short time. Because the hyperfine splitting frequency depends nonlinearly on the nuclear charge Z and on α, comparing the frequency of the hyperfine constant in hydrogen A_H to that of another alkali A_{Alkali} gives $\dot{\alpha}/\alpha$ according to [75]

$$\frac{d}{dt} \ln \frac{A_{Alkali}}{A_H} = \alpha \frac{d}{d\alpha} \ln \left(F_{rel}(\alpha Z) \right) \left(\frac{1}{\alpha} \frac{d\alpha}{dt} \right), \qquad (3)$$

where $F_{rel}(\alpha Z)$ is the Casimir correction factor [76]. Recently, the hyperfine structure frequency of ^{199}Hg$^+$ has been compared to that of hydrogen for 140 days to obtain $| \dot{\alpha}/\alpha | \leq 3.7 \times 10^{-14}$/year [75]. Optical atomic transition measurements have also been proposed for detecting changes in α [77]. Clearly, any improvement in the precision of laboratory frequency standards would tighten the limits on $\dot{\alpha}/\alpha$.

SUMMARY

Trapped and cooled ions are particularly suitable systems for precise and accurate spectroscopy. Frequency standards using trapped ions have stabilities and accuracies that are comparable to those of the best of Cs standards and are expected to continue to improve. Some directions toward improving trapped ion frequency standards use optical transitions and entangled states. In addition to improving technological applications, better frequency standards can test fundamental physics.

ACKNOWLEDGMENTS

We thank Lute Maleki, Travis Mitchell, John Prestage, Don Sullivan and Matt Young for valuable comments on this manuscript.

REFERENCES

1. Kinoshita, T., editor, *Quantum Electrodynamics*, World Scientific, Singapore, 1990.
2. Maleki, L., editor, *Proceedings of the Workshop on the Scientific Applications of Clocks in Space*, 1997, JPL Publication 97-15.
3. Bergquist, J., editor, *Proceedings of the 5th Symposium on Frequency Standards and Metrology*, Singapore, 1996, World Scientific.
4. Petley, B. W., *The Fundamental Physical Constants and the Frontier of Measurement*, A. Higler, Bristol, England, 1988.
5. Fisk, P. T. H., *Rep. Prog. Phys.* **60**, 761 (1997).

6. Itano, W. M., Bergquist, J. C., Bollinger, J. J., and Wineland, D. J., *Phys. Rev. ST* **59**, 106 (1995).
7. Bollinger, J. J., Heinzen, D. J., Itano, W. M., Gilbert, S. L., and Wineland, D. J., *IEEE Trans. Instrum. Meas.* **40**, 126 (1991).
8. Fisk, P. T. H., Sellars, M. J., Lawn, M. A., Coles, C., Mann, A. G., and Blair, D. G., *IEEE Trans. Instrum. Meas.* **44**, 113 (1995).
9. Ramsey, N. F., *Molecular Beams*, Oxford University Press, London, 1956.
10. Itano, W. M., Bergquist, J. C., Bollinger, J. J., Gilligan, J. M., Heinzen, D. J., Moore, F. L., Raizen, M. G., and Wineland, D. J., *Phys. Rev. A* **47**, 3554 (1993).
11. Wineland, D. J., Itano, W. M., Bergquist, J. C., Bollinger, J. J., Dietrich, F., and Gilbert, S. L., "High Accuracy Spectroscopy of Stored Ions," in *Proceedings of the 4th Symposium on Frequency Standards*, edited by Demarchi, A., pages 71–76, Heidelberg, 1989, Springer-Verlag.
12. Clairon, A., Ghezali, S., Santarelli, G., Laurent, P., Lea, S., Bahoura, M., Simon, E., Weyers, S., and Szymaniec, K., "Preliminary Accuracy Evaluation of a Cesium Fountain Frequency Standard," in Bergquist [3], pages 49–59.
13. Berkeland, D. J., Miller, J. D., Bergquist, J. C., Itano, W. M., and Wineland, D. J., *J. Appl. Phys.* **83**, 5025 (1998).
14. Berkeland, D. J., Miller, J. D., Bergquist, J. C., Itano, W. M., and Wineland, D. J., *Phys. Rev. Lett.* **80**, 2089 (1998).
15. Tjoelker, R., Prestage, J. D., and Maleki, L., "Record Frequency Stability with Mercury in a Linear Ion Trap," in Bergquist [3], pages 33–38.
16. Bollinger, J. J., Prestage, J. D., Itano, W. M., and Wineland, D. J., *Phys. Rev. Lett.* **54**, 1000 (1985).
17. Huang, X.-P., Bollinger, J. J., Mitchell, T. B., and Itano, W. M., *Phys. Rev. Lett.* **80**, 73 (1998).
18. Tan, J. N., Bollinger, J. J., and Wineland, D. J., *IEEE Trans. Instrum. Meas.* **44**, 144 (1995).
19. Prestage, J. D., Dick, G. J., and Maleki, L., *J. Appl. Phys.* **66**, 1013 (1989).
20. Fisk, P. T. H., Sellars, M. J., Lawn, M. A., and Coles, C., *IEEE Trans. Ultrason., Ferroel. and Freq. Control* **44**, 344 (1997).
21. Raizen, M. G., Gilligan, J. M., Bergquist, J. C., Itano, W. M., and Wineland, D. J., *J. Mod. Optics* **39**, 233 (1992).
22. Paul, W., *Rev. Mod. Phys.* **62**, 531 (1990).
23. Werth, G., *Hyperfine Interactions* **99**, 3 (1996).
24. Knöll, K. H., Marx, G., Hübner, K., Schweikert, F., Stahl, S., Weber, C., and Werth, G., *Phys. Rev. A* **54**, 1199 (1996).
25. Enders, K., Stachowska, E., Marx, G., Zölch, C., Georg, U., Dembczynski, J., and Werth, G., *Phys. Rev. A* **56**, 265 (1997).
26. Wada, M., Okada, K., Wang, H., Enders, K., Kurth, F., Nakamura, T., Fujitaka, S., Tanaka, J., Kawakami, H., Ohtoni, S., and Katayama, I., *Nucl. Phys. A* **626**, 365c (1997).
27. Tanaka, U., Imajo, H., Hayasaka, K., Ohmukai, R., Watanabe, M., and Urabe, S., *Phys. Rev. A* **53**, 3982 (1996).
28. Arbes, F., Benzing, M., Gudjons, T., Kurth, F., and Werth, G., *Z. Phys. D* **31**, 27

(1994).
29. Sunaoshi, H., Fukashiro, Y., Furukawa, M., Yamauchi, M., Hayashibe, S., Shinozuka, T., Fujioka, M., Satoh, I., Wada, M., and Matsuki, S., *Hyperfine Interactions* **78**, 241 (1993).
30. Matsubara, K., Tanaka, U., Imajo, H., Hayasaka, K., Ohmukai, R., Watanabe, M., and Urabe, S., *Appl. Phys. B* **67**, 1 (1998).
31. Tamm, C., Schnier, D., and Bauch, A., *Appl. Phys. B* **60**, 19 (1995).
32. Enders, V., Courteille, P., Huesmann, R., Ma, L. S., Neuhauser, W., Blatt, R., and Toschek, P. E., *Europhys. Lett.* **24**, 325 (1993).
33. Seidel, D. J., Williams, A., Berends, R. W., and Maleki, L., "The Development of the Ytterbium Ion Frequency Standard," in *1992 IEEE Frequency Control Symposium*, pages 70–75, New York, 1992, IEEE.
34. Prestage, J. D., Tjoelker, R. L., Dick, G. J., and Maleki, L., *J. Mod. Optics* **39**, 221 (1992).
35. Maleki, L., MS 298-100, JPL, 4800 Oak Grove Drive, Pasadena, CA 91109, private communication.
36. Poitzsch, M. E., Bergquist, J. C., Itano, W. M., and Wineland, D. J., *Rev. Sci. Instrum.* **67**, 129 (1996).
37. Fisk, P. T. H., National Measurement Laboratory, CSIRO Division of Telecommunications and Industrial Physics, PO Box 218, Lindfield NSW 2070, Sydney, Australia, private communication.
38. Tanaka, U., Kansai Research Center, Communications Research Laboratory, 588-2, Iwaoka, Nishi-ku, Kobe 651-24, Japan, private communication.
39. Itano, W. M., Lewis, L. L., and Wineland, D. J., *Phys. Rev. A* **23**, 1233 (1982).
40. Lee, W. D., Shirley, J. H., Lowe, J. P., and Drullinger, R. E., *IEEE Trans. Instrum. Meas.* **44**, 120 (1995).
41. Tjoelker, R., Prestage, J. D., and Maleki, L., "Long Term Stability of Hg^+ Trapped Ion Frequency Standards," in *Proceedings of the 1993 IEEE International Frequency Control Symposium*, pages 132–138, Piscataway, NJ, 1993, IEEE.
42. Drullinger, R. E., Shirley, J. H., and Lee, W., "NIST-7, the U.S. Primary Frequency Standard: New Evaluation Techniques," in *28th Annual PTTI Application and Planning Meeting*, pages 255–264, 1997.
43. Clairon, A., 1998, presented at the 1998 Conference on Precision Electromagnetic Measurements.
44. Taylor, P., Roberts, M., Gateva-Kostove, S. V., Clarke, R. B. M., Barwood, G. P., and Rowley, W. R. C., *Phys. Rev. A* **56**, 2699 (1997).
45. Taylor, P., Roberts, M., Barwood, G. P., and Gill, P., *Opt. Lett.* **23**, 298 (1998).
46. Tamm, C. and Engelke, D., "Optical Frequency Standard Investigations on Trapped ^{171}Yb Ions," in Bergquist [3], pages 283–288.
47. Barwood, G. P., Gill, P., Huang, G., Klein, H. A., and Rowley, W. R. C., *Opt. Commun.* **151**, 50 (1998).
48. Madej, A. A., 1998, these proceedings.
49. Madej, A. A., Siemsen, K. J., Whitford, B. G., Bernard, J. E., and Marmet, L., "Precision Absolute Frequency Measurements with Single Atoms of Ba^+ and Sr^+," in Bergquist [3], pages 165–170.

50. Urabe, S., Watanabe, M., Imago, H., Hayasaka, K., and Tanaka, U., "Observation of Doppler sidebands of a laser-cooled Ca^+ ion by using a low temperature-operated laser diode," in *Internation Workshop on Current Topics of Laser Technology*, page 55, Tokyo, 1998, Communications Research Laboratory.
51. Knoop, M., Vedel, M., and Vedel, F., *J. Phys. (Paris) II* **4**, 1639 (1994).
52. Plumelle, F., Desaintfuscien, M., and Houssin, M., *IEEE Trans. Instrum. Meas.* **42**, 462 (1993).
53. Peik, E., Hollemann, G., Abel, J., v. Zanthier, J., and Walther, H., "Single-ion Spectroscopy of Indium: Towards a Group-III Monoion Oscillator," in Bergquist [3], pages 376–379.
54. Nagourney, W., Burt, E., and Dehmelt, H. G., "Optical Frequency Standard Using Individual Indium Ions," in Bergquist [3], pages 341–346.
55. Bergquist, J. C., Itano, W. M., and Wineland, D. J., "Laser Stabilization to a Single Ion," in *Frontier in Laser Spectroscopy*, edited by Hänsch, T. and Inguscio, M., pages 359–376, Amsterdam, 1994, North Holland.
56. Young, B. C., Cruz, F. C., and Bergquist, J. C., in preparation.
57. Riehle, F., Schnatz, H., Lipphardt, B., Zinner, G., Kersten, P., and Helmcke, J., "Optical Frequency Standard Based on Laser-cooled Ca Atoms," in Bergquist [3], pages 277–282.
58. Zibrov, A. S., Fox, R. W., Ellingsen, R., Wiemer, C. S., Velichansky, V. L., Tino, G. M., and Hollberg, L., *Appl. Phys. B* **59**, 327 (1994).
59. Frech, B., Wells, J., Hollberg, L., Oates, C., Young, B. C., and Bergquist, J. C., in preparation.
60. Schnatz, H., Lipphardt, B., Helmcke, J., Riehle, R., and Zinner, G., *Phys. Rev. Lett.* **76**, 18 (1996).
61. Barwood, G. P., Edwards, C. S., Gill, P., Huang, G., Klein, H. A., and Rowley, W. R. C., *IEEE Trans. Instrum. Meas.* **44**, 117 (1995).
62. Lea, S. N. and Gill, P., "Proposed Infra-red to Visible Frequency Chain at NPL," in Bergquist [3], pages 507–508.
63. Bollinger, J. J., Itano, W. M., Wineland, D. J., and Heinzen, D. J., *Phys. Rev. A* **54**, R4649 (1996).
64. Wineland, D. J., Monroe, C., Itano, W. M., Leibfried, D., King, B., and Meekhof, D., *NIST J. Res.* **103**, 259 (1998).
65. Feynman, R. P., Vernon, F. L., and Hellwarth, R. W., *J. Appl. Phys.* **28**, 49 (1957).
66. Hellwig, H., *IEEE Trans. Ultrason., Ferroel. and Freq. Control* **40**, 538 (1993).
67. Vig, J. R., *IEEE Trans. Ultrason., Ferroel. and Freq. Control* **40**, 522 (1993).
68. Petley, B. W., *Proc. IEEE* **79**, 1070 (1993).
69. Detweiler, S., *Astrophys. J.* **234**, 1100 (1979).
70. Thorne, K. S., *300 Years of Gravitation*, pages 330–458, Cambridge University Press, Cambridge, 1987.
71. Taylor, J. H., *Proc. IEEE* **79**, 1054 (1991).
72. Vachaspati, T. and Vilenkin, A., *Phys. Rev. D* **31**, 3052 (1985).
73. Foster, R. S., Camilo, F., and Wolszczan, A., "High-precision Metrology from Pulsar J1713+0747," in *Marcel Grossmann Conference on General Relativity*, 1995.
74. Damour, T. and Dyson, F., *Nucl. Phys. B* **480**, 37 (1996).

75. Prestage, J. D., Tjoelker, R. L., Dick, G. J., and Maleki, L., *Phys. Rev. Lett.* **74**, 3511 (1995).
76. Casimir, H. B. G., *On the Interaction between Atomic Nuclei and Electrons*, Freeman, San Francisco, 1963, p. 54.
77. Dzuba, V. A., Flambaum, V. V., and Webb, J. K., "Space-time Variation of Physical Constants and Relativistic Corrections in Atoms," 1998, submitted to Phys. Rev. Lett.; preprint physics/9802029 on http://xxx.lanl.gov.

Precision Spectroscopy of Helium

Pablo Cancio[†], Maurizio Artoni[†], Giovanni Giusfredi[††],
Francesco Minardi[†], Francesco S. Pavone[†1] and Massimo Inguscio[†2]

[†]*European Laboratory for Non-linear Spectroscopy (LENS), Istituto Nazionale di Fisica Nucleare (INFN) and Istituto Nazionale di Fisica della Materia (INFM) Largo, E. Fermi 2, I-50125 Firenze, Italy*
[††]*Istituto Nazionale di Ottica (INO), Largo, E. Fermi 6, I-50125 Firenze, Italy*

Abstract. High-precision laser spectroscopy of helium is a powerful tool for testing QED calculations and for an accurate determination of the fine-structure constant α. Here, we report a new measurement of the fine-structure intervals of the $2\,^3P$ level of helium with an accuracy of kHz. They are measured from the frequency difference between $2\,^3S_1 \rightarrow 2\,^3P_J$ optical transitions observed with fluorescence-saturation spectroscopy in an atomic beam, in the absence of magnetic fields. A discussion of the systematic effects in the measurements is given, describing for the first time in helium the shift induced by mechanical effects of the light. Finally, a comparison with previously reported measurements is discussed.

INTRODUCTION

Different singlet and triplet levels of neutral helium (^{4}HeI) with $n > 2$ are accessible by one or two photon transitions from the $1S$ ground state or from the $2\,^1S_0$, $2\,^3S_1$ metastable states [1]–[9]. The energy levels are calculated by means of a power-series perturbation expansion of the fine-structure constant α, where the nonrelativistic terms are used as the zero-order, i.e. unperturbed, Hamiltonian [10]. Since the nonrelativistic eigenvalues are known within 1 part in 10^{15}, high-precision spectroscopy of the above-mentioned transitions can be used as a check for the calculated high-order radiative corrections, which are the limiting factor in the theoretical energy-level calculation. On the other hand, high-precision spectroscopy measurements of the fine-structure splittings of helium levels, when combined with an accurate theoretical determination from perturbation theory, lead to a new accurate value of α [11]. In this regard, we present a new accurate measurement of the fine-structure intervals of the $2\,^3P$ level by high-precision laser spectroscopy on the $2\,^3S_1 \rightarrow 2\,^3P$ transitions.

[1)] Present address: Collège de France et Ecole Normale Supérieure, 24 rue Lhomond, 75005 Paris, France
[2)] also: Dipartimento di Fisica, Università di Firenze, Largo, E. Fermi 2, I-50125 Firenze, Italy

Review of High-Precision Laser Spectroscopy Measurements in Helium

Among all precise helium energy-level measurements, the largest part of them refer to Lamb-shift determinations in the lower states because it scales with the principal quantum number as n^{-3}. The Lamb shift is obtained as the difference between the transition frequency and the well-known non-QED energy values of the levels involved. A theoretical prediction of Lamb shifts has been carried out considering the perturbing terms to order $O(\alpha^5 + \alpha^5 \log \alpha)$ a.u. [12,13] and to terms $O(\alpha^6)$ a.u. only for S levels [14]. The evaluation of the latter and of the Bethe logarithm [15] limits the accuracy of the calculated values.

Early experiments were performed using the two metastable $2S$ states. Starting from the triplet $2\,^3S_1$ state, our group performed the first pure frequency measurement in helium on the $2\,^3S_1 \to 3\,^3P_0$ transition at 389 nm [1] with an accuracy of 190 ppm in the Lamb-shift determination. Later, a $2\,^3S_1$ Lamb-shift value accurate to 15 ppm was measured by exciting the $2\,^3S_1 \to 3\,^3D_2$ transition [2]. The $2\,^1S_0$ state Lamb shift was investigated by laser spectroscopy of $2\,^1S_0 \to n\,^1D_2$ ($7 \geq n \geq 20$) two-photon transitions [3] and of $2\,^1S_0 \to n\,^1P_1$ ($7 \geq n \geq 74$) transitions [4], with accuracies of 53 ppm and 75 ppm, respectively. Theoretical Lamb shifts for $2S$ levels are only accurate to 8 MHz, but they are in quite good agreement with the measured values.

Recently, two groups have been able to perform laser spectroscopy on UV transitions from the $1S$ ground state. The $1\,^1S_0 \to 2\,^1P_1$ transition at 58.4 nm has been excited and its frequency measured with 45 MHz precision [5]. The two-photon transition $1\,^1S_0 \to 2\,^1S_0$ at 120 nm, which was measured within 48 MHz [6], is perhaps even more promising because of its 8 Hz natural linewidth. The ground-state Lamb shifts extracted from these experiments at the 10^{-3} level differ from each other by 1.8 times the combined standard deviations. The theoretical estimate reported in [16] carries an uncertainty of only 35 MHz, but uncalculated terms $O(\alpha^4)$ a.u. are expected to be of the order of 300 MHz.

Regarding the fine structure, the $2\,^3P$ level is the most interesting for a high-precision measurement linked with an accurate determination of α (Fig. 2). The group of Hughes at Yale has performed the first sub-MHz measurement, where an optical-microwave atomic-beam magnetic-resonance technique was used [7]. A more precise measurement was carried out using wavelength differences of the laser-excited $2\,^3S_1 \to 2\,^3P$ atomic-beam magnetic resonances [8], but there appeared an outstanding disagreement with previous results (more than five standard deviations in the best case, see table 3). Recently, another measurement for the $2\,^3P_0$-$2\,^3P_1$ interval, performed in the microwave domain [9], seems to be in good agreement with the Shiner et al. measurement, but at present its accuracy is limited to 13 kHz because of as yet uncontrolled systematic effects. All these measurements were performed in presence of moderate or high magnetic fields. From the theoretical point of view, kHz-level accuracy in the calculated fine-structure splittings can be attained by considering the perturbing terms up to $O(\alpha^4)$ a.u. [17] and up to $O(\alpha^5 + \alpha^5 \log \alpha)$ a.u. [18]. Actually, there are two predictions for the $2\,^3P$ intervals with accuracies of 15 kHz [16] and 20 kHz [18], limited respectively by $O(\alpha^5)$ a.u. and $O(\alpha^5 \log \alpha)$ a.u. contributions, but the values differ more than one standard deviation for the $2\,^3P_0$-$2\,^3P_1$ interval. The results of our experimental approach, based on heterodyne frequency measurements of the $2\,^3S_1 \rightarrow 2\,^3P$ optical transitions observed in an atomic beam in absence of any magnetic field, can help clarify the actual experimental and theoretical situation.

α and Helium Fine Structure

Among energy-dependent coupling constants of field-matter interactions, the fine-structure constant α is known with high accuracy from measurements of completely different physical systems. The consistency between these measurements gives strong support to the reliability of its value.

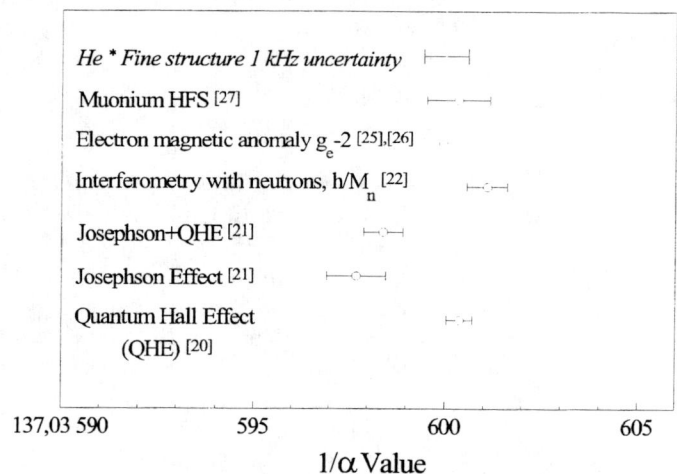

FIGURE 1. Determinations of α from different physical measurements.

Formally, we distinguish between QED-independent and QED-dependent determinations, depending on whether or not theoretical support from QED theory is needed for the measurement. The former includes measurements of the quantum Hall effect (QHE) [20], the Josephson effect [21], and the Planck constant to neutron mass ratio (h/M_n) [22]. The 24-ppb accurate QHE measurement is based on the determination of the von Klitzing constant R_K (equal to $\mu_0 c/2\alpha$ in SI units) from the measured quantized resistance. An accuracy of 56 ppb is inferred from the measured frequency-to-voltage ratio E_J in a Josephson junction: $\alpha^{-2} = (cE_J\mu_p)/(4\mu_B\gamma_p' R_\infty)$ where the limiting factor is the proton gyromagnetic ratio γ'_p value. A more accurate value (37 ppb) is determined by combining both measurements ($\alpha^{-3} = (E_J \mu_p R_K)/(2\mu_0 \mu_B \gamma'_p R_\infty)$ [21]. From the (h/M_n) measurement ($\alpha^2 = (2R_\infty/c)(M_n/m_e)(h/M_n)$), α was determined with an accuracy of 39 ppb. Finally, α can be calculated from h/M_{atom} ($\alpha^2 = (2R_\infty/c)(M_{atom}/m_e)(h/M_{atom})$) determined from an accurate atomic-recoil velocity measurement [23]: a 5 ppb accurate α value is anticipated soon from an experiment with cesium atoms [24]. On the other hand, the most accurate α value (4 ppb) comes from the comparison between the precisely measured electron magnetic anomaly $g_e - 2$ (3.9 parts in 10^{-9}) [25] and the calculated value from the QED theory up to order $O((\alpha/\pi)^4)$ [26]. However, this measurement is usually used as a "test of QED" with an assumed α value. Also, another QED-dependent determination is available from the muonium fine structure with an α value accurate to 58 ppb, thanks to the recently improved value of m_μ/m_e [27].

A plot of the above mentioned determinations shows the puzzling situation of Fig. 1, with a recommended α^{-1} value of 137.0359895 (61) (45 ppb). The situation can be clarified with a new determination stemming from a precision-spectroscopy measurement of the fine-structure energy of a simple atomic system combined with its calculated QED value. In this way, several α values were determined from fine-structure spectroscopic measurements on the ^2P level of hydrogen, but more accurate results were expected from precision spectroscopic measurements on the $2\,^3P$ level of helium [11]. In fact, an α value accurate to 38 ppb would be obtained with 1 kHz measurement and an accurate theoretical determination of the $2\,^3P$ fine-structure intervals.

A NEW MEASUREMENT FOR THE $2\,^3P$ FINE-STRUCTURE INTERVALS

Experiment

Our measurements are based on a heterodyne frequency experiment, as illustrated in Fig. 2. A first laser source, the so-called *master laser*, is frequency locked to a stable reference. A second laser, the *slave laser*, is successively tuned on two different fine-structure transitions and its frequency is measured with respect to that of the master laser. By subtraction of the two line-center frequencies the fine-structure interval is readily obtained. This procedure is used for the three measured intervals.

We observe the $2\,^3S_1 \rightarrow 2\,^3P_J$ ($J = 0, 1, 2$) transitions of Helium at 1083 nm by detecting the saturated fluorescence from a beam of metastable atoms. The measurements are performed in the absence of any bias magnetic field. The experimental setup is illustrated in Fig. 3, and the principal elements are described below.

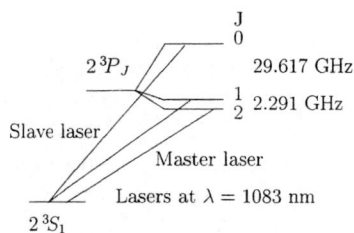

FIGURE 2. Scheme of helium energy levels relevant for our experiment.

Beam of metastable atoms

Our sample is a beam of helium atoms, excited to the metastable $2\,^3S_1$ level by electronic collisions in a DC discharge, similar to that described in [28]. From the detected fluorescence signal, we estimate a flux of metastable helium atoms of about 10^{15} atoms/s per steradian. Helium atoms are cooled to the temperature of liquid nitrogen before excitation, but they are subsequently heated by the longitudinal electronic collisions. This is confirmed by the recorded longitudinal-velocity distribution of the atomic flux, whose most probable value is about 2.5×10^3 m/s. The distribution is obtained from the Doppler profile of the $2\,^3S_1 \rightarrow 2\,^3P_2$ transition and is well fitted by a "generalised Maxwellian":

$$\frac{d\Phi}{dv} \propto \left(\frac{v}{u}\right)^\beta \exp\left(-\left(\frac{v}{u}\right)^2\right) \qquad (1)$$

with $\beta = 6.5(1)$ and $u = 1.56(7) \times 10^3$ m/s. The transverse velocity distribution gives rise to a typical Doppler-broadened linewidth of 120 MHz (FWHM).

After excitation, the metastable atoms fly downstream to the interaction region, where we realize a perturbation-free environment. A 1 mm thick high magnetic permittivity (mu-metal) enclosure screens the external electric and magnetic fields. Together with a pair of internal Helmholtz coils, it allows us to minimize the residual magnetic field to less than 0.1 μT in the 1 cm^3 interaction volume.

Laser sources

The 1083 nm radiation sources are accurately controlled Distributed Bragg Reflector (DBR) semiconductor lasers (SDL mod. 6702-H). The free-running linewidth of about 3 MHz is narrowed by a factor of 12 (i.e. to about 200 kHz) thanks to the extended-cavity configuration [30]. We stabilize the master laser frequency on the 1083 nm helium saturated absorption of an RF-discharge in a cell. For this purpose, the locking signal chosen depends on the fine-structure interval measured: we use the fine-structure transition that is not involved in the interval measurement (e.g. for the ν_{01} interval, the master laser was locked onto the

$2\,^3S_1 \to 2\,^3P_2$ transition). From the error signal, the short-term lock stability was measured to be 7 kHz in 0.5 s (root Allan variance), while the long-term frequency stability is assessed *a posteriori* to be on the order of 20 kHz/hour.

The slave laser is phase-locked to the master frequency by beating the two lasers on a fast photodiode and controlling their frequency difference using a microwave synthesizer as a local oscillator. In this way, the phase-lock ensures the slave laser to be a frequency shifted replica of the master laser on frequency intervals up to 40 GHz with a precision of 0.2 Hz in 1 s (root Allan variance); also, it allows a very accurate tuning of the slave laser across He* transitions by simply varying the microwave oscillator frequency.

After spatial filtering with a pinhole, the slave-laser beam shows a Gaussian-shaped intensity profile with a waist (radius at $1/e^2$ intensity) of 2.1(2) mm. Then, the linearly polarised slave laser interacts orthogonally with helium atomic beam in a saturation-spectroscopy configuration: a retro-reflecting dielectric mirror generates the standing wave that excites the helium fine-structure transitions. An accurately studied optical system before and after the interaction region ensures that both counter-propagating beams have plane-wave fronts. The laser power in the interaction region for each beam is 150(10) μW (i.e., 2.1 mW/cm^2) with a power imbalance less than 1 %.

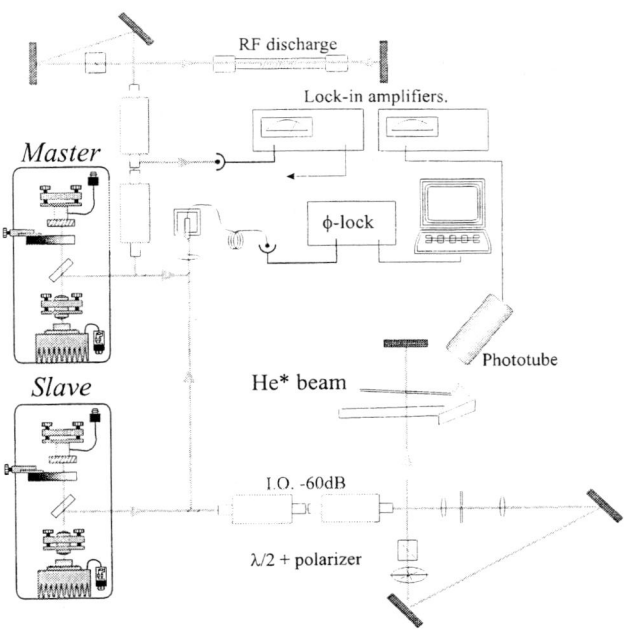

FIGURE 3. Experimental Setup.

Signal acquisition and analysis

Saturated atomic excitation is detected via fluorescence photons that are collected over a solid angle of $0.3 \times 4\pi$ steradian and then guided to a cooled photomultiplier with a Plexiglas lightpipe. The collection optics has been carefully designed to guarantee that all source points in the interaction region uniformly illuminate the photocathode.

The saturation dip is recorded by scanning the slave laser over fine-structure transition resonances. The dip is isolated from the residual gaussian Doppler pedestal by employing wavelength modulation of the laser at 3.23 kHz ("dither") and phase-sensitive third-harmonic demodulation. To acquire one spectrum we sweep the laser detuning twice at 101 points via computer control of the microwave synthesizer, with an integration time of 0.5 s/pt for a total scan of 50 MHz. The S/N ratios are 130, 50, 20 for the three $2\,^3S_1 \rightarrow 2\,^3P_J$ probed by transitions with $J = 2, 1, 0$ respectively.

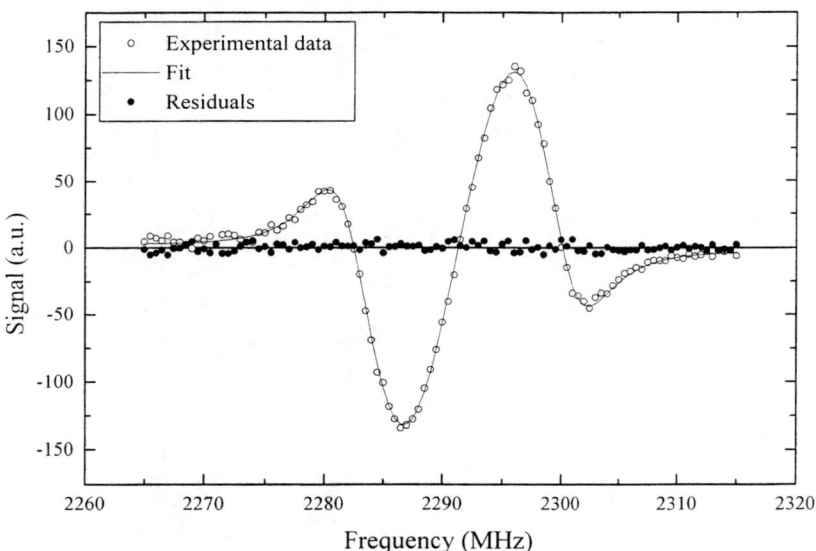

FIGURE 4. Spectrum, fit and residuals of the $2\,^3S_1 \rightarrow 2\,^3P_1$ transition.

The recorded spectra are fitted by a least-squares procedure to a function which takes into account possible background fluctuation, together with the saturation feature:

$$f(\nu) = a + b(\nu - \nu_0) + c \cdot \Gamma \left(\frac{i}{m}\right)^3 \frac{\left[\sqrt{\Gamma - i(\nu - \nu_0)^2 + m^2} - \gamma + i(\nu - \nu_0)\right]^3}{\sqrt{(\Gamma - i(\nu - \nu_0))^2 + m^2}} + c.c., \qquad (2)$$

where the baseline coefficients a and b, the modulation depth of the laser m, the line amplitude c, the linewidth Γ and the line-center frequency ν_0 are the six fitted parameters. The validity of this function is confirmed by the good agreement between the fitted values and the expected values of the c, m and Γ parameters, and by the flat residuals, as shown in Fig. 4 for the $2\,S \rightarrow 2\,^3P_1$ transition. Also, the χ^2/degrees of freedom ratio is close to one. From the fit, we measure the line center with a precision of 10 to 20 kHz, depending of the S/N ratio of the transition considered.

Statistical Results

To improve our precision, we take the weighted average of the several single frequency differences, measured over several months under different experimental conditions. So far, We have performed 230 measurements for the $2\,^3P_1 - 2\,^3P_2$ interval (ν_{12}), 190 measurements for the $2\,^3P_0 - 2\,^3P_1$ interval (ν_{01}), 230 measurements for the $2\,^3P_0 - 2\,^3P_2$ interval (ν_{02}), with the statistical result shown in the table 1.

TABLE 1. Statistical values and uncertainties of the three fine-structure intervals of the 2 P level of Helium.

Fine Structure interval	Statistical value [kHz]	Standard deviation [kHz]
$2\,^3P_1 - 2\,^3P_2$ (ν_{12})	2 290 971.8	2.4
$2\,^3P_0 - 2\,^3P_1$ (ν_{01})	29 616 931.3	1.8
$2\,^3P_0 - 2\,^3P_2$ (ν_{02})	31 907 907.9	3.4

The statistical uncertainty varies from 3.4 kHz for the worst case, to 1.9 kHz for the best case. These error values are directly correlated with the single-measurement reproducibility: as shown in Fig. 5, the single-measurement dispersion with respect to the mean value is largest for the ν_{02} interval. The main sources of statistical error are attributed to the master frequency long-term instability and to the fluctuations of the atomic beam velocity and intensity. The sum of the ν_{12} and ν_{01} mean values is 31907.904 (3) MHz, which is in good agreement with the ν_{02} mean value. This means that all measured intervals are affected by the same systematics uncertainties.

Systematic Effects

We performed checks in order to identify and eliminate (or at least minimize) systematic effects acting on the fine-structure intervals. In Table 2 we report the error budget, summarizing the corrections and uncertainties used to calculate the final values of the three measured intervals.

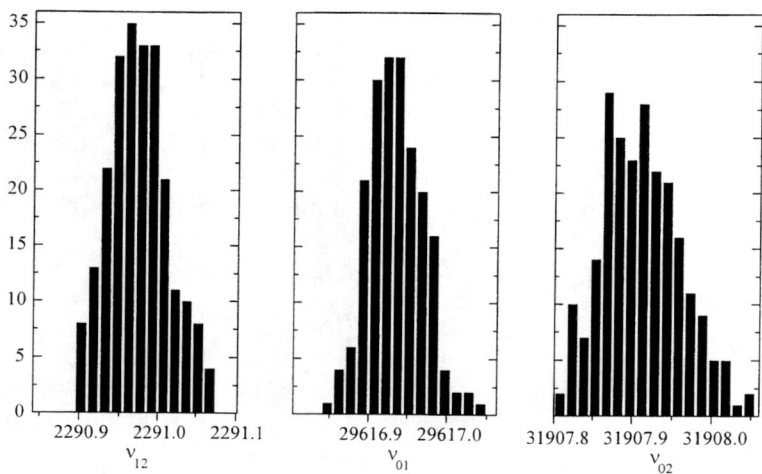

FIGURE 5. Statistical histogram of the 230, 195, 230 single measurements of the ν_{12}, ν_{01}, ν_{02} intervals, respectively. Bar-width 15 kHz.

First of all, we considered the systematic error due to a mutual misalignment of the counter-propagating laser beams, combined with the collimation of the atomic beam. This effect produces a shift of about 0.05 MHz for each transition, but it does not affect our fine-structure measurements because it cancels in subtracting the line-center frequencies. In any case, we randomize it by a periodic realignment of the retro-reflected beam. The second-order Doppler shift is about 14.4 (4) kHz for each transition, but, as above, it cancels by subtraction.

With our apparatus, we are able to control very precisely the magnetic field in the interaction region by means of the Hanle effect. We have found that a shift of the 0.05(1) MHz for the ν_{01} interval occurs if a 16 μT magnetic field is applied parallel to the atomic velocity and orthogonal to the pure (better than 99.9 %) π laser polarization. Since such a shift cannot be explained solely by the residual ellipticity of laser polarization, we suppose it to be due to a combination of optical pumping and Larmor precession [31]. Because of these significant Zeeman shifts, we believe that the absence of any magnetic field in the interaction region represents a major improvement over previous experiments. With a commercial magnetometer, we have measured a residual magnetic field of 0.1 μT in 1 cm^3 around the interaction region, and then minimised *in situ* with the Hanle effect. Therefore, assuming that the residual Zeeman effect is first-order in the field, we estimate an uncertainty of only 0.3 kHz in our experimental conditions. Also the mu-metal shield makes any systematic effect arising from spurious electric fields negligible.

Thanks to third-harmonic detection, the effect of the small residual amplitude modulation due to the dither is reduced to 0.1 kHz. The correction associated with the time-base calibration of the microwave synthesizer is $2.44(1) \times 10^{-7}$. It was determined by comparison with a Rb secondary frequency standard (stability better than 1 part in 10^{11}) traceable to Cs primary frequency standard.

TABLE 2. Budget of corrections and uncertainties of the three fine-structure intervals of the 2 P level of Helium.[a]

	ν_{12}	ν_{01}	ν_{02}
Statistical uncertainty	2.4	1.8	3.4
Second–order Doppler	<0.002	<0.002	<0.002
Residual magnetic field	<0.3	<0.3	<0.3
Residual amplitude modulation	<0.1	<0.1	<0.1
Synthesizer calibration	+7.23(3)	+7.23(3)	+7.23(3)
AC–Stark shift	+0.78(8)	-0.49(6)	+0.29(6)
Recoil–induced shift	+194(15)	+11.7(9)	+206(15)
Final correction and uncertainty	+202.0(15.2)	+18.4(2.0)	+213.0(15.4)

[a] All values are in kHz.

The shift induced by the radiation forces exerted on the atoms by the laser (the so-called recoil-induced shift) is the major systematic effect affecting our measurements, with values up to $0.1\,\Gamma$ for the *closed* transition $2\,^3S_1 \to 2\,^3P_2$. Since it has not been investigated before for high-precision saturation spectroscopic measurements on helium, we devote next section to a more detailed discussion.

Finally, we consider the AC-Stark shift induced by the presence of neighbouring energy levels. The AC-Stark shift correction for each interval is calculated from the difference between the correspondent transition AC-Stark shifts. Unfortunately, we are not able to measure them in the usual experimental manner, i.e. by linear extrapolation to null laser power of the transition frequency. A change in the laser power means a change of the Rabi frequency, which implies a change of the recoil-induced shift. As this shift is, at least, an order of magnitude larger than the AC-Stark shift under our experimental conditions, experimentally, the AC-Stark effect is swamped by the recoil-induced shift. As an experimental check we have measured the ν_{01} interval with double the laser power: the statistical value at 300 μW is 3 kHz less than the statistical value at 150 μW, contrary to what is expected in the case of a pure AC-Stark shift. On the other hand, the usual second-order perturbation theory shows that the interval corrections from this effect are on the order of a few kHz, which means that the AC-Stark shift cannot be neglected *a priori* to our accuracy level. In our analysis, we used the optical Bloch equations (OBE) to calculate the time evolution of the atomic system interacting with the laser standing wave. The OBE are solved by introducing the usual rotating wave approximation and an "adiabatic" elimination of the rapidly oscillating coherences, which we take at their steady-state values. Accordingly, these terms are expressed as algebraic functions of the density-matrix elements corresponding to the two levels connected by the resonant excitation. Finally, we calculate the shift of each fine-structure transition by numerical integration of the reduced OBE system, obtaining the following values:

$$\Delta f_{ls}(2\,^3S_1 \to 2\,^3P_2) = +0.239(54) \text{ kHz}$$
$$\Delta f_{ls}(2\,^3S_1 \to 2\,^3P_1) = -0.541(54) \text{ kHz}$$
$$\Delta f_{ls}(2\,^3S_1 \to 2\,^3P_0) = -0.050(21) \text{ kHz}.$$

The AC-Stark–shift corrections to the intervals, given in table 2, are calculated from these values.

Laser-Cooling–Induced Saturation-Dip Shift

The interaction of a resonant light beam with an atomic system affects not only its internal state, but also its center-of-mass motion, which is described as the mechanical effects of light on the atoms. The atomic cooling [29] and atomic trapping in potential wells [32,33] by exploiting the effects of the dissipative and conservative light forces, respectively, are well known. From the point of view of high-resolution spectroscopy, the mechanical effects of the light can produce line-shape asymmetries and, therefore, line-center shifts such as those observed for a two-level atomic system in a standing wave [34,35]. This is our case, where an atomic helium beam is orthogonally probed by the weak laser standing wave formed in the saturation spectroscopy experiment.

The effect can be qualitatively described as follow: we consider that the helium atoms contribute to the fluorescence saturation dip (i.e. $kv_z \lesssim \Gamma$). The mean radiation force over these atoms in the direction of the standing wave propagation is:

$$F_z = F_{Reactive}(z) + F_{Dissipative}(v_z)$$
$$= \frac{\hbar\delta}{2}\nabla\left\{ln\left[1 + \frac{1}{2}\frac{\Omega(z)^2}{\Gamma^2 + 4\delta^2}\right]\right\} + \frac{kv_z}{\Gamma}\frac{\hbar k\Gamma}{2}\frac{\delta\Gamma F^2(S)}{\Gamma^2 + 4\delta^2}, \quad (3)$$

where Ω is the Rabi frequency, $\delta = \omega - -\omega_0$ is the laser detuning, and $F(S)$ is a function of the saturation parameter S=$2\Omega^2/(\Gamma^2 + 4\delta^2) \simeq 1$. The reactive part of this force is a dipole force that channels the slow atoms ($kv_z \ll \Gamma$) in a periodic potential: the atoms localize in the nodes or antinodes of the standing wave, depending on whether the detuning is positive or negative, respectively. As a consequence, there is more fluorescence when $\delta < 0$ and less fluorescence when $\delta > 0$, which means an asymmetry in the fluorescence dip. The dissipative part of the force changes the transverse velocity distribution of the atoms: they are cooled or heated by a negative or positive laser detuning, respectively. As before, there is more fluorescence when the atoms are velocity damped ($\delta < 0$) and less fluorescence in the contrary case ($\delta > 0$). The overall effect is more fluorescence in the blue side of the dip and therefore a shift of the line center towards the blue. The shift scales with the number of the absorption-emission cycles, which establishes the process time scale.

We have studied the line-center shift for different interaction times from both experimental and theoretical side. Experimentally, this time is changed by varying the laser beam width in the direction of the atomic beam propagation. A variable slit in the laser beam path allows us to change the interaction time between 4τ and 32τ (τ is the lifetime of the $2\,^3P$ level) for the 2500 m/s longitudinal velocity atoms. Experimental results are shown in Fig. 6 (a) for the $2\,^3S_1 \rightarrow 2\,^3P_{1,2}$ transitions. Here, the horizontal scale is the slit width in mm instead of τ intervals since the interaction time is not the same for all the atoms; on the vertical scale, the line-center shift is given with respect to the line-center frequency obtained with an unobstructed laser beam in order to preserve the advantages of relative frequency measurements of our apparatus. For the $2\,^3S_1 \rightarrow 2\,^3P_2$ transition, the shift increases with the width of the interaction region, whereas it seems constant after several mm for the other transition. This is because of our experimental configuration, where π transitions between Zeeman sublevels are allowed by the laser: $2\,^3S_1 \rightarrow 2\,^3P_2$ is a *closed* transition and $2\,^3S_1 \rightarrow 2\,^3P_1$ is an *open* one, where the atoms are pumped to the $2\,^3S_1(m = 0)$ Zeeman sublevel after a few absorption-emission cycles and the effect "saturates." The situation is similar for the $2\,^3S_1 \rightarrow 2\,^3P_0$ transition, where $2\,^3S_1(m = \pm 1$ Zeeman sublevels are the dark states. On the theoretical side, such behaviour is analysed with a simple 1D model based on numerical integration of the optical Bloch equations of the levels involved in each transition and the time evolution of the kinematics (transverse velocity and position) of the atomic beam under the action of the radiation force $F = <d_{ud}> \nabla E(z,t)$. (Here, $<d_{ud}>$ is the dipole moment of the transition between upper and lower levels, and z is the laser beam propagation direction [36].) The actual experimental conditions for the laser and atomic beam characteristics are considered in the calculations. The fluorescence signal is calculated from the integrated population of the excited level for a laser detuning between -20 and $+20$ MHz. From this, we obtain the shift of the saturation dip center as is shown in Fig. 6 (b) for the $2\,^3S_1 \rightarrow 2\,^3P_{0,1}$ transitions. The good agreement between the numerical results and the experimental data (see Fig.6 (a) for the $2\,^3S_1 \rightarrow 2\,^3P_1$ transition) indicates that the effect is correctly described by our theoretical approach for the case of "open" transitions. We are also developing a theory for the $2\,^3S_1 \rightarrow 2\,^3P_2$ transition which reproduces the experimental result, but the comparison with experimental data is considerably more difficult because of the large sensitivity to atomic velocity.

With respect to the fine-structure interval measurements, the magnitude of these shifts under our experimental conditions ranges between $\Gamma/100$ and $\Gamma/10$, depending on whether the $2\,^3S_1 \rightarrow 2\,^3P_2$ transition is included in the interval or not, respectively. At first, we tried to cancel the shifts in the difference-frequency measurements by establishing experimentally the Rabi frequency conditions at which the recoil-induced shift was the same for both transitions of each interval. This meant changing the power and width of the laser beam for each single transition measurement, which could produce other unwanted systematic effects (for example, a change of the laser optical path between two consecutive measurements). Because of the good agreement between experiment and theory for the "open" transitions,

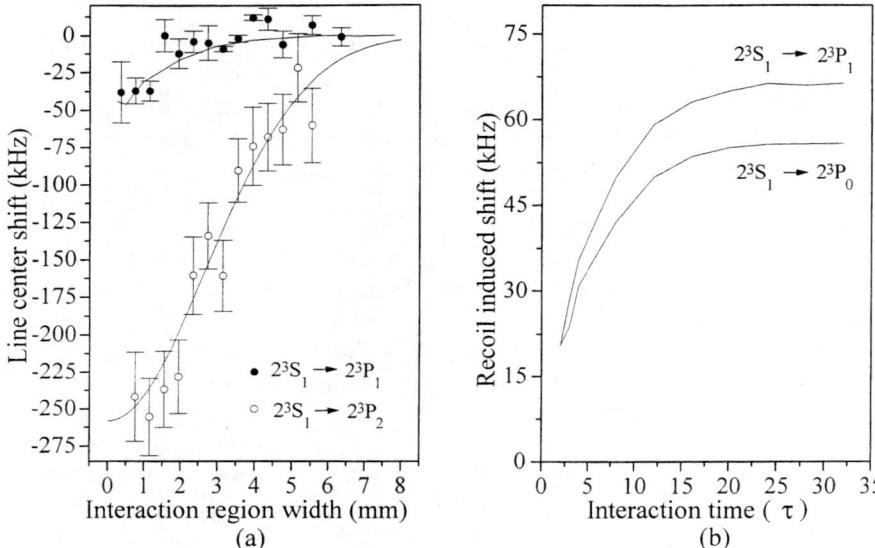

FIGURE 6. Recoil-induced shift vs. interaction time for helium fine-structure transitions; (a) experimental data and theoretical data for the $2\,^3S_1 \to 2\,^3P_1$ transition and experimental data and its fit to an inverted gaussian function for $2\,^3S_1 \to 2\,^3P_2$ (the vertical scale is the line-center frequency shift at each interaction-region width (delimited by the slit width) with respect to the line-center frequency at the maximum interaction-region width (without slit)); (b) theoretical data for $2\,^3S_1 \to 2\,^3P_J$ ($J = 0, 1$) transitions.

we take the calculated recoil-induced shifts at 8 mm beam width (or 32 τ) and 150 μW laser power to correct our measured ν_{01} interval. Another experimental test confirms this choice: we compare the difference between the ν_{01} interval measurements at two different interaction-region widths (0.95 mm and 8 mm), with the same calculated difference. The experimental result for the difference was +9.4(2.1) kHz, in good agreement with the theoretical one (+8.7(1.0) kHz). For the other two intervals, we correct the measured values using the difference between the calculated values of the $2\,^3S_1 \to 2\,^3P_{0,1}$ transitions and the shifted value for the $2\,^3S_1 \to 2\,^3P_2$ transition at zero interaction-region width, extrapolated from the fit of the experimental data to an inverted gaussian function (see Fig. 6 (a)). The poor S/N of the saturation dip at short beam widths introduces a large uncertainty in the experimental interval value, and the unknown functional behaviour of the effect produces a large error in the extrapolated value and, as a consequence, a large uncertainty in the intervals that include this transition. Summarising, the recoil-induced shifts are:

$$\Delta f_{rs}(2\,^3S_1 \to 2\,^3P_2) = +260(15) \text{ kHz}$$
$$\Delta f_{rs}(2\,^3S_1 \to 2\,^3P_1) = +66.2(8) \text{ kHz}$$

$$\Delta f_{rs}(2\,^3S_1 \to 2\,^3P_0) = +54.5(5) \text{ kHz},$$

where the shift uncertainties in the case of calculated values includes the computational error and the calculated error from the experimental uncertainties in the laser and atomic beam parameters.

Final Results and Discussion

All of the above systematic effects are summarised in table2. The corrected final values for the three measured intervals are:

$$\nu_{12} = 2,291,174(15) \text{ kHz}$$
$$\nu_{01} = 29,616,949.7(2.0) \text{ kHz}$$
$$\nu_{02} = 31,908,121(15) \text{ kHz},$$

where the accuracy for the ν_{12}, ν_{02} is limited by the recoil-induced shift correction. These results are compared with the previous ones as is reported in table 3.

TABLE 3. Comparison of this work with previous measurements and the theoretical prediction.[a]

	ν_{12}		ν_{01}		ν_{02}	
	V	D	V	D	V	D [b]
This work	2 291 174 (15)	—	29 616 949.7 (2.0)	—	31 908 121 (15)	—
Hughes et al.[a]	2 291 196 (21)	−22(26)	29 616 844 (21)	+102(21)	31 908 040 (20)	+81(25)
Shiner et al.[b]	2 291 173 (3)	+1(15)	29 616 962 (3)	−12(3)	31 908 135 (3)	−14(15)
Storry et al.[c]	—	—	29 616 966 (13)	−16(13)	—	—
Gabrielse et al.[d]	2 291 198 (8)	−24(16)	29 616 936 (8)	+14(8)	31 908 134 (8)	−13(16)
Theory[e]	2 291 180 (20)	−6(25)	29 616 974 (20)	−24(20)	31 908 154 (20)	−33(25)

[a] a: Ref. [7]; b: Ref. [8]; c: Ref. [9]; d: Ref. [37]; e: Ref. [19].
[b] V: Value; D: Difference; all values in kHz

With respect to a global comparison, our measurements are close to those of the Shiner group [8] values, to the unpublished data of the Gabrielse group [37], and to the Storry group [9] value for the ν_{01} interval. However, there is a large disagreement with the early measurements of Hughes et al. [7] and with the theoretical predictions [19] for the large frequency intervals (ν_{01}, ν_{02}). So far, our ν_{01} measurement is the most accurate value and a α value with a 38 ppb accuracy could be inferred from it together with a theoretical prediction at an accuracy of 1 kHz. This result could be used to complete and to clarify the situation with α determinations.

CONCLUSIONS

The precision spectroscopy of helium has been demonstrated to be a powerful tool for testing QED theory as applied to a multielectronic atomic system. Moreover, an accurate (38 ppb) determination of the α constant could be inferred from our precise measurement of the $2\,^3P_0$–$2\,^3P_1$ fine-structure splitting at the 2 kHz accuracy level. This splitting, together with the other two fine-structure intervals of the $2\,^3P$ level, were measured with an experimental approach that combines laser saturation spectroscopy, heterodyne pure frequency determination, and fluorescence detection. The absence of external perturbing fields, present in previous experiments, supports the reliability of our values and allows us to discriminate between contradictory results reported previously.

REFERENCES

1. Pavone, F. S., Marin, F., De Natale, P., Inguscio, M., and Biraben, F., *Phys. Rev. Lett.* **73**, 42 (1994).
2. Dorrer, C., Nez, F., de Beauvoir, B., Julien, L., and Biraben, F., *Phys. Rev. Lett.* **78**, 3658 (1997).
3. Lichten, W., Shiner, D., and Zhi-Xiang Zhou, *Phys. Rev. A* **43**, R1663 (1991).
4. Sansonetti, C. J., and Gillaspy, J. D., *Phys. Rev. A* **45**, R1 (1992).
5. Eikema, K. S., Ubachs, W., Vassen, W., and Hogervost, W., *Phys. Rev. A* **55**, 1866 (1997).
6. Bergeson, S. D. *et al.*, *Phys. Rev. Lett.* **80**, 3475 (1998).
7. Kponou, A., Hughes, V. W., Johnson, C. E., Lewis, S. A., and Pichanick, F. M. J., *Phys. Rev. A* **24**, 264 (1981);
   Frieze W, Hinds, E. A., Hughes, V. W., and Pichanick, F. M. J., *Phys. Rev. A* **24**, 279 (1981).
8. Shiner, D. L., Dixson, R., and Zhao, P., *Phys. Rev. Lett.* **72**, 1802 (1994);
   Shiner, D. L., and Dixson, R., *IEEE Trans. Instrum. Meas.* **44**, 518 (1995).
9. Storry, C. H., and Hessels, E. A., *Phys. Rev. A* **58**, R8 (1998).
10. Drake, G. W. F., *Long-Range Casimir Forces: Theory and Recent Experiments on Atomic Systems*, New York: Plenum, 1989.
11. Hughes, V. W., *Comm. At. Mol. Phys.* **1**, 5 (1969).
12. Araki, H., *Progr. Theoret. Phys. Japan* **17**, 619 (1957).
13. Sucher, J., *Phys. Rev.* **109**, 1010 (1958).
14. Pachucki, K., *J. Phys. B: At. Mol. Opt. Phys.* **31**, 2489 (1998).
15. Baker, J. D. *et al.*, *Bull. Am. Phys. Soc.* **38**, 1127 (1993).
16. Zhang, T., Yan, Z.-C., and Drake, G. W. F., *Phys. Rev. Lett.* **77**, 1715 (1996).
17. Douglas, M., and Kroll, N. M., *Ann. Phys.* **82**, 89 (1974).
18. Zhang, T., *Phys. Rev. A* **53**, 3896 (1996).
19. Zhang, T., and Drake, G. W. F., *Phys. Rev. A* **54**, 4882 (1996).
20. Jeffrey, A. M., Elmquist, R. E., Lee, L. H., Shields, J. Q., and Dziuba, R. F., *IEEE Trans. Instrum. Meas.* **46**, 264 (1997).
21. Cage, M. E., *et al.*, *IEEE Trans. Instrum. Meas.* **38**, 284 (1989).
22. Krüger, E., Nistler, W. and Weirauch, W., *Metrologia* **32**, 117 (1995).
23. Chu, S. *Frontiers in Laser Spectroscopy*. Varenna Summer School (1992).
24. Chu, S., private communication.

25. Van Dyck, R. S., Schwinberg Jr., P. B., and Dehmelt, H. G., *Phys. Rev. Lett.* **59**, 26 (1987).
26. Kinoshita, T., *IEEE Trans. Instrum. Meas.* **46**, 108 (1997).
27. Kawall, D. et al., *Proceedings of Workshop on Frontier Tests of QED and Physics of the Vacuum*, Bulgary, 1998.
28. Giusfredi, G., Godone, A., Bava, E., and Novero, C., *J. Appl. Phys.* **63**, 1279 (1988).
29. Aspect, A., et al., *Phys. Rev. Lett.* **57**, 1668 (1986).
30. Prevedelli, M., Cancio, P., Giusfredi, G., Pavone, F. S., and Inguscio, M., *Opt. Comm.* **125**, 231 (1996).
31. Röhricht, B., Eschle, P., Wigger, C., Dangel, S., Holzner, R., and Suter, D., *Phys. Rev. A* **50**, 2434 (1994).
32. Salomon, C., et al., *Phys. Rev. Lett.* **59**, 1659 (1987).
33. Chen, J., et al., *Phys. Rev. Lett.* **69**, 1344 (1992).
34. Prentiss, M. G., and Ezekiel, S., *Phys. Rev. Lett.* **56**, 46 (1986).
35. Grimm, R., and Mlynek, J., *Phys. Rev. Lett.* **63**, 232 (1989).
36. Cohen-Tannoudji, C., Dupont-Roc, J., and Grynberg, G., *Processus d'interaction entre photons et atomes*, InterEditions: Paris, 1988, ch. 5, pp. 354-363.
37. Wen, J., *Ph.D. thesis*, Advisor: Gabrielse G., Harvard University (1996).

Collective Collapse of a Bose-Einstein Condensate with Attractive Interactions

C. A. Sackett, J. M. Gerton, M. Welling, and R. G. Hulet[1]

Physics Department, MS 61
Rice University
Houston, TX 77251

Abstract. Bose-Einstein condensation (BEC) of atoms with attractive interactions is profoundly different from BEC of atoms with repulsive interactions, in several respects. We describe experiments with Bose condensates of ^{7}Li atoms, which are weakly attracting at ultralow temperature. We measure the distribution of condensate occupation numbers occurring in the gas, which shows that the number is limited and demonstrates the dynamics of condensate growth and collapse.

INTRODUCTION

The recent attainment of Bose-Einstein condensation (BEC) of dilute atomic gases [1–3] has enabled new investigations of weakly interacting many-body systems. ^{7}Li is unique among these gases in that the interactions are effectively attractive. These attractive interactions profoundly effect the nature of BEC. In fact, it was long believed that attractive interactions precluded the attainment of BEC in the gas phase [4,5]. It is now known that BEC *can* exist in a confined gas, provided the condensate number remains small [6]. These condensates are predicted to exhibit fascinating dynamical behavior, including soliton formation [7] and macroscopic quantum tunneling [8–11]. This paper reviews our work on BEC of ^{7}Li, including the measurement of limited condensate number, and the dynamics of condensate growth and collapse.

INTERACTIONS IN DILUTE GASES

One of the primary interests in dilute Bose-Einstein condensates is that the interactions are weak, facilitating comparison between theory and experiment. When

---

[1] This work is supported by the National Science Foundation, the Office of Naval Research, NASA, and the Welch Foundation.

the de Broglie wavelength Λ is much longer than the characteristic two-body interaction length, the effect of the interaction can be represented by a single parameter, the s-wave scattering length a [12]. The magnitude of a indicates the strength of the interaction, while the sign determines whether the interactions are effectively attractive ($a < 0$) or repulsive ($a > 0$). In the experiments, the density n is small enough that $n|a|^3 \ll 1$, so only binary interactions need be considered.

Photoassociative Spectroscopy

Although the interaction potentials for hydrogen and the alkali-metal atoms through francium are all qualitatively the same, in that they all have a repulsive inner-wall, a minimum that supports vibrational bound states (except for the triplet potential of hydrogen), and a long-range van der Waals tail, their respective scattering lengths differ enormously in magnitude and in sign. This variation arises because of differences in the proximity of the least-bound vibrational state to the dissociation limit. As with the familiar attractive square-well potential, a barely bound or barely unbound state leads to collisional resonances that produce very large magnitude scattering lengths. Therefore, small changes in the interaction potential may result in a large change in the magnitude, or even change the sign of a. In the past few years, photoassociative spectroscopy of ultracold atoms has proven to be the most precise method for determining scattering lengths [13]. In one-photon photoassociation, a laser beam is passed through a gas of ultracold atoms confined to a trap. As the laser frequency is tuned to a free-bound resonance, diatomic molecules are formed resulting in a detectable decrease in the number of trapped atoms. The intensity of the trap-loss signal is sensitive to the ground-state wavefunction, providing useful information for determining the ground-state interaction potential. The value of the scattering length is found by numerically solving the Schrödinger equation using this potential. This method has been used to find the scattering lengths for Li, Na, K, and Rb [13].

A more precise method for finding scattering lengths is to probe the ground state molecular levels directly. In particular, the scattering length is extremely sensitive to the binding energy of the least-bound molecular state. We have used two-photon photoassociation to directly measure this binding energy for both stable isotopes of lithium, the bosonic isotope ^{7}Li [14] and the fermionic isotope ^{6}Li [15]. In this method, a laser is tuned to the free-bound transition as in one-photon photoassociation, while the frequency of a second laser is tuned to resonance between the bound excited state and a bound ground state. The frequency difference between the two lasers gives the binding energy directly. This technique has resulted in the most precisely known atomic potentials. Table 1 gives the triplet and singlet scattering lengths for both isotopes individually, as well as for mixed isotope interactions. A summary of our scattering length measurements in lithium is given in Ref. [15]. Two-photon spectroscopy of the ground-state has also been used recently to find the scattering lengths of rubidium [16].

TABLE 1. Singlet and triplet scattering lengths in units of a_o, for isotopically pure and mixed gases of lithium isotopes [15]. The singlet scattering lengths were determined from one-photon photoassociative spectra, while the triplets were determined using the two-photon technique. The mixed case scattering lengths were calculated from knowledge of the $^6\text{Li}_2$ and $^7\text{Li}_2$ potentials

	^{6}Li	^{7}Li	^{6}Li/^{7}Li
a_T	-2160 ± 250	-27.6 ± 0.5	40.9 ± 0.2
a_S	45.5 ± 2.5	33 ± 2	-20 ± 10

Mean-Field Theory

The effects of interactions on the condensate have been studied using mean-field theory and neglecting inelastic collisions [17]. In this approximation, the interaction part of the Hamiltonian is replaced by its mean value, resulting in an interaction energy of $U = 4\pi\hbar^2 an/m$, where n is the density and m is the atomic mass [12]. For a gas at zero temperature, the net result of the interactions and the confining potential can be found by solving the non-linear Schrödinger equation for the wave function of the condensate, $\psi(r)$ [18]:

$$\left(-\frac{\hbar^2}{2m}\nabla^2 + V(r) + U(r) - \mu\right)\psi = 0. \quad (1)$$

Here μ is the chemical potential, and $V(r)$ is the confining potential provided by the trap. In a spherically symmetric harmonic trap with oscillation frequency ω, $V(r) = m\omega^2 r^2/2$. The interaction energy $U(r)$ is determined by taking $n(r) = |\psi(r)|^2$.

Implications of $a < 0$

Limited Condensate Number

For a dilute gas with $a > 0$, corresponding to repulsive interactions, it was shown long ago that the condensate will be stable and that its properties, such as its critical temperature T_c or its elementary excitation spectra, can be found from a perturbation expansion in the small parameter na^3 [12]. However, for $a < 0$ the situation is drastically different. Since U decreases with increasing n, an untrapped (homogeneous) gas is mechanically unstable to collapse. Therefore, it was believed that BEC was not possible in the gas phase. In a system with finite volume, however, the zero-point kinetic energy of the atoms provides a stabilizing influence. A numerical solution to Eq. (1) is found to exist only when N_0 is smaller than a limiting value N_m [19]. Physically, this limit can be understood as requiring that the interaction energy U be small compared to the trap level spacing $\hbar\omega$, so that the interactions act as a small perturbation to the ideal-gas solution. This condition implies that N_m is of the order $l_0/|a|$, where $l_0 = (\hbar/m\omega)^{1/2}$ is the length scale of the single-particle trap ground state [20]. It is at first surprising that N_m increases proportional to l_0, since it is known that BEC cannot occur in a homogeneous gas. However, the density of the condensate, N_0/l_0^3, tends to zero as $l_0 \to \infty$. This tradeoff between N_m and n is an important consideration when designing an experiment.

For condensate occupation numbers below N_m, ψ is determined using Eq. (1). It is found that for $N_0 \ll N_m$, ψ is closely approximated by the single-particle ground state, and as N_0 increases, the interaction energy causes the spatial extent of ψ to decrease. Note that even when a solution to Eq. (1) exists it represents only a metastable state of the trapped atoms [8–10,21], since the equilibrium state of lithium at low temperatures is a crystalline metal solid. Also, for temperatures $T > 0$, Eq. (1) must be modified to take into account the presence of thermally excited atoms, and N_m is slightly lower [21,22].

A variational method has been used to study the decay of condensates with attractive interactions [23,9,10], which we discuss here following the development of Stoof [10]. The ground-state solution to Eq. (1), ψ_0, satisfies an extremal condition

$$\langle \psi_0 | H | \psi_0 \rangle \leq \langle \psi | H | \psi \rangle \tag{2}$$

for any other function ψ. The energy operator H is given by

$$H = -\frac{\hbar^2}{2m} \nabla^2 + V(r) + \frac{U(r)}{2}, \tag{3}$$

where the factor of $1/2$ in the interaction term arises from the dependence of U on ψ. Because the solution to Eq. (1) for the ideal gas is a Gaussian function, it is reasonable to minimize $\langle H \rangle$ using the set of Gaussian trial wavefunctions

$$\psi(r; l) = \left(\frac{N_0}{\pi^{3/2} l^3} \right)^{1/2} \exp\left(-\frac{r^2}{2l^2} \right). \tag{4}$$

Evaluating $\langle H \rangle \equiv H(l)$ yields

$$H(l) = N_0 \frac{\hbar^2}{m} \left(\frac{3}{4l^2} + \frac{3l^2}{4l_0^4} - \frac{|a|}{\sqrt{2\pi}} \frac{N_0}{l^3} \right). \tag{5}$$

This function is plotted for three values of N_0, in Fig. 1. It is observed that for sufficiently small N_0, a local minimum exists near $l = l_0$, indicating that a metastable condensate is possible. For larger N_0, however, the minimum vanishes, and the system will be unstable. The condition for stability is $N_0 \leq 0.68\, l_0/|a|$, which is in reasonable agreement with the exact value obtained by numerical integration of Eq. (1), $N_m = 0.58\, l_0/|a|$ [19]. At very small l, the density is sufficiently high that Eq. (1) is no longer valid, so the divergence of H as $l \to 0$ is of no concern, since it means only that the true ground state of the system is not a dilute gas.

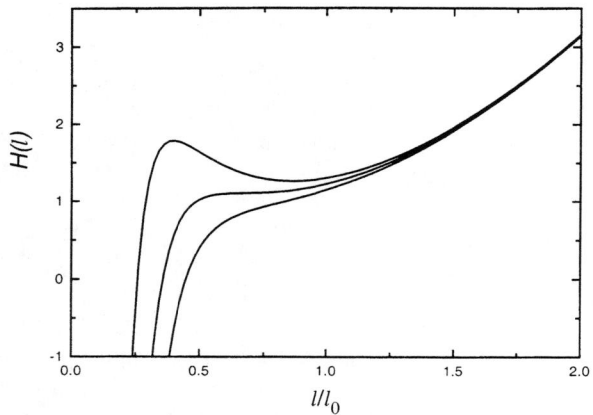

FIGURE 1. The condensate energy H, plotted in units of $N_0 \hbar^2 / m l_0^2$. The upper curve corresponds to $N_0 = 0.48 \, l_0/|a|$, the middle curve to $N_0 = 0.68 \, l_0/|a|$, and the lower curve to $N_0 = 0.87 \, l_0/|a|$. It is evident that a local minimum in H exists near $l = l_0$ if N_0 is sufficiently low, indicating that a metastable condensate can exist.

We have extended the variational calculation to the case of a cylindrically symmetric trap [20]. We find that N_m is determined by the direction of tightest confinement, so that the stability condition can be expressed as

$$N_m \approx \frac{l_{min}}{|a|}, \tag{6}$$

where l_{min} is the lesser of $l_{0\rho}$ and l_{0z}. This result is in agreement with those of Ueda et al. [11]. Therefore, we find that a spherically symmetric trap is optimal for most purposes, as it provides the highest density for a given N_m.

Condensate Collapse

Although a condensate can exist in a trapped gas, it is predicted to be metastable and to decay by quantum or thermal fluctuations [8–11]. The condensate has only one unstable collective mode, which in the case of an isotropic trap corresponds to the breathing mode [7,23]. The condensate therefore collapses as a whole, either by thermal excitation over, or by macroscopic quantum mechanical tunneling through the energy barrier in configuration space, shown in Fig. 1.

The rates of decay for both quantum tunneling and thermal excitation can be calculated within the formalism of the variational calculation [10] and are shown in Fig. 2. For large numbers of condensate atoms, these collective decay mechanisms are much faster than the decay caused by inelastic two and three-body collisions, since the energy barrier out of the metastable minimum vanishes as N_0 approaches N_m.

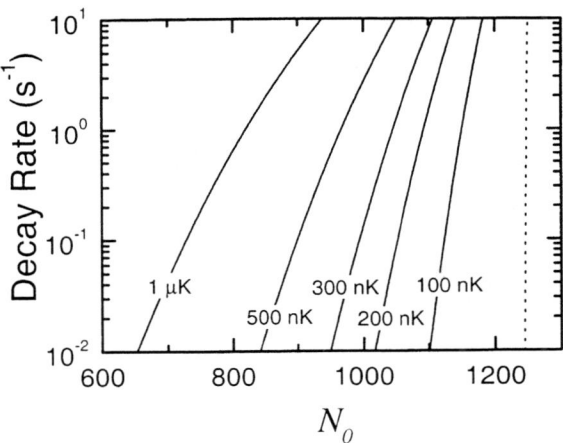

FIGURE 2. Decay rate of the condensate as a function of the number of condensate particles at several temperatures. The dotted line shows the macroscopic quantum tunneling rate.

Condensate Growth and Collapse During Evaporation

Experimentally, the condensate is formed by evaporatively cooling the gas. As the gas is cooled below the critical temperature for BEC, N_0 grows until N_m is reached. The condensate then collapses spontaneously if $N_0 \geq N_m$, or the collapse can be initiated by thermal fluctuations or quantum tunneling for $N_0 \simeq N_m$ [8–11,24]. During the collapse, the condensate shrinks on the time scale of the trap oscillation period. As the density rises, the rates for inelastic collisions such as dipolar decay and three-body molecular recombination increase. These processes release sufficient energy to immediately eject the colliding atoms from the trap, thus reducing N_0. The ejected atoms are very unlikely to further interact with the gas before leaving the trap, since the density of noncondensed atoms is low. As the collapse proceeds, the collision rate grows quickly enough that the density remains small compared to a^{-3} and the condensate remains a dilute gas [24,25]. However, the theories are not yet conclusive as to what fraction of the condensate atoms participates in the collapse, and of those participating, what fraction is eventually ejected.

Both the collapse and the initial cooling process displace the gas from thermal equilibrium. As long as N_0 is smaller than its equilibrium value, as determined by the total number and average energy of the trapped atoms, the condensate will continue to fill until another collapse occurs. This results in a cycle of condensate growth and collapse, which repeats until the gas comes to equilibrium with some $N_0 < N_m$. We have modeled the kinetics of the equilibration process by numerical solution of the quantum Boltzmann equation, as described in Ref. [24]. Fig. 3 shows a typical trajectory of N_0 in time, for our experimental conditions. In this

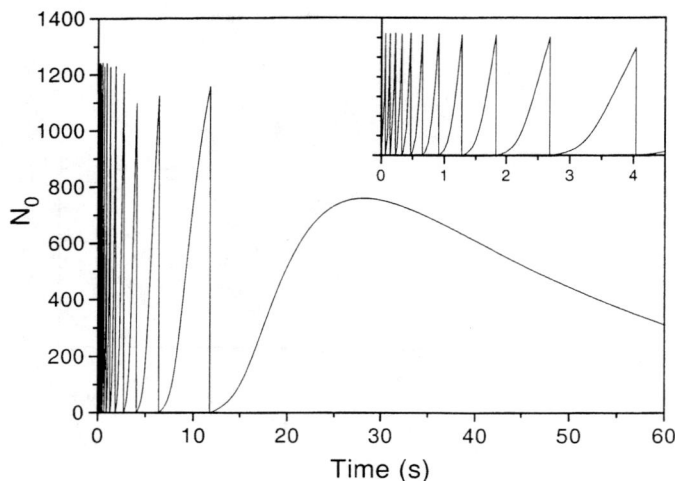

FIGURE 3. Numerical solution of the quantum Boltzmann equation, showing evolution of condensate occupation number. A trapped, degenerate ^{7}Li gas is cooled at $t = 0$ to a temperature of about 100 nK and a total number of 4×10^4 atoms. The gas then freely evolves in time. The inset shows an expanded view of the early time behavior on the same vertical scale.

calculation we assumed that N_0 is reduced to zero when a collapse occurs, on the basis of the model proposed in Ref. [24].

Relation to Other Collapse Phenomena

The non-linear Schrödinger equation (Eq. (1)) has been used to describe many wave-collapse phenomena occurring in classical wave physics. Some of these phenomena are the collapse of Langmuir waves in plasmas [26], and self-focusing of light waves propagating in a medium with a cubic non-linearity [27]. Because of this far-ranging applicability there is an extensive literature devoted to the solution of the non-linear Schrödinger equation under various conditions. Kagan *et al.* have begun to apply some of this accumulated experience to the difficult problem of describing the collapse of a condensate, including both growth and non-linear loss [25].

EXPERIMENT

Magnetic Trap

The apparatus used to produce BEC of ^{7}Li is described most completely in Ref. [20]. A Zeeman slower is used to slow an atomic beam of lithium atoms,

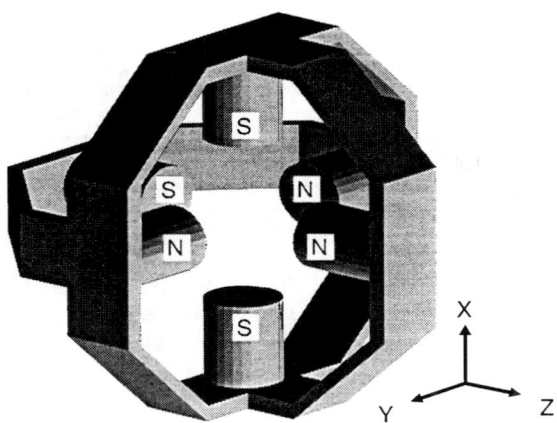

FIGURE 4. Diagram showing the orientation of the cylindrical trap magnets for the permanent magnet trap. The letters indicate the inner tip magnetizations of the NdFeB cylinder magnets. The tip-to-tip magnet spacing is 4.45 cm. The structure around the magnets is a magnetic stainless steel yoke which supports the magnets and provides low reluctance paths for the flux to follow between opposite signed magnets.

which are then directly loaded into a magnetic trap. There is no magneto-optical trap used in the experiment. The magnetic trap is unique in that it is made from permanent magnets, as shown in Fig. 4 [28]. By exploiting the enormous field gradients produced by rare-earth magnets, the resulting trap potential was made nearly spherically-symmetric with a large harmonic oscillation frequency of ∼150 Hz. As discussed above, N_m is limited by the tightest trap direction, so the condensate density is maximized for a spherically symmetric potential. In addition, by actively stabilizing the temperature of the magnets the fields are made highly stable, allowing for relatively repeatable and stable experimental conditions. The bias field at the center of the trap is 1004 G.

Evaporative Cooling

After about 1 s of loading, $\sim 2 \times 10^8$ atoms in the doubly spin-polarized $F = 2, m_F = 2$ state are accumulated. These atoms are then laser cooled to near the Doppler cooling limit of 200 μK. At this number and temperature, the phase space density, $n\Lambda^3$, is still more than 10^5 times too low for BEC. The atoms are cooled further by forced evaporative cooling [29]. The hottest atoms are driven to an untrapped ground state by a microwave field tuned just above the $(F = 2, m_F = 2) \leftrightarrow (F = 1, m_F = 1)$ Zeeman transition frequency of approximately 3450 MHz. As the atoms cool, the microwave frequency is reduced. The optimal frequency vs. time trajectory that maximizes the phase-space density of the trapped atoms

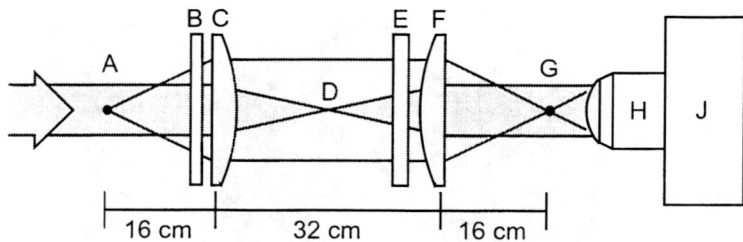

FIGURE 5. A schematic of the imaging system used for *in situ* phase-contrast polarization imaging. A linearly polarized laser beam is directed through the cloud of trapped atoms located at A. The probe beam and scattered light field pass out of a vacuum viewport B, and are relayed to the primary image plane G by an identical pair of 3-cm-diameter, 16-cm-focal-length doublet lenses C and F. The light is then re-imaged and magnified onto a camera J by a microscope objective H. The measured magnification is 19, and the camera pixels are 19 μm square. A linear polarizer E is used to cause the scattered light and probe fields to interfere, producing an image sensitive to the refractive index of the cloud.

is calculated ahead of time [30], and depends on the elastic collision rate and the trap loss rate. The elastic collision rate $n\sigma v$ is roughly 1 s^{-1}, with cross-section $\sigma = 8\pi a^2 \approx 5 \times 10^{-13}$ cm^2. The collision rate is approximately constant during evaporative cooling. We have recently measured the loss rate due to collisions with hot background gas atoms to be $< 10^{-4}$ s^{-1}, and the inelastic dipolar-relaxation collision rate constant to be 1.05×10^{-14} cm^3 s^{-1} [31]. From the low background collision loss rate, we estimate the background gas pressure in the apparatus to be $< 10^{-12}$ torr. Quantum degeneracy is typically reached after 200 seconds, with $N \approx 10^6$ atoms at $T \approx 700$ nK. Lower temperatures are reached by extending the cooling time or by the application of a short, deep cooling pulse.

Phase-Contrast Imaging

After evaporative cooling, the spatial distribution of the atoms is imaged *in situ* using an optical probe. Since the single-particle harmonic oscillator ground state of our trap has a Gaussian density distribution with a $1/e$-radius of only 3 μm, a high-resolution imaging system is required. Because the optical density of the atoms is sufficiently high to cause image distortions when probed by near-resonant absorption [32], we instead use a phase-contrast technique with a relatively large detuning from resonance $\Delta = \pm 250$ MHz. Our implementation of phase-contrast imaging, shown schematically in Fig. 5, is both simple and powerful. It exploits the fact that atoms in a magnetic field are birefringent, so that the light scattered by the atoms is polarized differently from the incident probe light. A linear polarizer decomposes the scattered and probe light onto a common axis, which causes them to interfere. Since the phase of the scattered light is equal to $\alpha/4\Delta$, where α is the

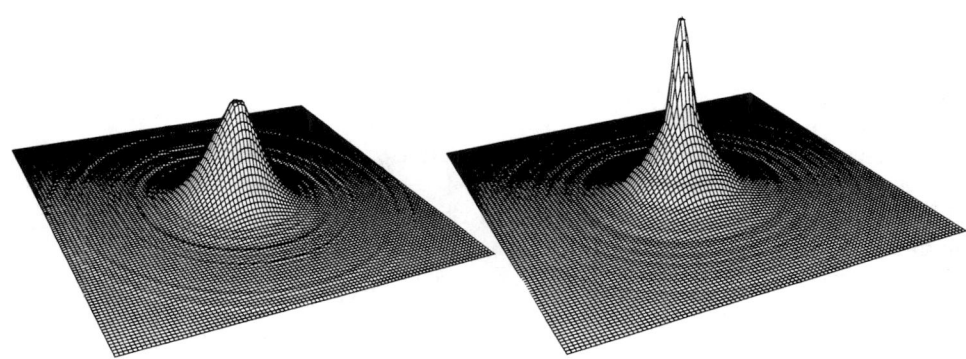

FIGURE 6. Phase-contrast images averaged around the cylindrical axis of the trap. For both cases, $N \approx 23{,}000$ atoms and $T \approx 190$ nK. For the image on the right, $N_0 \approx 1050$, while for the image on the left $N_0 \approx 65$. These images demonstrates our sensitivity to a small number of condensate atoms on a background of a large number of non-condensed atoms.

on-resonance optical density, the spatial image recorded on the CCD camera is a representation of the integrated atomic column density. Phase-contrast polarization imaging is described more fully in Ref. [20].

Fig. 6 shows two images obtained using phase-contrast polarization imaging. For these images, the trap symmetry is exploited by averaging the data around the cylindrical trap axis to improve the signal to noise. The total number of atoms is approximately the same for both images, but on the right T is slightly below T_c and a narrow condensate peak is clearly visible, while for the image on the left, $T \approx T_c$.

Data Analysis

Image profiles are obtained from the averaged data. These profiles are fit with a model energy distribution to determine N, T, and N_0. If the gas is in thermal equilibrium, then any two of N, T, or N_0 completely determine the density of the gas through the Bose-Einstein distribution function. However, if the gas is undergoing the growth/collapse cycles shown in Fig. 3, it certainly is not in thermal equilibrium and a more complicated function is required. Using the quantum Boltzmann equation model, we find that atoms in low-lying levels quickly equilibrate among themselves and the condensate, and that high-energy atoms are well thermalized among each other. Therefore, a three parameter function, including two chemical potentials corresponding to the two parts of the distribution, and a temperature given by the high-energy tail of the distribution, is sufficient to describe the expected non-equilibrium distributions and to determine N_0 [33]. The fits yield an average reduced χ^2 of very nearly 1, indicating that the model is consistent with the data within the noise level. The procedure was tested by applying it to simulated data generated by the quantum Boltzmann model, and also by comparing the analysis of experimental images of thermalized clouds using both equilibrium and nonequilibrium models. From these tests, the systematic error introduced by the nonequilibrium model is estimated to be not more than ±50 atoms. The most significant uncertainty in N_0 is the systematic uncertainty introduced by imaging limitations. While the imaging system is nearly diffraction limited, the resolution is not negligible compared to the size of the condensate, and imaging effects must be included in the fit [32]. Imaging resolution is accounted for by measuring the point transfer function of the lens system and convolving this function with the images. Uncertainties in the resolution lead to a systematic uncertainty in N_0 of ±20% [33].

EXPERIMENTAL RESULTS

In this section, we give our experimental results on the observation of limited condensate number [6], and on the collapse of the condensate [33].

Limited Condensate Number

We have measured N_0 for several thousand different degenerate distributions with T ranging between 80 and 400 nK, and for N between 2,000 and 250,000 atoms. In all cases, N_0 is found to be relatively small. The maximum N_0 observed is between 900 and 1400 atoms, depending on the assumed imaging resolution. This measurement is in very good agreement with the mean-field prediction of 1250 atoms.

In the analysis we have assumed that the gas is ideal, but interactions are expected to alter the size and shape of the density distribution. Mean-field theory predicts that interactions will reduce the $1/e$-radius of the condensate from 3 μm for low occupation number to ∼2 μm as the maximum N_0 is approached [34,7,10,21,22]. If the smaller condensate radius is used in the fit, the maximum N_0 decreases by ∼100 atoms.

Condensate Collapse

To explore the predicted collapse of the condensate, evaporative cooling is continued well into the degenerate regime, to $N \sim 4 \times 10^5$ atoms at a microwave frequency 100 kHz above the trap bottom. The frequency is then rapidly reduced to ∼10 kHz and raised again, leaving approximately 4×10^4 atoms. The frequency is swept quickly compared to the collision rate of ∼3 Hz, so that this "microwave razor" simply eliminates all atoms above a cutoff energy. It thereby creates a definite energy distribution at a specified time whose relaxation to equilibrium can be followed. Fig. 3 shows the expected trajectory of N_0 in time, for our experimental conditions. For this calculation, we have assumed that N_0 is reduced to zero following a collapse [24].

Although phase-contrast imaging can in principle be nearly nonperturbative, it is not possible to reduce incoherent scattering to a negligible level and simultaneously obtain low enough shot noise to measure N_0 accurately. Each atom therefore scatters several photons during a probe pulse, heating the gas and precluding the possibility of directly observing the evolution of N_0 in time as in Fig. 3. This limitation cannot be overcome by repeating the experiment and varying the delay time τ between the microwave razor and the probe, because the evolution of N_0 is made unrepeatable by random thermal and quantum fluctuations in the condensate growth and collapse processes, as well as experimental fluctuations in the initial conditions. Because of this, however, the values of N_0 occurring at a particular τ are expected to vary as different points in the collapse/fill cycle are sampled. We have observed such variations by measuring N_0 for many similarly prepared samples at several values of τ. Their measured distribution are shown as histograms in Fig. 7. For small τ, N_0 ranges from near zero to about 1200 atoms, as expected if the condensate is alternately filling to near the theoretical maximum and subsequently collapsing. At longer time delays, the histograms change shape, narrowing somewhat at $\tau = 30$ s, and having only small N_0 values at $\tau = 60$ s. The variations in N_0 are uncorrelated with changes in N, T, probe parameters, imaging model parameters, and goodness of fit. To our knowledge, no other explanation for variations of this magnitude has been proposed, so we consider the observation of these variations to strongly support the collapse/fill model.

The histogram data can be compared with the predictions of the quantum Boltzmann model. In the model trajectory shown in Fig. 3, three time domains can be discerned with which the data can be correlated. For $\tau \leq 20$ s, the condensate collapses frequently as the gas is equilibrating. Model histograms for delays of 5 and 10 seconds are similar to each other, and agree qualitatively with the experimentally observed distributions in being broadly spread between 0 and N_m. Around $\tau = 20\text{-}40$ s, equilibrium is reached and N_0 is stabilized for several seconds at a maximum value. As is observed in the data, N_0 declines at later times as atoms are lost through inelastic collisions.

The detailed shape of the model histogram for $\tau \leq 20$ s can be deduced from the dependence of N_0 on time as the condensate fills, since the probability of observing a particular N_0 value is proportional to $(dN_0/dt)^{-1}$ at that N_0 value. After a collapse, the condensate initially fills slowly because the stimulated Bose scattering factor is small. Subsequently, the growth rate increases until N_0 reaches ~ 100 atoms, when the growth becomes linear. This saturation occurs when the populations of low-lying energy levels in the trap become depleted. Condensate growth is then limited by the rate for collisions between high energy atoms which produce more low energy atoms, which yields a constant fill rate. Because of these effects, a

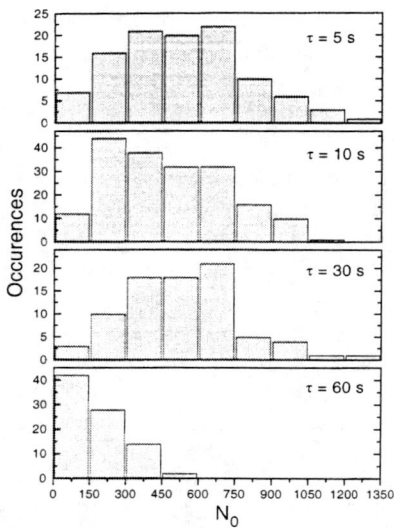

FIGURE 7. Frequency of occurrence of condensate occupation number. For each measurement, a nonequilibrium degenerate gas was produced, allowed to evolve freely for time τ, and then probed. The spread in N_0 values arises as the collapse/fill cycle is sampled at random points.

histogram based on our model is significantly peaked at small N_0, and lower but flat between $N_0 = 100$ and N_m.

The observed histograms differ quantitatively from the model predictions in several respects. There is no peak observed at low N_0; rather, a broad peak occurs at $N_0 = 200 - 700$ atoms. A possible explanation for this disagreement is that the condensate does not collapse to zero atoms. If this is the case, then the fact that we do observe some clouds with $N_0 \simeq 50$ atoms indicates that the condensate must collapse to a range of final values. Kagan *et al.* have observed the condensate to collapse to a nonzero value in numerical solutions of the NLSE [25]. However, while those authors found that close to 50% of the condensate was lost during a collapse, our data suggest that considerably smaller remainders are more likely, since a large fraction of our observations show $N_0 < 600$ atoms.

We also observe the frequency of occurrence to drop steadily as N_0 increases, rather than remaining flat up to $N_0 = N_m$ as predicted. This deviation might be explained in either of two ways. If the condensate growth does not saturate but continues to accelerate, then the probability of observing large N_0 values would decrease. This effect might be the result of the decreasing mean-field energy of the condensate with increasing N_0. Alternatively, if the condensate has a larger than expected rate for collapsing at relatively low N_0, then the probability of condensates surviving to large N_0 would decrease. This might be possible if the condensate is typically in a more excited state than expected from the temperature of the gas. Kagan *et al.* predict such excitations to occur during the growth and collapse

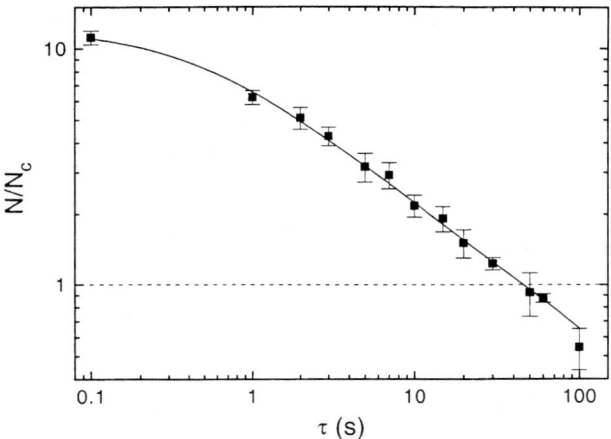

FIGURE 8. Relaxation of degenerate gas to equilibrium. Data were taken as in Fig. 7, but for each image the total number of atoms N and the temperature T were used to determine N/N_c, where $N_c = 1.2(kT/\hbar\omega)^3$. Points represent averages of several measurements, and error bars are standard deviations. The dashed line at $N/N_c = 1$ approximately denotes the point at which equilibrium is reached. The solid curve is a fit to the empirical form $A(1 + \kappa\tau)^\gamma$, yielding $A = 12 \pm 1$, $\kappa = 2.1 \pm .7$ s^{-1}, and $\gamma = -.55 \pm .03$.

processes, but quantitative estimates of the effect under our experimental conditions are not yet available. The comparison between theory and experiment at longer times is quite good.

The condensate growth and collapse cycle is driven by an excess of noncondensed atoms compared to a thermal distribution. This excess can be examined directly. From N and T, the critical number for the BEC transition, N_c, is calculated and the ratio N/N_c plotted as a function of delay time in Fig. 8. The ratio decays according to a power law, which signifies that a nonlinear process governs equilibration. This nonlinearity is reasonable since the rate of decay of the excess atoms should depend both on the excess number and on the collision rate, which in turn depends on N and T. Since $N_0 \ll N$, equilibrium is reached when $N/N_c \approx 1$, which occurs at $\tau \approx 40$ seconds. This time is consistent with the delay required to accurately fit the image data with an equilibrium model, and with the results of the quantum Boltzmann model. Comparison of Figs. 7 and 8 shows that the equilibration time is also consistent with the changing shape of the measured histograms. This further strengthens the conclusion that the variations in N_0 are related to the growth and collapse of the condensate during the equilibration process, since the distribution of N_0 values changes when the population imbalance driving condensate growth is eliminated.

CONCLUSIONS

These observations provide quantitative support for the applicability of mean-field theory to attractive gases. The measurements described here are the first indicator of the complex dynamics accompanying BEC in a gas with attractive interactions. We believe that they support the collective collapse/fill model as a useful framework for considering such systems. It is clear, however, that additional theoretical work is necessary to accurately describe the collapse in detail. Experimentally, we are pursuing more direct methods of observing the growth and collapse of the condensate, including minimally perturbative measurement techniques and controlled triggering of a collapse using an optically induced Feshbach resonance [35]. By such means, we hope to further our understanding of this novel and interesting state of matter.

REFERENCES

1. Anderson, M. H., Ensher, J. R., Matthews, M. R., Wieman, C. E., and Cornell, E. A., *Science* **269**, 198 (1995).
2. Bradley, C. C., Sackett, C. A., Tollett, J. J., and Hulet, R. G., *Phys. Rev. Lett.* **75**, 1687 (1995).
3. Davis, K. B., Mewes,M.-O., Andrews, M. R. , van Druten, N. J., Durfee, D. S. , Kurn, D. M., and Ketterle, W., *Phys. Rev. Lett.* **75**, 3969 (1995).
4. Bogolubov, N., *J. of Phys.* **XI**, 23 (1947).
5. Stoof, H. T. C., *Phys. Rev. A* **49**, 3824 (1994).
6. Bradley, C. C., Sackett, C. A., and Hulet, R. G., *Phys. Rev. Lett.* **78**, 985 (1997).
7. Dodd, R. J., Edwards, M., Williams, C. J., Clark, C. W., Holland, M. J., Ruprecht, P. A., and Burnett, K., *Phys. Rev. A* **54**, 661 (1996).
8. Kagan, Y., Shlyapnikov, G. V., and Walraven, J. T. M., *Phys. Rev. Lett.* **76**, 2670 (1996).
9. Shuryak, E., *Phys. Rev. A* **54**, 3151 (1996).
10. Stoof, H. T. C., *J. Stat. Phys.* **87**, 1353 (1997).
11. Ueda, M., and Leggett, A. J., *Phys. Rev. Lett.* **80**, 1576 (1998).
12. Huang, K., *Statistical Mechanics*, 2 ed. (John Wiley & Sons, New York, 1987).
13. Weiner, J., Bagnato, V. S., Zilio, S., and Julienne, P. S., review to be published.
14. Abraham, E. R. I., McAlexander, W. I., Sackett, C. A., and Hulet, R. G., *Phys. Rev. Lett.* **74**, 1315 (1995).
15. Abraham, E. R. I., McAlexander, W. I., Gerton, J. M., Hulet, R. G., Côté, R., and Dalgarno, A., *Phys. Rev. A* **55**, R3299 (1997).
16. Tsai, C. C., Freeland, R. S., Vogels, J. M., Boesten, H. M. J. M., Verhaar, B. J., and Heinzen, D. J., *Phys. Rev. Lett* **79**, 1245 (1997).
17. Dalfovo, F., Giorgini, S., Petaevskii, L. P., and Stringari, S., review to be published.
18. Lifshitz, E. M., and Pitaevskii, L. P., *Statistical Physics, Part 2* (Butterworth-Heineman, Oxford, 1980).

19. Ruprecht, P. A., Holland, M. J., Burnett, K., and Edwards, M., *Phys. Rev. A* **51**, 4704 (1995).
20. Sackett, C. A., Bradley, C. C., Welling, M., and Hulet, R. G., *Appl. Phys. B* **65**, 433 (1997).
21. Houbiers, M., and Stoof, H. T. C., *Phys. Rev. A* **54**, 5055 (1996).
22. Bergeman, T., *Phys. Rev. A* **55**, 3658 (1997).
23. Singh, K., and Rokhsar, D., *Phys. Rev. Lett.* **77**, 1667 (1996).
24. Sackett, C. A., Stoof, H. T. C., and Hulet, R. G., *Phys. Rev. Lett.* **80**, 2031 (1998).
25. Kagan, Y., Muryshev, A., and Shlyapnikov, G., *Phys. Rev. Lett* **81**, 933 (1998).
26. Zakharov, V. E., *Sov. Phys. JETP* **35**, 908 (1972).
27. Zakharov, V. E., *Sov. Phys. JETP* **41**, 465 (1976).
28. Tollett, J. J., Bradley, C. C., Sackett, C. A., and Hulet, R. G., *Phys. Rev. A* **51**, R22 (1995).
29. Ketterle, W., and van Druten, N. J., in *Advances in Atomic, Molecular, and Optical Physics* (Academic Press, San Diego, 1996), No. 37, p. 181.
30. Sackett, C. A., Bradley, C. C., and Hulet, R. G., *Phys. Rev. A* **55**, 3797 (1997).
31. Gerton, J. M., Sackett, C. A., Frew, B. J., and Hulet, R. G., to be published.
32. Bradley, C. C., Sackett, C. A., and Hulet, R. G., *Phys. Rev. A* **55**, 3951 (1997).
33. Sackett, C. A., Gerton, J. M., Welling, M., and Hulet, R. G., to be published.
34. Dalfovo, F., and Stringari, S., *Phys. Rev. A* **53**, 2477 (1996).
35. Fedichev, P. O., Kagan, Y., Shlyapnikov, G. V., and Walraven, J. T. M., *Phys. Rev. Lett.* **77**, 2913 (1996).

Interactions in Trapped Bose-Einstein Condensates

D. A. W. Hutchinson[1], R. J. Dodd[1], N. P. Proukakis[1,2], S. A. Morgan[1], S. Choi[1], M. Rusch[1], and K. Burnett[1]

[1] *Clarendon Laboratory, Department of Physics, University of Oxford, Oxford, OX1 3PU, United Kingdom.*

[2] *Foundation for Research and Technology Hellas, Institute of Electronic Structure and Laser, P.O. Box 1527, Heraklion 71 110, Crete, Greece.*

Abstract. This lecture describes recent work in our group on gapless mean-field theories for trapped Bose-Einstein condensates. We discuss the physical basis for these theories and compare them to the better known approaches. The proposed theories are based on suitable inclusion of the anomalous average of the Bose field operator. This leads to an effective interaction between two atoms which is both temperature and density dependent, as opposed to the widely used HFB-Popov approach for which it is constant. The predictions of these theories differ from the corresponding HFB-Popov ones by at most a few per cent for the lower-temperature gases studied in the laboratory at present. For systems that may well be studied in the next few years the effects can be much more profound.

INTRODUCTION

In this lecture we want to give an account of the physics of ultracold interactions relevant to the description of Bose-Einstein condensates produced using evaporative cooling. The range of physics that can be studied using these condensates is now very broad and we shall only be able to address a relatively narrow range of issues. We shall focus on the consistent inclusion of the effects of interactions in the theoretical description of condensates, and in particular on quantitative predictions of excitation frequencies at finite temperature. The theory we describe is best thought of in terms of an effective low-energy field theory which requires ultraviolet renormalisation. The infra-red divergence that usually plagues the use of such theories is absent in the case of the trapped gas. We shall discuss the region of validity for the mean-field theories that have been used with such success to date.

LOW-ENERGY INTERACTIONS

In the case of evaporatively cooled gases the de Broglie wavelength of the atoms is enormous in comparison with the range of the interatomic forces. This leads to the use of an effective contact interaction between the atoms which can be written in the form:

$$V(\mathbf{r} - \mathbf{r}') = U_0 \, \delta(\mathbf{r} - \mathbf{r}'). \tag{1}$$

Here U_0 is given in terms of the binary s-wave scattering length a by $U_0 = 4\pi\hbar^2 a/m$ [1]. It is crucial to realize that this effective interaction is strictly valid only when used to couple low-momentum states in vacuo. We use it as the starting point for the theory and shall see that the effect of scattering within a partially condensed gas is to produce a modified interaction strength. We shall first describe the basic features of the mean-field theory most beloved of trapped gas theorists and see where the contact interaction comes into the picture.

MEAN-FIELD THEORY OF CONDENSATES AND THEIR EXCITATIONS

Mean-field theory of the dilute Bose gas can be relied on when it is sufficiently dilute. This means that we require the scattering length to be much smaller than the interparticle spacing l_o, which is certainly true for the condensates produced using evaporative cooling. It is clearly not true for liquid helium. For a homogeneous gas, mean-field theory only holds for:

$$\frac{T - T_c}{T_c} \gg \frac{a}{l_o}. \tag{2}$$

Within this critical region fluctuations around the mean field cannot be treated using perturbation theory and the more sophisticated methods of the Renormalisation Group have to be deployed [2,3]. For the trapped gas, mean-field theory works right through the BEC transition. For a gas trapped in an harmonic potential this comes from the fact that the density of states is proportional to energy squared. This removes the infrared singularity that produces the breakdown of perturbation theory near the critical point in the uniform gas case.

The simplest mean-field theory of the trapped gases is of course Hartree-Fock theory where we assume that the atoms all share the same wavefunction, i.e. the condensate wavefunction $\Psi(\underline{r}, t)$. The evolution of the condensate is then described by a nonlinear Schrödinger equation called the Gross-Pitaevskii (GP) equation [4]. In free space this has the form:

$$i\hbar \frac{\partial \Psi}{\partial t} = -\frac{\hbar^2 \nabla^2 \Psi}{2m} + U_0 |\Psi|^2 \Psi, \tag{3}$$

where Ψ is normalized to the number of particles. For small k excitations this leads to the dispersion relation for longitudinal excitations in the condensate:

$$\omega = \sqrt{\frac{nU_o}{m}} k = ck. \tag{4}$$

The GP equation has been widely used to predict the properties of trapped condensates with near zero temperatures through the simple addition of the trapping potential

$$i\hbar \frac{\partial \Psi}{\partial t} = -\frac{\hbar^2 \nabla^2 \Psi}{2m} + V_{\text{trap}}(\mathbf{r})\Psi + U_0 |\Psi|^2 \Psi. \tag{5}$$

From the point of view of this lecture, it is the agreement between the predictions of this equation for low-temperature excitations and the experimental measurements that is particularly significant. To obtain the frequencies and form of the excitations one finds the solution to the normal modes of oscillation of the condensate given by linearizing the GP equation around the ground state. The set of equations one then has to solve is:

$$\left\{ -\frac{\hbar^2 \nabla_\mathbf{r}^2}{2m} + V_{\text{trap}}(\mathbf{r}) - \mu + U_0|\psi(\mathbf{r})|^2 \right\} \psi(\mathbf{r}) = 0 \tag{6}$$

$$\hat{\mathcal{L}}(\mathbf{r})u_j(\mathbf{r}) + U_0 \psi^2(\mathbf{r}) v_j(\mathbf{r}) = \hbar \omega_j u_j(\mathbf{r}) \tag{7}$$

$$\hat{\mathcal{L}}(\mathbf{r})v_j(\mathbf{r}) + U_0 \left[\psi^*(\mathbf{r})\right]^2 u_j(\mathbf{r}) = -\hbar \omega_j v_j(\mathbf{r}). \tag{8}$$

In these equations, the condensate wavefunction has been denoted by $\psi(\mathbf{r})$, μ represents the chemical potential of the assembly, $u_j(\mathbf{r})$ and $v_j(\mathbf{r})$ are the usual Bogoliubov functions [5] and the hermitian operator $\hat{\mathcal{L}}$ is defined by:

$$\hat{\mathcal{L}}(\mathbf{r}) = -\frac{\hbar^2 \nabla_\mathbf{r}^2}{2m} + V_{\text{trap}}(\mathbf{r}) - \mu + 2U_0|\psi(\mathbf{r})|^2. \tag{9}$$

This method is clearly equivalent to the Random Phase Approximation (RPA), as one would expect from a treatment that starts with the time-dependent Hartree-Fock approximation [6].

How accurate should we expect this approach to be? In atomic physics terms we would ask what the role of correlations between atoms will be. In particular, can these correlations change the form of the effective interaction between the particles? We should also include the effects of thermally excited atoms on the excitations. This is equivalent to including the effect of fluctuations in the field operators around their mean values. The standard approach, based on a finite-temperature version of the variational principle, results in the equations for the Hartree-Fock-Bogoliubov

(HFB) approximation. [7–9]. This again can be thought of as a finite-temperature version of RPA theory. The mean field now includes terms from the mean square of the fluctuating field, i.e. from the excited atoms, as well as from the condensate. The theory also takes account approximately of the correlations between particles. In the field-theory method these are given by the so-called anomalous average of the Bose field operator. In atomic physics terms they are the configuration-interaction terms from pairwise excitations of the condensate. These are pair correlations between atoms and they become crucial in the case of net attractive interactions where they may form bound pairs in the many-body system (in which case they may lead to a competing BCS-like phase transition, as discussed in [10]). These pair correlation terms produce a better approximation to the many-body wavefunction and hence their use in a variational calculation leads to a lower free energy than can be obtained if they are neglected.

The HFB has, however, a well-known problem: it predicts a gap in the single-particle excitation spectrum for a homogeneous gas [8]. This violates Goldstone's theorem [11] and is, therefore, a serious limitation of the theory. The simplest way out of this problem is to make the so-called Popov approximation [12–14], in which the correlation term does not explicitly appear. One can argue that this should be a good approximation as long as the nature of the interactions is not appreciably modified by the presence of the condensate. The effective interaction in vacuo is given by the contact interaction above. In the Popov approximation we have to assume that the pairwise interaction between atoms, i.e. the pair-correlation function, is not changed appreciably by the fact that the atoms are moving through a condensed gas. This is reasonable, as the correction is expected to be of the order $[a/l]^{3/2}$ for low temperatures in a three-dimensional uniform gas. We shall see that it is possible to calculate the change in the effective interaction directly for a trapped gas in a way which follows that done for the homogeneous case. What is more, this correction can be included in the theory of excitations in a way that does not produce a gap in the excitation spectrum.

Before we give a few of the details of this approach we should explain why it was felt necessary to look beyond the Popov approximation. We should first emphasize the fact that the HFB-Popov theory is gapless (see e.g. [14]), which implies that it can be considered to be a better theory for the elementary excitations of Bose-Einstein condensates than the fuller HFB approach. The HFB-Popov theory has therefore been employed in a variety of calculations [15–17]. Indeed, the frequencies of elementary excitations calculated via the HFB-Popov theory are in agreement with the experiments of Jin et al. [18] at reasonably low temperatures. At higher temperatures, however, ($T > 0.6T_c$, where T_c is the transition temperature), the theory appears to deviate from experimental results [16]. This may well be due to an inherent limitation of all the theories we are talking about here: they assume the condensate moves through a static thermal cloud [19–22]. In fact, the motion of the thermal cloud may be responsible for shifting the frequencies of the collective modes as discussed by Zaremba et al. [23]. We decided that we should check whether the other aspect of the problem, i.e. the proper inclusion of the pair correlations (and hence the effect of the condensate on the collisions) could significantly affect our predictions for the frequencies of the modes.

THE EFFECT OF THE CONDENSATE ON COLLISIONS

So what difference does the fact that the atomic interactions are taking place in the presence of other condensed and excited atoms make? [19,20] There are two principal effects on the colliding atoms: (i) the intermediate collisional states may be occupied, leading to a modification of the scattering amplitude via bosonic enhancement, and (ii) the spectrum of initial, intermediate and final states is altered, i.e. the atoms participating in the collisions are not bare atoms, but dressed ones, or quasiparticles. The medium therefore modifies the effective interaction experienced by a pair of colliding atoms from its value in vacuo [24].

The effects of the medium on a colliding pair lead to the replacement of the effective two-body interaction or T-matrix by its many-body equivalent. A detailed discussion of the two-body and many-body T matrix and their respective domains of validity can be found in [19,20,25,26]. In this lecture we just want to describe an approximation for the many-body T matrix which takes part of the effects discussed above into account. In formal terms this is achieved by consideration of the anomalous averages of the Bose field operator, which leads to a generalized effective interaction that is both temperature and density dependent. This effective interaction is shown to be consistent with analytical expressions in the homogeneous limit. It can also be thought of in terms of the effect of RPA correlations between the particles. [19,26]. The modifications, which are introduced over HFB-Popov, are found to be of the order of a few per cent for current experiments. The modifications due to the many-body T matrix can, however, be much more profound, e.g. in two-dimensional gases [27] or ones where the scattering-length interaction is attractive [10]. We shall now give briefly the form of the proposed theories and discuss the consequences for comparison with experiment.

SOME GAPLESS MEAN FIELD THEORIES

For a more detailed description of the theory of the excitations in an inhomogeneous Bose-condensed gas we refer the reader to the review article by Dalfovo et al. [28]. In this lecture we shall simply motivate the way in which we have modified the standard HFB approach. The modified self-consistent equations for the excitations can be summarized by the following generalized eigenvalue problem:

$$\left\{-\frac{\hbar^2\nabla_\mathbf{r}^2}{2m} + V_{\text{trap}}(\mathbf{r}) - \mu + \tilde{U}_{\text{con}}(\mathbf{r})|\psi(\mathbf{r})|^2 + 2\tilde{U}_{\text{exc}}(\mathbf{r})\tilde{n}(\mathbf{r})\right\}\psi(\mathbf{r}) = 0 \quad (10)$$

$$\hat{\mathcal{L}}(\mathbf{r})u_j(\mathbf{r}) + \tilde{U}_{\text{con}}(\mathbf{r})\psi^2(\mathbf{r})v_j(\mathbf{r}) = \hbar\omega_j u_j(\mathbf{r}) \quad (11)$$

$$\hat{\mathcal{L}}(\mathbf{r})v_j(\mathbf{r}) + \tilde{U}_{\text{con}}(\mathbf{r})\left[\psi^*(\mathbf{r})\right]^2 u_j(\mathbf{r}) = -\hbar\omega_j v_j(\mathbf{r}). \tag{12}$$

In these equations the modified version of the hermitian operator $\hat{\mathcal{L}}$ is defined by:

$$\hat{\mathcal{L}}(\mathbf{r}) = -\frac{\hbar^2 \nabla_{\mathbf{r}}^2}{2m} + V_{\text{trap}}(\mathbf{r}) - \mu + 2\tilde{U}_{\text{con}}(\mathbf{r})|\psi(\mathbf{r})|^2 + 2\tilde{U}_{\text{exc}}(\mathbf{r})\tilde{n}(\mathbf{r}). \tag{13}$$

To model the interactions between trapped atoms in the most general manner, we have allowed them to interact with each other via different effective interactions, depending on whether they are both in the condensate $[\tilde{U}_{\text{con}}(\mathbf{r})]$ or not $[\tilde{U}_{\text{exc}}(\mathbf{r})]$. The physical justification for such a distinction will be discussed shortly. The quantity $\tilde{n}(\mathbf{r})$ appearing in Eqs. (10)–(13) corresponds to the density profile of excited atoms, and is to be determined self-consistently [15] by a transformation to the quasiparticle basis. This is equivalent to putting population into the normal modes of vibration of the condensate in a self-consistent manner.

This set of equations predicts excitations at finite temperatures with no gap in the spectrum irrespective of the choice of potentials $\tilde{U}_{\text{con}}(\mathbf{r})$ and $\tilde{U}_{\text{exc}}(\mathbf{r})$ [29] and for that reason we term the theory gapless. One can see that the equations are gapless by noting that they support a zero-frequency mode with $[u_0(\mathbf{r}), v_0(\mathbf{r})] = [\psi(\mathbf{r}), -\psi^*(\mathbf{r})]$. There are, in effect, three different gapless theories (including HFB-Popov) that are special cases of the above set of equations. These are given by different choices of $\tilde{U}_{\text{con}}(\mathbf{r})$ and $\tilde{U}_{\text{exc}}(\mathbf{r})$, as follows:

- HFB-Popov corresponds to $\tilde{U}_{\text{con}}(\mathbf{r}) = \tilde{U}_{\text{exc}}(\mathbf{r}) = U_0$, where $U_0 = 4\pi\hbar^2 a/m$ is the usual dilute Bose gas effective interaction strength. This corresponds to assuming that the presence of the other atoms in the trap has no influence on the nature of the collisions.

- The next gapless theory (G1) is defined by $\tilde{U}_{\text{con}}(\mathbf{r}) = \tilde{U}(\mathbf{r})$ and $\tilde{U}_{\text{exc}}(\mathbf{r}) = U_0$, where $\tilde{U}(\mathbf{r}) = U_0\left[1 + \tilde{m}(\mathbf{r})/\psi^2(\mathbf{r})\right]$ and $\tilde{m}(\mathbf{r})$ represents the pair anomalous average [defined in Eq. (15)]. This takes into account in a simple way the momentum dependence of the effective interaction between particles. In this limit we are supposing that the somewhat larger relative momentum of a collision between excited and condensed atoms is sufficient to nullify the effect of the surrounding atoms. The form of $\tilde{U}(\mathbf{r})$ is justified later.

- The last gapless theory (G2) is defined by $\tilde{U}_{\text{con}}(\mathbf{r}) = \tilde{U}_{\text{exc}}(\mathbf{r}) = \tilde{U}(\mathbf{r})$, with $\tilde{U}(\mathbf{r})$ as above. This corresponds to the assumption that all pairwise interactions between atoms are modified in the same way, independent of the relevant momenta when they collide.

G1 and G2 extend beyond the usual HFB approximation and clearly correspond to the two extreme approximations for the momentum dependence of the many-body T matrix. The full HFB theory cannot be written in the (gapless) form of Eqs. (10)–(13), because it does not treat all condensate-condensate interactions consistently. The excited-state density $\tilde{n}(\mathbf{r})$ and the pair correlations (anomalous average) $\tilde{m}(\mathbf{r})$ are calculated simply by putting populations into the quasiparticle modes. This yields the following well-known expressions

$$\tilde{n}(\mathbf{r}) = \sum_j \left\{ \left[|u_j(\mathbf{r})|^2 + |v_j(\mathbf{r})|^2\right] N_Q(E_j) + |v_j(\mathbf{r})|^2 \right\}, \tag{14}$$

$$\tilde{m}(\mathbf{r}) = \sum_j u_j(\mathbf{r})v_j^*(\mathbf{r}) \left[2N_Q(E_j) + 1\right]. \tag{15}$$

Here, the quasiparticle populations $N_Q(E_j)$ are given by the Bose-Einstein distribution $N_Q(E_j) = 1/(e^{\beta E_j} - 1)$ with $\beta = 1/k_B T$. If we attempt to use these expressions directly we come up against the main problem with having started with the contact interaction, i.e. the expression for $\tilde{m}(\mathbf{r})$ is ultra-violet divergent. We can avoid this divergence by using the full pseudopotential [4,30,31]. We prefer, however, to continue using the contact interaction, and hence the same numerical methods. We can do this as long as we proceed in the following manner to renormalise the effective interaction in the trap. The term $\tilde{m}(\mathbf{r})$ represents the pair correlations between particles in the presence of the trapped condensate and excited atoms. Part of this correlation has, however, already been included in the contact interaction which represents the effect of the pair correlation in vacuo. We should therefore subtract from $\tilde{m}(\mathbf{r})$ its vacuum value. If we do this, we are calculating the change in the pair correlation in terms of the observed scattering length, and this justifies our renormalisation of $\tilde{m}(\mathbf{r})$. We shall show below why we should expect this procedure to produce a reasonable approximation to the full many-body T matrix by looking at the equivalent calculation in a homogeneous gas.

It is perhaps easiest to examine the nature of the effective interaction $\tilde{U}(\mathbf{r})$ in the context of the time-independent equation for the condensate mean field $\psi(\mathbf{r})$. In the usual form of HFB (see e.g. [14]) based on the contact interaction $V(\mathbf{r} - \mathbf{r}') = U_0 \delta(\mathbf{r} - \mathbf{r}')$, the condensate wavefunction is determined by

$$\left(-\frac{\hbar^2 \nabla^2}{2m} + V_{\text{trap}}(\mathbf{r}) - \mu\right)\psi(\mathbf{r}) + U_0 \left[|\psi(\mathbf{r})|^2 + 2\tilde{n}(\mathbf{r})\right]\psi(\mathbf{r}) + U_0 \tilde{m}(\mathbf{r})\psi^*(\mathbf{r}) = 0 \tag{16}$$

with $\psi(\mathbf{r})$ normalized to the total number of condensate particles, i.e. $\int d\mathbf{r} |\psi(\mathbf{r})|^2 = N_0$. The first term in this equation describes the 'free' evolution of the condensate mean field in a confining potential $V_{\text{trap}}(\mathbf{r})$. The first term in the square bracket arises from interactions between atoms in the condensate. The second term in the square brackets comes from the interaction between an atom in the condensate and the thermally excited ones. The final term (neglected in the HFB-Popov approximation) contains the effect of the simplest anomalous average of the Bose field, which represents pair correlations between atoms.

These pair correlations modify the scattering of two condensate atoms, producing an effective interatomic potential [20,26]. The form of the effective interaction in this case can be seen by grouping the final term with the expression $U_0 |\psi(\mathbf{r})|^2 \psi(\mathbf{r})$, to give:

$$U_0 \left[|\psi(\mathbf{r})|^2 \psi(\mathbf{r}) + \tilde{m}(\mathbf{r})\psi^*(\mathbf{r})\right] = U_0 \left[1 + \frac{\tilde{m}(\mathbf{r})}{\psi^2(\mathbf{r})}\right] |\psi(\mathbf{r})|^2 \psi(\mathbf{r}) = \tilde{U}(\mathbf{r})|\psi(\mathbf{r})|^2 \psi(\mathbf{r}). \tag{17}$$

Hence $\tilde{U}(\mathbf{r})$ can be thought of as an approximation to the full many-body interaction potential between two condensate atoms in the presence of the condensate and excited-state mean fields. Our definition of the effective interaction $\tilde{U}(\mathbf{r})$ is equivalent to the zero-energy, zero-momentum limit of the many-body T matrix that includes the effect of the mean field on the spectrum of the intermediate states in a collision. In using it we have avoided (or to be more precise ignored) a related problem associated with the well-known infrared divergences in the theory [19]. This issue will be addressed in future work of our group. $\tilde{U}(\mathbf{r})$ is an approximate extension of the existing homogeneous treatments [19,26] to the case of trapped gases.

Our discussion so far shows that $\tilde{m}(\mathbf{r})$ upgrades the effective interaction between two atoms to the many-body T matrix. However, in view of Eq. (16), this effective interaction only appears rigorously in the interactions between two condensate atoms. To obtain an effective interaction in collisions between condensed and excited atoms in a rigorous manner, one must deal explicitly with correlations of three particles, which requires a treatment beyond the usual mean-field approach [assumed in Eq. (16)] [20]. To second order in the interaction potential these three-particle correlations lead to the introduction of the two-body T matrix in the condensate-excited state interactions. At the next level of approximation they produce additional terms in Eq. (16), which correspond to many-body effects (i.e. a many-body T matrix) in the interactions between condensed and excited atoms. The novel theories discussed in this lecture differ in how the condensate-excited state interactions are approximated. We should expect the interaction between condensed and excited atoms to be similar to that between two condensate atoms, as long as the change in relative momenta of the colliding particles between the two cases is not too great. This leads to the use of $\tilde{U}(\mathbf{r})$ to describe all interactions, thus motivating the G2 theory. However, in the limit of high relative momenta the condensate-excited state interactions are best described by the two-body T matrix since many-body effects die out in this regime [29,26]. This motivates the G1 theory in which the condensate-condensate interactions are modeled by $\tilde{U}(\mathbf{r})$, whereas the condensate-excited state interactions are described in terms of U_0.

THE MANY-BODY T-MATRIX IN THE HOMOGENEOUS LIMIT

We now want to show that the effective interaction $\tilde{U}(\mathbf{r}) = U_0 \left[1 + \tilde{m}(\mathbf{r})/\psi^2(\mathbf{r})\right]$ of Eq. (17) is an approximation to the many-body T matrix. This is achieved by proving the equivalence of $\tilde{U}(\mathbf{r})$ to the homogeneous many-body T matrix (denoted here by $\tilde{\Gamma}_0$ as in [19]) in the zero-energy, zero-momentum limit.

It has been shown in [19,26] that $\tilde{\Gamma}_0$ is related to the zero-momentum limit of the vacuum scattering amplitude U_0 via

$$\tilde{\Gamma}_0 = \frac{U_0}{1 + \alpha(T)U_0}, \qquad (18)$$

where $\alpha(T)$ depends both on the energies of the intermediate states in a collision and on propagator factors for these states. Here we follow the notation of [19], which gives

$$\alpha(T) = \int \frac{d^3k}{(2\pi)^3} \left(\frac{1}{2E_k} \coth\frac{\beta E_k}{2} - \frac{1}{2\epsilon_k} \right). \qquad (19)$$

In this equation ϵ_k is the bare particle energy, and E_k corresponds to the quasi-particle energy given by:

$$E_k = \sqrt{\epsilon_k^2 + 2n_0 \tilde{\Gamma}_0 \epsilon_k}, \qquad (20)$$

where n_0 is the condensate density. $\tilde{\Gamma}_0$ rather than U_0 appears in this expression since the procedure is a self-consistent one. The term with $-(1/2\epsilon_k)$ corresponds to the scattering in vacuo and is subtracted so that we calculate the change in the effective interaction. It also has the welcome effect of removing the ultraviolet divergence.

We can now rewrite the expression for $\alpha(T)$ by noting that $\coth(\beta E_k/2) = (e^{\beta E_k} + 1)/(e^{\beta E_k} - 1) = 2N(E_k) + 1$, where $N(E_k) = [e^{\beta E_k} - 1]^{-1}$ is the Bose-Einstein distribution. Thus, rewriting the equation for $\tilde{\Gamma}_0$, we obtain:

$$\tilde{\Gamma}_0 = U_0 \left[1 - \tilde{\Gamma}_0 \int \frac{d^3k}{(2\pi)^3} \left(\frac{2N(E_k) + 1}{2E_k} - \frac{1}{2\epsilon_k} \right) \right]. \qquad (21)$$

The limit of $\tilde{U}(\mathbf{r})$ in a uniform gas yields exactly the same expression. To see this, we require the homogeneous expression for $\tilde{m}(\mathbf{r})$ which, after the subtraction of the ultraviolet divergent part, is given by:

$$\tilde{m}(\mathbf{r}) = \tilde{m} = \int \frac{d^3k}{(2\pi)^3} \left(u_k v_k [2N(E_k) + 1] - \lim_{k\to\infty} u_k v_k \right), \qquad (22)$$

where u_k and v_k are the Bogoliubov transformation factors [6]

$$u_k = \left[\frac{E_k + \epsilon_k + n_0\tilde{\Gamma}_0}{2E_k} \right]^{1/2}, \quad v_k = -\left[\frac{\epsilon_k + n_0\tilde{\Gamma}_0 - E_k}{2E_k} \right]^{1/2}. \qquad (23)$$

It is $\tilde{\Gamma}_0$ rather than U_0 that appears in these expressions, as in the quasiparticle energies, because they are evaluated self-consistently. A simple calculation then gives $u_k v_k = -n_0 \tilde{\Gamma}_0/2E_k$ enabling us to write $\tilde{m}$ as

$$\tilde{m} = -n_0 \tilde{\Gamma}_0 \int \frac{d^3\mathbf{k}}{(2\pi)^3} \left(\frac{2N(E_k)+1}{2E_k} - \frac{1}{2\epsilon_k} \right). \quad (24)$$

Since $n_0 = \psi^2$, comparison with Eq. (21) clearly shows that

$$\tilde{\Gamma}_0 = U_0 \left[1 + \frac{\tilde{m}}{\psi^2} \right], \quad (25)$$

which is the result we set out to prove. This expression corresponds formally to the zero-energy, zero-momentum limit of the homogeneous many-body T matrix, which takes into account the modification imposed by the medium on the energy spectrum. It is expected to be a very good approximation for most temperatures, except for a very small region near $T = 0$ [26].

PREDICTIONS FOR EXCITATION FREQUENCIES

Having discussed the theoretical basis of the theories we now turn to the predictions of G1 and G2 and their comparison with HFB-Popov. We have found that as far as density profiles or excitation frequencies are concerned, the predictions of G1 lie close to those of HFB-Popov. We have tested this in predictions of excitations in one-dimensional harmonic traps, and in spherically symmetric and anisotropic TOP traps. The deviation of G2 from HFB-Popov is more significant, but is still only of order a few per cent. This justifies the use of the simpler Popov theory for modeling the trapped gas for many situations. The G2 theory is in better agreement with the measurements at JILA for the lower lying $m = 2$ mode. Unfortunately the new theory's prediction of the $m = 0$ mode is further from measurement than the previous theories.

As mentioned earlier, a gapless theory should have a solution with precisely zero excitation energy, and this is the case for all the theories considered here. Another important issue is whether any of the theories considered satisfy the generalized Kohn theorem for atoms in parabolic traps [32]. This says that there should be a mode of oscillation corresponding to the rigid motion of the entire system (i.e. condensate and excited atoms) at the trap frequency. In 3D simulations, we have found this theorem to be satisfied by the theories presented here to better than 1% (except near T_c but even here the difference is never greater than a couple of percent). A full theory of excitations should, however, recover the Kohn mode exactly. These theories do not do so because they ignore the motion of the excited atoms that is driven by the oscillation of the condensate. [15,16,19–23] The condensate is assumed to move in a static thermal cloud, which cannot be strictly correct even if the coupling between the two clouds is modest. For the Kohn theorem to be completely satisfied, one must take into account properly the dynamics of the thermal cloud. This has been done by Zaremba et al. [23] using the assumption of local thermodynamic equilibrium for the excited states. This is a reasonable assumption for experiments in the hydrodynamic limit, i.e. when the excited atom cloud is sufficiently dense for the collisional mean free path to be much smaller than the wavelength of the excitations, leading to rapid local equilibration of the system. Nonetheless, such an approach is not expected to be of much use in the case of very dilute condensates such as those of [18]. One should not be too discouraged, however, if the Kohn theorem is only approximately satisfied, because this need not necessarily have a large effect on the excitation frequencies of other modes [33]. This appears reasonable since the low-temperature predictions of the HFB-Popov theory agree with experimental data [18] within the few percent experimental uncertainty [16]. Bijlsma and Stoof [34] have examined the nature of the coupling of the excitations in the collisionless limit and they find significant shifts in the modes: sufficient to explain the discrepancies with the theory described here.

SO WHEN IS THE EFFECTIVE INTERACTION REALLY IMPORTANT?

The correction we have discussed here has modest effects on the predictions for excitations in the three-dimensional gas with repulsive interactions. This will not be so for a two-dimensional gas. In that case the change in the effective interaction is absolutely crucial. The low-energy limit of the T matrix in vacuo [27] is now zero in the uniform gas limit and one has to include $\tilde{m}(\mathbf{r})$ to obtain a sensible expression for the energy of the particles in the condensed gas. This has important consequences for the Kosterlitz-Thouless phase transition that one observes in a two-dimensional system.

One should also expect the effects of the many-body T matrix to be profound in the case of attractive interactions where a transition related to the BCS phenomena may occur. This comes about if particles in the gas can form bound pairs due to the combined effects of the attractive forces and the presence of the surrounding particles. This has been discussed in detail by Stoof [10].

The third case in which we have to consider the effects of $\tilde{m}(\mathbf{r})$ is close to the critical temperature in a uniform gas. The zero-energy, zero-momentum limit of the many-body T matrix is known to vanish at the transition temperature [19,26,35], consistent with the nature of a second order phase transition. This will not be the case for trapped condensates where collisions do not occur precisely at zero momentum. In the case of a one-dimensional trap, our expression for $\tilde{U}(\mathbf{r})$ predicts a decrease of the effective interaction (with respect to the value in vacuum, or U_0) of the order of $4-5\%$, with the lower prediction being made by G2.

DISCUSSION

In this paper we have described how one can look at the effects that the presence of the trapped atoms has on the atomic interactions that determine the mean-fields present. We believe we have made an important step in the calculation of excitation frequencies for a trapped gas that go beyond HFB-Popov and HFB. The new features of these theories arise from the effect of the medium on the colliding atoms and are taken into account using an effective many-body interaction. Our theories are in agreement with the requirements recently laid forward by Giorgini [22] for a mean-field theory beyond HFB-Popov. For many of the cases being studied at the moment the change in the effective interactions is rather modest. We have set up a computationally based formalism that can treat the problem in a quantitative manner and the future programme of our group will involve looking at cases where the effects are important. The theories we have discussed in this paper (including HFB-Popov) have the inherent limitation of treating the thermal cloud statically. This is the limitation that we have also addressed in other work in our group, but we do not have time to discuss this here.

ACKNOWLEDGMENTS

We are indebted to Charles Clark, Mark Edwards, Allan Griffin, Henk Stoof and Eugene Zaremba for stimulating discussions. This research was funded by the Engineering and Physical Sciences Research Council of the United Kingdom and by the European Union under the TMR network 'Coherent Matter wave Interactions' ERB-FMRX-CT-0002.

REFERENCES

1. Huang, K., and Yang, C. N., *Phys. Rev.* **106**, 1135 (1957); Huang, K., *Statistical Mechanics*, 2nd ed. (Wiley, New York, 1987).
2. Rasolt, M., Stephen, M. J., Fisher, M., and Weichman, P., *Phys. Rev. Lett.* **53**, 798 (1984).
3. Bijlsma, M. J., *Quantum degeneracy in a Bose gas*, Doctoral Thesis, University of Eindhoven.
4. Lifshitz, E. M., and Pitaevskii, L. P., *Statistical Physics Part 2*, Landau and Lifshitz Course of Theoretical Physics, Vol. 9 (Pergamon, Oxford, 1980).
5. Fetter, A. L., *Ann. Phys.* **70**, 67 (1972).
6. Nozieres, P., and Pines, P., *The Theory of Quantum Liquids*, Vol. II, (Addison-Wesley, 1990).
7. Kobe, D. H., *Ann. Phys.* **47**, 15 (1968); Goble, G. W., and Kobe, D. H., *Phys. Rev.* **10**, 851 (1974); Huse, D. A., and Siggia, E. D., *J. of Low Temp. Phys.* **46**, 137 (1982) and references therein.
8. Girardeau, M., and Arnowitt, R., *Phys. Rev.* **113**, 755 (1959); Wentzel, G., *Phys. Rev.* **120**, 1572 (1960); Hohenberg, P. C., and Martin, P. C., *Ann. Phys.* **34**, 291 (1965).
9. Blaizot, J. P., and Ripka, G., *Quantum Theory of Finite Systems*, MIT Press (1986).
10. Stoof, H. T. C., *Phys. Rev. A* **49**, 3824 (1994).
11. Goldstone, J., *Nuovo Cim.* **19** 154 (1961).

12. Popov, V. N., *Functional Integrals and Collective Modes*, (Cambridge University Press, New York, 1987), Ch. 6.
13. Shi, H., Verechaka, G., and Griffin, A., *Phys. Rev.* **B 50**, 1119 (1994).
14. Griffin, A., *Phys. Rev. B* **53**, 9341 (1996).
15. Hutchinson, D. A. W., Zaremba, E., and Griffin, A., *Phys. Rev. Lett.* **78**, 1842 (1997).
16. Dodd, R. J., Edwards, M., Clark, C. W., and Burnett, K., to appear in *Phys. Rev. A* (1998).
17. Giorgini, S., Pitaevskii, L. P., and Stringari, S., *J. Low Temp. Phys.* **109**, 309 (1997).
18. Jin, D. S., Matthews, M. R., Ensher, J. R., Wieman, C. E., and Cornell, E. A., *Phys. Rev. Lett.* **78**, 764 (1997).
19. Shi, H., PhD thesis, Univerity of Toronto (1997); Shi, H., and Griffin, A., *Phys. Rep.* to be published).
20. Proukakis, N. P., Burnett, K., and Stoof, H. T. C., *Phys. Rev. A* **57**, 1230 (1998).
21. Minguzzi, A., and Tosi, M. P., *J. of Phys: Condens. Matter* (to be published) (cond-mat/9709323).
22. Giorgini, S., preprint (cond-mat/9709259).
23. Zaremba, E., Griffin, A., and Nikuni, T., preprint (cond-mat/9705134).
24. Proukakis, N. P., and Burnett, K., *Phil. Trans. R. Soc. Lond. A* **355**, 2235 (1997).
25. Stoof, H. T. C., Bijlsma, M., and Houbiers, M., *J. Res. Natl. Inst. Stand. Technol.* **101**, 443 (1996).
26. Bijlsma, M., and Stoof, H. T. C., *Phys. Rev. A* **55**, 498 (1997).
27. Fisher, D. S., and Hohenberg, P. C., *Phys. Rev. B*, **37**, 4936 (1988).
28. Dalvovo, F., Giorgini, S., L. P. Pitaevskii and Stringari, S., to appear in Reviews of Modern Physics.
29. Stoof, H. T. C., Private Communication.
30. Abrikosov, A. A., Gor'kov, L. P., and Dzyaloshinskiĭ, I. E., *Quantum Field Theoretical Methods in Statistical Physics* (Pergamon, Oxford, 1965).
31. Mohling, F., and Sirlin, A., *Phys. Rev.* **118**, 370 (1960).
32. See Dobson, J. F., *Phys. Rev. Lett.* **73**, 2244 (1994) and references therein.
33. Zaremba, E., Private Communication.
34. Bijlsma, M. J. and Stoof, H. T. C., private communication.
35. Bijlsma, M., and Stoof, H. T. C., *Phys. Rev. A* **54**, 5085 (1996).

Atomic Ion Crystals in Non-Neutral Plasmas*

J. J. Bollinger, T. B. Mitchell, X.-P. Huang, W. M. Itano,
J. N. Tan[†], B. M. Jelenković[‡], and D. J. Wineland

*National Institute of Standards and Technology
Boulder Colorado 80303*

Abstract. Laser-cooled ions in a trap can be strongly coupled and form crystalline states. We describe experimental studies that measure the spatial correlations of the ion crystals formed in Penning traps. Both Bragg scattering of the cooling-laser light and spatial imaging of the laser-induced ion fluorescence are used to measure these correlations. In spherical plasmas with more than 2×10^5 ions, body-centered-cubic (bcc) crystals, the predicted bulk structure, are the only type of crystals observed. We are able to phase-lock the orientation of the ion crystals to a rotating electric-field perturbation. With this "rotating wall" technique and stroboscopic detection, images of individual ions in a Penning trap are obtained. The rotating-wall technique also provides a precise control of the time-dilation shift due to the plasma rotation, which is important for Penning trap frequency standards.

INTRODUCTION

Trapped ions are a good example of a one-component plasma (OCP). An OCP consists of a single charged species immersed in a neutralizing background [1]. In an ion trap, the trapping fields provide the neutralizing background [2]. Examples of OCP's include such diverse systems as the outer crust of neutron stars [3], electrons on the surface of liquid helium [4], and elemental metals. The thermodynamic properties of the classical OCP of infinite spatial extent are determined by its Coulomb coupling constant [1]

$$\Gamma \equiv \frac{1}{4\pi\epsilon_0} \frac{e^2}{a_{WS} k_B T}, \qquad (1)$$

which is a measure of the ratio of the Coulomb potential energy of nearest neighbor ions to the kinetic energy per ion. Here, ϵ_o is the permittivity of the vacuum, e is the

---

*) Work of the U.S. Government. Not subject to U.S. copyright
†) Present address: Physics Dept., Harvard Univ., Cambridge, MA 02138
‡) On leave from the Institute of Physics, University of Belgrade, Belgrade, Yugoslavia

charge of an ion, k_B is Boltzmann's constant, T is the temperature, and a_{WS} is the Wigner-Seitz radius, defined by $4\pi(a_{WS})^3/3 = 1/n_o$ where n_o is the ion density. For low-temperature ions in a trap, n_0 equals the equivalent neutralizing background density provided by the trapping fields. Plasmas with $\Gamma > 1$ are called strongly coupled. The onset of fluid-like behavior is predicted [1] at $\Gamma \approx 2$, and a phase transition to a body-centered cubic (bcc) lattice is predicted [1,5] at $\Gamma \approx 170$. From a theoretical perspective, the strongly coupled OCP has been used as a paradigm for condensed matter for decades. However, only recently has it been realized in the laboratory [6].

Experimentally, freezing of small numbers ($N < 50$) of laser-cooled atomic ions into Coulomb clusters was first observed in Paul traps [7-9]. With larger numbers of trapped ions, concentric shell structures were observed directly in Penning [10] and linear Paul [11,12] traps. The linear Paul traps provided strong confinement in the two dimensions perpendicular to the trap axis and very weak confinement along the trap axis. This resulted in cylindrically shaped plasmas whose axial lengths are large compared to their cylindrical diameters. Cylindrical-shell crystals which are periodic with distance along the trap axis were observed. The diameter of these crystals was limited to $\sim 10 a_{WS}$ in Ref. [11] and $\sim 30 a_{WS}$ in Ref. [12], presumably due to rf heating [13] which is produced by the time-dependent trapping fields and increases with the plasma diameter. These plasma diameters appear to be too small to observe the 3-D periodic crystals predicted for the infinite, strongly coupled OCP. Strong coupling and crystallization have also been observed with particles interacting through a screened Coulomb potential. Examples include dusty plasma crystals [14] and colloidal suspensions [15,16].

In this manuscript we summarize recent progress on the study of strongly coupled ion OCP's in Penning traps. Because Penning traps use static fields to confine charged particles, there is no rf heating. This has enabled ion plasmas which are large in all three dimensions to be laser-cooled. For example, we have laser-cooled $\sim 10^6$ Be$^+$ ions in an approximately spherical plasma with diameter $\sim 200 a_{WS}$. With these large ion plasmas we have used Bragg scattering of the cooling laser light to detect the formation of bcc crystals [17,18], the predicted state for a bulk OCP with $\Gamma > 170$. In addition, we have studied the spatial correlations in planar, disk-like plasmas with axial thickness $\lesssim 10 a_{WS}$. These plasmas consist of extended, two-dimensionally periodic lattice planes. The importance of the plasma boundary in this case results in different crystalline structures depending on the details of the plasma shape.

A potential drawback of the Penning trap versus the rf trap is that the ions rotate about the trap magnetic field, and this has previously prevented the imaging of the ion crystals as done in Paul traps. This is because the rotation, created by the $\mathbf{E} \times \mathbf{B}$ drift due to the radial electric and the trap magnetic fields is, in general, not stable. For example, fluctuations in the plasma density or shape produce fluctuations in the ion space charge fields which change the plasma rotation. With a technique first employed by the non-neutral plasma physics group at the University of California at San Diego [19], we are able to phase-lock the rotation of the laser-cooled ion

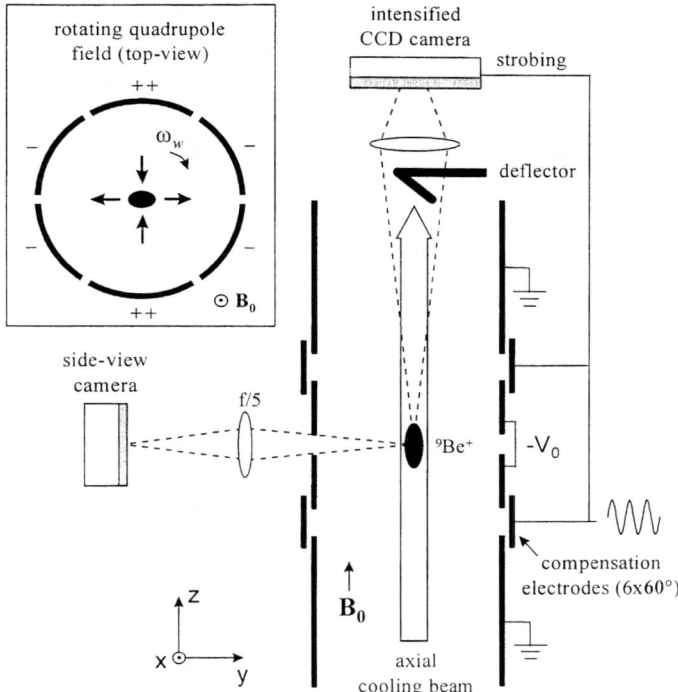

FIGURE 1. Schematic view of the cylindrical trap with real space imaging optics for the side-view camera and Bragg diffraction detection system for the axial cooling beam. Cross section of the rotating quadrupole field (in the x-y plane) is shown in the insert. From Ref. [20].

crystals to a rotating electric field perturbation [20,21]. The success of this "rotating wall" technique enables us to strobe the cameras recording the ion fluorescence synchronously with the plasma rotation and obtain images of individual ions in the plasma crystals [22].

Figure 1 is a schematic of the cylindrical Penning trap we use to confine ^{9}Be$^+$ ions. The trap consists of a 127 mm long vertical stack of cylindrical electrodes with an inner diameter of 40.6 mm, enclosed in a room temperature, 10^{-8} Pa vacuum chamber. The uniform magnetic field $B_o = 4.46$ T is aligned parallel to the trap axis within 0.01° and produces a ^{9}Be$^+$ cyclotron frequency $\Omega = 2\pi \times 7.61$ MHz. A quadratic, axially symmetric potential $(m\omega_z^2/2e)[z^2 - r^2/2]$ is generated near the trap center by biasing the central electrodes to a negative voltage $-V_o$. At $V_o = 1$ kV, the single-particle axial frequency $\omega_z = 2\pi \times 799$ kHz and the magnetron $\mathbf{E} \times \mathbf{B}$ drift frequency $\omega_m = 2\pi \times 42.2$ kHz. The trapped Be$^+$ ions are Doppler laser-cooled by two 313 nm laser beams. The principal cooling beam (waist diameter ~ 0.5 mm, power ~ 50 μW) is directed parallel to $\mathbf{B}_o$. A second, typically weaker cooling beam with a much smaller waist (~ 0.08 mm) is directed perpendicularly to $\mathbf{B_o}$ (not shown in Fig. 1). This beam can also be used to vary the plasma rotation frequency by applying a torque with radiation pressure. With this configuration ion temperatures close to the 0.5 mK Doppler laser-cooling limit are presumably achieved. However, experimentally we have only placed a rough 10 mK upper bound on the ion temperature [23]. For a typical value of $n_0 = 4 \times 10^8$ cm^{-3}, this implies $\Gamma > 200$.

Two types of imaging detectors were used. One is a charge-coupled device (CCD) camera coupled to an electronically gateable image intensifier. The other is an imaging photomultiplier tube based on a microchannel-plate electron multiplier and a multielectrode resistive anode for position sensing. For each detected photon, the position coordinates are derived from the current pulses collected by the different electrodes attached to the resistive anode. This camera therefore provides the position and time of each detected photon. However, in order to avoid saturation, we placed up to 20 dB of attenuation in front of this camera to lower the detected photon count rate to less than ~ 300 kHz.

In thermal equilibrium, the trapped ion plasma rotates without shear at a frequency ω_r where $\omega_m < \omega_r < \Omega - \omega_m$ [24,25]. For the low temperature work described here, the ion density is constant and given by $n_o = 2\epsilon_o m \omega_r (\Omega - \omega_r)/e^2$. With a quadratic trapping potential the plasma has the simple shape of a spheroid, $z^2/z_o^2 + r^2/r_o^2 = 1$, where the aspect ratio $\alpha \equiv z_o/r_o$ depends on ω_r [23,25]. This is because the radial binding force of the trap is determined by the Lorentz force due to the plasma's rotation through the magnetic field. Thus low ω_r results in a "disk-like" plasma (an oblate spheroid) with large radius. As ω_r increases, r_o shrinks and z_o grows, resulting in an increasing α. However, large ω_r ($\omega_r > \Omega/2$) produces a large centrifugal acceleration which opposes the Lorentz force and disk-like plasmas are once again obtained for $\omega_r \sim \Omega - \omega_m$. In our work, torques from a laser or a rotating electric field are used to control ω_r and therefore the plasma density and shape. The plasma shape is observed by imaging the ion fluorescence scattered perpendicularly to $\mathbf{B_o}$ with an f/5 objective. (See Fig. 1.) All possible values of ω_r from ω_m to $\Omega - \omega_m$ have been accessed using either torque [21,26,27]. Azimuthally segmented compensation electrodes located between the main trap electrodes are used to apply the rotating electric-field perturbation. Both a rotating quadrupole (see inset in Fig. 1) and rotating dipole field (not shown in Fig. 1) have been used to control ω_r. Below we explain how the rotating quadrupole field provides precise control of ω_r.

BRAGG SCATTERING

BCC Crystals

An infinite OCP with $\Gamma \gtrsim 170$ is predicted to form a bcc lattice. However, the bulk energies per ion of the face-centered-cubic (fcc) and hexagonal-close-packed (hcp) lattices differ very little from bcc ($< 10^{-4}$) [28]. Because some of the fcc and hcp planes have lower surface energies than any of the bcc planes, a boundary can have a strong effect on the preferred lattice structure. One calculation [28] estimates that the plasma may need to be $\gtrsim 100 a_{WS}$ across its smallest dimension to exhibit bulk behavior. For a spherical plasma this corresponds to $\sim 10^5$ ions.

We used Bragg scattering to measure the spatial correlations of approximately spherical plasmas with $N > 2 \times 10^5$ trapped Be^+ ions [17,18]. The cooling-laser beam directed along the trap axis was used for Bragg scattering as indicated in Fig. 1. First the plasma shape was set to be approximately spherical. (In early experiments this was done with the perpendicular laser beam; more recent experiments used the rotating wall.) The parallel laser beam was then tuned approximately half a linewidth below resonance, and a Bragg scattering pattern recorded ($\sim$ 1-30 s integration). The plasma was then heated and recooled, and another Bragg scattering pattern was recorded. Because the 313 nm wavelength of the cooling laser is small compared to the inter-ion separation ($\sim 10 - 20$ μm), Bragg scattering occurs in the forward (few degree) scattering direction. In order for a diffracted beam to form, the incident and scattered wave vectors $\mathbf{k}_i$ and $\mathbf{k}_s$ must differ by a reciprocal lattice vector (Laue condition) [29]. In a typical x-ray crystal diffraction case, satisfying the Laue condition for many reciprocal lattice vectors requires that the incident radiation have a continuous range of wavelengths. Here the Laue condition is relaxed because of the small size of the crystal, so a crystalline Bragg scattering pattern is frequently obtained even with monochromatic radiation.

Figure 2(a) shows a time-averaged diffraction pattern obtained on a spherical plasma with $N \sim 7.5 \times 10^5$. The multiple concentric rings are due to Bragg scattering off different planes of a crystal. A concentric ring rather than a dot pattern is observed because the crystal was rotating about the laser beam. In general, many different patterns were observed, corresponding to Bragg scattering off crystals with different orientations. Figure 3 summarizes the analysis of approximately 30 time-averaged patterns obtained on two different spherical plasmas with $N > 2 \times 10^5$. It shows the number of Bragg peaks as a function of the momentum transfer $q = |\mathbf{k}_s - \mathbf{k}_i| = 2k \sin(\theta_{scatt}/2)$ ($\simeq k\theta_{scatt}$ for $\theta_{scatt} \ll 1$), where $k = 2\pi/\lambda$ is the laser wave number and θ_{scatt} is the scattering angle. The density dependence of the Bragg peak positions is removed by multiplying q by a_{WS}, which was determined from ω_r. The positions of the peaks agree with those calculated for a bcc lattice, within the 2.5% uncertainty of the angular calibration. They disagree by about 10% with the values calculated for an fcc lattice. The ratios of the peak positions of the first five peaks agree within about 1% with the calculated ratios for a bcc lattice. This provides strong evidence for the formation of bcc crystals in spherical plasmas with $N > 2 \times 10^5$ ions. This result is significant because it is evidence for bulk behavior in a laboratory, strongly coupled OCP.

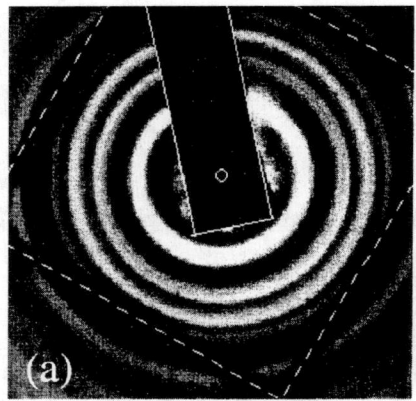 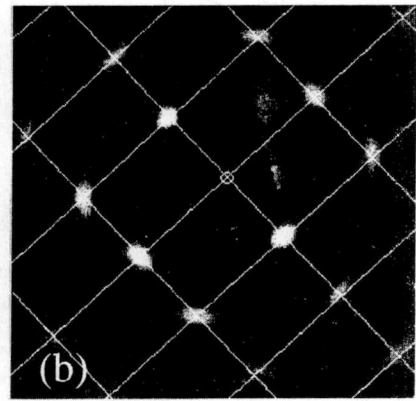

FIGURE 2. Bragg diffraction patterns from a plasma phase locked to a rotating quadrupole field ($\omega_r = 2\pi \times 140$ kHz, $n_0 \approx 4.26 \times 10^8$ cm^{-3}, $\alpha \approx 1.1$). (a) 1 s time-averaged pattern. The long rectangular shadow (highlighted by solid lines) is from the deflector for the incident beam; four line shadows (highlighted by dashed lines) that form a square are due to a wire mesh at the exit window of the vacuum chamber. The small open circle near the center of the figure marks the position of the undeflected laser beam. (b) Time-resolved pattern obtained nearly simultaneously with (a) by strobing the camera with the rotating field (integration time ≈ 5 s). A spot is predicted at each intersection of the rectangular grid lines for a bcc crystal with a [110] axis aligned with the laser beam. The grid spacings were determined from the n_0 calculated from ω_r and are not fitted. From Ref. [21].

Rotating Wall

By strobing the camera recording the Bragg scattering pattern synchronously with the plasma rotation, we should be able to recover a dot pattern from the time-averaged concentric ring pattern in Fig. 2(a). Initially we used the time-dependence of the Bragg scattered light to sense the phase of the plasma rotation [18,30]. More recently we used a rotating electric field perturbation to phase-lock the ion plasma rotation [20,21].

Consider the rotating quadrupolar perturbation shown in the inset of Fig. 1. This z-independent perturbation produces a small distortion in the shape of the spheroidal plasma. In particular, the plasma acquires a small elliptical cross section normal to the z-axis. (In our work the distortion created by the rotating quadrupole field was typically less than 1% of the plasma diameter.) The elliptical boundary rotates at the applied rotating wall frequency ω_w. An ion near the plasma boundary experiences a torque due to this rotating boundary. If the ion is rotating slower than ω_w, the torque will speed it up. If it is rotating faster than ω_w, the torque will slow it down. Through viscous effects, this torque is transmitted to the plasma interior. Therefore, if other external torques are small, the rotating wall perturbation will

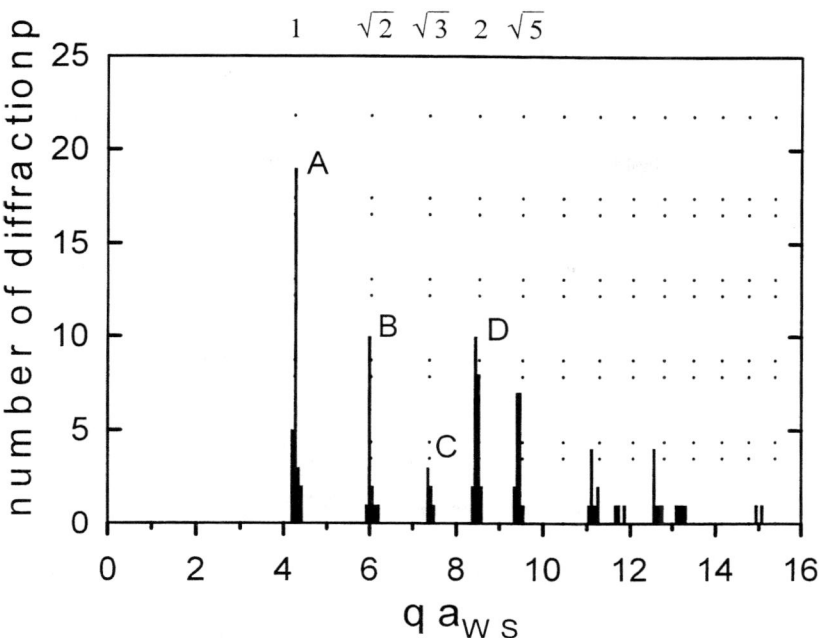

FIGURE 3. Histogram showing the numbers of peaks (not intensities) observed as a function of qa_{WS} (defined in the text) for 30 time-averaged Bragg scattering patterns obtained on two different spherical plasmas with $N > 2 \times 10^5$. The dotted lines show the expected peak positions for a bcc crystal, normalized to the center of gravity of the peak at A (corresponding to Bragg reflections off {110} planes). From Ref. [18].

make ω_r equal ω_w. Crystallized plasmas behave more like a solid than a liquid or gas. Because the viscosity is high, the whole plasma will tend to rotate rigidly with its boundary. In particular, the orientation of the ion crystals could phase-lock to the rotating quadrupolar perturbation if the frequency difference between ω_r and ω_w is small.

To check for phase-locked control of ω_r, we strobed the camera recording the Bragg scattering pattern in Fig. 2(a) with the synthesizer used to generate the rotating wall signal. Specifically, once each $2\pi/\omega_w$ period, the rotating wall signal gated the camera on for a period of time $\lesssim 0.02(2\pi/\omega_w)$. The resulting Laue dot pattern in Fig. 2(b) shows that the plasma rotation was phase-locked to the rotating electric-field perturbation. The dot pattern provides detailed information on the number and orientation of the crystals which contributed to the Bragg scattering signal. For example, the pattern in Fig. 2(b) was due to a single bcc crystal with a [110] axis aligned along the laser beam. For phase-locked operation of the rotating wall, other external torques must be small. For example, a misalignment of the trap magnetic field with the trap electrode symmetry axis of $> 0.01°$ prevented phase-locked control of the plasma rotation. In our work, alignment to $\lesssim 0.003°$ was obtained by minimizing the excitation of zero-frequency plasma modes [26,27].

In addition to the rotating quadrupole perturbation, phase-locked control was also achieved with a uniform rotating electric field (a "dipole" field) [21]. In fact under many circumstances a uniform oscillating field worked equally well. In these cases the co-rotating component of the oscillating field controlled the plasma rotation while the perturbing effects due to the counter-rotating component were minimal. The simplicity of the oscillating dipole field makes it a convenient tool for controlling ω_r. However, in a quadratic trap, control of ω_r with a uniform rotating or oscillating electric field requires an effect which breaks the separation of center-of-mass and internal degrees of freedom of the plasma. In our work this is done by other-mass ions which experience a different centrifugal potential than the $^9Be^+$ ions.

REAL-SPACE IMAGES

Bragg scattering measures the Fourier transform of the spatial correlations of the trapped ions. It provides a picture of these correlations in reciprocal-lattice space. With phase-locked control of ω_r, real-space imaging of individual ions in a Penning trap becomes possible. To obtain real-space images with high resolution, we replaced the Bragg scattering optics (see Fig. 1) with optics, starting with an f/2 objective, which formed a real, top-view image of the ion plasma. The combined resolution limit of the optics and camera was less than 5 μm near the optimal object plane of the f/2 objective. This is less than the ~ 10 μm resolution limit required to resolve individual ions. However, the depth of field of an f/2 objective for 10 μm resolution is ~ 80 μm. Therefore, in order to resolve individual ions, the ions needed to reside within ± 40 μm of the optimal object plane. For disk-like plasmas with $2z_o \lesssim 80$ μm, all of the ions within the plasma were resolvable. For plasmas with $2z_o > 80$ μm, the cooling-laser beam directed perpendicularly to $\mathbf{B_o}$ was used to illuminate a section of the plasma within the depth of field.

Figure 4 shows side-view and top-view images of an approximately spherical plasma with $N \sim 1.8 \times 10^5$. The fluorescence from the perpendicular laser beam used to highlight a small region of the plasma is clearly visible. In the top-view image a square grid of dots is observed near the plasma center. The measured spacing between nearest neighbor dots is 12.8 ± 0.3 μm, in good agreement with the 12.5 μm spacing expected for viewing along a [100] axis of a bcc crystal with density determined by the ω_r set by the rotating field. Real-space imaging provides direct information on the location and size of the crystals. In Fig. 4 the crystal was located in the radial center of the plasma and was at least 230 μm across, or at least 1/4 of the plasma diameter.

For disk-like plasmas with $2z_o \lesssim 80$ μm, all of the ions within the plasma are resolved without the use of the perpendicular laser beam. Disk-like plasmas are

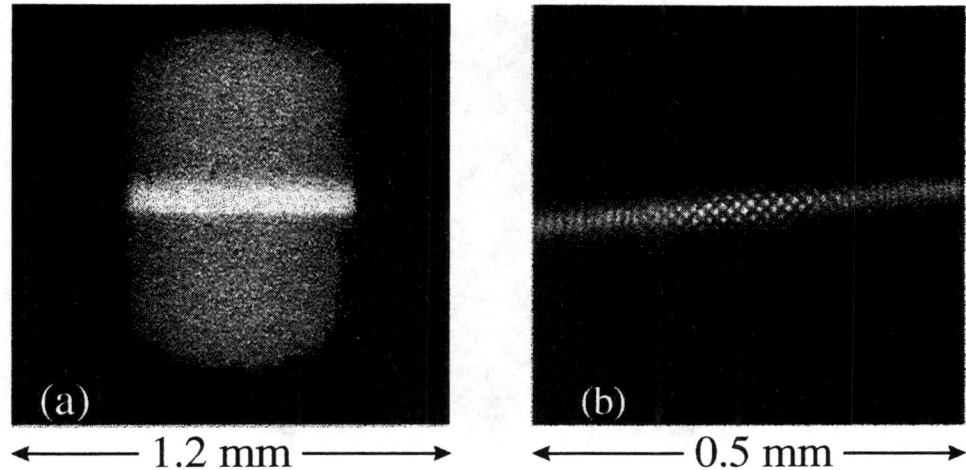

FIGURE 4. Real-space images of an $N \sim 1.8 \times 10^5$ ion plasma phase-locked with an oscillating dipole field at $\omega_r = 2\pi \times 120$ kHz. (a) Time-averaged side-view image showing the overall plasma shape. The bright line of fluorescence through the plasma center is due to a laser beam directed perpendicularly to $\mathbf{B_o}$. The plasma shape is approximately spherical. The presence of heavier-mass ions, which centrifugally separate from the $^9\text{Be}^+$ ions, produces the straight vertical boundaries in the image. (b) Strobed top-view image, obtained simultaneously with (a), showing the presence of a bcc crystal in the plasma center. The distance scales in (a) and (b) are different as noted.

obtained with ω_r slightly greater than ω_m. For small plasmas ($N \lesssim 2000$ ions) we were able to use the rotating-dipole electric field to lower ω_r and obtain a single plane while maintaining long-range order in the top-view images. Figure 5 shows a top-view image of such a plasma. Near the plasma center a 2-D hexagonal lattice is observed, the preferred lattice for a 2-D system. Here each dot is the image of an individual ion.

Starting with a single plane like that shown in Fig. 5, we studied the structural phase transitions that occur as ω_r is increased [22]. With increasing ω_r, the radial confining force of the Penning trap increases, which decreases r_o. At a particular point, there is a structural phase transition near the plasma center from a single, hexagonal lattice plane to two lattice planes where the ions form a square grid in each plane. Further increases in ω_r increase the number of ions per unit area of each plane as well as the spacing between the planes. During this process the square lattice planes smoothly change into rhombic lattice planes and eventually there is a sudden transition to hexagonal lattice planes. Further increases in ω_r eventually produce a structural transition to three square lattice planes, and the basic pattern repeats. The structural phases from one to five lattice planes were carefully measured [22]. Agreement with the predictions of a 2-D planar model,

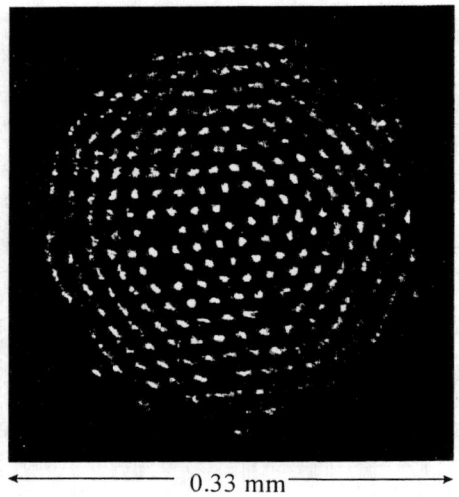

←——— 0.33 mm ———→

FIGURE 5. Strobed top-view image of a small $N \sim 300$ ^{9}Be$^+$ ion plasma phase-locked with a rotating dipole field at $\omega_r = 2\pi \times 65.7$ kHz. Heavier-mass ions are located outside the ^{9}Be$^+$ ions.

which is confined by a quadratic potential in 1-D, but infinite and homogeneous in the other two dimensions, is good [22,31,32].

DISCUSSION

With Bragg scattering and spatial imaging, we have measured the correlations in both disk-like and spherical, strongly coupled ^{9}Be$^+$ ion plasmas. For finite plasmas, the disk-like geometry permits a more detailed comparison with theoretical calculations. We have measured the preferred lattice structures for up to five lattice planes in disk-like plasmas and obtain good agreement with theory. By increasing the number of planes (by adding more ions to the plasma), the transition from surface-dominated to bulk behavior in the planar geometry can be studied. Ions in a trap have been proposed as a register for a quantum computer [33]. Work in this area has focused on a string of a few ions in a linear Paul trap [34]. A single lattice plane of ions as in Fig. 5 could provide a 2-D geometry of trapped ions for studies of quantum computing or entangled quantum states.

In spherical plasmas with more than 2×10^5 ions, we have observed the formation of bcc crystals, the predicted state for the infinite strongly coupled OCP. The crystals occupied the inner quarter of the plasma diameter. Outside the crystal there was a complicated transition to shell structure. In this system we have not observed the thermodynamic liquid-solid phase transition predicted for the bulk OCP. Our measurements have concentrated on the correlations obtained at the coldest temperatures (therefore maximum Γ) where the ion fluorescence is maxi-

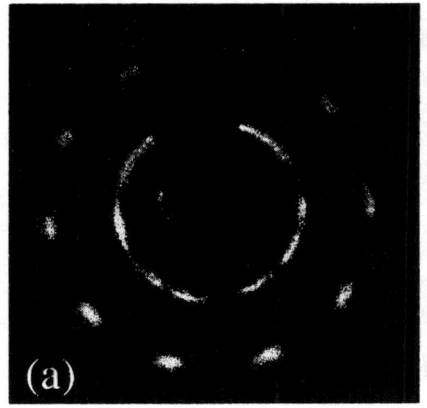 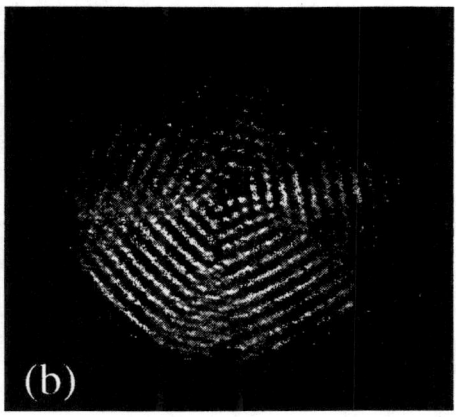

FIGURE 6. Five-fold Bragg scattering and real-space patterns obtained by strobing the intensified CCD camera synchronously with the rotating electric field perturbation. (a) Bragg scattering pattern obtained on an $N \sim 1.2 \times 10^5$ ion plasma phase-locked with a rotating dipole field at $\omega_r = 2\pi \times 166.84$ kHz. Here $V_o = 500$ V and $\alpha = 2.6$. (b) Real-space image of a disk-like plasma consisting of 4 horizontal planes in the plasma center. The rotating dipole field was used to set $\omega_r = 2\pi \times 74.35$ kHz.

mum. The phase transition may take place in the present system, but we have experimentally missed detecting it, or possibly larger crystals (for example, where the number of ions in the crystal is large compared to the number of ions in the shells) may be required in order for a sharp phase transition to be exhibited.

We have observed structures for which we do not have any current theoretical understanding. Figure 6(a) shows an approximate five-fold Bragg scattering pattern that was observed a number of times under different experimental circumstances. A five-fold Bragg scattering pattern is characteristic of a quasi-crystal. However, more sets of dots would be present in a true quasi-crystalline Bragg scattering pattern. We now think that the five-fold Bragg scattering pattern of Fig. 6(a) is due to a structure like that shown in Fig. 6(b). Figure 6(b) is a top-view image of a disk-like plasma which consisted of four horizontal planes. Even though it is difficult to distinguish individual ions in this figure, it is possible to see that there are five distinct regions where the ions resided in vertical planes. The planes from these different regions form a five-sided structure that would produce a Bragg scattering pattern like Fig. 6(a). (With the small crystals and forward Bragg scattering angles of this work, each set of vertical planes produces two Bragg peaks.) Once formed, this five-fold "twinning" was stable and persisted for reasons which we do not understand.

In addition to enhancing studies of Coulomb crystals, the phase-locked control of ω_r has improved the prospects of a microwave frequency standard based on a hyperfine-Zeeman transition of ions stored in a Penning trap. This is because the time-dilation shift due to the plasma rotation is one of the largest known systematic shifts in such a standard. Reference [35] discusses the potential frequency stability and accuracy of a microwave frequency standard based on 10^6 trapped ions. For ions such as ^{67}Zn$^+$ and ^{201}Hg$^+$, fractional frequency stabilities $\lesssim 10^{-14}/\tau^{1/2}$ with time-dilation shifts due to the plasma rotation of $\sim$few$\times 10^{-15}$ are possible. Here τ is the measurement time in seconds. With phase-locked operation of the rotating wall, we think it should be possible to stabilize and evaluate the rotational time-dilation shift within 1%. Therefore the inaccuracy due to this shift would contribute a few parts in 10^{-17}.

ACKNOWLEDGEMENTS

We gratefully acknowledge the support of the Office of Naval Research. We thank Fred Walls and Fred Moore for their comments and careful reading of the manuscript.

REFERENCES

1. Ichimaru, S., Iyetomi, H., and Tanaka, S., *Phys. Rep.* **149**, 91 (1987).
2. Malmberg, J. H., and O'Neil, T. M., *Phys. Rev. Lett.* **39**, 1333 (1977).
3. Van Horn, H. M., *Science* **252**, 384 (1991).
4. Grimes, C. C., and Adams, G., *Phys. Rev. Lett.* **42**, 795 (1979).
5. Pollock, E. L., and Hansen, J. P., *Phys. Rev. A* **8**, 3110 (1973); Slattery, W. L., Doolen, G. D., and DeWitt, H. E., *ibid.* **21**, 2087 (1980); *ibid.* **26**, 255 (1982); Ogata, S., and Ichimaru, S., *ibid.* **36**, 5451 (1987); Stringfellow, G. S., and DeWitt, H. E., *ibid.* **41**, 1105 (1990); Dubin, D. H. E., *ibid.* **42**, 4972 (1990).
6. Schiffer, J., *Science* **279**, 675 (1998).
7. Diedrich, F., Peik, E., Chen, J. M., Quint, W., and Walther, H., *Phys. Rev. Lett.* **59**, 2931 (1987).
8. Wineland, D. J., Bergquist, J. C., Itano, W. M., Bollinger, J. J., and Manney, C. H., *Phys. Rev. Lett.* **59**, 2935 (1987).
9. Strongly coupled clusters of highly charged, micrometer-sized aluminum particles were previously observed in Paul traps. See Wuerker, R. F., Shelton, H., and Langmuir, R. V., *J. Appl. Phys.* **30**, 349 (1959).
10. Gilbert, S. L., Bollinger, J. J., and Wineland, D. J., *Phys. Rev. Lett.* **60**, 2022 (1988).
11. Birkl, G., Kassner, S., and Walther, H., *Nature (London)* **357**, 310 (1992).
12. Drewsen, M., Brodersen, C., Hornekaer, L., Hangst, J. S., and Schiffer, J. P., to appear in *Phys. Rev. Lett.*.
13. Walther, H., *Adv. At. Opt. Phys.* **31**, 137 (1993).
14. See, for example, Melzer, A., Homann, A., and Piel, A., *Phys. Rev. E* **53**, 2757 (1996); Thomas, H., and Morfill, G. E., *Nature (London)* **379**, 806 (1996).
15. Murray, C. A., and Grier, D. G., *American Scientist* **83**, 238 (1995).

16. Vos, W. L., Mehens, M., van Kats, C. M., and Bösecke, P., *Langmuir* **13**, 6004 (1997).
17. Tan, J. N., Bollinger, J. J., Jelenković, B., and Wineland, D. J., *Phys. Rev. Lett.* **72**, 4198 (1995).
18. Itano, W. M., *et al.*, *Science* **279**, 686 (1998).
19. Huang, X.-P., Anderegg, F., Hollman, E. M., Driscoll, C. F., and O'Neil, T. M., *Phys. Rev. Lett.* **78**, 875 (1997).
20. Huang, X.-P., Bollinger, J. J., Mitchell, T. B., and Itano, W. M., *Phys. Rev. Lett.* **80**, 73, (1998).
21. Huang, X.-P., Bollinger, J. J., Mitchell, T. B., Itano, W. M., and Dubin, D. H. E., *Phys. Plasmas* **5**, 1656 (1998).
22. Mitchell, T. B., *et al.*, "Direct observations of structural phase transitions in planar crystallized ion plasmas," submitted to *Nature (London)*.
23. Brewer, L. R., *et al.*, *Phys. Rev. A* **38**, 859 (1988).
24. Davidson, R. C., *Physics of Nonneutral Plasmas*, New York: Addison-Wesley Publishing, 1990, ch. 3, pp. 39-75.
25. O'Neil, T. M., and Dubin, D. H. E., *Phys. Plasmas* **5**, 2163 (1998).
26. Heinzen, D. J., Bollinger, J. J., Moore, F. L., Itano, W. M., and Wineland, D. J., *Phys.Rev. Lett.* **66**, 2080 (1991).
27. Bollinger, J. J., *et al.*, *Phys. Rev. A*, **48**, 525 (1993).
28. Dubin, D. H. E., *Phys. Rev. A* **40**, 1140 (1989).
29. Ashcroft, N. W. and Mermin, N. D., *Solid State Physics*, Philadelphia, PA: Saunders College, 1976, ch. 6, pp. 95-110.
30. Tan, J. N., Bollinger, J. J., Jelenković, B., Itano, W. M., and Wineland, D. J., in *Physics of Strongly Coupled Plasmas*, Proceedings of the International Conference, Binz, Germany, Kraeft, W. D., and Schlanges, M., Eds., Singapore: World Scientific, 1996, pp. 387-396.
31. Schiffer, J. P., *Phys. Rev. Lett.* **70**, 818 (1993).
32. Dubin, D. H. E., *Phys. Rev. Lett.* **71**, 2753 (1993).
33. Cirac, J. I., and Zoller, P., *Phys. Rev. Lett.* **74**, 4091 (1995).
34. Wineland, D. J., *et al.*, "Experimental issues in coherent quantum-state manipulation of trapped ions," *Journal of Research of NIST* **103**, 259 (1998).
35. Tan, J. N., Bollinger, J. J., and Wineland, D. J., *IEEE Trans. Instrum. Meas.* **44**, 144 (1995).

Novel Dipole-Force Atom Traps

I. Manek[a], U. Moslener[a], Yu. B. Ovchinnikov[a,1], P. Rosenbusch[b], A. I. Sidorov[c], G. Wasik[d], M. Zielonkowski[a,e], and R. Grimm[a]

[a] *Max-Planck-Institut für Kernphysik, Heidelberg, Germany*
[b] *Sussex Centre for Optical and Atomic Physics, Brighton, UK*
[c] *School of Physics, University of Melbourne, Australia*
[d] *Institute of Physics, Jagiellonian University, Krakow, Poland*
[e] *Physikalisches Institut der Universität, Heidelberg, Germany*

Abstract. We discuss our recent experimental demonstrations of three novel atom traps which are all based on the optical dipole force in far-detuned laser light. In the gravito-optical surface trap (GOST), Cs atoms are trapped and cooled by means of an evanescent-wave mirror very closely above a dielectric surface. This trap allows for optical cooling at very high densities and provides good starting conditions for evaporative cooling to quantum degeneracy and the realization of a 2D quantum gas. The conical atom trap (CAT) uses a hollow blue-detuned laser beam in combination with gravity to trap atoms in a dark spatial region. This trap is experimentally simple and provides high loading efficiency, tight confinement, efficient sub-Doppler cooling, and strong suppression of collisional losses. The standing-wave red-detuned (STAR) trap allows one to confine atoms in a shallow optical potential at a very low photon scattering rate, and is thus an interesting tool for studies of the atomic ground-state dynamics in external electro-magnetic fields. We have studied atomic spin polarization in this trap.

INTRODUCTION

Atom traps have become very important tools for many experiments in atomic physics and related research [1–3]. Various trapping schemes have been developed to accumulate, cool, and store atoms, and to study different types of interactions. Applications can be found in many different respects, like high-resolution spectroscopy, atom optics and interferometry, quantum effects of the atomic motion, ultracold collisions, quantum statistics, and Bose-Einstein condensation.

Atom traps can be divided into three main classes: radiation-pressure traps, magnetic traps, and optical dipole-force traps. The first class is based on the scattering force exerted on the atoms in near-resonant laser light; the well-known magneto-optical trap (MOT) [4] is by far the most important example. The main

---

[1]) now at National Institute of Standards and Technology, Gaithersburg, USA.

advantage of radiation-pressure traps is the high capture velocity and potential depth, because of which the MOT is widely used as a source for cold atoms. In such a trap, however, the atoms are strongly influenced by the trapping light, as each atom typically scatters about 10^7 photons per second, the atomic levels exhibit light shifts of the order of the natural linewidth, and the near-resonant light has dramatic consequences for their collisonal properties. Magnetic traps [5] are based on the ground-state Zeeman interaction in an inhomogeneous magnetic field and thus avoid any optical interaction. Recently, they became very popular for evaporative cooling [6] and Bose-Einstein-condensation [7–9]. Magnetic traps do not allow for trapping of all ground-state sublevels, and possible trap geometries are restricted in practice by the necessity to use coils or permanent magnets.

Optical dipole-force traps [10], which are based on the off-resonant interaction with laser light, offer very interesting features for experiments on ultracold atomic samples. Photon scattering from the trapping light is drastically reduced as compared to radiation-pressure traps. Moreover, dipole force traps do not depend on the particular ground-state hyperfine or magnetic sublevel. This allows for sub-Doppler [11] and sub-recoil [12,13] cooling and facilitates experiments at zero or arbitrary magnetic fields. As a further advantage, many different trapping geometries can be realized, which allows for many different potential shapes ranging from highly anisotropic potentials to optical lattices.

The optical dipole force can attract atoms into the light field or repel them out of it, depending on the sign of the resonance detuning: In "red" and "blue" detuned traps, the potential minima are located at the intensity maxima or minima of the laser field, respectively. The most simple dipole-force trap is a tightly focused, red-detuned laser beam [14]. In order to obtain very low photon scattering rates in red-detuned traps, one has to use very far detuned fields [15–18]. Blue-detuned traps [19,12,20] are more difficult to be realized experimentally, but offer the great advantage of storing the atoms in dark regions, which further reduces the perturbing interaction with the trapping light.

In these proceedings, we discuss three novel dipole-force traps: The gravito-optical surface trap (GOST) and the conical atom trap (CAT), which are both blue detuned, and the standing-wave red-detuned (STAR) trap.

Dipole force potential

The optical dipole force can be derived from a potential, which in the far-detuned case corresponds to the energy of the light-shifted ground state. All our experiments are peformed on cesium atoms with linearly polarized light in the vicinity of the D_2 resonance line at a wavelength of 852.1 nm. The optical detuning greatly exceeds the excited-state hyperfine splitting, so that the ground-state potentials do not depend on the Zeeman sublevel. In this case, the dipole force potential can be written as

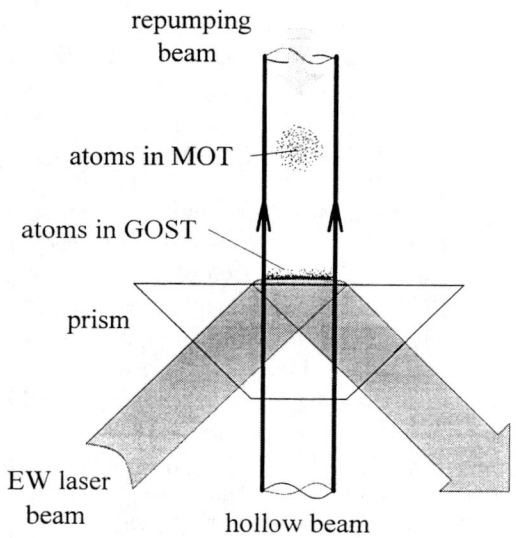

FIGURE 1. Illustration of the gravito-optical surface trap.

$$U_F(z) = \frac{\Gamma}{\delta_F} \frac{\lambda^3}{8\pi^2 c} I(\mathbf{r}). \qquad (1)$$

Here the natural linewidth is $\Gamma = 2\pi \times 5.3\,\text{MHz}$, and δ_F is the relevant optical detuning for the lower (F=3) or upper (F=4) hyperfine ground state. The detunings are related by $\delta_4 - \delta_3 = 2\pi \times 9.2\,\text{GHz}$.

GRAVITO-OPTICAL SURFACE TRAP (GOST)

The GOST, schematically shown in Fig. 1 and described in more detail in Ref. [21], facilitates storage and efficient cooling of a dense atomic gas closely above an evanescent-wave (EW) atom mirror [22–24]. The atoms bounce on the surface like on a trampoline [25,26] with their vertical motion being confined by the repulsive EW field in combination with gravity. In our experiment, the EW is produced on the flat horizontal surface of a fused-silica prism by total internal reflection of a 60-mW beam delivered by a laser diode.

Horizontal confinement is provided by the conservative optical dipole potential of a hollow, cylindrical laser beam, far blue-detuned by about 0.3 nm ($\sim 100\,\text{GHz}$) from the atomic resonance. The upward directed hollow beam (HB) used for this purpose has a ring-shaped profile with an inner and outer $1/e$ diameter of 700 μm and 750 μm, respectively, and a total power of 120 mW. This beam is generated

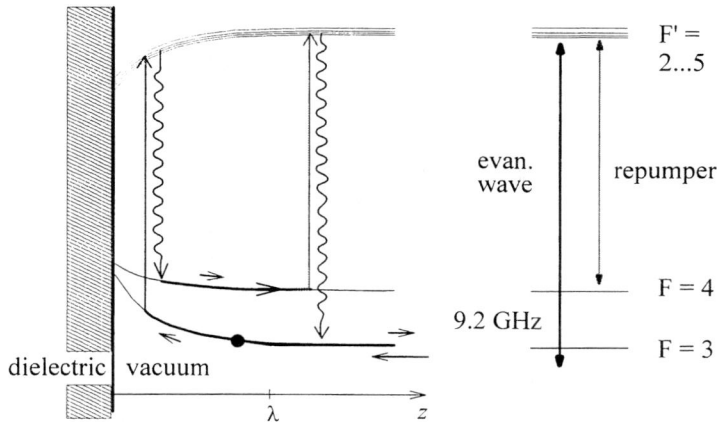

FIGURE 2. Illustration of the EW cooling cycle together with the relevant level scheme of the Cs D_2 line. The dot indicates an atom approaching the dielectric surface and moving along the optical ground-state potentials.

with the help of an *axicon* as described in Ref. [27]. In the dark central region, the intensity is as low as 10^{-3} of the peak intensity in the ring.

The GOST is loaded from a standard magneto-optical trap (MOT), which is placed right into the center of the HB about $800\,\mu$m above the prism surface. The MOT is loaded from the low-velocity tail of an effusive Cs atomic beam with about $\sim 3 \times 10^5$ atoms. The atoms are transferred into the GOST by turning off the MOT after a short sub-Doppler cooling phase at large detuning.

Evanescent-wave Sisyphus cooling

The key to prepare a very cold and dense surface gas in the GOST is evanescent-wave Sisyphus cooling [28–31]. This sub-Doppler cooling mechanism is based on *inelastic reflections* of the atoms from the EW. As a very important feature for attaining very high densities, EW cooling allows to keep the atoms predominantly in the absolute ground state($F = 3$), thus avoiding trap losses by light-assisted collisions [28,29].

The EW cooling mechanism, illustrated in Fig. 2, is based on the ground-state hyperfine structure of cesium. We can model the atom as a three-level scheme [29] with two ground states ($F = 3, 4$) separated by $\delta_{\text{hfs}}/2\pi = 9.2\,\text{GHz}$ and one excited state, for which the much smaller hyperfine splitting can be neglected. The EW has an intensity of $I(z) = I_0 \exp(-2z/\Lambda)$ with $I_0 = 2.9 \times 10^4\,\text{mW/cm}^2$ and $\Lambda = 0.30\mu$m, and its frequency is on the blue side of both transitions with detunings $\delta_3 = \delta_{\text{ew}}$ (for $F=3$) and $\delta_4 = \delta_{\text{ew}} + \delta_{\text{hfs}}$ (for $F=4$). In the experiment, we chose $\delta_{\text{ew}}/2\pi = 1.0\,\text{GHz}$. The optical ground-state potentials, given by Eq. 1, exponentially decrease with the distance z from the surface and depend on the hyperfine state F via the detuning δ_F.

An inelastic reflection takes place when the atom enters the EW in the $F=3$ state and, by scattering an EW photon during the reflection process, is pumped into the less repulsive $F=4$ state (see Fig. 2). In such a spontaneous transition the potential energy is reduced according to the ratio of the detunings by $\delta_{\rm ew}/(\delta_{\rm ew}+\delta_{\rm hfs})$ (≈ 0.1 for $\delta_{\rm ew}/2\pi = 1.0\,{\rm GHz}$) and the atom loses energy in an elementary Sisyphus process [28–31]. The mean energy loss $\Delta E_\perp$ from the motion perpendicular to the surface per inelastic reflection is found to be $\Delta E_\perp/E_\perp = -\frac{2}{3}\delta_{\rm hfs}/(\delta_{\rm ew}+\delta_{\rm hfs})$ (≈ 0.6), where $E_\perp = mv_\perp^2/2$ is the kinetic energy of the incoming atom. The probability for spontaneously scattering an EW photon during the reflection is given by $p_{\rm sp} = m\Lambda\Gamma v_\perp/\hbar\delta_{\rm ew}$ ($\approx 0.033 \times v_\perp/({\rm cm/s})$) as far as $p_{\rm sp} \ll 1$. With the branching ratio $q = 0.25$ of the excited state to the $F=4$ ground state, the probability for a cooling reflection is $p_{\rm cool} = q\,p_{\rm sp}$ ($\approx 0.0084 \times v_\perp/({\rm cm/s})$). The cooling cycle can be closed by repumping the atom into the $F=3$ state shortly after the reflection (see Fig. 2). In the experiment, we used a weak repumping beam that was directed downward and tuned close to resonance with the $F=4\rightarrow F'=4$ transition.

An atom repeatedly bouncing in the field of gravity on a horizontal surface with a time $t_{\rm r} = 2v_\perp/g$ between two successive reflections loses its vertical kinetic energy with an average rate

$$\beta = p_{\rm cool}\frac{\Delta E_\perp}{E_\perp}t_{\rm r}^{-1} = \frac{q}{3}\frac{\delta_{\rm hfs}}{\delta_{\rm ew}}\frac{mg\Lambda}{\hbar(\delta_{\rm ew}+\delta_{\rm hfs})}\Gamma. \qquad (2)$$

For our experimental parameters we obtain $\beta \approx 1/(400\,\mu{\rm s})$. The final temperature attainable with EW cooling is recoil-limited to a value of the order of $10\,\hbar^2 k^2/mk_B \approx 2\,\mu{\rm K}$ [29].

Experimental results

With EW cooling, the storage of atoms in the GOST is essentially limited by collisions with the residual gas in the vacuum chamber. At a background pressure of 4.2×10^{-10} mbar we observe an exponential decay of the number of stored atoms (initially 2×10^5) with a $1/e$ lifetime of 6.0 s. The measurement is performed by a recapture method: After storage times of up to 25 s, the atoms remaining in the GOST are retrapped into the MOT and then detected by fluorescence imaging. We do not observe any non-exponential particle loss, as it would result from inelastic ultracold collisions between trapped atoms.

For measuring the vertical and horizontal temperatures $T_{\rm v}$ and $T_{\rm h}$ in the GOST we use a time-of-flight method: For a short time interval $\Delta t_{\rm ew}$ or $\Delta t_{\rm hb}$, we switch off either the EW or the HB, respectively. This leads to a loss of atoms from the GOST either by hitting the prism surface ($\Delta t_{\rm ew} \neq 0$) or by escaping horizontally out of the trap ($\Delta t_{\rm hb} \neq 0$). With the assumption of a Boltzmann distribution in

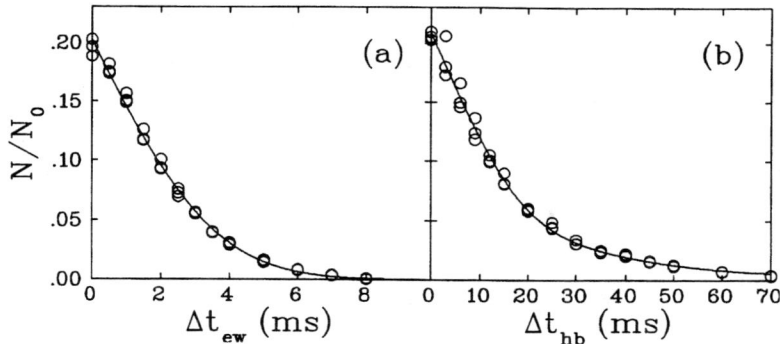

FIGURE 3. Time-of-flight measurements of the vertical (a) and horizontal (b) temperature in the GOST after a storage time of 4 s. The number N of atoms recaptured from the GOST into the MOT relative to the initial number N_0 in the MOT is plotted as a function of the switch-off time Δt_{ew} (evanescent wave) or Δt_{hb} (hollow beam).

phase space and a free ballistic flight it is straightforward to calculate the number of remaining atoms as a function of Δt_{ew} or Δt_{hb}. A fit to the experimental data then yields the vertical and horizontal temperature. Fig. 3 shows an example for such a pair of measurements of T_{v} and T_{h}, which was made after 4 s of storage and cooling in the GOST.

The vertical measurement displayed in Fig. 3(a) yields $T_{\text{v}} = (3.0 \pm 0.1)\,\mu\text{K}$. At this low temperature the mean time between two bounces is 2.2 ms, corresponding to 450 bounces per second. The mean probability for an incoherent bounce is $p_{\text{sp}} = 3.5\%$ and thus for a cooling bounce $p_{\text{cool}} = 0.9\%$. According to the Boltzmann distribution, the vertical dependence of the atomic density obeys the *barometric equation* $n(z) = n_0 \exp(-mgz/k_B T_{\text{v}})$, where in our case the $1/e$ height is as low as $k_B T_{\text{v}}/mg = 19\,\mu\text{m}$.

The horizontal measurement shown in Fig. 3(b) yields a temperature $T_{\text{h}} = (3.1 \pm 0.3)\,\mu\text{K}$, which is equal to the vertical one within the experimental uncertainty. As the anisotropic EW cooling mechanism acts only vertically, it seems very surprising that we find equal horizontal and vertical temperatures. However, as an atom performs about 100 bounces between two cooling reflections, even a weak motional coupling effect can completely redistribute the energy between the different degrees of freedom. In our experiment, the most likely cause of this coupling is the weak diffusive component of the EW reflection that was observed in Ref. [32]. The total photon scattering rate in the GOST is about 30 photons per atom and second and thus more than five orders of magnitude less than in a typical MOT.

Under our present loading conditions, we obtain number densities n_0 of up to about $2 \times 10^{10}\,\text{cm}^{-3}$ at a temperature of $3\,\mu\text{K}$ corresponding to a mean sample height of $19\,\mu\text{m}$. This gives a maximum phase-space density of $1/7 \times (1.3 \times 10^{-5})$, where the degeneracy factor $1/7$ takes into account the distribution among the seven magnetic sublevels of the $F = 3$ ground state. This phase-space density is already comparable to the best values measured in an optimized Cs MOT [33]. Much higher number and phase-space densities in the GOST can be obtained in a straightforward way by increasing the number of atoms loaded into it. In contrast to a MOT, the GOST offers an enormous potential for future experiments at high densities as losses by hyperfine-changing and excited-state collisions are dramatically suppressed as a result of the predominant population of the absolute ground state $F = 3$ [29]: In the GOST, the fraction of atoms in the upper hyperfine state ($F = 4$) is in the range of 10^{-4}–10^{-5}, and the fraction in the optically excited state is as low as $\sim 1.5 \times 10^{-6}$. Furthermore, the density and temperature-limiting reabsorption of spontaneously emitted photons is strongly reduced as compared to a MOT, because the flat geometry of the GOST offers a large surface for photons to escape.

Towards quantum degeneracy in the GOST

In near-future experiments we plan to load the GOST with 10^7 to 10^8 Cs atoms, and we expect to reach the same temperature as before at 100 to 1000 times higher number densities (10^{12} - 10^{13} atoms/cm^3). With an expected phase-space density of about 10^{-3} and an elastic collision rate of $\sim 100\,\text{s}^{-1}$, the GOST will then offer very promising starting conditions for evaporative cooling [6] with the aim to reach quantum degeneracy.

A particular motivation for these planned experiments are the *unusual quantum gas properties of cesium*, revealed in a beautiful series of experiments at the ENS Paris [34–37] and studied theoretically in Refs. [38,39]. On one hand, cesium has a large cross section for elastic scattering, which is resonantly enhanced both in the upper ($F=4$) and in the lower hyperfine state ($F=3$). On the other hand, both states show an unusually fast dipolar relaxation, which for Cs unlike other alkali atoms (Rb, Na, Li) leads to severe losses from magnetic traps. According to the present state of knowledge, it appears to be impossible to produce a Bose-Einstein condensate of Cs atoms in a magnetic trap by evaporative cooling in the $F=4$ state [35] and, at least, very difficult in the $F=3$ state [36,37].

As an optical dipole-force trap, the GOST does not suffer from dipolar relaxation losses as it stores atoms in all magnetic sublevels. A polarized sample can be trapped in the *absolute ground state $F=3, m_F=3$*, which is impossible in a magnetic trap. Collisional losses can then only occur as a result of three-body recombination, which is relevant only at very high densities [18,40]. The GOST is therefore very interesting for producing a Bose-Einstein condensate of Cs as it can take full advantage of the large cross section for elastic scattering without being limited by collisional losses.

In the GOST, we plan to evaporatively cool a polarized sample of Cs atoms in the absolute ground state $F=3, m_F=3$, as discussed in more detail in [46]. A very important issue for achieving BEC by evaporative cooling in the GOST is photon scattering from the trapping light, which heats and depolarizes the sample. We estimate that it is necessary to reduce the photon scattering rate in the GOST to values well below $1\,\text{s}^{-1}$. In order to fulfill this requirement, we plan to implement a second stage of the GOST, where the detunings of the hollow beam and of the evanescent wave will be substantially increased after the initial Sisyphus pre-cooling phase. For implementing evaporative cooling we plan to ramp down the EW potential by continuously increasing the optical detuning.

The s-wave scattering length of the state $F = m_F = 3$ is predicted to be negative [38]. Therefore, at low magnetic field, a Bose-Einstein condensate can only be expected with a relatively small particle number [9]. In general, however, the scattering length depends on the magnetic field and can exhibit Feshbach resonances [41,42]. In the GOST, one is completely free to apply a variable homogeneous magnetic field and thus to study magnetic-field dependencies of collisonal processes and mean-field interactions without changing the trap characteristics. A Feshbach resonance may be used to tune the scattering length to positive values and thus to produce a stable Bose-Einstein condensate of cesium.

The highly anisotropic potential of the GOST is also very prospective for experiments to realize a *two-dimensional quantum gas* of ultracold atoms. This issue is discussed in Ref. [46]. An interesting option in this respect is to create a narrow potential well of optical wavelength size by using a second, red-detuned evanescent-wave as suggested in Ref. [45]. Such an experiment would show interesting analogies to experiments on H atoms adsorbed on the surface of liquid He [43,44].

CONICAL ATOM TRAP (CAT)

We have studied another gravito-optical trap which does not use an evanescent wave. In our CAT, described in more detail in Ref. [47], the atoms are trapped in an intense blue-detuned conical hollow laser beam which propagates vertical in the field of gravity. The gravito-optical confinement principle of the CAT is the same as in the GOST and in previously demonstrated light-sheet traps [19,12], but the experimental implementation of the conical trapping field based on axicon optics [27] is considerably simpler. Besides this practical advantage, the interesting features of the trap are high loading efficiency in a funnel-like geometry, tight confinement in a dark spatial region, and a strong suppression of losses caused by ultracold collisions due to the predominant population of the absolute ground state ($F = 3$).

The CAT is schematically shown in Fig. 4. The conical trapping beam with an opening angle of $\theta \approx 150$ mrad is realized in a simple optical setup using two identical axicons: In a telescope-like arrangement, the axicons (base angle ~ 10 mrad) are separated by 2.5 m and transform a Gaussian laser beam ($1/e$-diameter 4 mm) into a collimated hollow beam. This tubular beam (outer diameter 34 mm, inner diameter 30 mm) is then focused with a spherical lens (focal length 200mm) to generate the conical beam with its apex located in the focal plane. The 250-mW beam, derived from a single-mode Ti:Sapphire laser, is linearly polarized and di-

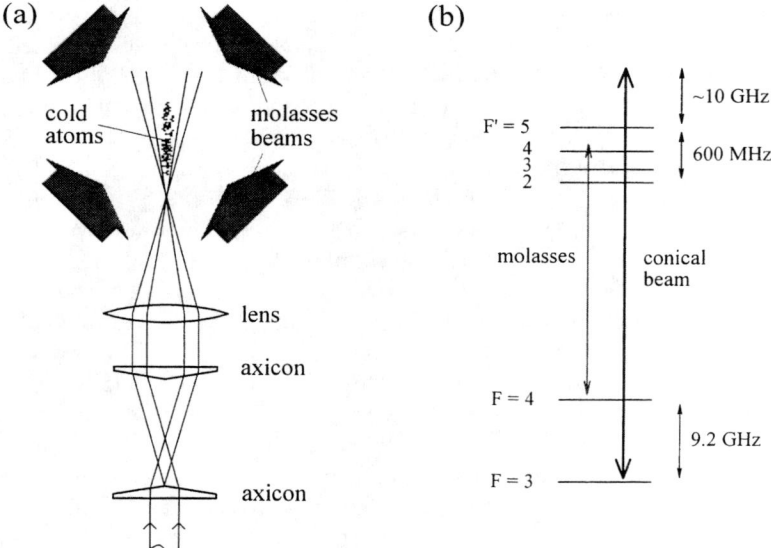

FIGURE 4. (a) Illustration of the conical atom trap (CAT). The conical hollow beam, directed upwards against gravity, is generated with two axicons and one spherical lens. Cooling is provided by a 3D optical molasses in combination with inelastic reflections from the conical field. (b) Level scheme of the Cs atom (level spacings not to scale). The blue-detuned conical hollow beam repels atoms in both hyperfine ground states and also acts as a slow repumper for the molasses, which cools atoms in the $F=4$ ground state.

rected upwards. In the focal plane, the beam profile is roughly Gaussian with a diameter of about 100 μm. Within a few millimeters, the beam profile evolves into a ring resembling a higher-order Laguerre-Gaussian mode [20].

We have operated the CAT with blue detunings $\delta_3/2\pi$ in the range of 3 GHz to 30 GHz ($\delta_4/2\pi$ of 9 GHz to 39 GHz), where the conical field acts repulsively for both hyperfine ground states. The CAT is loaded by releasing about 10^6 Cs atoms from a standard magneto-optical trap (MOT) situated inside the conical light beam about 5 mm above the apex.

For cooling, we use an optical molasses in combination with "reflection cooling"; the latter is analogous to the cooling in the GOST, but much less efficient as the reflecting laser field in the CAT is much less steep (about 100 times) as compared to the evanescent wave. The molasses is realized with the laser beams of the MOT tuned about 125 MHz to the red side of the closed $F=4 \rightarrow F'=5$ transition, corresponding to a blue detuning of 125 MHz with respect to the open $F=4 \rightarrow F'=4$ transition. No additional repumping field from the $F=3$ ground state is present. Let us consider an idealized cooling cycle: An atom enters the conical field in the strongly repulsive lower hyperfine state ($F=3$). During the reflection process it is

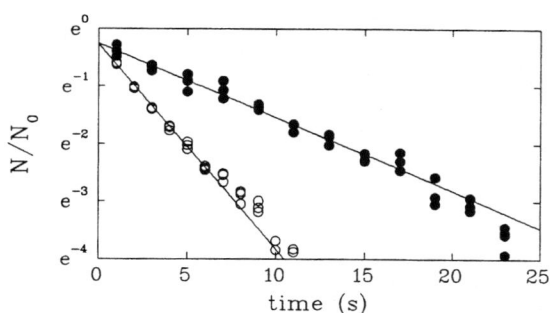 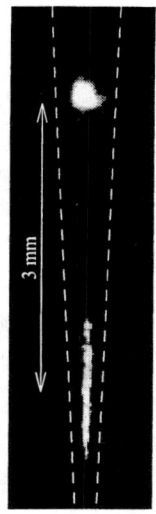

FIGURE 5. Experimental results on the storage of Cs atoms in the CAT. Left-hand side, measurements of the storage time; at a background pressure of 2.5×10^{-10} mbar (●) an exponential fit to the data gives a lifetime of 7.8 s, and at 6.9×10^{-10} mbar (○) a lifetime of 2.8 s is observed. Right-hand side, fluorescence image of atoms in the CAT (lower, cigar-shaped blob) combined with an image of atoms in the MOT (upper spot). The dashed lines indicate the conical field.

pumped into the upper hyperfine level, which is less repulsive by the factor δ_4/δ_3, and thus loses a considerable part of its potential energy. The atom leaves the reflecting laser field with reduced kinetic energy. In the $F=4$ ground state, the atom then experiences the optical molasses and is cooled there for a short time involving a few spontaneous emission events only until it is depumped again into the lower ground state ($F=3$). It remains in this dark level until it is again reflected and repumped from the intense conical beam.

We have demonstrated the storage of Cs atoms in the CAT with a lifetime of several seconds, limited by collisions with the residual gas in our vacuum chamber. In this experiment, the laser frequency was set to $\delta_3/2\pi \approx 3$ GHz ($\delta_4/2\pi \approx 12$ GHz). For measuring the storage time we use the same retrapping method as applied for the GOST. The experimental results are shown in Fig. 5 for two different values of the residual gas pressure in our vacuum chamber. The data clearly shows that collisions with the residual gas are the predominant loss mechanism of atoms out of the CAT. The measurement also demonstrates the very high transfer efficiency from the MOT into the CAT of ∼80%, as obtained from the extrapolation of the exponential fits to $t=0$.

A fluorescence image of the atomic cloud in the CAT is shown in Fig. 5 (lower, elongated blob) combined with a fluorescence image of the original MOT (upper, round spot) taken 100 ms before the transfer into the CAT. The fluorescence light emitted by the atoms in the CAT is at least 1000 times weaker as from the MOT. Therefore the CAT picture has been taken by shortly tuning the molasses beams close to resonance and turning on the repumping field. The atoms in the CAT form a sample with the form of a "cigar" of about $100\,\mu$m in diameter (full width at half maximum) and with a length of about 1 mm. The vertical distribution of the trapped atoms is determined by their temperature and the shape of the potential well that results from the optical dipole force counteracted by gravity. The center of the sample is located about 2.5 mm above the apex. With nearly 10^6 atoms in the CAT the peak density is on the order of $\sim 10^{11}\,\text{cm}^{-3}$.

The temperature of atoms stored and cooled in the CAT is determined by time-of-flight fluorescence imaging. After turning off the conical field, the suddenly released atoms were detected with a near-resonant 5-ms light pulse after 50 ms of free ballistic expansion. The temperatures derived from corresponding CCD pictures are between $10\,\mu$K and $15\,\mu$K.

For the storage of atoms at high densities, a very important feature of the CAT is the predominant population of the lower hyperfine level ($F=3$). In this absolute ground state, no internal energy can be released in an ultracold binary collision so that trap losses by inelastic processes are strongly suppressed as compared to a MOT. ¿From our experimental data we estimate that the population of the F=4 state is reduced at least a factor of 30 and the relative excited-state population in the CAT is below 10^{-4}.

We also tested the CAT in a pure *reflection-cooling mode* [28,29], without any molasses applied. In addition to the intense conical beam, we used a weak beam coming from above to pump the atoms from the $F=4$ ground state into the $F=3$ state. The beam was resonant with the $F=4 \leftarrow F'=4$ transition and had an intensity of a few $10\,\mu$W/cm^2. The cooling then results from inelastic reflections from the optical walls of the CAT, analogous to EW cooling in the GOST. This mechanism, however, can only provide efficient cooling if the number of spontaneous emissions per single reflection process is $\lesssim 1$ [28,29]. While this condition is easily fulfilled in the case of the very steep optical potentials of an evanescent wave, the much less steep CAT potentials require much larger detunings. Therefore, we observed reflection cooling at detunings $\delta_3/2\pi \simeq 30\,\text{GHz}$ ($\delta_4/2\pi \simeq 40\,\text{GHz}$). In this case, we obtained trapping with lifetimes of up to a few seconds, close to residual-gas limited conditions. When we blocked the weak repumping beam to disable reflection cooling, the trapped atoms were lost much faster (within about 100 ms) due to heating out of the conical field by photon scattering. The low potential depth resulting from the large detunings limited the transfer from the MOT to about 10 %. Reflection cooling alone does thus not lead to a very efficient and stable operation of the CAT, but in combination with an optical molasses excellent performance is obtained.

A possible application of the CAT could be photoassociation spectroscopy [48], as its geometry allows one to match a tightly focused, intense photoassociation beam to the dense cigar-shaped trapped atom cloud. For spectroscopic measurements [49], the level shifts induced by the trapping light in the CAT are much less than in a MOT. Another interesting application of the conical hollow beam is to serve as a funnel for loading surface traps like the GOST.

STANDING-WAVE RED-DETUNED (STAR) TRAP

With the motivation to study spin polarization of optically trapped atoms we have realized another kind of atom trap [50]: Our STAR trap consists of a simple standing-wave laser beam, far detuned below resonance and oriented vertically in the field of gravity [51], and combines several advantages for precision measurements concerning the atomic ground-state dynamics: In comparison with a usual far-off-resonance trap obtained by tight focusing of a single laser beam [15], a similar number of atoms can be stored in a much shallower optical potential of much larger volume. The low potential depth strongly reduces the heating and depolarization caused by scattering trap photons, and the low density minimizes unwanted collisional effects leading to trap losses and possible line broadening [52]. The STAR trap also represents a 1D optical lattice [53,54], but in our experiments the periodic structure and the confinement in wavelength-size potential wells plays no explicit role. We just use the high vertical potential gradient to keep the atoms from falling down in the field of gravity.

The STAR trap is realized with the linearly polarized 220-mW beam of a Ti:Sapphire laser operated at a wavelength of $\lambda = 858.2$ nm, i.e. detuned by $\Delta\lambda_2 = 6.1$ nm to the red side of the D_2 line of Cs at 852.1 nm. The beam passes through the vacuum chamber from below and is retro-reflected from a mirror above the chamber. In the interaction region, its $1/e$ radius is $r = 0.25$ mm and the peak intensity is $I_0 = 4.6$ W/mm^2. The ground-state optical potentials are independent of the magnetic substate and the maximum potential depth is calculated to $U_0 = k_B \times 17\,\mu$K $= 1.4$ neV. The maximum vertical dipole force exceeds the gravitational force as well as the about three times weaker maximum horizontal dipole force by about three orders of magnitude.

For magnetic Stern-Gerlach (SG) selection of atoms in the STAR trap, we simply use the anti-Helmholtz coils of the MOT to generate the required inhomogeneous magnetic field, see Fig. 6(a). Along the symmetry axis of the magnetic field (z-axis), the gradient is 25 G/cm. In addition we apply a homogeneous bias field $B_0 \simeq 1$ G in z-direction to shift the magnetic field zero horizontally out of the trap volume. In this way, we define the z-axis as quantization axis and exclude Majorana spin flips of trapped atoms. The m_F-dependent potentials for the $F=4$ ground state resulting from both the magnetic potential and the optical potential are plotted in Fig. 6(b). With the magnetic gradient applied, the STAR trap confines atoms only in the substate $m_F = 0$. All other atoms ($m_F \neq 0$) are horizontally pulled out of the trap by the SG force and then fall down in the field of gravity.

The STAR trap is loaded from a standard MOT, the center of which is carefully placed onto the axis of the intense standing-wave beam. A short sub-Doppler cooling phase in the MOT at large detunings provides sufficiently low temperatures

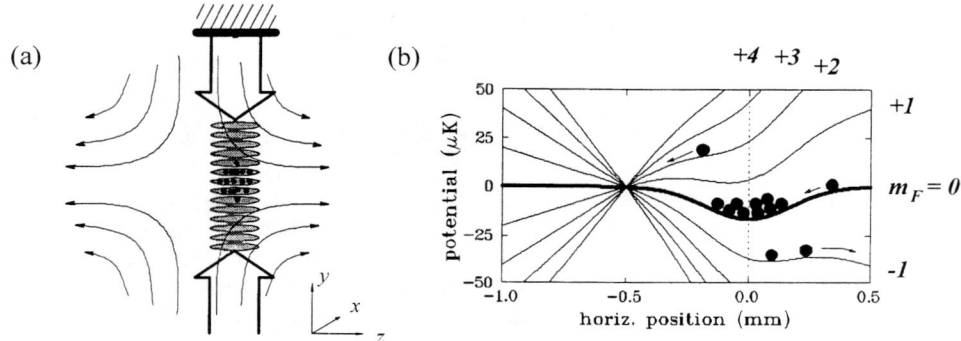

FIGURE 6. (a) Illustration of the STAR trap with the magnetic quadrupole field used for SG selection of atoms in the $m_F=0$ state. (b) Horizontal motional potentials for atoms in the different magnetic substates.

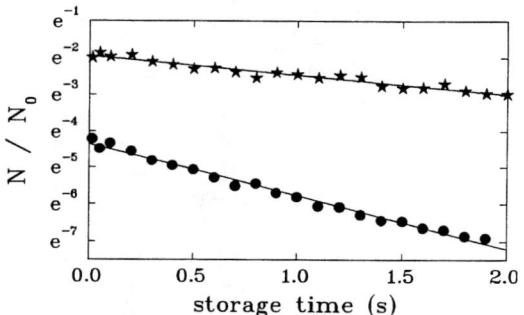

FIGURE 7. Fraction of Cs atoms remaining in the STAR trap after variable storage time without (stars) and with (dots) continuous SG selection; the solid lines represent exponential fits to the experimental data.

($\sim 10\,\mu$K) for loading the shallow standing-wave potentials. The lifetime of atoms in the STAR trap is measured with the same recapture method, as used for the GOST and CAT. The result is shown in Fig. 7 (stars). The observed transfer from the MOT into the shallow potential of the STAR trap is about 14% ($\sim 10^5$ atoms). The exponential decay shows a $1/e$ lifetime of $\tau = 1.9 \pm 0.2$ s, which is faster than the decay time of about 7 s observed for the residual-gas limited storage in the GOST and in the CAT under the same vacuum conditions. We attribute the additional loss to heating out of the shallow trap by photon scattering and presumably also by intensity fluctuations of the trapping light [55].

Our SG selection with the anti-Helmholtz coils now allows us to directly measure the trap-induced decay of spin polarization. About 20 ms after finishing the loading sequence, we turn on the bias field $B_0 = 1$ G and a few ms later the magnetic quadrupole field for continuous SG selection. An atom in the trapped $m_F = 0$ state that changes its magnetic quantum number by scattering a linearly polarized trap photon and emitting a circularly polarized photon is immediately lost from the trap. The resulting increase of the trap decay as compared to the trap without magnetic gradient therefore represents a direct measure for the depolarizing effect of the trap light [56]. The number N_{SG} of atoms remaining in the STAR trap with permanent SG selection is also plotted in Fig. 7 (dots). With the magnetic gradient, the initial number of atoms in the STAR trap is $N_{SG}/N_0 \simeq 1.3\%$, i.e. about ten times less as compared to the magnetic-field free case. This reduction of the initially stored atom number is due to the fact that all atoms with $m_F \neq 0$ are immediately expelled from the trap. With SG selection, an exponential decay of the number of trapped $m_F = 0$ atoms is observed with a lifetime of $\tau_{SG} = 0.7 \pm 0.1$ s.

¿From the measurements of the storage times τ_{SG} and τ with and without SG selection, we can directly derive the trap-induced depolarization rate (the rate of pumping from $m_F = 0$ into $m_F \neq 0$), for which we obtain $\Gamma_{depol} = 1/\tau_{SG} - 1/\tau = 0.9 \pm 0.2 \, \text{s}^{-1}$; the corresponding lifetime of spin polarization is $1/\Gamma_{depol} = 1.1 \pm 0.2$ s. ¿From this measurement and the calculated branching ratio of 1/3 for m_F-changing scattering processes [57], we can deduce an average photon scattering rate of $\Gamma_{sc} = 3\Gamma_{depol} \simeq 2.5 \, \text{s}^{-1}$. In comparison, the peak photon scattering rate (at the standing-wave nodes and in the center of the beam) is calculated to $\sim 5 \, \text{s}^{-1}$. This number is consistent with the observed average rate because of the spatial distribution of atoms in the trap. The observed decay of spin polarization can thus fully be attributed to the expected depolarization by trap photon scattering.

In a further experiment, we demonstrate the spin manipulation of trapped atoms by a 'fictitious magnetic field' [58,59], induced by an off-resonant circularly polarized laser beam. This field, which results from m_F-dependent light shifts, affects internal spin dynamics in the same way as if a real magnetic field of magnitude [59]

$$B_{fict} = \frac{\Gamma}{\Delta_1} \frac{\lambda_0^3}{16\pi^2 \, c \, \mu_B} I_1 \qquad (3)$$

was applied in laser-beam direction. Here λ_0 is the transition wavelength, I_1 represents the laser intensity, and Δ_1 denotes the detuning with respect to the line center of the hyperfine-split excited state. In our experiment, the fictitious field beam is provided by a diode laser, propagates in x-direction, has a central intensity of $\sim 500 \, \text{mW/cm}^2$, and is detuned by $\Delta_1/2\pi = +3 \, \text{GHz}$. The maximum fictitious field strength, calculated from Eq. 3, is about 50 mG.

Our measurement of spin precession, see timing scheme in Fig. 8(a), is based on SG preparation of a trapped $m_F = 0$ sample and subsequent SG analysis, each accomplished by application of the magnetic gradient in a 50-ms time interval. In the middle of the 150-ms delay time in between SG preparation and probing, the spins are rotated by a short pulse of circularly polarized light. At the time of this pulse the bias field B_0, serving as holding field for the spins, is reduced from its full value of 1 G to $\sim 200 \, \text{mG}$. Spin precession takes place around the total field

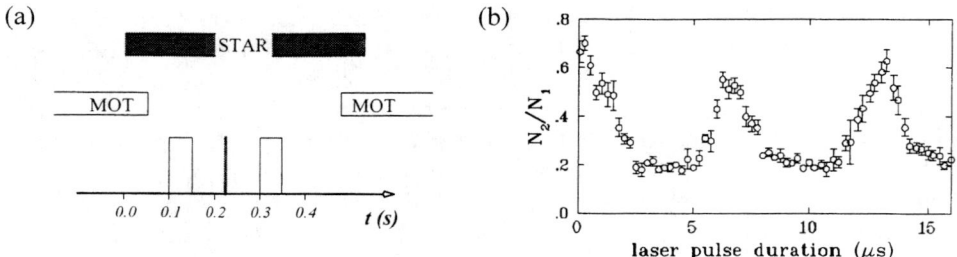

FIGURE 8. Observation of spin precession in a 'fictitious magnetic field'. (a) Timing scheme, and (b) relative number of trapped atoms N_2/N_1 measured as a function of the fictitious-field pulse duration.

$\mathbf{B}_{\text{tot}} = B_{\text{fict}}\hat{\mathbf{x}} + B_0\hat{\mathbf{z}}$, which is tilted against the holding field (quantization axis) by about 15°. After the SG analysis, the $m_F=0$ atoms remaining in the STAR trap are detected by recapture into the MOT. The observed number N_2 is normalized to a reference measurement of the atom number N_1 performed with SG preparation but without the SG analysis. The relative number N_2/N_1 directly shows the fraction of atoms in the $m_F = 0$ state at the time of the SG analysis.

The experimental results are shown in Fig. 8(b). Without the fictitious-field pulse we observe $N_2/N_1 \approx 0.7$, which means that about 30% of the atoms are pumped into $m_F \neq 0$ states in the course of a measurement. With the duration of the fictitious-field laser pulse, the $m_F = 0$ population is observed to oscillate between a maximum of ~ 0.65 and a minimum of ~ 0.3 with a period of 13 µs. This oscillation can be directly interpreted as the Larmor precession in the field $B_{\text{tot}} \simeq 220$ mG. The two revivals of the $m_F = 0$ population in Fig. 8(b) are observed without any significant loss of contrast. This shows that the fictitious field is sufficiently homogeneous, i.e. the laser intensity I_1 is nearly constant over the trapped atom sample. A calculation of the precession of an atom initially prepared in the state $F = 4, m_F = 0$ in a magnetic field [57] agrees well with the observed shape of the experimental curves.

A very interesting application of the STAR trap would be the measurement of permanent electric dipole moments of paramagnetic atoms [60]. For such an experiment, exceptionally long lifetimes of spin polarization could be realized in a straightforward way by using much further detuned laser light as compared to the present experiment. By using, e.g., a Nd:YAG laser operating at the 1.35 µm line, the detuning could be increased by a factor of 100. To obtain the necessary potential depth the standing wave could be generated in a build-up cavity, which would not require much experimental effort. By this and other possible improvements, it seems realistic to obtain lifetimes of spin polarization of the order of several minutes. Another important issue for this application is the dephasing of spins in an inhomogeneous environment limiting the coherence time. In order to overcome this problem, spin-echoes induced by a fictitious magnetic field [59] could be used.

CONCLUSION

We have presented three examples highlighting the unique features of optical dipole-force traps for experiments with ultracold atoms. Interesting new trap geometries can be realized, as demonstrated by our two novel blue-detuned traps, the GOST and the CAT. The GOST allows one to store and cool atoms at high densities closely above an evanescent-wave mirror. We are currently exploring the possibility to create a quantum-degenerate gas of cesium atoms, which may exhibit interesting 2D properties. The CAT provides high loading efficiency, tight confinement, efficient cooling, and storage in a dark spatial region, combined with experimental simplicity. The fact that any ground-state sublevel can be stored in a dipole-force trap opens up further new applications. In our STAR trap, we have demonstrated the manipulation of spin polarization and a first magnetic resonance experiment with ultracold trapped atoms.

ACKNOWLEDGMENTS

We thank J. Söding, M. Weidemüller, M. DeKieviet, and E. A. Hinds for stimulating discussions. We are indebted to D. Schwalm for discussions, encouragement, and support. Yu. B. O. gratefully acknowledges a fellowship by the Alexander von Humboldt-Stiftung, A. I. S. acknowledges an Australian Research Fellowship, and G. W. thanks the Polish Committee for Science for travel support. This work was supported by the Deutsche Forschungsgemeinschaft in the frame of the Gerhard-Hess-Programm.

REFERENCES

1. *Laser Manipulation of Atoms and Ions*, Proceedings of the International School of Physics "Enrico Fermi," Course CXVIII, edited by Arimondo, E., Phillips, W. D., and F. Strumia (North Holland, Amsterdam, 1992).
2. Metcalf, H., and van der Straten, P., *Phys. Rep.* **244**, 203 (1994).
3. Adams, C., and Riis, E., *Prog. Quant. Electr.* **21**, 1 (1997).
4. Raab, E., Prentiss, M., Cable, A., Chu, S., and Pritchard, D., *Phys. Rev. Lett.* **59**, 2631 (1987).
5. Bergeman, T., Erez, G., and Metcalf, H. J., *Phys. Rev. A* **35**, 1535 (1987).
6. Ketterle, W., and van Druten, N. J., *Adv. At. Mol. Opt. Phys.* **37**, 181 (1996).
7. Anderson, M. H., Ensher, J. R., Matthews, M. R., Wieman, C. E., and Cornell, E., *Science* **269**, 198 (1995).
8. Davis, K. B., Mewes, M.-O., Andrews, M. R., van Druten, N. J., Durfee, D. S., Kurn, D. M., and Ketterle, W., *Phys. Rev. Lett.* **75**, 3969 (1995).
9. Bradley, C. C., Sackett, C. A., Tollet, J. J., and Hulet, R. G., *Phys. Rev. Lett.* **75**, 1687 (1995); Bradley, C. C., Sackett, C. A., and Hulet, R. G., *Phys. Rev. Lett.* **78**, 985 (1997).

10. Grimm, R., Weidemüller, M., and Ovchinnikov, Yu. B., *Adv. At. Mol. Opt. Phys.*, in preparation.
11. Boiron, D., Michaud, A., Fournier, J. M., Simard, L., Sprenger, M., Grynberg, G., and Salomon, C., *Phys. Rev. A* **57**, R4106 (1998).
12. Lee, H. J., Adams, C. S., Kasevich, M., and Chu, S., *Phys. Rev. Lett.* **76**, 2658 (1996).
13. Kuhn, A., Perrin, H., Hänsel, W., and Salomon, C., in *OSA TOPS on ultracold atoms and BEC*, edited by Burnett, K., (Optical Society of America, Washington D. C., 1996).
14. Chu, S., Bjorkholm, J. E., Ashkin, A., and Cable, A., *Phys. Rev. Lett.* **57**, 314 (1986).
15. Miller, J. D., Cline, R. A., and Heinzen, D. J., *Phys. Rev. A* **47**, R4567 (1993).
16. Adams, C. S., Lee, H. J., Davidson, N., Kasevich, M., and Chu, S., *Phys. Rev. Lett.* **74**, 3577 (1995).
17. Takekoshi, T., and Knize, R. J., *Opt. Lett.* **21**, 77 (1996).
18. Stamper-Kurn, D. M., Andrews, M. R., Chikkatur, A. P., Inouye, S., Miesner, H.-J., Stenger, J., and Ketterle, W., *Phys. Rev. Lett.* **80**, 2027 (1998).
19. Davidson, N., Lee, H. J., Adams, C. S., Kasevich, M., and Chu, S., *Phys. Rev. Lett.* **74**, 1311 (1995).
20. Kuga, T., Torii, Y., Shiokawa, N., and Hirano, T., *Phys. Rev. Lett.* **78**, 4713 (1997).
21. Ovchinnikov, Yu. B., Manek, I., and Grimm, R., *Phys. Rev. Lett.* **79**, 2225 (1997).
22. Cook, R. J., and Hill, R. K., *Opt. Commun.* **43**, 258 (1982).
23. Balykin, V. I., Letokhov, V. S., Ovchinnikov, Yu. B., and Sidorov, A. I., *JETP Lett.* **45**, 353 (1987); *Phys. Rev. Lett.* **60**, 2137 (1988).
24. Dowling, J. P., and Gea-Banacloche, J., *Adv. At. Mol. Opt. Phys.* **37**, 1 (1996).
25. Kasevich, M., A., Weiss, D. S., and Chu, S., *Opt. Lett.* **15**, 607 (1990).
26. Aminoff, C. G., et al., *Phys. Rev. Lett.* **71**, 3083 (1993).
27. Manek, I., Ovchinnikov, Yu. B., and Grimm, R., *Opt. Commun.* **147**, 67 (1998).
28. Ovchinnikov, Yu. B., Söding, J., and Grimm, R., *JETP Lett.* **61**, 21 (1995).
29. Söding, J., Grimm, R., Ovchinnikov, Yu. B., *Opt. Commun.* **119**, 652 (1995).
30. Ovchinnikov, Yu. B., Laryushin, D. V., Balykin, V. I., Letokhov, V. S., *JETP Lett.* **62**, 113 (1995); Laryushin, D. V., Ovchinnikov, Yu. B., Balykin, V. I., Letokhov, V. S., *Opt. Commun.* **135**, 138 (1997).
31. Desbiolles, P., Arndt, M., Szriftgiser, P., and Dalibard, J., *Phys. Rev. A* **54**, 4292 (1996).
32. Landragin, A., Labeyrie, G., Henkel, C., Kaiser, R., Vansteenkiste, N., Westbrook, C. I., and Aspect, A., *Opt. Lett.* **21**, 1591 (1996).
33. Townsend, C. G., Edwards, N. H., Cooper, C. J., Zetie, K. P., Foot, C. J., Steane, A. M., Szriftgiser, P., Perrin, H., and Dalibard, J., *Phys. Rev. A* **52**, 1423 (1995).
34. Arndt, M., Ben Dahan, M., Guéry-Odelin, D., Reynolds, M., and Dalibard, J., *Phys. Rev. Lett.* **79**, 625 (1997).
35. Söding, J., Guéry-Odelin, D., Desbiolles, P., Ferrari, G., and Dalibard, J., *Phys. Rev. Lett.* **80**, 1869 (1998).
36. Guéry-Odelin, D., Söding, J., Desbiolles, P., and Dalibard, J., *Opt. Express* **2**, 323 (1998).

37. Guéry-Odelin, D., Söding, J., Desbiolles, P., and Dalibard, J., preprint 1998.
38. Kokkelmans, S. J. J. M. F., Verhaar, B. J., and Gibble, K., *Phys. Rev. Lett.* **81**, 951 (1998).
39. Leo, P. J., Tiesinga, E., Julienne, P. S., Walter, D. K., Kadlecek, S., and Walker, T. G., *Phys. Rev. Lett.* **81**, 1389 (1998).
40. Burt, E. A., Ghrist, R. W., Myatt, C. J., Holland, M. J., Cornell, E. A., and Wieman, C. E., *Phys. Rev. Lett.* **79**, 337 (1997).
41. Tiesinga, E., Verhaar, B. J., and Stoof, H. T. C., *Phys. Rev. A* **47**, 4114 (1993).
42. Inouye, S., Andrews, M. R., Stenger, J., Miesner, H.-J., Stamper-Kurn, D. M., and Ketterle, W., *Nature* **392**, 151 (1998).
43. Matsubaru, A., Arai, T., Hotta, S., Korhonen, J. S., Suzuki, T., Masaike, A., Walraven, J. T. M., Mizusaki, T., and Hirai, A., *Physica B* **194**, 899 (1994).
44. Safonov, A. I., Vasilyev, S. A., Yashnikov, I. S., Lukashevich, I. I., and Jaakola, S., *JETP Lett.* **61**, 1032 (1995).
45. Ovchinnikov, Yu. B., Shul'ga, S. V., and Balykin, V. I., *J. Phys. B: At. Mol. Opt. Phys.* **24**, 3173 (1991).
46. Engler, H., Manek, I., Moslener, U., Nill, M., Ovchinnikov, Yu. B., Schlöder, U., Schünemann, U., Zielonkowski, M., Weidemüller, M., and Grimm, R., *Appl. Phys. B*, in press.
47. Ovchinnikov, Yu. B., Manek, I., Sidorov, A. I., Wasik, G., and Grimm, R., *Europhys. Lett.* **43**, 510 (1998).
48. Walker, T., and Feng, P., *Adv. At. Mol. Opt. Phys.* **34**, 125 (1994); Weiner, J., *ibid.* **35**, 45 (1995).
49. Tabosa, J. W. R., Chen, G., Hu, Z., Lee, R. B., and Kimble, H. J., *Phys. Rev. Lett.* **66**, 3245 (1991).
50. Zielonkowski, M., Manek, I., Moslener, U., Rosenbusch, P., and Grimm, R., submitted to *Europhys. Lett.*
51. Basically the same trap configuration has been used recently in three other experiments reported at this conference: S. Chu, talk; M. Kasevich, talk; D. S. Weiss, poster B7.
52. Bijlsma, M., Verhaar, B. J., and Heinzen, D. J., *Phys. Rev. A* **49**, R4285 (1994).
53. Haycock, D. L., Hamann, S. E., Klose, G., and Jessen, P. S., *Phys. Rev. A*, **55**, R3991 (1997).
54. Friebel, S., D'Andrea, C., Walz, J., Weitz, M., and Hänsch, T. W., *Phys. Rev. A* **57**, R20 (1998).
55. Savard, T. A., O'Hara, K. M., and Thomas, J. E., *Phys. Rev. A* **56**, R1095 (1997).
56. The relaxation of hyperfine population due to photon scattering in a far-off-resonance trap has been investigated by Cline, R. A., Miller, J. D., Matthews, M. R., and Heinzen, D. J., *Opt. Lett.* **19**, 207 (1994).
57. Zielonkowski, M., doctoral thesis, Heidelberg University, in preparation.
58. Cohen-Tannoudji, C. and Dupont-Roc, J., *Phys. Rev. A* **5**, 968 (1972).
59. Zielonkowski, M., Steiger, J., Schünemann, U., DeKieviet, M., and Grimm, R., *Phys. Rev. A* **58** (1998), in press.
60. Khriplovich, I. B., and Lamoreaux, S. K., *CP Violation without Strangeness: Electric Dipole Moments of Particles, Atoms, and Molecules* (Springer, Berlin, 1997).

Single Atoms in a MOT

Dieter Meschede, Bernd Ueberholz, Victor Gomer, Svenja Knappe,
Uwe Reiter, Harald Schadwinkel, Frank Strauch

Institut für Angewandte Physik, Universität Bonn
Wegeler Str. 8, D-53115 Bonn, F.R.Germany

Abstract. We are experimenting with individual neutral cesium atoms stored in a magneto-optical trap. The atoms are detected by their resonance fluorescence, and fluorescence fluctuations contain signatures of the atomic internal and external degrees of freedom. This noninvasive probe provides a rich source of information about atomic dynamics at all relevant time scales.

INTRODUCTION

Individual and distinguishable particles are truly at the heart of our theoretical description of physical processes at atomic scales. This is due not only to the purported transparency of the system, but also to our fascination with the control of a single atomic object, a possibility that goes beyond our immediate experience. While transient observations of single particles go back to the detection of energetic charged particles in cloud chambers [1], continuous and repeated interaction with individual particles became possible only once long-time confinement of ions in electromagnetic traps was achieved. A whole branch of atomic physics was opened when in 1978 W. Neuhauser and colleagues [2] observed the light field scattered from a single Ba^+ ion in a Paul trap. Extensions of this method have spread and many beautiful experiments have now illustrated quantum mechanics at a most elementary level of description, including for instance the occurrence of quantum jumps [3] in the absorptive interaction of individual particles with electromagnetic fields.

From the beginning, resonance fluorescence has played an important role as an efficient way of detecting isolated and confined particles, and today even certain molecules adsorbed on suitable surfaces have been shown to fluoresce with sufficient intensity to be unambiguously detected [4]. More recently the advances of laser cooling have also brought neutral atoms into the realm of single-particle manipulations, even though the confinement forces are most feeble in this case.

The invention of the magneto-optical trap (*MOT*) [5] was perhaps the most important experimental step toward routine trapping and cooling of neutral atoms.

The identification of individual atoms in a MOT was then a matter of tightly controlling the flux of atoms into the trap, employing sensitive photon counting detectors, and carefully suppressing ambient light [6–8]. In the neutral-atom trap, near-resonant scattering of light contributes to the restoring force and is always available as a sensitive and noninvasive probe.

EXPERIMENTAL

Single Atom Trapping and Detection. All our experiments on single atoms have been carried out with cesium atoms. They are stored in a usual 6-beam MOT operated with two diode lasers stabilized to the trapping and repumping wavelengths near 852 nm. Our MOT differs from the standard setup only through the application of strong magnetic field gradients up to 800 G/cm, which dramatically reduce the capture cross section for atoms from the vapor phase and lead to better localization but do not change the dynamical characteristics, for instance the temperature of the trapped sample [9]. The loading rate of individual atoms from a room-temperature reservoir is furthermore controlled by a gate valve that allows us to match the average loading time to typical dwelling times, which in our UHV chamber with residual gas pressure below 10^{-10} mbar range from seconds to minutes. It is therefore possible to adjust experimental conditions such that at any given time a small and readily countable number of neutral atoms reside in the MOT.

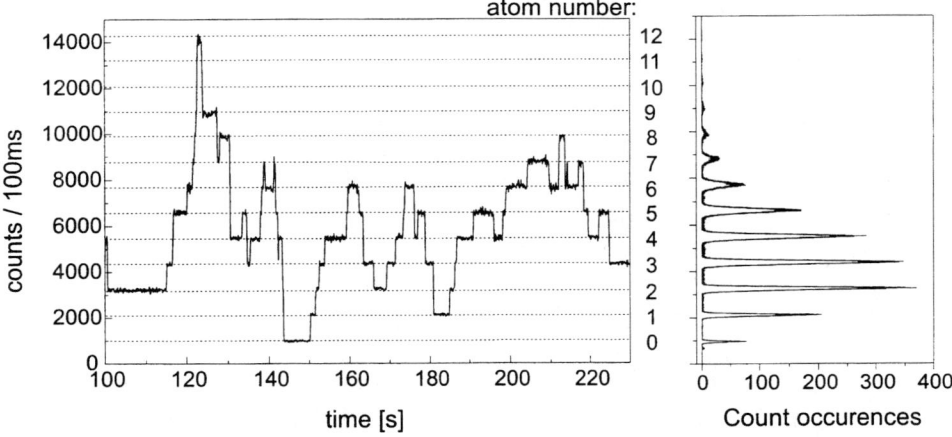

FIGURE 1. Random arrival and departure of individual atoms in a MOT (up to 12 atoms). Right side: Histogram of the distribution of photon count rates.

An example of the time evolution of the photon counting rate is shown in Fig. 1. The random arrival and departure of atoms is observed by well-resolved steps in the fluorescence intensity. We use avalanche photodetectors (*APD*) in the single-photon counting mode with measured photon detection efficiency better than 50% and dark count rates below 15 s^{-1}. The scattering rate of a single atom depends on the detuning δ from the resonance frequency (δ/Γ = -8...-1, in terms of the natural linewidth $\Gamma = 2\pi \times 5.2$ MHz). While collecting 4.5% of the total fluorescence, the detectors have a count rate that varies from 3 to 20 kHz. The histogram in Fig. 1 shows the distribution of count rates. The width of each peak is nearly shot-noise limited, i.e. it is essentially equal (within better than 5%) to the square root of the number of counts in a given integration time.

Photon Correlations. The fluorescence fluctuations constitute our prime experimental source of information about atomic motion.

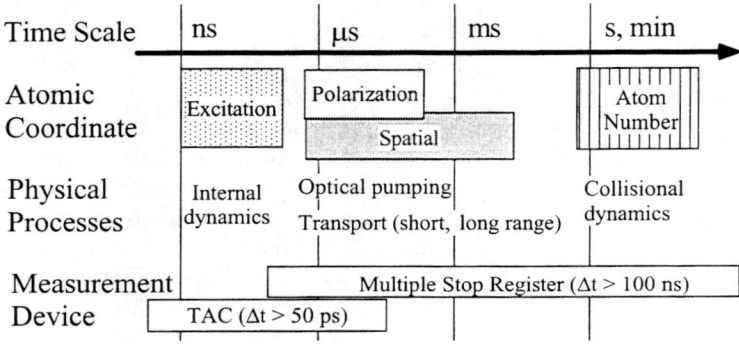

FIGURE 2. Overview: Dynamical time scales of atoms stored in a MOT. TAC: Time to Amplitude Converter.

We analyze radiative fluctuations by means of correlation functions. Related methods were used in [10,11] who investigated radiative fluctuations from a MOT and an optical lattice containing many atoms.

The normalized intensity-intensity correlation function is classically defined in terms of the fluorescence intensity $I(t)$ through [12]

$$g^{(2)}(\tau) = \frac{\langle I(t)I(t+\tau)\rangle}{\langle I(t)\rangle^2} \qquad (1)$$

At very long delays $\tau \to \infty$, correlations always vanish: $\langle I(t)I(t+\tau)\rangle \to \langle I(t)\rangle\langle I(t)\rangle$, and hence we have $g^{(2)}(\tau) \to 1$. It is thus convenient to present and discuss the physical information in the form $g^{(2)}(\tau) - 1$.

In our experiment we detect individual photons rather than a continuous intensity. In photon language $g^{(2)}(\tau)$ describes the probability of detecting a second photon after a first photon was detected. One can detect just the "next" second photon ("single-stop method") and thus measure the probability for a certain waiting time, which coincides with the correlation function $g^{(2)}(\tau)$ at count rates much smaller than the reciprocal delay time of interest. Application of the famous Hanbury-Brown-Twiss technique [13] then allows us to measure $g^{(2)}(\tau)$ with sub nanosecond time resolution. However, at longer delay times this method is impaired by a systematic error [14]. To completely eliminate this problem we have recorded the arrival time of every individual photon and then analyzed correlations on a computer by a multiple-stop technique that does not need any corrections. In parallel we have always monitored the total count rate as a measure of the number of trapped atoms.

While information on the internal dynamics is in principle already contained in the total fluorescence trace of Fig. 1 (albeit with a much better time resolution), further information can be gathered by conditioning the signal, for instance through polarization or spatial analysis. Finally, it is interesting to note that we can decode the whole dynamics of a single atom in the MOT with the one experimental method.

RESULTS

Nanosecond Dynamics: Rabi Oscillations

When a photon emitted by an atom is detected, the atom is instantaneously 'prepared' in its ground state. To emit a second photon the atom must be excited again, and this requires a finite amount of time — an atom cannot emit two photons at the same time. This phenomenon causes $g^{(2)}(0) = 0$; it is called 'photon antibunching' and has received much interest in the past [15–17] since it is regarded as an important manifestation of the quantum nature of light. All classical fields have auto correlations $g^{(2)}(0) \geq g^{(2)}(\tau)$, and a value $g^{(2)}(0) - 1 < 0$ is indeed a signature for a pure quantum effect. The first detected photon is followed by a transient oscillation, which is a direct record of Rabi oscillations. For us it provides an excellent opportunity to elucidate a few more experimental details concerning photon correlation measurements with one or more atoms.

An original record of photon coincidences measured by the TAC in Fig. 3 shows the photon antibunching effect for 1 to 3 stored cesium atoms. The sequence of the traces is determined by the frequency of trap occupation by 1, 2, or 3 atoms, respectively. The total number of coincidences does not drop to zero at $\tau = 0$ even for a single atom due to random coincidences coming from stray light. These can be measured independently, however, (at 0 atoms in the trap) and subtracted, showing that the minimum agrees indeed with $g^{(2)}(0) = 0$ for 1 atom. If more than one atom is stored in the trap, the number of random coincidences is enhanced due to efficient near-resonant scattering from different and distinguishable atoms. In contrast to a frequent assumption in textbooks (e.g. [18]), emission does not occur into a single spatial mode of the radiation field for which a transition from antibunching to photon bunching ($g_N^{(2)}(0) = 2 - 2/N$) is predicted for an increasing number of atoms N. It is in principle possible to observe this transition even with only two atoms by reducing the observation-angle aperture or by improving atomic

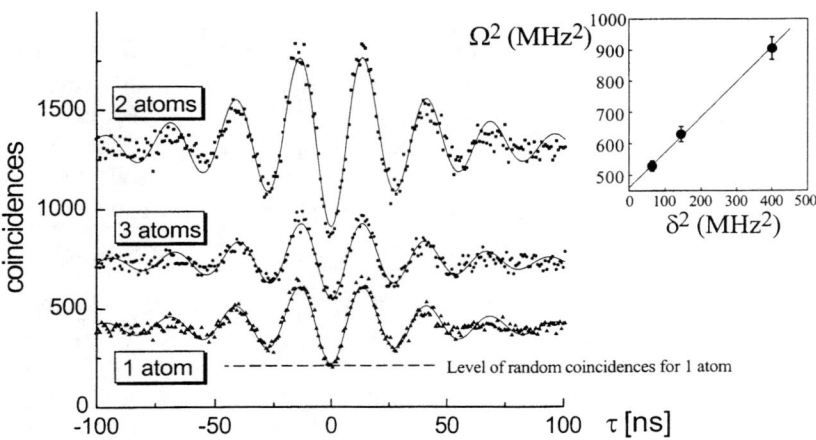

FIGURE 3. Raw data for photon correlations of neutral cesium atoms stored in a MOT at laser detuning $\delta = -20$ MHz. The trap occupation times by 1,2 and 3 atoms were 136, 147 and 69 min, respectively. Solid lines represent the theoretical prediction for a two-level atom. The inset shows a measurement of the Rabi frequency Ω vs. laser detuning δ.

localization. In our case, however, stronger localization leads to collisional loss of the atoms, and major reductions in the aperture result in an insufficient signal-to-noise ratio for the coincidence rate. Therefore the one-atom photon antibunching signature survives for N atoms according to

$$g_N^{(2)}(\tau) = \frac{1}{N}\left\{g^{2)}(\tau) + (N-1)\right\}. \qquad (2)$$

We find it surprising that a simple theoretical description in terms of a two-level atom [19] reproduces the observed Rabi oscillations quite well. In former experiments, great care was taken to properly prepare a simple light field as well as a suitable two-level model atom. In our MOT in contrast there is an unknown, acoustically jittering and thermally drifting interference pattern of 6 light beams, and a trapped cesium atom that offers 27 different transitions between magnetic sublevels interacting with the light fields. We interpret this observation as a strong tendency fort the cesium atoms to spend most of their time in a situation where extreme magnetic sublevels are coupled by the light field, acting again like a two-level system. In this model the observed oscillation frequency $\Omega/2\pi$ is related to the resonant Rabi frequency $\Omega_R = d \cdot E/\hbar$, giving the coupling strength of induced dipole moment d and field amplitude E, through $\Omega^2 = \Omega_R^2 + \delta^2 - (\Gamma/4)^2$. The absolute value can be determined with excellent precision (see inset of Fig. 3), and it agrees well with an independent measurement from power broadening and also with the rule of thumb that Ω_R should correspond to an intensity of 4 times one of the trapping laser beams.

One may speculate about the possibility of observing cooperative effects of the atoms in their radiative properties. Such cooperative effects are expected if the atoms approach each other within $\lambda/2\pi$. However, the corresponding fraction of the total trapping volume ($< 10^{-4}$) is too small to permit observations of any significant deviation from single-atom behaviour.

Microsecond Dynamics: Optical Pumping

As it moves through the trap, the cesium atom experiences continuous changes of the interference pattern and tries to adjust its magnetic orientation to the equilibrium value with regard to the local polarization state by optical pumping. This mechanism is the origin of sub-Doppler cooling [20], which is most prominent in the so-called $lin \perp lin$ configuration. It is a consequence of the entanglement of internal and external atomic degrees of freedom.

The orientation of an atom determines the polarization of its radiation field, and hence polarization-sensitive photon correlations will display the dynamic behaviour of the orientation. We have carried out such measurements both for auto correlations, and also for cross correlations of orthogonal polarization states according to $g^{(2)}_{\alpha\beta}(\tau) = \langle I_\alpha(t) I_\beta(t+\tau) \rangle / \langle I_\alpha(t) \rangle \langle I_\beta(t) \rangle$, which is a minor generalization of Eq. (1). The indices (α, β) represent circular $(+,-)$ or linear (h,v) polarization components.

Measurements. The results for the polarization-resolved correlation measurements are shown in Fig. 4. For very short time scales Rabi oscillations are observed again but will be disregarded now.

Three independent $\sigma^+\sigma^-$ standing waves of the MOT have two relative time phases, and from an interferometric measurement we know that they are subject to continuous drift at the ms scale. Our integration time is much longer and therefore presents an average over all possible phase relations. A striking feature of the measurement is the strong contrast for auto ($g^{(2)}_{++}(0) - 1 \simeq 1$) and cross correlation ($g^{(2)}_{+-}(0) - 1 \simeq -0.4$) for circular polarization components, while no significant correlation is observed for linear polarizations, $g^{(2)}_{vh}(0) - 1 \simeq 0$! The contrast for circular light is straightforwardly interpreted from a qualitative point of view: the large total angular momentum of a cesium atom ($F = 4, 5$) lends inertia to its orientation, and our measurements show consistently that it takes a few microseconds to turn a cesium atom around, depending on the laser detuning (the time constant increases with decreasing detuning). Auto and cross correlation seem to represent identical dynamical behaviour, for instance both the time constant of the decay of correlations as well as a slight but reproducible periodic modulation seem to agree as one might intuitively expect, because in a given direction of observation an atom can only emit either one of two orthogonal polarization states ($I_{tot} = I_+ + I_-$). We therefore expect a sum rule $g^{(2)}_{++}(\tau) + g^{(2)}_{+-}(\tau) = 2g^{(2)}_{ii}(\tau)$, where i stands for total intensity. We have explicitly tested this prediction (lower right of

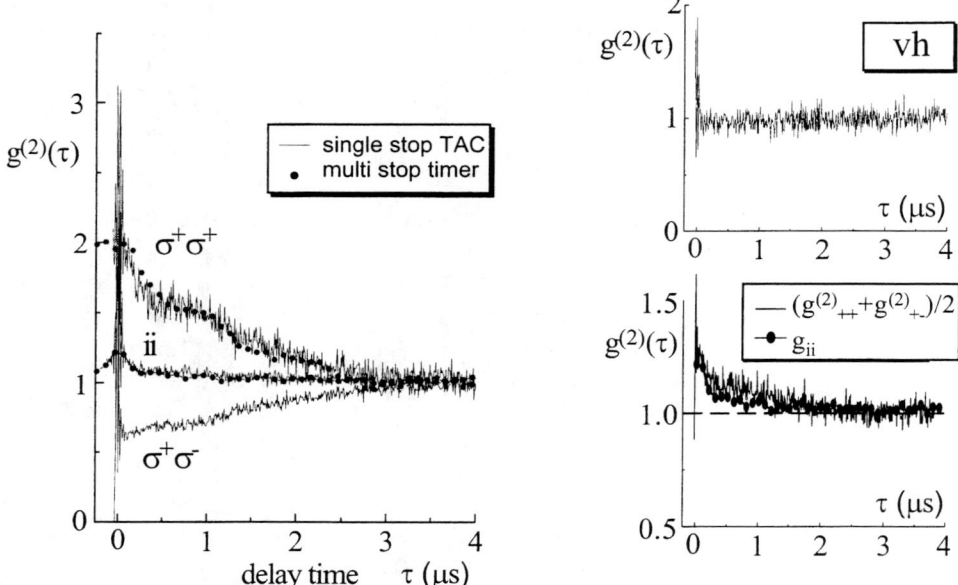

FIGURE 4. Normalized auto correlation and cross correlation for circularly polarized light (left), for linearly polarized light (upper right), and a comparison of the sum rule with an actual measurement.

Fig. 4) by directly measuring the intensity-intensity auto correlation function much beyond the nanosecond scale, where it shows Rabi oscillations, and comparing it to the values expected from polarization-resolved analysis. The measurement shows good agreement, an unexpectedly strong contrast $g_{ii}^{(2)}(0) - 1 \simeq 0.25$, and a decay time significantly shorter than for polarization correlations.

A qualitative argument for the time constant of the polarization correlations may be taken from the typical time it takes a cesium to undergo reorientation: for this process to take place an atom has to cross one unit cell of the light field interference pattern of length λ. In our experiment the atomic velocity corresponds to a temperature not too far from the Doppler limit, and the characteristic time may indeed be estimated from $\tau \simeq (kv_D)^{-1} \simeq 1\mu s$ with the Doppler velocity of cesium $v_D \simeq 0.1 m/s$.

A Simplified Model: Diffusive Transport in 1D Optical Lattices . A more detailed interpretation of our data is not only complicated by the atomic multilevel structure and its interaction with various states of the light-field polarization and intensity, but also by the entanglement of atomic excitation and atomic motion. In order to gain physical insight, let us outline a simplified one-dimensional model in which we first consider the treatment of atomic motion and subsequently discuss the implications for photon correlation measurements [21].

Atomic motion in the MOT is frequently described in terms of a damping constant α and a spring constant κ [5] which depend on the light forces exerted by the trapping laser beams. Diffusive motion in the trap can be described by the theory of Brownian motion, accounting for stochastic heating processes in terms of a Fokker-Planck equation [22,21] and yielding a time-dependent distribution function $f(x, x_0, t)$ that describes the diffusion of an atom initially at position x_0.

The polarization state of the atomic radiation field is determined by the atomic orientation, which is in turn closely related to the local polarization state of the driving light field. It is subject to change when an atom moves on the scale of the laser wavelength. For such short distances, or equivalently for short times, we can completely neglect the restoring force and consider diffusive motion in potential-free space with diffusion constant $D = kT/\alpha$, since the MOT operates in the overdamped regime ($\alpha \gg \sqrt{\kappa m}$).

The light-field pattern is created in our case by 3 interfering pairs of the trapping laser beams. Although the detailed light-field structure depends on relative time phases of the standing waves and is not known, a periodic structure (optical lattice) of both the intensity and the polarization topography does exist in the MOT and can be considered as constant on the *ms* time scale, much longer than the time it takes an atom to cross a unit cell of the lattice.

In our measurements we analyze the time-dependent conditional probability of observing a second photon with certain polarization properties, once a first is observed. Our model assumes that the intensity as well as the polarization of the emitted radiation are completely and instantaneously determined by the driving light field at the position of the atom. In other words we assume a linear dielectric scattering object resembling a $J = 0 \to J = 1$ dipole transition. The atom diffuses away from its initial position, causing modifications in its orientation and hence in the polarization state of its radiation.

Three types of different light fields that play the role of one-dimensional models for optical lattices come to mind as limiting situations: The *lin* $\perp$ *lin* standing wave consists of two counter-propagating waves with equal amplitude and orthogonal linear polarizations, and the $\sigma^+\sigma^-$ configuration is similarly constructed from two counter-propagating waves of the same handedness. The resulting standing wave field may be decomposed into two circular (*lin* $\perp$ *lin*) or linear ($\sigma^+\sigma^-$) polarization components, each with sinusoidal spatial variation ($I_\alpha = I_{\alpha 0} \sin^2 kz$). These models have no modulation of the total intensity; they are pure polarization gradient lattices. We may also add the case of a pure intensity optical lattice with sinusoidal intensity variation for a light field composed from two counter-propagating waves with identical linear polarization (*lin*||*lin*).

In all three limiting cases an auto correlation function of the form

$$g^{(2)}_{\alpha\alpha}(\tau) = 1 + \frac{1}{2}e^{-2k^2 D\tau} \tag{3}$$

is found [21] ($\alpha = +$ for *lin* $\perp$ *lin*, $\alpha = x$ for $\sigma^+\sigma^-$ and $\alpha = +, x, i$ for *lin*||*lin*).

While for circular polarization the gross structure agrees with our observation, it differs significantly in several important details: The contrast of the auto correlation function ($g^{(2)}_{++}(0) - 1 \simeq 1$) much exceeds the value 1/2 predicted from Eq. (3). Furthermore, it does not give any indication why we observe strong correlations for circularly polarized light fields but vanishing contrast for linear polarization components, and there is no modulation of the decaying transient in the model. While it may be exaggerated to expect such interpretation power from an oversimplified one-dimensional model, let us briefly examine several possible extensions of the model holding the promise to improve the theoretical description:

- *Dimensions.* By construction our model accounts for one-dimensional situations only, and it is known that a simple description of the polarization variations in a MOT is complicated [23]. For the 'classical', i.e. dielectric, scatterer we can of course analyze the influence of a 2D or 3D geometry. We find [21] that in 3D, the linear correlations are still more prominent than in 1D, in contrast to our observation.

- *Atoms with large angular momentum.* The orientation of an atom for a given polarization state of the local field depends strongly on the total angular momentum. For instance, even relatively small amounts of circular polarization cause the high-angular-momentum cesium atom to spend most of its time in the outermost magnetic sublevels, favouring the emission of circularly polarized light. Also, the spatial dependence of the circular components of the radiation field changes more rapidly than suggested in our model. Replacing the linear $\sin^2 kx$ model by steeper functions rapidly enhances the contrast (for instance $g^{(2)}_{\alpha\alpha}(0) - 1 = 1$ for a square function), in accordance with our observations. This mechanism also reduces the contrast for linear polarization correlations.

- *Diffusion in a periodic potential.* A light-induced periodic potential influences atomic diffusion motion and produces a periodic equilibrium distribution, which enhances contributions of the potential minima to the photon correlations. Preliminary results show that the periodic structure may even mimic transient oscillatory behaviour as observed in our measurement. We conclude therefore that our observation is not necessarily an indication of contributions by atoms oscillating in optical micropotentials, such as reported in [11].

Millisecond Dynamics: Global Motion in the Trap

We have interpreted the polarization correlations in the preceding paragraph as a mapping of atomic motion within the polarization unit cell of an optical lattice. This method is inappropriate for analyzing transport over larger distances. We have therefore generated an intermediate image of the trapping volume, which allows us either to insert apertures or to direct different parts of the trapping volume onto separate APDs. This arrangement corresponds to a crude 2 pixel camera and allows us to detect atomic motion in the trap at large. In Fig. 5 we

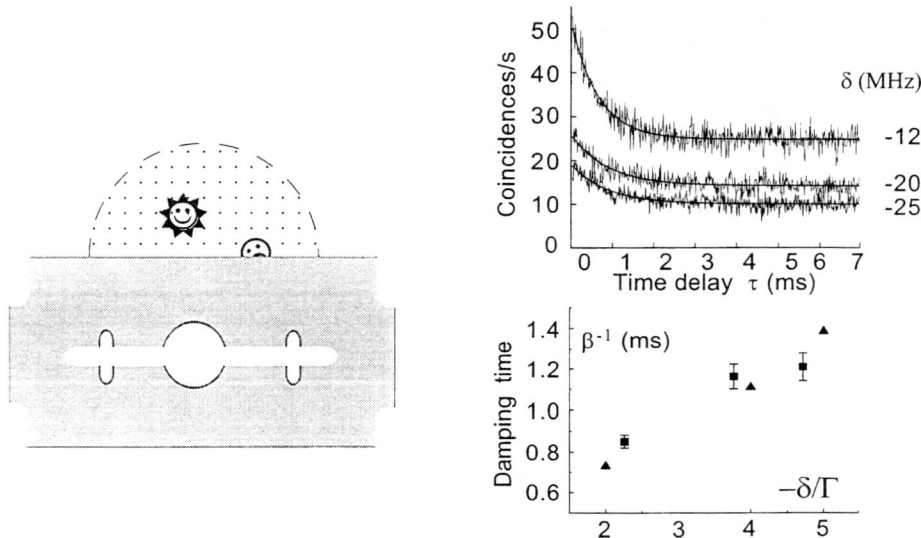

FIGURE 5. Left: An aperture is inserted into an intermediate image of the trapping volume, or alternatively different parts of the trapping volume are imaged onto separate photo detectors. Upper right: Difference of coincidences from two different detectors as a function of trapping laser detuning. Lower right: Damping-time constants determined from correlation measurements in comparison with results from [24].

show experimental records of spatially resolved coincidences. A theoretical solution of the Fokker-Planck equation resembling an exponential decay is fitted to the data and produces a value for the damping constants that is in good agreement with a measurement by [24] if the results are scaled by the gradient of the magnetic quadrupole field as expected from theory.

The spatial resolution of this method is determined by the resolving power of the imaging system used to collect atomic radiation. Therefore, it allows the extension of the polarization method of the preceding section, which may be interpreted as a mapping of atomic motion at sub-wavelength scales to longer distances. This method may have a potential for identifying less common types of diffusive motion [25].

COUNTING COLD COLLISIONS

The slowest time scale in the MOT is set by the arrival and departure of atoms. The arrival of atoms in the trap is determined by the light-force configuration of the MOT and the supply of cesium atoms only, and it is independent of the atomic trap occupation number N. We observe in Fig. 6 that atom captures are always single-particle events, in stark contrast to departure, where already on this short stretch of the time record we observe 5 departures of two atoms simultaneously, which is orders of magnitude above the statistical probability for occasional coincidence of two independent one-atom losses within the integration time of 100 ms. Cold

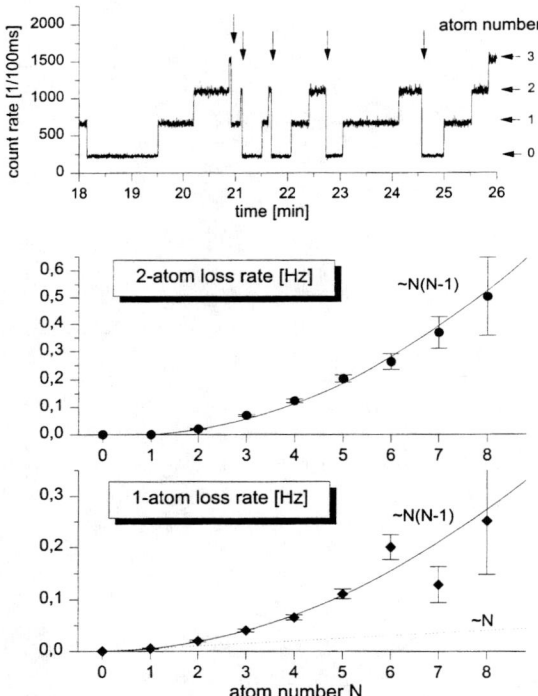

FIGURE 6. One- and two-atom loss rates as a function of atom number N.

collisions have been intensely studied with a large number of atoms N in the MOT [26], and a typical analysis matches the observed loss of atoms to the differential equation that is used to describe the decay of the number of trapped atoms once the trap has been turned off:

$$\frac{d}{dt}N = -\alpha N - \frac{\beta}{V}N^2, \qquad (4)$$

where N is treated as continuous variable, α is the loss rate due to collisions with residual gas, and β is the coefficient of cold intra-trap collisions of cesium atoms. For the interpretation of the β-coefficient, several loss mechanisms have been described that can generate sufficient kinetic energy to eject atoms out off the trap, including fine-structure and hyperfine-structure changing collisions, radiative escape, and others [26].

In our experiment N is a digital number, and the rate equation (4) has to be replaced by an appropriate nonlinear statistical equation. It is clearly seen from the chart record of Fig. 6 that we can unambiguously distinguish one-atom loss events from two-atom ones, i.e. with 100% contrast. By evaluating the one- and two-atom loss rates as a function of the initial atom occupation number, we have obtained preliminary data for the cold-collision cross sections in our trap. The two-atom loss rate beautifully follows a square law ($\propto N(N-1)$) as is expected for a two-body collision.

The lower trace of Fig. 6 shows the one-atom loss rate as a function of trapped atom number. Quite unexpectedly it exhibits a contribution $\propto N(N-1)$ as well, indicating 'soft' cold collisions resulting in loss of only one atom. This unique ability (impossible in experiments with many atoms)to separately detect all losses from the trap (also losses due to collisions with background gas for $N = 1$!) allows us to distinguish between different loss channels.

A PHASE STABLE MOT FOR SINGLE ATOMS

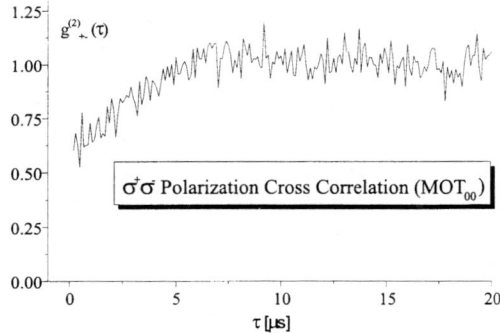

FIGURE 7. First measurement of the cross correlation function for a single atom stored in a MOT with zero time phases.

All experiments described above have been carried out in a MOT without stabilization of the time phases of the 3 interfering pairs of light beams. In order to improve and simplify experimental conditions, we have recently developed a concept [27] and experimentally realized a MOT with intrinsically stable and *a priori*-known time phases. Meanwhile we have obtained initial results of single atoms radiating in a phase-stable MOT with zero-time phase delays between all three standing waves. It is interesting to note that in such a light field there exist exclusively linear polarizations. Nevertheless we again observe strong cross correlations for circularly polarized light fields and negligible correlations for linearly polarized radiation.

REFERENCES

1. Wilson, C. R. T., Philos. Trans. R. Soc. London **189**, 265 (1897).
2. Neuhauser, W., Hohenstatt, M., Toeschek, P., Dehmelt, H., *Phys. Rev. Lett.* **41**, 233 (1978).
3. Nagourney, W., et al., *Phys. Rev. Lett.* **56**, 2797 (1986); Sauter, Th., et al., *Phys. Rev. Lett.* **57**, 1696 (1986); Berquist, J. C., et al., *Phys. Rev. Lett.* **57**, 1699 (1986).
4. Basché, Th., Moerner, W. E., Orrit, M., and Talon, H., *Phys. Rev. Lett.* **69**, 1516 (1992).
5. Raab, E. L., Prentiss, M., Cable, A., Chu, S., Pritchard, D. E., *Phys. Rev. Lett.* **59**, 2631 (1987); for a recent review see Townsend, C. G., Edwards, N. H., Cooper, C. J., Zetie, K. P., and Foot, C. J., Steane, A. M., Szriftgiser, P., Perrin, H., and Dalibard, D., *Phys. Rev.* **A52**, 1423 (1995).
6. Hu, Z., and Kimble, H. J., *Opt. Lett.* **19**, 1888 (1994).
7. Ruschewitz, F., Bettermann, D., Peng, J. L., and Ertmer, W., *Europhys. Lett.* **34**, 651 (1996).
8. Haubrich, D., Schadwinkel, H., Strauch, F., Ueberholz, B., Wynands, R., and Meschede, D., *Europhys. Lett.* **34**, 663 (1996).
9. Höpe, A., Haubrich, D., Müller, G., Kaenders, W. G., and Meschede, D., *Europhys. Lett.* **22**, 669 (1993); Haubrich, D., Höpe, A., Meschede, D., *Opt. Commun.* **102**, 225 (1993).
10. Bali, S., Hoffmann, D., Siman, J., and Walker, T., *Phys. Rev.* **A 53**, 3469 (1996).
11. Jurczak, C., Desruelle, B., Sengstock, K., Courtois, J.-Y., Westbrook, C. I., and Aspect, A., *Phys. Rev. Lett.* **77**, 1727 (1996).
12. For a rigorous quantum-mechanical treatment see: Glauber, R. J. *Phys. Rev.* **130**, 2529 (1963); *Phys. Rev.* **131**, 2766 (1963).
13. Hanbury Brown, R., and Twiss, R. Q., *Nature* **177**, 27 (1956).
14. Coates, P. B., *J. Sci. Instr.*, Ser. 2, **1**, 878 (1968).
15. Kimble, H. J., Dagenais, M., and Mandel, L., *Phys. Rev. Lett.* **39**, 691 (1977).
16. Rateike, F.-M., Leuchs, G., and Walther, H., results cited by Cresser, J. D. *et al.*, in "Dissipative Systems in Quantum Optics", edited by R. Bonifacio, *Topics in Current Physics*, vol. **27**, Berlin: Springer, 1982), p. 21.
17. Diedrich, F., and Walther, H., *Phys. Rev. Lett.* **58**, 203 (1987).
18. Loudon, R., *The quantum theory of light*, Oxford: Oxford University Press, second edition, 1983.
19. Kimble, H. J., and Mandel, L., *Phys. Rev.* **A13**, 2123 (1976).
20. Dalibard, J., and Cohen-Tannoudjii, C. *J. Opt. Soc. Am.* **B 6**, 2023 (1989).
21. Gomer, V., Strauch, F., Ueberholz, B., Knappe, S., and Meschede, D., *Phys. Rev.*

A58, September (1998); Gomer, V., Ueberholz, Strauch, F., B., Knappe, S., Frese, D., Kuhr, S., and Meschede, D., submitted to Appl. Phys. B;
22. Pathria, R. K., *Statistical Mechanics*, Oxford: Pergamon Press 1985, corrected edition.
23. An attempt was made by: Hopkins, S., and Durrant, A., *Phys. Rev.* **A56**, 4012 (1997).
24. Drewsen, M., Laurent, Ph., Nadir, A., Santerelli, G., Clairon, A., Castin, Y., Grison, D., and Salomon, C., *Appl. Phys.* **B 59** 283 (1994).
25. Katori, H., Schlopf, S., and Walther, H., *Phys. Rev. Lett.* **79**, 2221 (1997).
26. Julienne, P. S., Smith, A. M. and Burnett, K., in *Advances in Atomic, Molecular and Optical Physics*, edited by D. R. Bates and B. Bederson (Academic Press, San Diego, 1993), Vol. 30, 141; Walker, T. and Feng, P., in *Advances in Atomic, Molecular and Optical Physics*, edited by B. Bederson and H. Walther (Academic Press, San Diego, 1994), Vol. 34, 125; Weiner, J., in *Advances in Atomic, Molecular and Optical Physics*, edited by B. Bederson and H. Walther (Academic Press, San Diego, 1995), Vol. 35, 45.
27. Rauschenbeutel, A., Schadwinkel, H., Gomer, V., and Meschede, D., *Opt. Commun.* **148**, 45 (1998).

Atomic Collisions at Sub-Microkelvin Temperatures

Daniel J. Heinzen

Dept. of Physics, The University of Texas, Austin, TX 78712

Abstract. With the advent of evaporative cooling methods, atomic collisions can be studied at microkelvin and even sub-microkelvin temperatures. At these energies, resonance and threshold phenomena become very pronounced. Near zero-energy resonances lead to large variations in collision cross sections with energy. Magnetically tunable Feshbach resonances have been observed, which allow elastic collision cross sections and scattering lengths to be tuned over large ranges with small changes in an applied magnetic field. These resonance effects are extremely sensitive to small variations in atomic interaction parameters, including scattering lengths. The combined results of ultracold atomic collision and photoassociation spectroscopy experiments is leading to a complete and consistent picture of collisions of ultracold alkali atoms.

INTRODUCTION

Over the past several years evaporative cooling methods have led to the production of trapped alkali gases with sub-microkelvin temperatures and, most dramatically, to dilute gas Bose-Einstein condensation [1–3]. Atomic collisions are very important to the physics of these cold gases. They determine the stability of the gas with respect to inelastic collisional losses such as three-body molecular recombination. Elastic collisions set the time scale for thermalization of the gas. Also, in a Bose-condensate the amplitudes for elastic scattering in the gas add coherently, and give rise to an effective self energy $U = 4\pi\hbar^2 na/m$, where n is the density of the gas, a is the two-body scattering length, and m is the atomic mass [4]. This self-energy plays a crucial role in determining such properties as the stability and excitation spectrum of the condensate [4].

At microkelvin temperatures, resonance and threshold effects become very pronounced in atomic collisions [5,6]. Large, resonant cross sections for s-wave scattering can arise from the presence of a molecular bound state very near threshold [7]. Shape [8,9] and Feshbach resonances may also occur. In such cases the collision cross sections exhibit a dramatic dependence on collision energy. Magnetically tunable Feshbach resonances have been discovered [10–12], which allow the low-energy elastic cross section and scattering length to be tuned over a wide range with a small change in magnetic field about some resonance field value B_0. These resonance effects are pronounced at ultracold temperatures because thermal averaging effects are substantially reduced.

The precise energies E_{res} of collision resonances or resonant field values B_0 for tunable Feshbach resonances are very sensitive to atomic interaction parameters, including the singlet and triplet scattering lengths a_S and a_T, and ground state Van der Waals interaction coefficient C_6. These parameters have recently been determined for the various alkali pairs through a combination of cold atom photoassociation and cold collision experiments. These determinations have allowed the precise locations of collision resonances to be predicted, considerably aiding their experimental observation. Conversely, precisely measured values of E_{res} or B_0 can be inverted to provide the atomic interaction parameters. For several of the alkali atom collision pairs, a very complete and consistent picture of their cold collision properties has been obtained.

ULTRACOLD ALKALI ATOM COLLISIONS

The most important ultracold collision parameter is the scattering length $a = -\lim_{k \to 0}(\delta_0(k)/k)$, where δ_0 is the s-wave phase shift. It can be given a simple geometrical interpretation as shown in Fig. 1. At short range the zero-energy radial wavefunction u(R) exhibits a large number of oscillations associated with the strongly attractive part of the interaction potential $V(R)$. However at large R it tends to a constant times $(R-a)$, which is simply the first part of the first cycle of a sine wave of wavelength $2\pi/k \to \infty$. The scattering length a is the displacement of the zeros of this sine wave from those which would occur for $V = 0$. The scattering length depends critically on the positions of the highest bound states of the potential, as shown in Fig. 2. Here, v_D is the number of bound states in the potential, with an integer value corresponding to the case in which the last bound state lies precisely at the dissociation limit. A fractional value interpolates linearly between integer values with the integrated WKB phase across the potential [13]. The value of a depends only on the fractional part of v_D. It is for this reason that determination of scattering lengths is somewhat difficult. An uncertainty of a fraction of an cm^{-1} in the binding energy of the highest state can correspond to an uncertainty in v_D greater than 1, and to complete uncertainty in a.

Since the alkali atoms in a collision pair have valence electron spin $\mathbf{s}_i$ ($i = 1, 2$), with $s_i = 1/2$, and nuclear spin $\mathbf{i}_i$, the interaction potentials must account for the spin structure. At short range, the electronic exchange interaction dominates, so that total electron spin S, with $\mathbf{S} = \mathbf{s}_1 + \mathbf{s}_2$, and nuclear spin I, with $\mathbf{I} = \mathbf{i}_1 + \mathbf{i}_2$, are the good quantum numbers. This gives rise to the familiar singlet and triplet interaction potentials. However at long range, the atomic hyperfine interaction dominates over the exchange interaction, and the atomic hyperfine quantum numbers f_1 and f_2, are good, where $\mathbf{f}_i = \mathbf{s}_i + \mathbf{i}_i$. Neither set of quantum numbers is good at all internuclear distances; only the total spin $\mathbf{F} = \mathbf{S} + \mathbf{I}$ of both atoms is

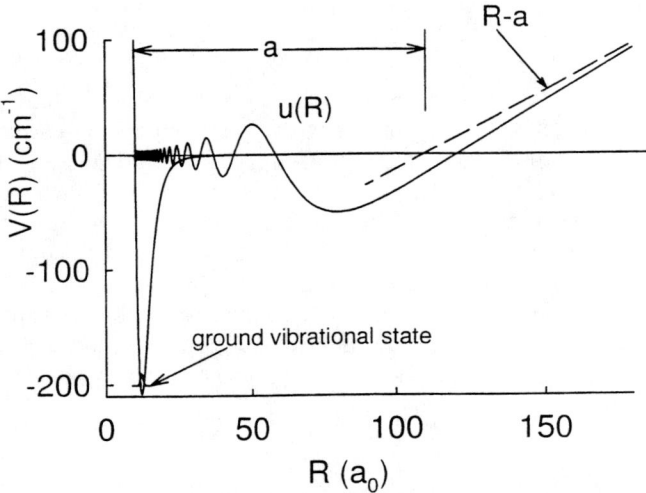

FIGURE 1. Approximate interaction potential and zero-energy wavefunction $u(R)$ for triplet collisions of two ^{87}Rb atoms, illustrating the definition of the scattering length a.

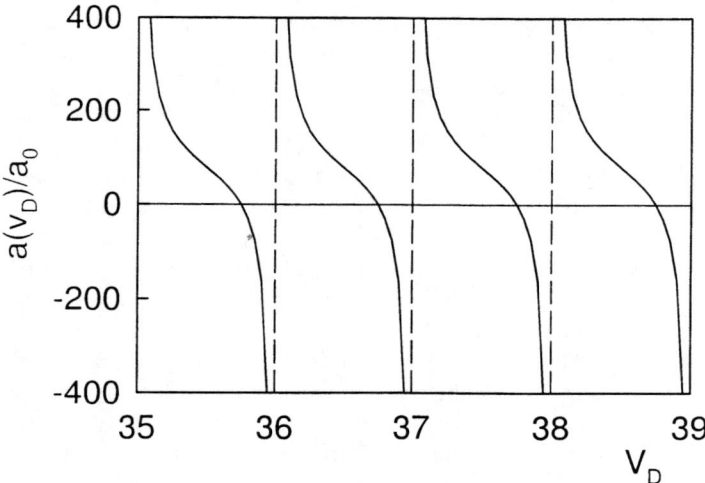

FIGURE 2. Scattering length a for triplet collisions of Rb atoms as a function of the vibrational quantum number at dissociation v_D.

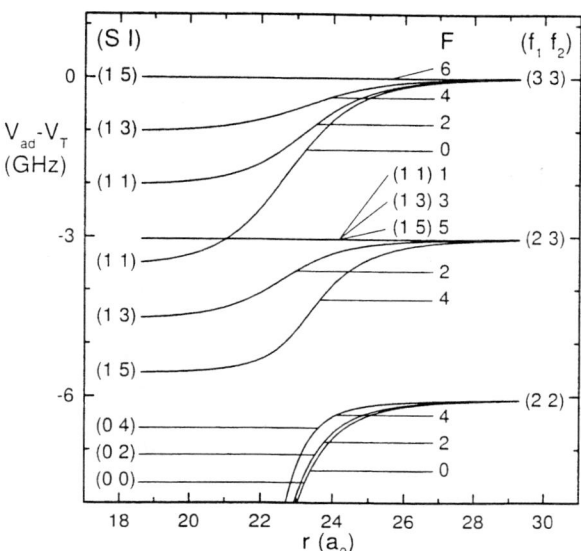

FIGURE 3. Adiabatic potential curves for two interacting, ground state ^{85}Rb atoms. In this figure, the energy of the pure triplet, $F = 6$ potential has been subtracted off, so that only the difference between the potential curves is visible.

good at all distances. Diagonalizing the complete interaction Hamiltonian at each internuclear separation results in adiabatic interaction potentials with a structure like that shown in Fig. 3 for the case of ^{85}Rb$_2$ [14]. Note that in this figure, the energy of the pure triplet $F = 6$ state has been subtracted from each curve in order to make it easier to see the splitting between curves. If it were included, all curves would show a very strong attractive interaction. The change in spin coupling occurs over the range of a few a_0, at an internuclear separation of about 24 a_0 for the case illustrated. The motion of alkali atoms through the recoupling region is neither perfectly adiabatic nor diabatic; in general motions on different channels with the same values of F and M_F are coupled together. To parameterize the interactions it is sufficient to provide the singlet and triplet scattering lengths a_S and a_T, defined as the scattering lengths that would be computed for singlet and triplet collisions of hypothetical atoms with zero hyperfine interaction, and additional parameters which characterize the hyperfine, dispersion, and long range part of the exchange interaction. From this parameter set, it is possible to compute the scattering lengths which apply to collisions of atoms in any given $|f_1, m_{f1}; f_2, m_{f2}\rangle$ combination using coupled channels methods [5,15].

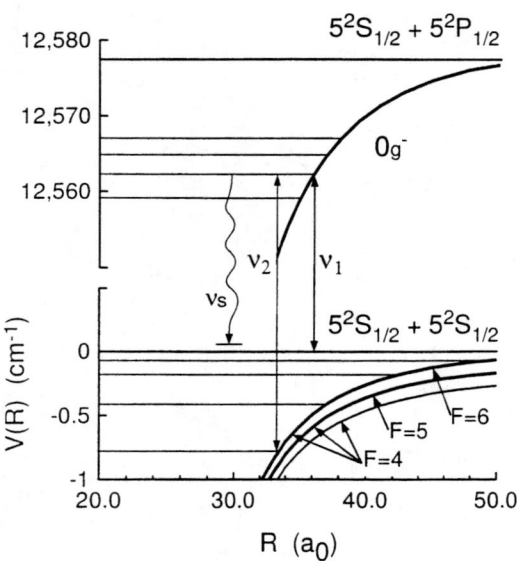

FIGURE 4. Photoassociation spectroscopy of ultracold ^{85}Rb($f = 3$) atoms. For two-color spectroscopy, colliding atoms are irradiated by laser fields of fixed frequency ν_1 and variable frequency ν_2. Optical double resonance (free-bound-bound) signals occur when the frequency difference $\nu_2 - \nu_1$ coincides with the binding energy of a ground state vibrational level.

PHOTOASSOCIATION SPECTROSCOPY OF ULTRACOLD ATOMS

In order to determine alkali scattering lengths, information must be obtained on the phase of the low energy collisional wavefunctions at long range, or equivalently, on the locations on the highest bound states of the potential. Cold atom photoassociation spectroscopy [16–18] provides one of the best ways to do this, as shown in Fig. 4. In single-color experiments, atoms collide in the presence of a laser field of frequency ν_1. As this laser is tuned, free-bound transitions are resonantly excited. This excitation may be detected through photoionization of the excited state molecule followed by detection of the resulting molecular ion [17]. Alternatively, the loss of atoms from the trap may be used as a measure of the photoassociation rate [18]. This trap loss occurs because the excited molecule decays by spontaneous emission predominantly to free states with kinetic energies that are larger than the trap depth.

Single-color photoassociation spectroscopy has been used to determine alkali atom scattering lengths through the analysis of the photoassociation line shapes and intensities. The line strengths are proportional to the free-bound Franck-Condon factors $|\int u_g(R) u_e(R) dR|^2$, where u_g and u_e are the free and excited bound state vibrational wavefunctions, respectively. Since the excited potentials and wavefunc-

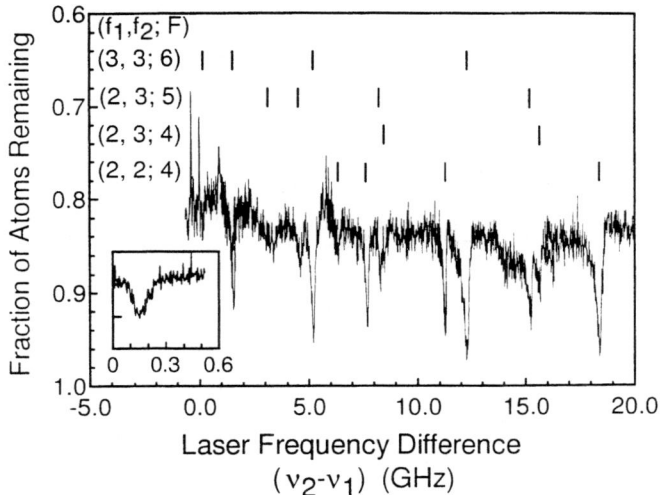

FIGURE 5. Two-color photoassociation spectrum of ^{85}Rb$_2$, with ν_1 tuned into resonance with a $0_g^-(J=2)$ level at 12563.1 cm^{-1}. Double resonance signals are observable as a decrease in the trap loss. The positions and quantum numbers of the levels inferred from the analysis are also indicated. The inset at the lower left shows the highest level observed, with a binding energy of only 0.16 GHz.

tions can be precisely determined through spectroscopic analysis, the photoassociation line intensities provide information on u_g, and in turn on the scattering length. This type of analysis has been carried out for Li [19], Na [20], and Rb [8,9,21,22] atoms.

A more straightforward method to determine scattering lengths is to directly measure the highest bound level energies of the ground electronic state with two-color photoassociation spectroscopy. In this case, the atoms collide in the presence of two lasers fields of frequency ν_1 and ν_2. The frequency ν_1 is fixed on resonance to a particular photoassociation line, and ν_2 is scanned to the blue of ν_1. When the frequency difference $\nu_2 - \nu_1$ matches the binding energy of a high-lying vibrational state, the trap loss decreases. This occurs because the second laser power broadens the excited level, reducing the absorption rate of photons on the free-bound transition. Results of this type were first obtained for triplet ^{7}Li$_2$ by the group of Hulet at Rice [23]. In collaboration with the group of Verhaar at Eindhoven, our group at Texas has obtained the result for ^{85}Rb$_2$ shown in Fig. 5 [14]. In this case, the atoms were initially spin polarized in their $f = 3, m_f = 3$ state, so that with a two-photon transition only $F = 4, 5,$ or 6 two-atom states can be excited. The levels can be assigned to the $F = 4, 5,$ and 6 potentials illustrated in Fig. 3. The $F = 6$ and $F = 5$ states are pure triplet states, whereas the $F = 4$ states have mixed singlet-triplet character. A coupled channels version of inverse perturbation analysis was developed to analyze this data, and excellent agreement was obtained

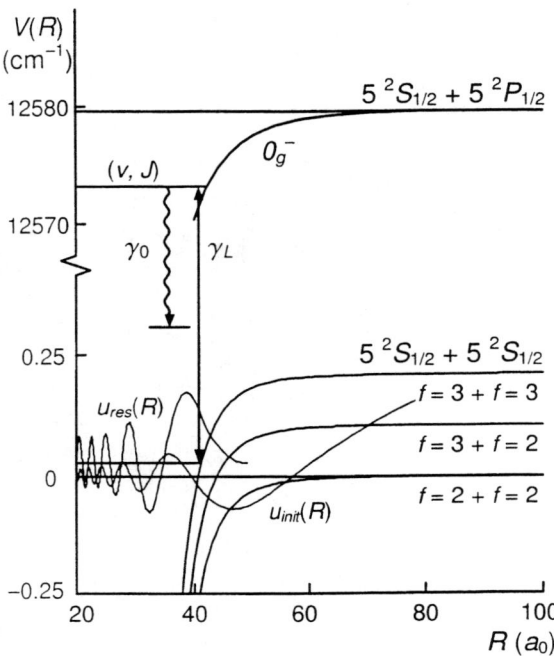

FIGURE 6. Photoassociation method for the detection of a Feshbach resonance in collisions of ultracold ^{85}Rb($f = 2$) atoms.

between the best-fit theoretical and experimental level energies. This result also led to a precise determination of both triplet and singlet scattering lengths for ^{85}Rb.

FESHBACH RESONANCES OF ULTRACOLD, COLLIDING ATOMS

One of the most exciting developments during the past year was the observation of magnetically tunable Feshbach resonances by several groups [10–12]. The origin of these resonances can be understood with reference to Figs. 3 and 6, which illustrate a particular case that has been studied in ^{85}Rb. Free, ground-state ^{85}Rb atoms collide in the $|f = 2, m_f = -2\rangle + |f = 2, m_f = -2\rangle$ entrance channel. The entrance channel has a total angular momentum projection quantum number $M_F = -4$, equal to the sum of the two atomic m_f values. It is coupled to other $M_F = -4$ channels at small internuclear distance by the electronic-exchange interaction. The other $M_F = -4$ potential curves all correlate to the higher energy $(f = 2) + (f = 3)$

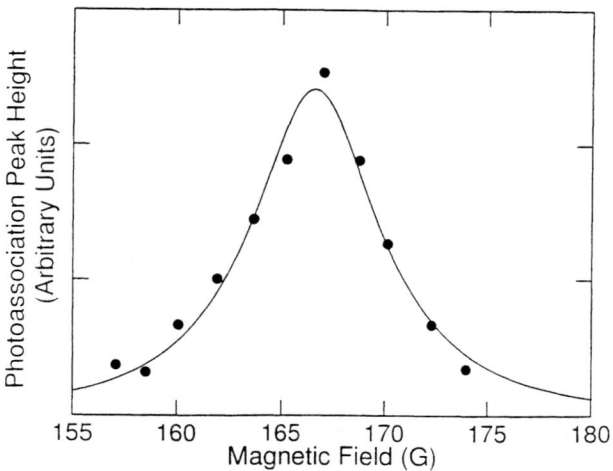

FIGURE 7. Height of an $F = 4, M_F = -4, J = 0$ peak in an ^{85}Rb$_2$ photoassociation spectrum, as a function of magnetic field, with the atoms colliding in their lower hyperfine (f=2) state. The resonance in the peak height as a function of magnetic field is due to a magnetically tunable Feshbach resonance.

or $(f = 3) + (f = 3)$ dissociation limits. They support multichannel quasi-bound states at positive energies, where we take the zero of energy to be the threshold of the entrance channel. If the energy of the incoming atoms matches the energy of one of these states, a Feshbach resonance occurs in which a large wavefunction amplitude builds up in the quasi-bound state. Physically, the resonance occurs because atoms may make a transition during a collision to the quasi-bound level, and oscillate for some number of periods before making a transition back to the initial channel. In some cases the energy of a Feshbach resonance can be tuned to zero with a magnetic field, because the quasi-bound state and threshold energy Zeeman shift at different rates [6]. In that case the resonance has a large effect on ultracold collisions. These tunable resonances are of particular interest to dilute gas BEC, since they allow the scattering length to be tuned. The resonance in the scattering length has the dispersive form $a = a^{(0)}[1 - \Delta/(B - B_0)]$ [24], where $a^{(0)}$ gives the "background" value of the scattering length, and B_0 and Δ are constants that characterize the resonance field value and width.

The group of Ketterle at M.I.T. has observed tunable Feshbach resonances in an optically trapped ^{23}Na BEC, and actually observed the change in self-energy that occurs as a result of the change in scattering length [10]. In addition, they observed an enhancement in the inelastic collisional loss rates of the atoms. Feshbach resonances should enhance inelastic collision rates, but the observed enhancements are so large that they are not presently understood.

In collaboration with the group of Verhaar, our group at Texas has studied the

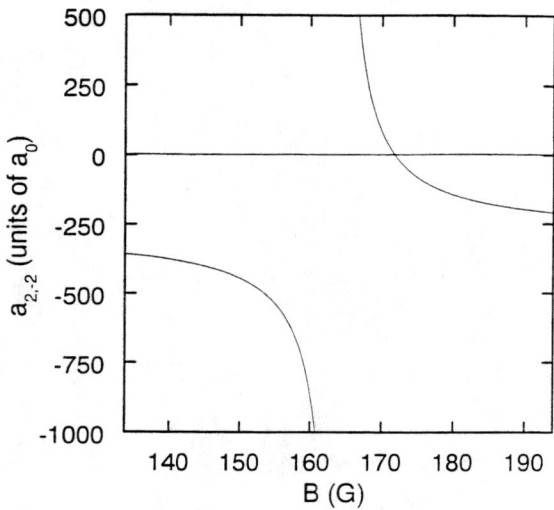

FIGURE 8. Calculated value of the $^{85}\text{Rb}_2(f = 2, m_f = -2)$ scattering length as a function of magnetic field that follows from the width and position of the Feshbach resonance shown in Fig. 7.

^{85}Rb resonance illustrated in Fig. 6, using photoassociation as a probe [11]. In order to detect the resonance, we drive photoassociation transitions to the excited $^{85}\text{Rb}_2$ 0_g^- bound molecular vibrational state at an energy 5.9 cm^{-1} below the $5^2S_{1/2} + 5^2P_{1/2}$ dissociation limit. The experiment is carried out in a magnetic field between 70 G and 200 G. The photoassociation spectrum shows a Zeeman splitting, and we are able to isolate a single spectral component which originates from the s-wave, $M_F = -4$, collisional resonance state. We observe the resonance through an enhancement in the photoassociation rate as the field is tuned. This enhancement occurs because the photoassociation rate is proportional to the square of the wavefunction overlap between the collisional state and the excited state, and therefore shows an enhancement when the Feshbach resonance is tuned near zero energy.

We take repeated scans over the $M_F = -4$ photoassociation peak at a laser intensity of 0.1 W/cm^2, and show the peak height as a function of magnetic field in Fig. 7. The photoassociation signal reaches a maximum at about 167 G, and shows a resonance that is about 6 G wide. We do not observe the change in scattering length directly, but can model the data shown in the figure to determine the resonance parameters that are consistent with our observed photoassociation resonance structure. We find that these data imply that $a^{(0)} = -295 \pm 80$ a_0, $\Delta = 8.2 \pm 3.8$ G, and $B_0 = 164 \pm 7$ G. A plot of the predicted scattering length as a function of magnetic field is shown in Fig. 8.

A different study of this resonance has been carried out by the JILA BEC group [12]. They measure the elastic collision rate of $^{85}\text{Rb}_2(f = 2, m_f = -2)$ atoms as

TABLE 1. Scattering lengths for Rb atoms at zero magnetic field, in atomic units, and indicating the method used: elastic collision measurement (EL), inelastic collision measurements (IN), one-color photoassociation spectroscopy (1PA), two-color photoassociation spectroscopy (2PA), or Feshbach resonance measurements (FB). The last column gives values for the $|f = i - 1/2, m_f = -(i - 1/2)\rangle$ state.

Atom	Triplet	Singlet	Lower Hyperfine
^{85}Rb	$-1000 < a < -60$ (1PA [21])	$4500 < a < \infty$ or	-450 ± 140 (2PA [22])
	-400 ± 100 (1PA [8])	$-\infty < a < -1200$	
	-440 ± 140 (2PA [14])	(2PA [14])	
	-369 ± 16 (FB & 1PA [12])	2400^{+600}_{-350}	
		(FB & 1PA [12])	
^{87}Rb	$85 < a < 140$ (1PA [21])	93 ± 5 (2PA [22])	106 ± 6 (2PA [22])
	109 ± 10 (1PA [9])	88 ± 14 (IN & 1PA [25])	87 ± 11 (EL [26])
	102 ± 6 (2PA [22])		103 ± 5 (IN & 1PA [27])
	106 ± 4 (FB & 1PA [12])	90 ± 1 (FB & 1PA [12])	

a function of temperature and magnetic field strength. They observe directly the large change in elastic cross section that occurs as the field is tuned through the resonance value at low temperature. A maximum in the cross section occurs near $B = B_0$, and a zero in the cross section occurs at $B = B_0 + \Delta$. A variation in the elastic collision rate of more than four orders of magnitude is observed in this experiment. Also, the field measurements in this experiment are very precise, so that very accurate resonance parameters $\Delta = 11.6 \pm 0.5$ G, and $B_0 = 155.2 \pm 0.4$ G are obtained.

The positions and widths of the Feshbach resonances are very sensitive to the potential parameters. For the case of the ^{85}Rb resonance, the value of B_0 depends mostly on the sum of v_{DS} and v_{DT} and on C_6, whereas Δ depends mostly on $v_{DS} - v_{DT}$ [11]. These measurements provide similar information to that provided by the very high molecular bound state energies, but with better accuracy than can easily be obtained with photoassociation in a laser-cooled gas.

CONCLUSION

A wide variety of cold collision studies and photoassociation measurements have been carried out on the alkali atoms. Together with coupled channels collision and bound state theory, these measurements are leading to a complete and consistent understanding of ultracold alkali atom collisions and interactions. For example, Table 1 lists the determinations of the Rb scattering lengths which have been carried out. Very good agreement is obtained across a wide range of measurements. These experiments show a range of interesting resonance and threshold phenomena, including low-energy shape and Feshbach resonances. In turn, these phenomena play important roles in trapped ultracold alkali gases and Bose-Einstein condensates.

ACKNOWLEDGEMENTS

Work on ultracold Rb collisions at the University of Texas was carried out by John Miller, Robert Cline, Jeff Gardner, Chin Tsai, Riley Freeland, and Philippe Courteille, and in collaboration with Boudewijn Verhaar, Hugo Boesten, Johnny Vogels, and Frank van Abeelen at the Eindhoven University of Technology. We gratefully acknowledge the support of this work by the National Science Foundation, the R. A. Welch Foundation, and the NASA Microgravity Research Division.

REFERENCES

1. Anderson, M. H., Ensher, J. R., Matthews, M. R., Wieman, C. E., and Cornell, E. A., *Science* **269**, 198-201 (1995).
2. Davis, K. B., Mewes, M.-O., Andrews, M. R., van Druten, N. J., Durfee, D. S., Kurn, D. M., and Ketterle, W. *Phys. Rev. Lett.* **75**, 3969-3973 (1995).
3. Bradley, C. C., Sackett, C. A., and Hulet, R. G., *Phys. Rev. Lett.* **78**, 985-989 (1997).
4. Dalfovo, F., Giorgini, S., Pitaevskii, L. P., and Stringari, S., to be published in *Rev. Mod. Phys.* **1**, vol. 2.
5. Julienne, P. S., Smith, A. M., and Burnett, K., *Adv. At. Mol. Phys.* **30**, 141-198 (1993).
6. Tiesinga, E., Verhaar, B. J., and Stoof, H. T. C., *Phys. Rev. A* **47**, 4114-4121 (1993).
7. Arndt, M., Ben Dahan, M., Guéry-Odelin, D., Reynolds, M. W., and Dalibard, J. *Phys. Rev. Lett.* **79**, 625-628 (1997).
8. Boesten, H. M. J. M., Tsai, C. C., Verhaar, B. J., and Heinzen, D. J., *Phys. Rev. Lett.* **77**, 5194-5197 (1996).
9. Boesten, H. M. J. M., Tsai, C. C., Gardner, J. R., Heinzen, D. J., and Verhaar, B. J., *Phys. Rev. A* **55**, 636-640 (1997).
10. Inouye, S., Andrews, M. R., Stenger, J., Miesner, H.-J., Stamper-Kurn, D. M., and Ketterle, W., *Nature* **392**, 151-154 (1998).
11. Courteille, P., Freeland, R. S., Heinzen, D. J., van Abeelen, F. A., and Verhaar, B. J., *Phys. Rev. Lett.* **81**, 69-72 (1998).
12. Roberts, J. L., Claussen, N. R., Burke, Jr., J. P, Greene, C. H., Cornell, E. A., and Wieman, C. E., *Phys. Rev. Lett.* **81**, 5109-5112 (1998).
13. LeRoy, R. J., and Bernstein, R. B., *J. Chem. Phys.* **52**, 3869-3879 (1970).
14. Tsai, C. C., Freeland, R. S., Vogels, J. M., Boesten, H. M. J. M., Verhaar, B. J., and Heinzen, D. J., *Phys. Rev. Lett.* **79**, 1245-1248 (1997).
15. Stoof, H. T. C., Koelman, J. M. V. A., and Verhaar, B. J., *Phys. Rev. B* **38**, 4688-4697 (1988).
16. Lett, P. D., Julienne, P. S., and Phillips, W. D., *Ann. Rev. Phys. Chem.* **46**, 423-452 (1995).
17. Lett, P. D., Helmerson, K., Phillips, W. D., Ratliff, L. P., Rolston, S. L., and Wagshul, M. E., *Phys. Rev. Lett.* **71**, 2200-2203 (1993).
18. Miller, J. D., Cline, R. A., and Heinzen, D. J., *Phys. Rev. Lett.* **71**, 2204-2207 (1993).

19. Abraham, E. R. I., McAlexander, W. I., Gerton, J. M., Hulet, R. G., Côté, R., and Dalgarno, A., *Phys. Rev. A* **55**, 3299-3302 (1997).
20. Tiesinga, E., Williams, C. J., Julienne, P. S., Jones, K. M., Lett, P. D., and Phillips, W. D., *J. Res. Nat. Inst. Stand. Technol.* **101**, 505-520 (1996).
21. Gardner, J. R., Cline, R. A., Miller, J. D., Heinzen, D. J., Boesten, H. M. J. M., and Verhaar, B. J., *Phys. Rev. Lett.* **74**, 3764-3767 (1995).
22. Vogels, J. M., Tsai, C. C., Freeland, R. S., Kokkelmans, S. J. J. M. F., Verhaar, B. J., and Heinzen, D. J., *Phys. Rev. A* **56**, 1067-1070 (1997).
23. Abraham, E. R. I., McAlexander, W. I., Sackett, C. A., and Hulet, R. G., *Phys. Rev. Lett.* **74**, 1315-1318 (1995).
24. Moerdijk, A.-J., Verhaar, B. J., and Axelsson, A., *Phys. Rev. A* **51**, 4852-4861 (1995).
25. Burke, Jr., J. P., Bohn, J. L., Esry, B. D., and Greene, C. H., *Phys. Rev. A* **55**, 2511-2514 (1997).
26. Newbury, N. R., Myatt, C. J., and Wieman, C. E., *Phys. Rev. A* **51**, 2680-2683 (1995); erratum, *Phys. Rev. A* **55**, 3279 (1997).
27. Julienne, P. S., Mies, F. H., Tiesinga, E., and Williams, C. J., *Phys. Rev. Lett.* **78**, 1880-1883 (1997).

Ultracold Collisions: Exploring the Quantum Threshold Regime

Paul S. Julienne, Eite Tiesinga, Paul Leo, and Carl J. Williams

Atomic Physics Division
National Institute of Standards and Technology
Gaithersburg, MD 20899

Abstract. This talk reviews some of the basic concepts in cold collisions, and gives examples of calculations directed towards understanding the following areas: (1) collisions of Na atoms trapped in Zeeman components of the $F = 1$ hyperfine sublevel, and their control by magnetically-induced Feshbach resonances; (2) elastic and inelastic collisions of Cs atoms in the double spin-polarized $F = 4$, $M = 4$ level; and (3) stimulated Raman photoassociation of two colliding atoms in a Bose-Einstein condensate (BEC) to make translationally cold molecules in a selected target level.

INTRODUCTION

The topic of atomic collisions at temperatures of a few μK and below continues to develop as an important subject, due to its relevance to a number of forefront areas in atomic physics opened up by continuing progress in the cooling and trapping of atoms. This talk will review some of the basic concepts in cold collisions, and give examples of calculations directed towards understanding the following areas: (1) collisions of Na atoms trapped in Zeeman components of the $F = 1$ hyperfine sublevel, and their control by magnetically-induced Feshbach resonances; (2) elastic and inelastic collisions of Cs atoms in the double spin-polarized $F = 4$, $M = 4$ level; and (3) stimulated Raman photoassociation of two colliding atoms in a Bose-Einstein condensate (BEC) to make translationally cold molecules in a selected target level. We are interested in particular in collisions in the lower range of current cooling methods, that is, at temperatures on the order of 1 μK or lower that apply to an evaporatively cooled gas or a BEC.

BACKGROUND

A collision of two atoms is described by a wavefunction that consists of a plane wave plus a scattered wave. These are usually resolved into "partial waves" associated with the quantized relative angular momentum quantum numbers $\ell =$

$0, 1, 2, \ldots$, or s-, p-, d-, $\ldots$ waves. The scattering amplitudes for elastic or inelastic processes vary as some power of k^ℓ, where $\hbar k$ is collision momentum, according to the threshold laws for cold collisions. As collision momentum $\hbar k \to 0$, only s-waves can give nonvanishing contributions to cross sections σ or rate coefficients $K = \langle \sigma v \rangle$, where the brackets imply an average over the distribution of relative velocity v. As $k \to 0$ the elastic collision cross section for two identical bosons is $\sigma_{el} = 8\pi A^2$, where A is the scattering length, and the corresponding rate coefficient $\to 0$. The sign of the scattering length is also crucial for determining the stability and mean field energy of a Bose-Einstein condensate. The scattering length is extremely sensitive to the details of the interatomic potential(s), and must be measured since accurate ab initio calculations of A are not possible for the alkali species. In contrast to elastic collisions, the cross sections for exoergic inelastic processes vary as $1/k$ and $K \to$ a constant value as $k \to 0$.

Ground state alkali atoms have hyperfine structure due to the coupling of the spin $S = \frac{1}{2}$ electron to the nucleus with spin I to give a resultant $F_< = I - \frac{1}{2}$ or $F_> = I + \frac{1}{2}$. Ground state ^{23}Na or ^{87}Rb atoms have $F = 1$ and 2, and the ^{133}Cs atom has $F = 3$ or 4. Each F level is split into $2F + 1$ Zeeman sublevels when a magnetic field B is applied. The levels $F = F_>$, $M = F_>$ and $F = F_<$, $M = -F_<$ are weak-field-seeking levels near $B = 0$ that have the lowest energy when the field is lowest. These are the levels that can be evaporatively cooled and Bose-condensed in traps with a spatial minimum in B.

When two ground state alkali atoms collide, the two unpaired electron spins can couple to give molecular states of $^1\Sigma_g^+$ or $^3\Sigma_u^+$ symmetry, which are characterized by Born-Oppenheimer molecular potential energy curves $V_1(R)$ and $V_3(R)$ respectively. These describe the chemical bonding interactions that occur when the two atoms are close together at small R and the weak van der Waals interaction varying as R^{-6} at large R. The full theoretical treatment that we use to describe ground state collisions must not only take into account the influence of these molecular potentials but also the effect of electron and nuclear spins and their R−dependent interactions. For example, the atoms separate to individual Zeeman sublevels, which are split by hyperfine and magnetic field terms in the Hamiltonian. Our full theory uses an expansion of the wavefunction in a basis that includes the electronic, spin, and rotational degrees of freedom, collectively designated α, and then solves the coupled Schrödinger equations for the radial motion. One useful selection rule to keep in mind for cold collisions is that for weak magnetic fields the resultant total electron and nuclear spin angular momentum $\mathbf{f} = \mathbf{F_1} + \mathbf{F_2}$ of the two atoms 1 and 2 is conserved to a very good approximation, as well as ℓ, and the space projection m of total spin. Many properties of cold collisions follow from these simple selection rules. For example, using the notation $\{f\ell\}$ for the approximately conserved quantum numbers, s−wave collisions of doubly spin polarized atoms only gives rise to a $\{f = 2F_>, s\}$ state, which is only weakly coupled to other channels with a consequently small inelastic relaxation rate. Weak-field collisions of two ^{23}Na or ^{87}Rb $F = 1$ atoms are characterized by two scattering lengths, $\{0s\}$ and

{2s}. Only the {2s} one is important for collisions of two $F = 1, M = -1$ atoms, but both contribute to collisions of unpolarized atoms or to spinor condensates. This {2s} channel is strongly coupled to {2s} channels from $F = 1$ and $F = 2$ atoms or two $F = 2$ atoms, so collisions of $F = 1, M = -1$ atoms depend on both the V_1 and V_3 molecular potentials, and can be affected by Feshbach resonances due to the presence of closed channels contributing to the collision.

SODIUM $F = 1$ COLLISIONS

We have previously used high resolution photoassociation spectroscopy to determine the zero-field scattering length $A_{1,-1}$ for two $F = 1, M = -1$ ^{23}Na atoms [1]. This is based on an analysis of data for the 0_g^- excited pure long range state of the Na$_2$ dimer. In photoassociation two colliding ground state atoms are excited to a molecular dimer excited electronic state by a tunable laser. Detecting product ions (due to a second color) or loss of atoms from the trap as a function of frequency of the photoassociation laser maps out a spectrum that measures the positions of the vibrational and rotational levels of the upper state. From the positions of the levels, we were able to determine the atomic transition dipole (and atomic lifetime) from the strength of the long range potential, and also detect the presence of small relativistic finite-speed-of-light effects on the potential [2]. From the relative intensities and line shapes of different spectral features, we were able to deduce that the low B-field scattering length of $F = 1, M = -1$ ^{23}Na atoms is $+52 \pm 5$ a$_0$ (a$_0$ = 0.0529177 nm) [1]. The basic reason why such spectra can be used to determine scattering lengths is that the intensity of the transition is determined by a Franck-Condon factor. Thus, the photoassociation transition probes the shape of the ground state wavefunction in the region of internuclear separation where the molecule is in resonance with the light, that is, the region of the Condon point [3].

Recently the MIT group [4] has developed methods to trap $F = 1$ ^{23}Na atoms optically instead of magnetically. This means that any Zeeman sublevel can be trapped, not just the low field $F = 1, M = -1$ level. Since F is not a good quantum number when the field strength is increased to the regime that the Zeeman shift is not small compared to the hyperfine splitting, we use the designation a, b, c to label the levels that correlate with the respective $F = 1, M = 1, 0$, and -1 levels at small field. The MIT group was able to trap a Bose-Einstein condensate of a level atoms and independently vary the magnetic field up to above 0.1 T. In this way they could observe the variation of scattering length as the magnetic field was tuned to position a molecular Feshbach resonance near the zero collision energy of the very cold atoms [5]. The Feshbach resonance is a closed channel molecular quasibound state that is made to pass through the collision threshold energy as the field is changed so as to vary the position of the molecular bound level relative to the energy of the separated atoms in the a level. By making small modifications in the Na$_2$ V_1 and V_3 molecular potentials obtained in [1], we can get excellent agreement between the calculated and observed positions of the measured resonances. The MIT group observed two resonances in $a + a$ collisions at 85.3 ± 2 mT and 90.7 ± 2 mT, with widths on the order of 0.01 mT and 0.1 mT respectively [5]. Using potentials that give an $A_{1,-1}$ scattering length of 55 a_0, we calculate resonances at 86.7 mT and 91.6 mT, in excellent agreement with those measured, and the resonance widths are comparable to the measured ones. Our revised potentials predict $A_{2,2} = 65$ a_0 for the doubly spin-polarized state; $A_{2,2}$ is also the scattering length for the V_3 potential. For low B field, the two resonances are $\{2s\}$ and $\{4s\}$ components of the $v = 14$ level of the $^3\Sigma_u^+$ state with a binding energy on the order of 4 GHz, and they increase in energy as B increases as they become levels with both electron spins up. Since the energy of two a atoms decreases as B increases, these bound levels cross from below into the scattering continuum as B increases and become unbound Feshbach resonance states that influence the scattering length for the collision of two a atoms. We also predict two Feshbach resonances in collisions of two c level atoms at 117.9 mT and 151 mT, with widths of 0.1 mT and 10 mT respectively. The lower resonance, due to a $v = 15$ $^3\Sigma_u^+$ level which moves through threshold from above, has also been observed by the MIT group at 119.5 ± 5 mT [6], in agreement with our calculations. The upper $c + c$ resonance, which has not yet been observed, is associated with a low B $\{4s\}$ $v = 14$ level which moves through threshold from below.

The properties of spinor condensates [7] made from mixtures of the $M = $ -1, 0, and 1 components of $F + 1$ ^{23}Na atoms depend on the scattering length differences among the various M-components. At low B-field these differences depend on the difference between the $\{2s\}$ and $\{0s\}$ scattering lengths. We calculate $A(2s) - A(0s) = +5 \pm 2$ a_0, as compared to a value of 3.5 ± 1.5 measured by the MIT group from experiments on spinor condensates.

We have recently started an analysis of the bound state spectra of the Na$_2$ dimer obtained in the dissociation threshold region by the group of Tiemann in Hannover [8]. The experiment was done by recording the emission spectrum from the $v = 120$ level of the excited $A^1\Sigma_u^+$ state, which decays to ground state levels near the dissociation limit; the $v = 120$ level was pumped in a molecular beam using a multicolor pumping scheme. Calculated levels agree well with the measured levels close to threshold, but there are discrepancies on the order of 0.01 cm^{-1}, or 300 MHz, for levels on the order of several GHz or more from threshold. We assume this is due to remaining uncertainties in the molecular potentials. Although the photoassociation and Feshbach data strongly constrain the phase very close to threshold, the energy-variation of the phase which determines bound state positions away from threshold depends on details of the potentials that are still in need of minor adjustment. The hope is that analysis of the data can be used to produce more accurate potentials over their whole range. In any case, we now have an excellent understanding of threshold collisions of Na atoms.

CESIUM $F = 4$ COLLISIONS

Collisions of cool Cs ground state atoms are of interest because of applications involving atomic clocks, optical lattices, or Bose-Einstein condensates. The Cs atom has two hyperfine ground state manifolds with $F = 3$ and 4. Recent experiments in Paris [9,10] have measured both the elastic and inelastic collision rate coefficients for Cs atoms in the doubly spin-polarized level with $F = 4$, $M = 4$. Both results were surprising: the elastic collision cross section was observed to be very close to its maximum permissible value for s-waves and the inelastic rate was about three orders of magnitude larger than had been previously predicted. We have carried out new calculations on the elastic and inelastic collisions of Cs $F = 4$, $M = 4$ atoms, and find that both these observations can be understood if the scattering length $A_{4,4}$ has a large magnitude and if the second-order spin-orbit contribution to the spin-dipolar relaxation is taken into account [11].

Arndt et al. [9] measured thermalization rates in an evaporatively cooled Cs gas, and interpreted their result to show that the elastic scattering cross section of $F = 4$, $M = 4$ Cs atoms was equal (within experimental error limits) to the unitarity limit for s-wave scattering, $\sigma_{4,4}(k) = 8\pi/k^2$, over a temperature range between about 5 and 50 μK. Thus, the cross section is still increasing with decreasing k even for collision energies in the μK range, and has not yet reached the low T limiting behavior where $\sigma_{4,4}(k) = 8\pi A_{4,4}^2$ is a constant independent of k. We have calculated $\sigma_{4,4}(k)$, including s-, d-, and g-wave contributions, between 0.1 and 300 μK. Since the $V_3(R)$ molecular potential is not known with sufficient accuracy to predict a reliable scattering length, we took the best potential available, using the long range potential parameters from Marinescu and Dalgarno [12], and varied the uncertain inner part of the potential to produce different scattering lengths in the $k \to 0$ limit. Agreement with the $\sigma_{4,4}$ reported by Arndt et al. is found if $|A_{4,4}|$ is assumed to be larger than about 600 a_0. In this case, the threshold limit where $A_{4,4}$ is a constant independent of k is not reached until T is less than 1 μK.

It is necessary to consider carefully the effective spin-dipolar interaction Hamiltonian in order to understand the anonymously large inelastic relaxation rate coefficient, on the order of 10^{-12} cm^3/s, measured by Söding et al. [10]. The weak dipolar interaction between the two unpaired electrons in the different Cs atoms is proportional to α^2 (α is the fine structure constant) and varies as $1/R^3$. This interaction splits the degeneracy of the $|\Omega| = 1$ and $\Omega = 0$ projections on the interatomic axis of the spin $S = 1$ state. A splitting of these spin projections can also be produced by a second-order spin-orbit interaction mediated through distant electron states (such as $^3\Pi_u$) [13]. This latter contribution to the effective spin-spin interaction, which increases exponentially with decreasing R as the charge clouds of the atoms overlap and a chemical bond is formed, plays a dominant role in a high Z atom such as Cs.

Relaxation of doubly spin-polarized Cs at room temperature is also controlled by the same spin Hamiltonian as the collision at 1 μK. By using a semiclassical theory and adjusting the strength of the second-order spin-orbit contribution to the spin Hamiltonian, retaining the exponential form calculated by Mies et al. [13], Leo et al. [11] were able to explain the measured room temperature relaxation rate. This Hamiltonian, without any additional adjustments, was used in a full quantum scattering calculation at low T of the inelastic relaxation rate coefficient, including all spins and partial waves through $\ell = 4$. Excellent agreement was found with the measured rate coefficient of Söding et al. between 10 and 100 μK, as long as the magnitude of the scattering length $A_{4,4}$ is taken to be larger than around 500 a_0.

Our conclusion is that the recent measurements on elastic and inelastic collisions of Cs $F = 4$, $M = 4$ atoms can be understood as long as the $A_{4,4}$ scattering length is large in magnitude. We can not give precise bounds, but a magnitude of 600 a_0 or larger seems likely. We are unable to deduce the sign of the scattering length from this analysis. In any case, Bose-Einstein condensation of this level seems to be ruled out due to the rapid inelastic spin relaxation [9], which is several orders of magnitude faster than the corresponding relaxation rate for ^{87}Rb at comparable density. We are currently exploring collisions of two $F = 3$ Cs atoms. Unfortunately collisional heating in weak magnetic fields [14] seems to be unusually high because of a low-field d-wave resonance in this case. Additional experimental and theoretical work on the Cs system is needed to clarify the remaining options for cooling and condensation, and to assess the role of collisions in optical lattices.

STIMULATED RAMAN MOLECULE FORMATION IN A BEC

One potentially important use of collisions in a Bose-Einstein condensate is molecule production using a stimulated Raman process. One-color photoassociation in a magneto-optical trap (MOT) is known to lead to the formation of diatomic molecules as the excited molecular vibrational level excited by the photoassociation laser decays back to the molecular ground state. A part of this emission results in hot atoms (or a dissociated molecule) and another part results in dimer formation. The branching ratios between hot atom and dimer production are controlled by Franck-Condon factors. Thorsheim et al. [15] calculated such branching ratios for the A-X band system of the Na$_2$ molecule and showed that photoassociation is usually not very efficient for molecule formation. Band and Julienne [16] proposed a two-color scheme that should be more efficient, involving excitation to a double-excited molecular state that preferentially decayed to a few vibrational levels of the molecular ground state. The experimental realization of cold molecule production in MOTs has recently been demonstrated experimentally for Cs$_2$ [17] and K$_2$ [18]. In both cases sufficient molecules for detection were produced by the spontaneous decay following one-color photoassociation.

The question we are interested in here is whether a two-color stimulated process can be used efficiently to convert trapped cold atoms into translationally cold molecules. Vardi et al. [19] made a theoretical study of such processes in a MOT using a time-dependent wavepacket formalism. Molecule formation is relatively slow since the molecules will be photodissociated by the same lasers that produce them, and a long time, several minutes, would be needed to recombine all the atoms to ground state molecules using a sequence of short pulses and waiting for the molecules to leave the trap between pulses.

There are reasons to think that the two-color process may be especially useful in a BEC. First, a condensate will have a prominent photoassociation spectrum. Burnett et al. [20] showed that the spectrum of light scattering by a condensate in the far red detuning region (detuning Δ larger than a few GHz from atomic resonance) is dominated by a series of one-color photoassociation resonances. If Γ_{atom} represents the free atom light scattering rate, proportional to I/Δ^2 where I is the laser intensity, then the light scattering rate due to collisions of atom pairs is $\Gamma_{pair} = \Gamma_{atom} C n \left(\frac{\lambda}{2\pi}\right)^3 f_{res}$, where C is a constant on the order unity, n is atomic density, λ is the wavelength of the light, and f_{res} is a photoassociation resonance enhancement factor. If the light is tuned to photoassociation resonance, $f_{res} \gg 1$ is on the order of the molecular vibrational level spacing divided by the natural linewidth, whereas $f_{res} \ll 1$ if the light is tuned between photoassociation resonances. Since $n \left(\frac{\lambda}{2\pi}\right)^3$ is on the order of unity or larger for typical condensate densities, one-color photoassociation resonances should be prominent features of large detuning light scattering in a condensate.

A semianalytic theory of two-color photoassociation line shapes has been worked out by Bohn and Julienne [21]. In the two-color process two frequencies of light are absorbed by the colliding atoms to produce a target molecular vibrational level by a stimulated Raman process that involves an intermediate excited molecular level. Assuming that the target molecular level decays to some products that are detected, the theory uses a time-independent scattering formalism to calculate the rate coefficient for making the product states as a function of the intensities and frequencies of the two lasers and the collision energy of the atom pair. If these parameters are properly chosen, the theory suggests that the molecule production can be made very efficient, with the inelastic S-matrix element for molecule production approaching the unitary limit where one molecule is made per collision. This large efficiency is the consequence of a dark state Raman resonance which strongly suppresses excited state spontaneous decay by the interference of two paths to spontaneous decay [21]: one is the direct path following one-color photoassociation; the other is a three-photon path to the intermediate excited level involving one photoassociation photon and two photons of the second color that couples the intermediate level to the target level.

Although the semianalytic theory suggests that molecule formation can be efficient at MOT temperatures, it does not specifically look into the case of nK near-threshold collisions that would apply in the case of a BEC. Therefore, we have set up a model calculation to test the possibility of efficient molecule formation by stimulated Raman photoassociation for BEC conditions [22]. The model sets up an S-matrix calculation for a five-channel model for a collision in two radiation fields. The channels represent: (1) the colliding ground state atoms, (2) the excited molecular intermediate state optically coupled to the ground state by the first color, (3) spontaneous decay from this excited state, (4) the ground molecular state to be formed, optically coupled to the channel (2) by the second color, and (5) the decay of the target molecular level to the detected products. Channel (3) is represented by a complex potential, leading to loss of unitarity of the S-matrix. Channel (5) is represented by an artificial channel which simulates the loss of the molecules from the trap as they exit, since they are assumed to be untrapped. The coupling to the artificial channel is adjusted so that the decay width of the target molecular level in channel (4) represents the loss rate of molecules due to their finite residence time in the trap. The model calculates two collision rate coefficients, K_{loss} for channel (3) due to spontaneous emission loss from the excited state, and K_{mol} for channel (5) representing molecules in the target vibrational level that leave the trap.

The detuning of the two laser frequencies can be chosen to exactly match the position of the target Raman resonance to the collision energy near threshold. A typical loss time for molecules from a condensate gives a width of the Raman resonance on the order of 1 kHz, or about 50 nK. Our model calculations show that even for a few nK collision energy, K_{mol} remains large, with values on the order of 10^{-10} cm^3/s to 10^{-9} cm^3/s. The excited state spontaneous decay can also be suppressed by up to two orders of magnitude near the Raman resonance, so that K_{loss} can be much smaller than K_{mol}. With condensate densities on the order of 10^{15} cm^{-3}, the characteristic time $(nK_{\text{mol}})^{-1}$ for converting atoms to the target molecule level is on the order of microseconds. Thus, if excited state spontaneous decay can indeed be strongly suppressed, it may be possible to convert a significant fraction of condensed atoms to molecules in specific target vibrational-rotational levels in a series of short (microsecond) laser pulses. At this stage, the theory is only suggestive, and needs to be tested experimentally.

There are a number of specific molecular band systems that could be tried. The intensities and frequencies of the two colors will depend on the system in mind. Intensities will scale according to the free-bound and bound-bound Franck-Condon factors involved. The necessity of locking the two Raman frequencies to a few kHz will be much easier to achieve experimentally if the frequencies of the two colors do not differ by more than a few GHz, thereby implying experiments in which the target level is a high-lying vibrational level near the dissociation limit of the ground state molecule. Such levels will have favorable Franck-Condon factors for both steps. If collisions involve the lower of the two ground state hyperfine levels, then both the $^1\Sigma_g^+$ and $^3\Sigma_u^+$ states are produced in the collision, and target levels of either symmetry are possible. If the upper doubly spin-polarized hyperfine level is chosen, then only the $^3\Sigma_u^+$ state is possible. In the one color photoassociation experiments done so far, $^3\Sigma_u^+$ molecular levels were produced for Cs$_2$ and $^1\Sigma_g^+$ ones for K$_2$. We have assumed that the molecules formed are untrapped and rapidly leave the trap. However, it should be possible to produce $^3\Sigma_u^+$ molecular levels in Zeeman sublevels that remain trapped. This raises the intriguing question as to whether it would be possible to form molecular condensates *in situ* from an atomic one using a Raman pulse.

CONCLUSION

The field of cold collisions continues to be a very active one, driven partly by the stunning successes of BEC experiments. We are now at a point where a very precise determination of threshold scattering properties in the nK region is possible by combining data from MOT and FORT photoassociation studies, trapped atom BEC collision data, and data from more conventional molecular spectroscopy. Quantitative predictive theoretical models are now available for Li, Na, and Rb, and work is in progress on K and Cs. We are getting a qualitative understanding of Cs collisions, and hopefully a quantitative understanding will soon be available. The problem is that an accurate determination of the very large scattering lengths that occur for the ^{133}Cs species is difficult, and it is necessary to use as many different experimental constraints as possible. Recent work by Kokkelmaans *et al.* [23] should help to clarify the situation. We have in progress a collaborative effort with the JILA group [24] analyzing the photoassociation spectra of the 0_g^- and 1_u band systems of the K_2 molecule by the University of Connecticut group [25]. Accurate scattering lengths for the K system should soon be available.

One important prospect for the future is the extension of cooling and trapping techniques to other parts of the periodic table and to molecular species. Several posters at this meeting by John Doyle's group illustrate the prospects for helium buffer gas loading methods for cooling and trapping Cr atoms [26] or CaH molecules [27]. We have already seen that translationally cold molecules have been produced by two groups [17,18]. It is only a matter of time until such molecules will be trapped. Whether it will prove possible to produce molecule efficiently from a BEC remains to be seen. In any case, the subject of cold collisions between molecules or between an atom and a molecule, will become an important one to investigate if molecule trapping develops as an experimental field.

Finally, although this talk does not discuss the issue of collision of mixed species, such as Na and Cs or K and Rb, the interest in mixed-species traps and sympathetic cooling makes such collisions important ones to investigate. Photoassociation spectroscopy of such mixed species should be an important aid for understanding threshold ground state collisions for mixed systems just as it has been for the homonuclear cases.

ACKNOWLEDGMENTS

This work was supported in part by grants from the Office of Naval Research and the Army Research Office.

REFERENCES

1. Tiesinga, E., Williams, C. J., Julienne, P. S., Jones, K. M., Lett, P. D., and Phillips, W. D., *J. Res. Natl. Inst. Stand. Technol.* **101**, 505 (1996).

2. Jones, K. M., Julienne, P. S., Lett, P. D., Phillips, W. D., Tiesinga, E., and Williams, C. J., *Europhys. Lett.* **35**, 85 (1996).
3. Julienne, P. S., *J. Res. Nat. Inst. Stand. Technol.* **101**, 487 (1996).
4. Stamper-Kurn, D. M., Andrews, M. R., Chikkatur, A. P., Inouye, S., Miesner, H.-J., Stenger, J., and Ketterle, W., Phys. Rev. Lett. **80**, 2027 (1998).
5. Inouye, S., Andrews, M. R., Stenger, J., Miesner, H.-J., Stamper-Kurn, D. M., and Ketterle, W., *Nature* **392**, 152 (1998).
6. Ketterle, W., private communication (1998).
7. Stenger, J., Inouye, S., Stamper-Kurn, D. M., Miesner, H.-J., Chikkatur, A. P., and Ketterle, W., preprint (1998).
8. Elbs, M., Laue, T., and Tiemann, E., private communication (1998).
9. Arndt, M., Ben Dahan, M., Guery-Odelin, D., Reynolds, M. W., and Dalibard, J., *Phys. Rev. Lett.* **79**, 625 (1997).
10. Söding, J., Guery-Odelin, D., Desbiolles, P., Ferrari, G., and Dalibard, J., *Phys. Rev. Lett.* **80**, 1869 (1998).
11. Leo, P. J., Tiesinga, E., Julienne, P. S., Walter, D. K., Kadlecek, S., and Walker, T. G., *Phys. Rev. Lett.* **81**, 1389 (1998).
12. Marinescu, M., Babb, J. F., and Dalgarno, A., *Phys. Rev. A* **50**, 3096 (1994).
13. Mies, F. H., Williams, C. J., Julienne, P. S., and Krauss, M., *J. Res. Natl. Inst. Stand. Technol.* **101**, 521 (1996).
14. Dalibard, J., private communication (1998).
15. Thorsheim, H. R., Weiner, J., and Julienne, P. S., *Phys. Rev. Lett.* **58**, 2420 (1987).
16. Band, Y. B., and Julienne, P. S., *Phys. Rev. A* **51**, R4317 (1995).
17. Fioretti, A., Comparat, D., Crubellier, A., Dulieu, O., Masnou-Seeuws, F. and Pillet, P., *Phys. Rev. Lett.* **80**, 4402 (1998).
18. Nikolov, A. N., Eyler, E. E., Wang, X., Wang, H., Li, J., Stwalley, W. S., and Gould, P. L., Poster B33, International Conference on Atomic Physics, Windsor, Ontario (1998).
19. Vardi, A., Abrashkevich, D., Frishman, E., and Shapiro, M., *J. Chem. Phys.* **107**, 6166 (1997).
20. Burnett, K., Julienne, P. S., and Suominen, K.-A., *Phys. Rev. Lett.* **77**, 1416 (1996).
21. Bohn, J. L., and Julienne, P. S., *Phys. Rev. A* **54**, R4637 (1996).
22. Julienne, P. S., Burnett, K., Band, Y. B., and Stwalley, W. C., *Phys. Rev. A* **58**, R797 (1998).
23. Kokkelmans, S. J. J. M. F., Verhaar, B. J., and Gibble, K., *Bull. Am. Phys. Soc.* **43**, 1327 (1998).
24. Burke, Jr., J. P., Bohn, J. L., and Greene, C. H., private communication (1998).
25. Wang, H., Gould, P., Stwalley, W. C., private communication (1998).
26. Weinstein, J. D., deCarvalho, R., Kim, J., Patterson, D., Friedrich, B., and Doyle, J. M., Poster B21, International Conference on Atomic Physics, Windsor, Ontario (1998).
27. Weinstein, J. D., deCarvalho, R., Guillet, T., Friedrich, B., and Doyle, J. M., Poster B22, International Conference on Atomic Physics, Windsor, Ontario (1998).

BEC: the Alkali Gases from the Perspective of Research on Liquid Helium

Anthony J. Leggett

Department of Physics
University of Illinois at Urbana-Champaign
1110 West Green Street
Urbana, IL 61801-3080

Abstract.
I consider the analogies and differences between, on the one hand, the BEC alkali gases and, on the other, the superfluid phases of the two helium isotopes, ^{4}He and ^{3}He, with particular reference to the phenomena of metastable superflow in ^{4}He and the "internal" Josephson effect in ^{3}He.

INTRODUCTION

Very soon after the experimental discovery in 1938 of superfluidity in ^{4}He-II, that is liquid ^{4}He below the lambda-temperature (2.17 K at SVP), it was suggested by Fritz London that this temperature simply marked the onset of Bose-Einstein condensation (BEC) in the liquid, and that the superfluid properties of the He-II phase were due to the existence of a Bose condensate. Over the sixty years since London's suggestion, at the hands of London himself, Tisza, Landau, Bogoliubov, Feynman and others, a well-developed theory of superfluidity has taken shape on the basis of the idea of BEC, and London's hypothesis has long been almost universally accepted in the condensed-matter community. However, despite the widespread confidence that BEC is indeed occurring in He-II, it has proved extremely difficult to establish this directly. The most promising method of detecting the existence of a condensate, that is of a macroscopic occupation of a single one-particle state, was long thought to be high-energy neutron scattering [1]; however, in real life this technique is complicated by problems of experimental resolution and final-state effects, and a recent review [2] by one of its chief practitioners concludes that the data, while undoubtedly consistent with a condensate fraction of approximately 9%, does not unambiguously establish it. Very recently, following direct measurements of the condensate in the alkali gases, Wyatt [3] has re-examined his earlier data on the spectrum of atoms evaporated from the surface of liquid He-II under the impact of a pulse of phonons, and concludes that these indicate the existence of a condensate. However, there are no analogs, for He-II, of the in-situ measurements of the condensate profile and magnitude now achieved [4] [5] in the alkalis; the reason for this state of affairs will become clear below.

The above history is probably familiar to most people in the audience. What may be less familiar to some is the fact that the heavy (mass-4) isotope of liquid helium is not the only one to show "BEC"—at least if the definition of the latter is generalized in a natural way. In fact, while the atoms of the lighter isotope (^{3}He) obey Fermi rather than Bose statistics and the system thus cannot undergo BEC in the original literal sense, it too undergoes a transition to an apparently superfluid phase (or rather phases) below about 3 mK, and a natural representation [6] of these phases is in terms of formation of *pairs* of fermions (Cooper pairs), which when regarded as single compound objects essentially obey Bose statistics, accompanied by their immediate Bose condensation. Although direct evidence for this (pseudo-) BEC (in the sense of macroscopic occupation of a single state in *momentum* space) is even harder to come by than in ^{4}He (and even the evidence for superfluidity is considerably more circumstantial), there is a sense in which the onset of BEC is manifested quite spectacularly in this system; I return to this below. Moreover, as we shall see, it displays certain phenomena, notably an "internal Josephson effect," that are very closely analogous to some predicted (though not yet seen) in the BEC alkali gases. It is for this reason that I have used in my title the words "liquid helium" rather than "liquid ^{4}He."

COMPARISON WITH BEC IN ALKALI GASES

Anyone reviewing the history of the last six decades of experimental work on He-II—let alone on the superfluid phases of ^{3}He—and that of the last three years on the BEC alkali gases, might well conclude at first sight that the two systems have virtually nothing in common. However, it is important to appreciate that the apparent glaring differences have little to do with the basic physical principles involved—which are essentially the same in the two cases—and much to do, first, with the very different regimes in parameter space that they occupy and, second and perhaps even more important, with the very different situation with respect to control and diagnostic techniques. I will now try to list some of these differences and briefly discuss their implications, concentrating in the case of liquid helium for the moment principally on ^{4}He.

The most glaringly obvious difference between liquid He-II and the BEC alkali gases relates to the density and hence to the relative effect of interactions. Typical atomic densities n of bulk liquid He are in the range $2 - 3 \times 10^{22}$ atoms/cc; thus the mean interatomic spacing is of order $3 - 3.5 \text{Å}$, and since the hard-core radius of the He atom is about 2.5Å and the minimum of the van der Waals potential is at about 2.8Å, this means that the atoms feel one another's potential fields virtually all the time. In fact, if one could meaningfully introduce the concept of an *s-*

wave scattering length a_s for liquid He, the value of the dimensionless number na_s^3 would certainly be of order 1. By contrast, as you all know, the maximum density attained to date in the alkalis is of order 10^{15} atoms/cc, and even though the s-wave scattering length may be much greater than the atomic "size," the parameter na_s^3 is never much greater than about 10^{-4} and may be considerably smaller than this. It should be mentioned that a situation not too different from this can in fact be attained with ^{4}He atoms, if we study, as done by Reppy and co-workers, not the bulk liquid but a submonolayer coverage of a large surface such as is provided by the pores in various kinds of porous glass such as Vycor; see Ref. [7]. In this case the "gas" of helium atoms is of course two- rather than three-dimensional, and considerations connected with the topology of the relevant surfaces may play an important role.

One immediate consequence of this state of affairs relates to the theoretical tractability of the interatomic interactions. In the case of bulk liquid ^{4}He, no systematic expansion in a small parameter is available, and thus one is faced with the choice of doing numerical computations for a relatively small number of particles (see e.g., Ref. [8]) or making phenomenological *Ansätze* for the many-body wave function whose accuracy is often debatable. By contrast, there is considerable confidence in the theoretical community working on the BEC alkali gases that perturbation theory in the interatomic interaction (or more precisely in the renormalized interaction expressed by the s-wave scattering length) will converge rapidly and give quantitatively reliable results, and the comparisons so far carried out of the results of such calculations with experiments would seem consistent with this optimism. One particular consequence of the weakness of the interactions in the BEC gases is that it is usually not necessary to distinguish between the density of condensate (n_o) and the superfluid density (ρ_s)—a distinction that is very important in the case of liquid He. On the experimental side, the small effect of the interatomic interactions in the BEC gases allows a very high compressibility and hence a sensitivity to external perturbations which is many orders of magnitude greater than that of liquid He; more on this below.

A second important difference relates to the nature of the "confinement" in the two cases. In the case of liquid helium this must be achieved by a physical (solid) vessel, whereas in the case of the BEC alkali gases it is achieved by magnetic and, so far less commonly, laser trapping. This difference has at least two important consequences. In the first place, in the case of liquid helium the potential energy $V(\mathbf{r})$ felt by an atom is very weakly varying in the "free space" inside the vessel and then rises very sharply at the walls (over a distance of the order of a few angstroms) to an effectively infinite value. Combined with the very low compressibility of ^{4}He (by "gas" standards!) this means that the profile of the liquid is very rigid and in particular insensitive to the onset of BEC.

A further consequence of the differences in both density and confinement mechanism between liquid He and the BEC alkali gases is the very different role played by the kinetics and by conservation laws. In the case of liquid helium, the atoms are continually jostling up against one another's potential hard cores, and in so far as we can define a "collision time" at all it would be of the order of 10^{-12} s, many orders of magnitude smaller than any characteristic time related to the macroscopic behavior (say, the time taken by a sound wave to propagate across the sample). By contrast, the characteristic time for an alkali gas atom to undergo an irreversible collision process with a second atom may well be long compared to its transit time across the potential well. This has an immediate consequence for nonequilibrium processes such as the decay of a metastable (e.g., circulating, see below) state by thermal activation. In the case of liquid He, the kinetics is so fast that the rate of any such process is overwhelmingly determined by energetic considerations, and kinetic effects enter at most as a small correction. In the alkalis quite the opposite may be the case, i.e. the kinetics itself may be the main bottleneck. A further difference relevant to such processes is that in the case of helium, unless the container is prepared with quite extraordinary care (something which is virtually never done in practice!) its surface will carry all sorts of rugosities, phonons, etc., as a result of which the system-container interaction fails to conserve angular momentum, energy or indeed just about anything other than particle number; thus, the overwhelmingly natural description of the system in statistical mechanics is by a canonical ensemble. By contrast, in the absence of a deliberate choice by the experimenter, the interaction with the container "walls" or indeed with the outside world generally, of an alkali gas in a magnetic or laser trap is extraordinarily weak, and it is entirely arguable that the more natural description is by a microcanonical (constant-energy) ensemble. (This conclusion strictly speaking would hold only for a trap so deep that evaporation is negligible, and in the absence of appreciable recombination; however, it is clear that even for a realistic experimental situation the difference with the case of helium is dramatic).

To my mind, however, the single most dramatic difference between the alkali gases and liquid helium—and certainly the one which, more than anything else, is likely to permit (to an extent has indeed already permitted) a whole class of experiments to be done which have not even a qualitative analog in helium—is the availability, in the case of the alkalis, of a whole set of manipulation and detection techniques which can be finely tuned not only in space but, even more importantly, in time. As we all know, laser beams are used on an entirely routine basis not only to cool the alkali gases but to provide "interesting" potential wells, to move the atoms around and to image them. Why are similar operations not routinely done with liquid helium? The answer is very simple and, alas, apparently at least at present ineluctable: Almost all these applications require relatively high-intensity and monochromatic light beams operated close to the fundamental resonance, and for the He atom the Lyman-α line lies at around 20 eV, a region of the spectrum in which at present no practical (portable) lasers exist! Thus any optical investigations of liquid helium must be carried out in the weak-field limit, and the results of such investigations have been, to put it mildly, unspectacular; in particular, no dramatic change, and in fact very little change at all, is seen at the superfluid transition.

However, the feature of the laser-induced potential which has been really crucial to novel types of experiments in the BEC alkali gases is not so much its strength, but the fact that it can be switched on and off in a short time (presumably in effect of the order of the inverse linewidth of the principal transition, namely $\sim 10^{-7}$ s), which is smaller by several orders of magnitude than the characteristic timescales of the intrinsic system dynamics (e.g., the period of oscillation in the harmonic well, which is typically at least of the order of 1 ms). Thus, for example, the effective potential in which the system moves can be changed effectively "instantaneously," and moreover without producing appreciable unwanted effects such as heating. No such possibility is realistic for liquid helium; while it certainly is possible to change thermodynamic parameters such as pressure over a timescale ($\sim$ 1 ms) which is comparable to (though not much smaller than) characteristic dynamical timescales, attempts to do so in the existing literature (as e.g., in Ref. [9]) have usually resulted in the production of much unwanted noise, etc. In addition, as we have already seen, any such "manipulation" of the potential, etc. is much less specific in liquid helium. As a result, it seems very doubtful that the literal analog of (e.g.) the experiment of Ref. [10] would ever be practical in that system.

A final difference between the alkali gases and liquid ^{4}He is of course the existence in the former of an "internal" degree of freedom, namely the hyperfine-Zeeman index. However, while there is indeed no analog of this in ^{4}He, the lighter isotope, ^{3}He, possesses a nuclear spin of 1/2, and, more importantly in the present context, the complexes that undergo (pseudo-) BEC in that system, namely the Cooper pairs, themselves possess a non-zero total spin. Consequently, while some phenomena associated with the hyperfine degree of freedom in the alkali gases, such as the (expected) "internal" Josephson effect, have no obvious analog in ^{4}He, there do exist rather close analogs in the superfluid phases of ^{3}He; cf. below.

RELEVANCE OF LIQUID-HE RESEARCH TO ALKALI BEC

Given the above major differences between liquid helium and the BEC alkali gases as regards both the physical parameter regime and the kinds of questions that can feasibly be investigated experimentally, this audience might perhaps be tempted to ask whether there is really anything much they can learn from the research of the last sixty years on the former system. I think the answer is certainly yes, for both obvious reasons and some rather more subtle ones. As regards the obvious reasons, the first is that while liquid ^{4}He is certainly nothing like a weakly interacting Bose gas, attempts to understand the qualitative features of its behavior have led, over the years, to extensive studies of the latter system, in particular the classic early work of Bogoliubov and its extension in the 50's and 60's by Yang, Lee, Huang, Girardeau and others. Although the systems explicitly considered in this work are almost invariably translation-invariant, the extension to the case, realistic for the alkali gases, of a finite trap is relatively straightforward, and this old work forms the conceptual underpinning of our current quantitative understanding of the static properties of these systems.

With regard to the dynamic properties, a little more caution is necessary. It is by now widely appreciated that at zero temperature the Gross-Pitaevskii (GP) equation (nonlinear Schrödinger equation, Hartree equation for condensed bosons) which was originally proposed as a qualitative description of the dynamics of liquid ^{4}He, actually gives a rather good *quantitative* account of the condensate dynamics in a dilute BEC alkali gas. The degree of quantitative agreement is actually rather surprising (to me at least), and I suspect may reflect the fact that the situation mostly studied in existing experiments, namely one in which the external potential is nearly harmonic and one is well into the "Thomas-Fermi" limit, may be actually rather pathological from the point of view of the generic problem: in particular it seems that the real excitation of quasiparticles out of the condensate, which would automatically invalidate the use of the GP equation, is probably very much less than one would expect in the generic case. Be that as it may, at finite temperature the GP equation, to the extent that it remains valid, needs to be supplemented with a description of the dynamical behavior of the normal (noncondensed) component, and here it is already less obvious that standard results from the theory of ^{4}He can be taken over as they stand. The most obvious difference between the two systems in this context (apart from geometry) is that most of the experiments on superfluid ^{4}He which can be compared quantitatively with the theory are performed in the hydrodynamic limit, where the frequency of oscillation is much smaller than the rate, e.g., of conversion, by collisions, of condensate into normal component. By contrast, the collective excitations observed experimentally in the alkali gases may, depending on temperature, density and well parameters, lie in either the hydrodynamic or the opposite, "collisionless" limit, or somewhere in between; correspondingly, the details of the kinetics of (*inter alia*) normal-condensate conversion, which has a relatively minor effect in ^{4}He, may here play a major role. In view of Keith Burnett's talk at this conference I do not discuss this further.

Let me now turn to some rather less obvious reasons why I believe that existing research on the helium liquids may be relevant to the study of the alkali gases. Before doing so, it may be helpful to make a general remark about the field of helium research. As we have seen, there is a whole class of experiments, involving spatial interference on a microscopic scale, that have been or can be conducted in the alkali gases and have no obvious practical analog in liquid helium. On the other hand, there is also a class of experiments, involving in one sense or another interference on a *macroscopic* scale, that have been the motivation for much of all the research conducted on liquid helium but that are yet to be observed in the alkali gases. Such phenomena include the two major manifestations of "superfluidity," namely the Hess-Fairbank effect (see below) and the metastability of superflow, and, later in the historical sequence, the "external" Josephson effect in both helium isotopes and the "internal" effect in superfluid ^{3}He. (I do not include "vortices" in the list, despite the popularity of this topic in the helium literature, because I believe it is most naturally regarded as a special case of the phenomenon of metastable superflow). In trying to explain these phenomena, condensed matter physicists have accepted from an early stage that the strength of the inter-atomic interaction and the unknown details of, e.g., the precise potential exerted by the container is likely to make even an approximate solution of the many-body Schrödinger equation problematic, and they have therefore tended to concentrate on phenomenological *Gestälte* and on qualitative and/or topological arguments whose validity is independent of the microscopics. The overall attitude to what constitutes a satisfactory "explanation," etc., that results from this approach is, of course, very different from that which has traditionally been embraced by practitioners of atomic physics or quantum optics, fields in which one is used to starting with a reasonably simple and well-isolated system described by a Hamiltonian that is presumed exactly known and making a series of controlled, e.g., perturbative, approximations. What I find fascinating about the BEC alkali gases is that some problems in these systems, and I believe particularly attempts to replicate or expand the kind of macroscopic interference phenomena routinely seen in superfluid helium, seem to lie precisely on the border between the "natural" areas of applicability of the very different philosophies developed in condensed matter physics on the one hand and in atomic physics and quantum optics on the other, and I believe that such problems will be a major area of (hopefully fruitful!) cross-fertilization between the two fields. I now turn to a description of some of these problems in a little more detail.

Historically, the term "superfluidity" was originally used to describe the ability of liquid ^{4}He, when in the so-called He-II phase realized below the λ-point, to flow through narrow capillaries without apparent friction with the walls. From a modern point of view, however, "superfluidity" is a complex of phenomena of which the two most striking are the Hess-Fairbank effect and the metastability of circulating currents. To discuss these, we consider an annular geometry such that the width d is much less than the radius R, and define the classical moment of inertia, $I_{cl} \equiv NmR^2$ (N = no. of atoms) and also a characteristic angular velocity $\omega_c \equiv \hbar/mR^2$ where m is the mass of the ^{4}He atom. The *Hess-Fairbank effect* [11] is, from the theoretician's viewpoint, the analog for an electrically neutral system of the Meissner effect in superconductors. If we fill the annulus with liquid helium at a temperature $T > T_\lambda$ and rotate it with some angular velocity $\omega \ll \omega_c/2$, the helium will behave just like any other liquid, i.e. it will follow the rotation of the container. But if, while continuing to rotate the container, we now lower the temperature through T_λ, the helium will appear to slow down and, as T approaches zero, to come to rest in (approximately) the frame of the laboratory! It is clear that this behavior cannot be explained by the hypothesis that the viscosity vanishes (since the liquid has come *out* of equilibrium with its immediate environment, namely the container walls), but that the non-rotating state must be the *thermodynamic equilibrium* one. Further investigation shows that if the angular velocity ω of the container is larger ($\gtrsim \omega_c$), then the final angular velocity of the helium takes the "quantized" value $n\omega_c$, where n is the nearest integer to ω/ω_c; the behavior described above corresponds of course to the special case $n = 0$.

A phenomenon that at first sight resembles the Hess-Fairbank effect but is actually very different in its conceptual implications is that of *metastable superflow*. Suppose we start above the λ-point and rotate the container with a "large" angular velocity, $\omega \gg \omega_c$. If we subsequently cool, still rotating, through the λ-temperature, then according to the considerations of the previous paragraph the liquid will adopt the quantized value $n\omega_c$ of angular velocity "as close as possible" to ω; for large n the fractional change will be very small and practically unobservable, so that to all intents and purposes the helium will appear to behave like any other liquid. However, if we now stop the rotation of the container, we find that the helium appears to continue to rotate for as long as we care to observe it! More precisely, the behavior is as if some finite fraction of the liquid, which is a function of temperature and tends to 1 as $T \to 0$, continues to rotate while the remainder comes to rest in the laboratory frame; by ramping T up and down (but taking care never to go above T_λ) we can decrease or increase reversibly the "rotating fraction." In this situation there is a rigorous proof that the circulating state cannot be the absolute minimum of the free energy, so we have here an example of a *metastable* state, whose lifetime is, however, greater than that of the Universe.

As remarked earlier, our present theoretical understanding of these phenomena takes as its central premise the hypothesis that the λ-point is simply the temperature of onset of BEC, so that below T_λ a finite fraction of the atoms (not necessarily 100%, even at $T = 0$) are in a *single* one-particle state, the "condensate." It is absolutely crucial to the theory that any attempt to create *more than one* condensate, i.e. to populate more than one single-particle state macroscopically, costs a thermodynamically extensive amount of energy; for a system of spinless atoms such as ^{4}He with repulsive interactions, this crucial feature is guaranteed by the "Fock" term in the mean-field energy i.e., the fact that, e.g., for $k, k' \to 0$, atoms in *different* plane-wave states repel twice as strongly as atoms in *the same* state k.

The currently accepted theory of superfluidity in ^{4}He is actually quite sophisticated, and involves the careful definition of a number of concepts such as superfluid velocity and normal density in a way which is not restricted to a weakly interacting system. However, for the purposes of comparison with the alkalis it is convenient to pretend that liquid ^{4}He were indeed a dilute gas with weakly repulsive interactions. How then would we explain the Hess-Fairbank effect and the metastability of circulating currents?

The explanation of the Hess-Fairbank effect is very straightforward and can be seen by treating the system to a first approximation as a free Bose gas. When the walls rotate, we must do the thermodynamics in the rotating frame, and in this frame the energy of a single-particle state with angular momentum $\ell\hbar$ and transverse quantum number $n_\perp$ is of the form (for $d \ll R$)

$$E(\ell, n_\perp) = \frac{\ell^2 \hbar^2}{2mR^2} - \ell\hbar\omega + E_\perp(n_\perp), \quad \ell \text{ integral.} \tag{1}$$

In the normal state the particle distribution is a smooth function of E, with a characteristic scale $\sim kT$, and provided only that $kT \gg \hbar^2/mR^2$ (a condition well fulfilled for helium under all realistic conditions) any sums over ℓ may be replaced by integrals, and in particular the total angular momentum is given to a very good approximation by $L = I_{cl}\omega$. The situation changes drastically at the onset of BEC, because now the system must choose a unique value of ℓ (as well as of $n_\perp$) for the condensate, namely that which minimizes the ℓ-dependent term in (1); evidently the correct choice is the nearest integer to ω/ω_c where, as above, $\omega \equiv \hbar/mR^2$. The resulting contribution of the N_c condensate atoms to the angular momentum is, in the limit of the weak interaction, simply $N_c \ell \hbar$, while the normal (uncondensed) particles contribute the smooth function $(1 - N_c/N)I_{cl}\omega$. In particular, for $\omega \ll \omega_c/2$ the condensate contribution vanishes and the total angular momentum is reduced from its normal state value by the factor $(1 - N_c/N)$ (ρ_n/ρ in the standard helium notation), in agreement with experiment. Note that, provided the interactions are repulsive, they strengthen rather than spoil this argument and in particular ensure, by the argument above about "double condensates," that the transition between different values of ℓ is discontinuous rather than smooth. (The case of attractive interactions is more interesting in this respect and needs separate consideration).

The argument regarding the metastability of circulating currents is a little more subtle. Suppose for simplicity that the helium has initially been set into rotation above the λ-point at an angular velocity close to ω_c, and then cooled through the transition, so that the relevant value of ℓ taken up by the condensate is 1 (p-state); the rotation of the container is then stopped, and the experimental observation is then that the p-state has an astronomical degree of metastability. It is clear that this phenomenon, unlike the HF effect, cannot be understood by simple analogy with atomic physics. For consider an electron in an atom in a p-state. There are two obvious ways for it to return to the ground (s-) state, which in practice would be realized respectively by induced and spontaneous emission processes (see below for the relevant sense of "induced"):

(a) It can remain in (approximately) a pure state but allow this to evolve smoothly from a p-state to an s-state, i.e. the wave function $\psi(\theta, t)$ is schematically of the form

$$\psi(\theta, t) = a(t)\psi_p(\theta) + b(t)\psi_s(\theta),$$

where $\mid a(t) \mid^2 + \mid b(t) \mid^2 = 1$, and $b(t)$ increases smoothly from 0 at $t = 0$ to 1 at $t = \infty$. I call this the "coherent path."

(b) It can go through a *mixed* state, such that at time t the diagonal elements of the density matrix in the (s, p) representation are given by

$$\rho_{pp}(t) = \mid a(t) \mid^2, \quad \rho_{ss}(t) = \mid b(t) \mid^2$$

with $a(t)$ and $b(t)$ given as above, but the off-diagonal elements $\rho_{sp}(t), \rho_{ps}(t)$ are zero. I call this the "Fock path." It is clear that in the case of the BEC gas there are exact analogs of each of these processes: in the "coherent" case the many-body wave function is, schematically, simply the expression (2a) raised to the N-th power, while in the "Fock" case the density matrix evolves in such a way that at time t there are approximately $N_p(t)$ atoms definitely in the p-state and $N_s(t) \equiv N - N_p(t)$ definitely in the s-state, where $N_p(t) \cong N \mid a(t) \mid^2$. I.e., to the extent that we can neglect the difference of $N(t)$ from an integer, the Fock path is *schematically* represented by the many-body wave function (apart from normalization)

$$\psi(t) = (a_p^+)^{N_p(t)} (a_s^t)^{N_s(t)} |0\rangle ,\qquad(3)$$

where $|0\rangle$ indicates the vacuum.

Now it is clear from the linearity of the Schrödinger equation that both for the single electron, and for the BEC gas in the absence of interactions, the expectation value $E(t)$ of the (system) energy relative to the groundstate is simply of the form $(E_p - E_s) \mid a(t) \mid^2$, and is therefore a monotonically decreasing function of time. There is therefore no energy barrier against decay of the p-state, and the free Bose gas should *not* be able to sustain metastable circulating currents, any more than can a single electron in an atom.

Repulsive interactions, however, change the situation qualitatively. This is easily seen as regards the Fock path: the Hartree term in the interaction energy is insensitive to the state, but the Fock term contributes an extra repulsive energy of the form $V_0 N_p(t) N_s(t)$, and provided that V_0 is greater than $\hbar^2/NmR^2$ (a condition well fulfilled for helium in practical geometries) this provides an energy barrier, which is extensive in N, against the decay of the p-state. A similar remark applies to the coherent path; there is a repulsive term that, locally, is proportional to $\mid \psi_p(\theta) \mid^2 \mid \psi_s(\theta) \mid^2$, and thus does not give zero when averaged over θ, and in fact the dependence of $E(t)$ on $N_p(t) \equiv N \mid a(t) \mid^2$ along the coherent path turns out to be identical to that along the Fock path, so that in particular the energy barrier is the same. We note for future reference that this identity applies only in rather simple geometries such as the uniform annular one we are considering: in the generic case the barrier along the coherent path is lower than that along the Fock path. Also, it is worth noting that while the above argument, which is based in effect on a simple "Gross-Pitaevskii" description of the coherent state, gets things qualitatively right, it does not quantitatively reproduce the critical velocity for superflow which was originally derived by Landau [12] and is believed to be the physically correct result.

Let's now consider the application of these ideas to the BEC alkali gases; we assume that in this case a suitably "rotating" state can be achieved, above T_c, over a time scale short compared to the sample lifetime, for example by rotating an appropriate weakly asymmetric magnetic trap configuration. (In view of the long collision times and the "non-physical" nature of the confinement, this assumption is not entirely trivial). As regards the HF effect, the argument given above for helium should essentially go through; in particular, the ratio $(\hbar^2/mR^2)/kT$, though larger than for helium, is small enough that above the condensation temperature the system should still behave "normally." The actual *kinetics* of the process by which the system achieves its new equilibrium in the BEC phase is however a more delicate question, and is only just beginning to be studied [13].

As regards the question of metastability of supercurrents, there are several complications. In the first place, as was shown by Ho [14] many years ago in the context of spin-polarized hydrogen, any degeneracy associated with an internal (e.g., Zeeman) degree of freedom immediately invalidates the above argument for metastability. This is most easily seen by tracing out the "Fock" path, but associating with the particles in the s-state an internal degree of freedom *different* from the original one of the particles in the p-state. Because the internal degree of freedom now in effect "tags" the two sets of particles, they are distinguishable and there is no Fock term in their interaction and thus no energy barrier against decay, (at least in the approximation used above). Ho [14] gives a parallel argument for the coherent path. Of course, this complication could be removed by application of a magnetic field sufficient to produce a Zeeman splitting large compared to $\hbar^2/mR^2$—a not very stringent condition.

A more intriguing question is the following: The established theory [15] of the decay, due to thermal fluctuations, of circulating supercurrents at velocities close to critical in superfluid ^{4}He always assumes that the system follows the *coherent* path, i.e. that we always deal with a *single* condensate the detailed form of whose wave function evolves continuously in time from the metastable state to a more stable one. But is it in fact obvious that the kinetics will allow this, or is the system constrained to follow, rather, what we have called the Fock path? In the case of an electron in an atom, we have seen that, crudely speaking, the coherent path is followed in induced radiation processes, i.e. these in which the electromagnetic field can be treated as a c-number, but the Fock path where the decay is spontaneous. Since in the case of the macroscopic Bose condensate there is no obvious "classical field" around, one might at first suspect that the second option is the appropriate one. As noted above, this question probably has no substantial experimental consequences in the case of a uniform geometry, but for anything more complicated, such as the geometries in fact likely to be used if the experiment is actually done on the alkali gases, the energy barriers are different on the two paths and, since these usually enter in exponential form, the consequences may be substantial. It is amusing that while, as remarked, the traditional theory of superfluidity in ^{4}He has implicitly assumed without argument that it is the coherent path that is followed, recent treatments of related problems in the BEC in alkali gases (e.g., Ref. [16]) have tended to use without argument the master equation, which is equivalent to assuming the exact opposite! Hopefully there is scope for some cross-fertilization here, and the outcome may affect our thinking about superfluid ^{4}He as well as about the alkalis.

Finally, I would like to say a word about another problem in which one can draw a close parallel between phenomena known to occur in superfluid helium and one that some confidently expected to occur in the BEC alkali gases, namely the Josephson effect. The original effect predicted by Josephson in 1962 in the context of superconductivity relates of course to a situation where a (pseudo)-Bose condensate can occupy either or both of two *spatially* separated regions, and there is a weak matrix element for tunneling between them. Such a situation can be and has been realized not only in metallic superconductors but in both isotopes of superfluid liquid helium, and there seems little doubt that it could be realized, by appropriate manipulation of laser barriers etc., also for the BEC alkali gases (cf. Ref. [17]); however, it is probable [18] that the conditions for observation of anything that would be reasonably called a "Josephson effect" are extremely stringent. A much more promising version, for the BEC alkali gases, is one in which the two "bulk" states in question are not spatially distinct but are different hyperfine-Zeeman states, and the coupling is provided by a Raman laser pair or something similar. It does not seem widely known in the BEC community that a situation rather closely analogous to this exists in the superfluid phases of liquid ^{3}He, and has been extensively explored both experimentally and theoretically, so I would like to describe the relevant phenomenon briefly and then compare it with the analogous effects expected in the BEC alkali gases. For a detailed account I refer, e.g., to Ref. [19], ch. 8; here, for simplicity, I concentrate on a special case, namely the so-called longitudinal resonance of the A phase of superfluid ^{3}He.

Consider the generic problem of a Bose condensate of N atoms constrained to occupy one or both of two distinct states 1,2, created (destroyed) by operators $a_j^+(a_j), j = 1, 2$; thus $N \equiv a_1^+ a_1 + a_2^+ a_2$. The Hamiltonian of the system includes *inter alia* a "tunneling" term of the form

$$H_T = -\frac{1}{2}\hbar\omega_R(a_1^+ a_2 + H.c.), \qquad (4)$$

where ω_R is the "Rabi frequency" (I here use deliberately the standard notation of atomic physics). In addition there will be, in general, a "single-particle splitting" term of the form

$$H_v = -\delta(a_1^+ a_1 - a_2^+ a_2) \equiv -\delta \Delta \hat{N} \qquad (5)$$

and also terms of higher order than quadratic in the a's. Of the latter, the most important for our purposes are those which depend only on the quantity $\Delta \hat{N}$; we will keep only these, and moreover make a Taylor expansion around $\Delta N = 0$ and incorporate any linear term in H_v. Thus, the final term we consider is of the form

$$H_c = \frac{1}{2}\kappa\frac{(\Delta \hat{N})^2}{N}, \qquad (6)$$

where the factor of N is introduced for subsequent convenience.

We will assume in the following that κ is positive, although the opposite case is also of interest.

It is convenient to decompose the operators a_j into "number" and "phase" operators (a process which involves some subtleties [20], which however, are believed to have little effect in the limit of large N):

$$a_j \equiv (\hat{N}_j)^{1/2} \exp i\hat{\phi}_j, \quad \Delta\hat{\phi} \equiv \hat{\phi}_1 - \hat{\phi}_2, \quad \Delta\hat{N} \equiv \hat{N}_1 - \hat{N}_2. \qquad (7)$$

The operators $\Delta \hat{N}, \Delta \hat{\phi}$ effectively obey a canonical commutation relation:

$$\left[\Delta \hat{N}, \Delta \hat{\phi}\right] = -2i \tag{8}$$

and in terms of them the complete Hamiltonian we are considering reads:

$$\hat{H} = -N\hbar\omega_R \left(1 - \left(\Delta \hat{N}\right)^2 / N^2\right)^{1/2} \cos\Delta\hat{\phi} - \delta\Delta\hat{N} + \frac{1}{2}\kappa \frac{\left(\Delta\hat{N}\right)^2}{N}. \tag{9}$$

For the case of the BEC alkali gases, this Hamiltonian has the following implementation (in the "rotating" frame): The states 1,2 are different hyperfine-Zeeman states in the same trap, ω_R is the (single-atom) Rabi frequency, δ is the detuning from resonance and κ is related to the difference in effective interactions (scattering lengths) between the different species; specifically, we have

$$\kappa \sim K \left(a_s^{(11)} + a_s^{(22)} - 2a_s^{(12)}\right) \bar{n} \cdot 4\pi\hbar^2/m, \tag{10}$$

where K is a (possibly N-dependent) geometrical factor of order unity and $\bar{n}$ is some reference total density, say at the trap center. In the case of superfluid ^{3}He-A, where the Cooper pairs have a total spin of 1, the states 1,2 correspond to the two different "allowed" spin orientations $S_z = +1$ and $S_z = -1$. (The $S_z = 0$ Zeeman substate is not allowed in the A phase). The quantity δ is then simply the Zeeman energy (if any) in an external magnetic field, and the quantity κ is the inverse of the spin susceptibility per atom. The "tunneling" term is in this case provided by the only term in the original many-body Hamiltonian which fails to conserve $\Delta \hat{N}$ (i.e. the z-component of total spin), namely the nuclear dipole-dipole interaction; strictly speaking this term does not necessarily have the precise $\Delta \hat{N}$-dependence given in (9), but as we shall see this is irrelevant in practice. The "Rabi frequency" ω_R is to a first approximation independent of N.

In analyzing the general behavior of a system described by the Hamiltonian (9) it is convenient to distinguish three different principal regimes, which I have elsewhere [21] suggested should be designated respectively as the "Rabi," "Josephson," and "Fock" regimes. The classification is most clear-cut for the "symmetric" case $\delta = 0$, so I will confine myself to that case here.

(1) The Rabi regime, $\kappa \ll \omega_R$

In this regime the effect of the $\Delta \hat{N}$-dependence in the last term in (9) is negligible compared to that in the coefficient of $\cos\Delta\phi$ in the first term. The dynamics of the Bose condensate is then simply that of a standard two-state system without interactions; this remains true for finite δ.

(2) The Josephson regime, $\omega_R \ll \kappa \ll N^2\hbar\omega_R$

Quite generally, for $\omega_R \ll \kappa$ (whether or not the second inequality is satisfied) the effect of the $\Delta \hat{N}$-dependence of the coefficient of $\cos \Delta\phi$ can be neglected relative to that in the κ-term *provided* $\Delta N/N$ is not too close to 1 and we are not in certain very small regions of the parameter space (for the behavior in those "special" regimes, see Ref. [22]). Under these conditions the Hamiltonian (9), with $\delta = 0$, reduces simply to that of a *simple quantum pendulum*, with $\Delta\phi$ corresponding to the angle of the problem with the vertical and ΔN to its angular momentum. In the "Josephson" limit, $\kappa \ll N^2\hbar\omega_R$, the problem further simplifies to that of the *classical* pendulum (in this limit $\Delta\phi$ and ΔN can be regarded as classical quantities and the commutator in (8) replaced by a Poisson bracket). This "pendulum" analogy has proved extremely useful in visualizing the behavior of the quantity ΔN (the longitudinal component of spin) in ^{3}He-A, e.g., following a sudden stepping of the external magnetic field; see e.g., Ref. [19], section 8.3.3. In particular, one predicts that in the absence of dissipation an initial "large" value of ΔN does not reverse itself, as it would in the Rabi regime, but rather undergoes oscillations of limited amplitude around its original value (just as a violently struck pendulum rotates, speeding up and slowing down slightly, but in the absence of dissipation never comes to rest). This behavior has been amply confirmed experimentally [23]; it is the precise analog of the "self-trapping" effect predicted [24] to occur for a BEC alkali gas.

(3) The Fock regime, $N^2\hbar\omega_R \lesssim < \kappa$.

In this regime the quantum fluctuations of the pendulum around its mean position are of order unity, so that it can no longer be treated classically and, in the extreme Fock limit, the groundstate corresponds to zero angular momentum and thus to a $\Delta\phi$ that is completely undefined. In the alkali-gas case this limit corresponds to having a definite number of atoms in each of the two hyperfine states separately, with no atom "simultaneously" in both states.

An interesting difference between superfluid ^{3}He-A and the BEC alkali gases lies in the regimes that can be explored in the two cases. In the case of ^{3}He-A, under normal conditions one is always in the Josephson regime; the Rabi regime is completely inaccessible, since the nuclear dipole-dipole interaction is fixed and cannot be increased, and the Fock regime, while in principle realizable in very small samples, has probably never been attained in any existing experiment on ^{3}He. (It is, of course, routinely attained in another "BEC" system, namely mesoscopic superconducting grains.) By contrast, in the case of the alkali gases one can trivially obtain the Fock regime simply by turning off any Raman lasers, and there seems no good reason why one should not also obtain the Rabi regime, although at a "typical" condensate density of the order of 10^{15} atoms/cc this may require fairly high laser power. Moreover, the possibility of varying the quantity ω_R with time so as to make transitions between the different regimes leads to a whole series of interesting problems concerning the transient response of the system that have no experimentally realizable analog in superfluid ^{3}He.

Finally, it is worth mentioning that in liquid ^{3}He the expectation value of the nuclear dipole energy, or more precisely of a quantity very intimately related to it, may be measured directly from the transverse NMR spectrum [25]. Since the elementary interaction g_D between two neighboring nuclei is only of order $10^{-7}K$, i.e. many orders of magnitude smaller than the thermal energy ($\sim 10^{-3}$K), we find that in the normal state the expectation value is of order g_D^2/kT and hence almost unobservably small. However, in the BEC state all Cooper pairs have to behave identically, not only as regards their center-of-mass motion but also as regards their spin configuration, and the effect is as if g_D were multiplied by a factor N, thus completely swamping the thermal energy. As a result, the dipole energy takes a value much larger than in the normal phase, and one which is easily observable (for details of the argument, see Ref. [25]. Thus, the spectacular shift of the transverse NMR frequency in ^{3}He-A could be regarded as direct evidence for the onset of BEC! Related effects have been proposed theoretically in the alkali gases (see e.g., Ref. [26] but have so far not been observed (though cf. the elegant results on phase coherence reported in Ref. [27].

This work was supported by the National Science Foundation under grants nos. DMR-96-14133 and PHY94-07194. I am grateful to the Institute for Theoretical Physics, Santa Barbara for hospitality while some of it was being performed, and to Gordon Baym for valuable comments on the draft manuscript.

REFERENCES

1. Hohenberg, P. C., and Platzman, P. M., *Phys. Rev.* **152**, 198 (1966).
2. Sokol, P. E., in *Bose Einstein Condensation*, ed. A. Griffin, D. W. Snoke and S. Stringari, Cambridge UK: Cambridge University Press, 1995.
3. Wyatt, A. F. G., *Nature* **391**, 56 (1998).
4. Andrews, M. R., Mewes, M-O., van Druten, N. J., Durfee, D. S., Kurn, D. M., and Ketterle, W., *Science* **273**, 84 (1996).
5. Hau, L. V., Busch, B. D., Chien, L., Dutton, Z., Burns, M. M., and Golovchenko, J. A., *Phys. Rev.* A**58**, R54 (1988).
6. Leggett, A. J., *Rev. Mod. Phys.* **47**, 331 (1975).
7. Reppy, J. D. *J. Low Temp. Phys.* **87**, 279 (1995).
8. Ceperley, D. M., *Rev. Mod. Phys.* **67**, 279 (1995).
9. Hendry, P. C., Lawson, N. S., Lee, R. A. M., McClintock, P. V. E., and Williams, C. H. D., *Nature* **368**, 315 (1994).
10. Andrews, M. R., Townsend, C. G., Miesner, H.-J., Durfee, D. S., Kurn, D. M., and Ketterle, W., *Science* **275**, 637 (1997).
11. Hess, G. B., and Fairbank, W. M., *Phys. Rev. Lett.* **19**, 216 (1967).
12. Landau, L. D., *J. Phys. (USSR)* **5**, 71 (1941).
13. Jaksch, D., Gardiner, C. W., and Zoller, P., *Phys. Rev. A* **56**, 575 (1997): cf. Kagan, Yu. M., Svistunov, B. V., and Shlyapnikov, G. V., *Zh. Eksp. Teor. Fiz.* **74**, 523 (1992) [*JETP* **75**, 387 (1992)].
14. Ho, T. L., *Phys. Rev. Lett.* **49**, 1837 (1982).
15. Langer, J. S., and Fisher, M. E., *Phys. Rev. Lett.* **19**, 560 (1967).
16. Anglin, J., and Zurek, W. H., quant-ph/9804035.
17. Wallis, H., Röhrl, A., Naraschewski, M., and Schenzle, A., *Phys. Rev. A* **55**, 2109 (1997).
18. Zapata, I., Sols, F., and Leggett, A. J. *Phys. Rev. A* **47**, R28 (1998).
19. Vollhardt, D., Wölfle, P., *The Superfluid Phases of Helium 3*, London: Taylor and Francis, 1990, ch. 8.
20. Carruthers, P., and Nieto, M. M., *Phys. Rev. Lett.* **14**, 387 (1965).

21. Leggett, A. J., *Remarks at ITP BEC Workshop, Santa Barbara, CA*, April 1998 (unpublished).
22. Smerzi, A., Fantoni, S., Giovanazzi, S., and Shenoy, S. R., *Phys. Rev. Lett.* **79**, 4950 (1997).
23. Wheatley, J. C., *Prog. Low Temp. Phys.* **Vol. VIII A**, 1 (1978).
24. Milburn, G. J., Corney, J., Wright, E. M., and Walls, D. F., *Phys. Rev. A* **55**, 4318 (1997).
25. Leggett, A. J., *J. Phys. C* **6**, 3187 (1973).
26. Javanainen, J., *Phys. Rev. A* **54**, 4629 (1996).
27. Hall, D. S., Matthews, M. R., Wieman, C. E., and Cornell, E. A., cond-mat/9805327.

Quantum Communication and Computation

H. Briegel,*† W. Dür,* S. Van Enk,‡ J. I. Cirac,* and P. Zoller*

*Institut für Theoretische Physik, Universität Innsbruck, A-6020 Innsbruck, AUSTRIA
†Departamento de Fisica Aplicada, Universidad de Castilla-La Mancha, 13071 Ciudad Real, SPAIN
‡Norman Bridge Laboratory of Physics 12-33, California Institute of Technology, Pasadena CA 91125

Abstract. We show how to build quantum networks based on atoms trapped in cavities that are connected via optical fibers. This networks can be used for quantum communication and to scale up current models of quantum computers.

INTRODUCTION

Quantum entanglement has been a focus of fundamental debate since the original paper of Einstein, Podolsky and Rosen (EPR) [1] and the work of Bell [2], discussing its implications for fundamental issues related to the concepts of physical reality and locality. Only during the last few years has it been recognized that this feature of quantum mechanics may also have important applications in the fields of communication and computation. In particular, it has been shown that by making use of quantum entanglement one can perform certain computational tasks, such as factoring, much faster than without it. Furthermore, quantum entanglement provides the basis for provably secure secret communication and cryptography.

Bell states (or EPR states) play a central role in quantum communication and computation. These are quantum states of two two-level systems (or qubits) that represent maximally entangled states. Denoting by $|0\rangle$ and $|1\rangle$ the computational basis in a qubit, that is, two orthogonal states of a two-level system, the states

$$|\Phi^{\pm}\rangle = 1/\sqrt{2}(|0\rangle|0\rangle \pm |1\rangle|1\rangle) \tag{1a}$$

$$|\Psi^{\pm}\rangle = 1/\sqrt{2}(|0\rangle|1\rangle \pm |1\rangle|0\rangle), \tag{1b}$$

are the Bell states. Ekert [3] showed that if two distant partners share pairs of entangled particles in one of these states (EPR pairs) they can establish a secret key (sequence of random numbers) by performing certain measurements. Once they share such a secret key, they can utilize it to encode/decode secret measurements without any possibility of eavesdropping by a third partner. Experimental results on quantum key distribution have been reported (see, for example, [4]). Another remarkable application of entanglement in the context of communication is teleportation, in which an *unknown* state of a quantum system may be swapped between distant observers by means of a local measurement. This scheme utilizes Bell states and can also be used for secret communication since one can teleport one message from one place to another without being intercepted by an eavesdropper. Quantum teleportation, as proposed by Bennett *et al.* [5] has quite recently been realized experimentally with polarization states of photons [6,7].

For practical applications, it is crucial how close the experimentally established entanglement between particles come to the ideal quantum case. In general, these correlations are built up via some noisy channel such as an optical fiber through which (pairs of) polarized photons are sent. Due to the noise introduced by the channel, the resulting EPR pair will be imperfect and needs to be described by a density operator rather than a pure state. A central quantity is the *fidelity F* of the state, which measures the overlap of this density operator with the ideal EPR pair. Only if the fidelity is large enough will the thus created pair be useful for genuine quantum communication. For increasing the fidelity of an imperfect pair, procedures have been developed that are summarized under the term 'entanglement purification' [8,9]. Entanglement purification as proposed by Bennett and coworkers, is a scheme for distilling a few pairs of high fidelity (strong correlations, large entanglement) out of many pairs with less entanglement.

In this paper we outline some of the steps needed in creating a distant EPR pair of arbitrarily high fidelity over a noisy channel. The discussion will be illustrated with a quantum optical implementation, although the arguments can easily be adapted to other systems. The scheme combines elements that have been experimentally realized or can be expected to be realized in the near future. For example, remarkable experimental progress has been reported recently on secret key distribution for quantum cryptography [10,11], on teleportation of the polarization state of a single photon [6], and on the creation of entanglement between different atoms [12]. In addition, the first steps towards the implementation of quantum logical operations, which are the building blocks of quantum computing, have been demonstrated [13].

ENTANGLEMENT DISTRIBUTION

Operations for creating entangled states of two distant systems A and B can be divided into two groups: (1) local operations which allow us to manipulate particles locally (in each location separately), and (2) transfer operations for transmitting the states of the particles from one place to the other.

A scheme in which both elements can be realized has been introduced by Cirac *et al.* [14], using long-lived states of atoms as the physical basis of qubits, and photons as a means for transferring these qubits from one atom to another, see Fig. 1. To allow for a controlled transfer of photons, the atoms are embedded in high-finesse optical cavities that are connected by an optical fiber. For every transmission of a photon, appropriately tailored Raman pulses are applied to the atoms at the sending and the receiving time. These pulses map the qubit from the atomic state to a specific photon wave packet and *vice versa*.

This scheme, ideally, realizes the transfer operation

$$[\alpha|0\rangle_A + \beta|1\rangle_A]|0\rangle_B \longrightarrow |0\rangle_A[\alpha|0\rangle_B + \beta|1\rangle_B] \qquad (2)$$

where an unknown superposition of internal states $|0\rangle$ and $|1\rangle$ of atom A is transferred to a distant atom B. The idea is that if atom A is initially in state $|0\rangle_A$, then it does not interact with the laser pulse, and nothing occurs. However, if it is in the state $|1\rangle_A$, then it is transferred to the state $|0\rangle_A$ leaving a photon in the cavity. This photon leaks out of the cavity, and via the fiber enters the second cavity, where induces the transition $|0\rangle_B \to |1\rangle_B$. The shape of the laser pulses are selected so that the photon is not reflected at the second cavity.

If atom A is itself entangled to other atoms in the same cavity or at other nodes of a network, the coefficients α and β in (2) are no longer complex numbers but denote unnormalized states of the other atoms. Thus the transmission (2) can be used not only to transfer single atomic states, but also to transfer *entanglement*. For instance, starting from single-particle states, an EPR pair can be created by a two-step process

$$\frac{1}{\sqrt{2}}[|0\rangle_A + |1\rangle_A]|0\rangle_{A_2}|0\rangle_B$$
$$\downarrow$$
$$\frac{1}{\sqrt{2}}[|0\rangle_{A_2}|0\rangle_A + |1\rangle_{A_2}|1\rangle_A]|0\rangle_B$$
$$\downarrow$$
$$|0\rangle_{A_2}\frac{1}{\sqrt{2}}[|0\rangle_A|0\rangle_B + |1\rangle_A|1\rangle_B] \equiv |0\rangle_{A_2}|\Phi^+\rangle_{AB} \,. \qquad (3)$$

Here, the first arrow refers to a *local* CNOT operation between two atoms A and A_2 in the first cavity. The second arrow transfers the state of A_2 to B, thereby transferring the entanglement between the atoms A and A_2 to an entanglement between atoms A and B. At the end of this composite transformation, the state of the auxiliary atom A_2 is the same as at the start and *factors out*. Thus, an ideal EPR pair $|\Phi^+\rangle_{AB}$ is created between A and B.

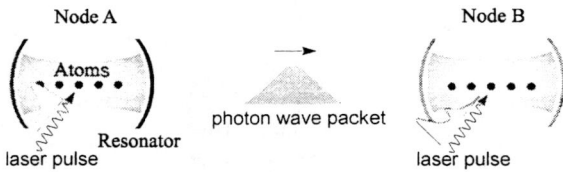

FIGURE 1. Scheme for transferring the state of one atom in the first node (cavity) to a second node.

NOISY CHANNELS AND ERROR CORRECTION

A realistic model must account for imperfections in the transfer of the atomic state from cavity A to B. The optical-cavity–fiber system, together with the laser pulses, constitutes what we abstractly call a noisy quantum channel. For example, photoabsorption and depolarization processes in an optical fiber will lead to transfer errors. Generally speaking, the coupling of the compound atom-cavity-fiber system with the environment, e.g. with the cavity walls, and with the radiation field of the free space results in an entanglement of the final atomic states in (2) and (3) with the environment. Losses may also occur by incoherent scattering on the surface of the cavity mirrors and at the coupling segments between the cavities and the fiber. Examples for local errors are imperfectly designed laser pulses or spontaneous emission processes during the gate operation. Generally speaking, there is a finite probability that the atom in B will not be excited, even though A was excited.

Under these circumstances, it is essential to have a means for *error correction* (for a review, see [15]). In the quantum optical implementation discussed here, an efficient error correction scheme [16] can be realized with the help of auxiliary atoms in both cavities. Before the transmission of a certain atomic state as in (3), the auxiliary atoms in the first cavity are used to *encode* the atomic state into a local three-particle entangled state. After *transmitting* photons to two different atoms on the receiver side, one obtains a certain multiparticle entangled state of 5 atoms in total, distributed over both cavities. By finally *measuring* states of certain atoms in both cavities (the 'syndrome,' in the language of error correction theory), one is able to detect a photon loss without destroying the initial coherence of the atomic state that was to be sent. The transmission may then be repeated.

The result of this whole process is a two-particle entangled state that can be described by a density operator of the form

$$\rho_{AB} = F|\Phi^+\rangle\langle\Phi^+| + \epsilon_1|\Psi^+\rangle\langle\Psi^+| \\ +\epsilon_2|\Phi^-\rangle\langle\Phi^-| + \epsilon_3|\Psi^-\rangle\langle\Psi^-|, \tag{4}$$

which is the realistic version replacing the state $|\Phi^+\rangle$ in the ideal process (3). The states $|\Phi^\pm\rangle = (|00\rangle_{AB} \pm |11\rangle_{AB})/\sqrt{2}$ and $|\Psi^\pm\rangle = (|01\rangle_{AB} \pm |10\rangle_{AB})/\sqrt{2}$ are the four Bell states. For a stationary channel such as the optical fiber, the size of the positive parameters $\epsilon_1, \epsilon_2, \epsilon_3$ in (4) is essentially limited by the imperfection of local gate operations and measurements. The overlap $F = \langle\Phi^+|\rho_{AB}|\Phi^+\rangle = 1-\epsilon_1-\epsilon_2-\epsilon_3$ of the state (4) with the ideal result $|\Phi^+\rangle$ is a measure for the fidelity of the real pair.

There is no room here for giving any details of the procedure that determines the actual values of the parameters ϵ_i in (4). Roughly speaking, the noise introduced in creating (4) multiplies with every imperfect operation that is involved in the process. For a protocol involving 5 imperfect operations, one can expect a fidelity of the order of 95% as long as the error probability for each operation can be controlled on the per cent level.

QUANTUM REPEATERS

When building up quantum correlations via an optical fiber, there is the fundamental problem that the absorption probability for the photon grows exponentially with the length of the fiber. By using the error correction methods described in the previous section, one can, in principle, establish EPR correlations over arbitrary distances. However, since the probability for a transmission error grows exponentially with the distance, so will the average number of repetitions required for a single successful transmission. For distances corresponding to more than a few absorption lengths of the fiber (10 km), the scheme soon becomes impractical [17].

In *classical* digital communication, amplifiers (repeaters) are placed at certain positions in the fiber, at which the signal is restored to its initial strength and shape. In quantum communication, the signals used are states of single photons. Owing to fundamental principles, these states cannot be amplified. In fact, they cannot be genuinely attenuated either, since any photon can only be absorbed as a unity. At any time, therefore, the signal is either still there 'in its full strength,' or it is gone. Guided by the ideas of classical communication, however, we can divide the channel into N segments with connection points (i.e. auxiliary nodes) in between. We then create N elementary EPR pairs of fidelity F_1 between the nodes A & C_1, C_1 & C_2, ... C_{N-1} & B, as in Fig. 2(a). The number N is chosen such that $F_{\min} < F_1 < F_{\max}$, where $F_{\max}$ is the maximal attainable fidelity after purification, and $F_{\min}$ is the minimum fidelity required to be able to purify. Subsequently, we connect these pairs by making Bell measurements at the nodes C_i and classically communicating the results between the nodes as in the schemes for teleportation [5] and entanglement swapping [5]. Unfortunately, with every connection the fidelity F' of the resulting pair will decrease: on the one hand, the connection process involves imperfect operations that introduce noise; on the other hand, even for perfect connections, the fidelity decreases. Both effects lead to an exponential decrease of the fidelity F_N with N of the final pair shared between A and B. Eventually, the value of F_N drops below a certain value $F_{\min}$ and therefore it will not be possible to increase the fidelity by purification (e.g. with the aid of many similar pairs that are constructed in parallel). The only way to circumvent this limitation is to connect a smaller number $L \ll N$ of pairs so that $F_L > F_{\min}$ and purification is possible. The idea is then to purify, connect the resulting pairs, purify again, and continue in the same vein. The way in which these alternating sequences of connections and purifications is done has to be properly designed so that the number of resources needed does not grow exponentially with N and thus with the length l of the channel ($N \propto l$).

Our proposal, the *nested purification protocol* [17], consists of connecting and purifying the pairs simultaneously in the following sense. For simplicity, assume that $N = L^n$ for some integer n. On the first level, we simultaneously connect the pairs (initial fidelity F_1) at all the checkpoints except at $C_L, C_{2L}, \ldots, C_{N-L}$. As a result, we have N/L pairs of length L and fidelity F_L between A and C_L, C_L and C_{2L}, and so on. To purify these pairs, we need a certain number M of copies that we construct in parallel fashion. We then use these copies on the segments A and C_L, C_L and C_{2L}, etc., to purify and obtain one pair of fidelity $\geq F_1$ on each segment. This last condition determines the (average) number of copies M that we need, which will depend on the initial fidelity, the degradation of the fidelity under connections, and the efficiency of the purification protocol. The total number of elementary pairs involved in constructing one of the more distant pairs of length L is LM. On the second level, we connect L of these more distant pairs at every checkpoint C_{kL} ($k = 1, 2 \ldots$) except at $C_{L^2}, C_{2L^2}, \ldots, C_{N-L^2}$. As a result, we have N/L^2 pairs of length L^2 between A and C_{L^2}, C_{L^2} and C_{2L^2}, and so on of fidelity $\geq F_L$. Again, we need M parallel copies of these long pairs to repurify up to the fidelity $\geq F_1$. The total number of elementary pairs involved in constructing one pair of length L^2 is thus $(LM)^2$. We iterate the procedure to higher and higher levels, until we reach the n-th level. As a result, we have obtained a final pair between A and B of length N and fidelity $\geq F_1$. In this way, the total number R of elementary pairs will be $(LM)^n$. We can re-express this result in the form

$$R = N^{\log_L M + 1}, \qquad (5)$$

which shows that the resources grow polynomially with the distance N. A similar formula was obtained in [18] for the overhead required in propagating the concatenated quantum code. Note that R depends only on L and M. In order to evaluate M, one needs to know the specific form of the error mechanisms involved in the purification and connections, which in turn depend on the specific physical implementation of the quantum network. In general, we have only limited knowledge of these details. In order to estimate M, we have chosen a generic error model for imperfect operations and measurements. The details of this error model will be given elsewhere.

Figure 3 shows the curves for connection and purification for a certain set of parameters in the specific error model here considered. The purification curve has two intersection points with the diagonal, which are the fixed points of the corresponding map. The upper point, $F_{\max} < 1$ is an attractor and gives the maximum value of the fidelity beyond which no pair can be purified. Note also the existence of the minimum value $F_{\min} > 1/2$. Together, they define the interval within which purification is possible. The connection curve, which looks like a simple power in Fig. 3, stays below the diagonal for all values of F between $1/4$ and 1. The offset of this curve at $F = 1$ from the ideal value $F' = 1$ quantifies the amount of noise that is introduced through imperfect operations in the connection

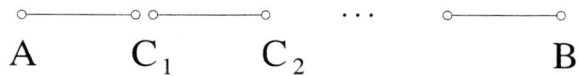

A C_1 C_2 B

FIGURE 2. EPR pairs of atoms are simultaneously created across different segments of a compound fiber. After connecting the pairs, one obtains a single distant EPR pair between A and B.

process.

With the above results, we can now analyze the nested purification protocol. Let us consider a given level k in this protocol, where we have N/L^{k-1} pairs of fidelity

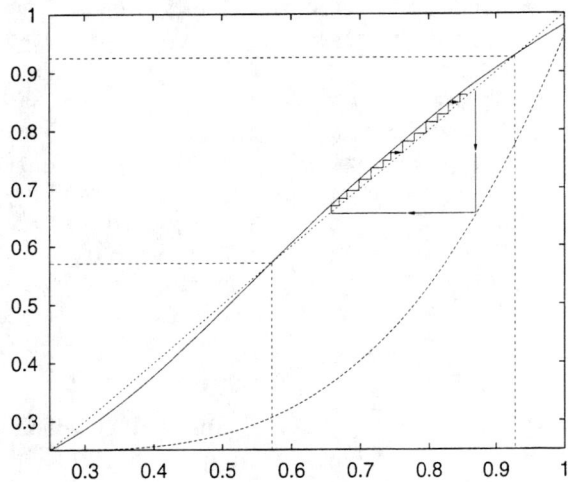

FIGURE 3. Typical purification loop for connecting and purifying EPR pairs.

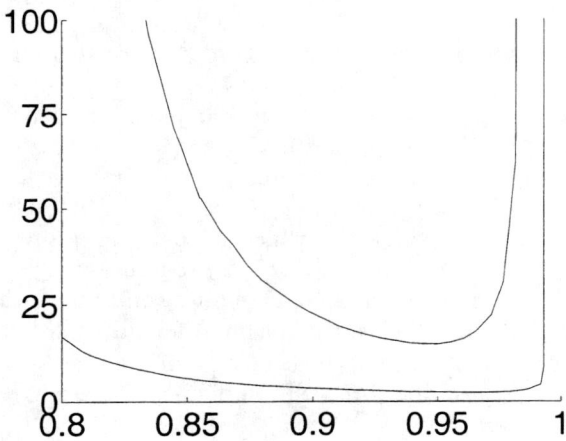

FIGURE 4. M (see text) versus working fidelity F. (a) Realization of the repeater with the aid of the purification schemes of Bennett *et al.* (upper curve) and Deutsch *et al.* (lower curve). The error probabilities of all operations are 0.5% (error parameters 0.995), and $L = 2$. (b) Lower curve in (a) for different error probabilities. From bottom to top: 0, 0.25, 0.5, 0.75, 1 percent probability

F each. The two-step process of connection–purification can now be visualized as follows (see Fig. 3). Starting from F, the fidelity F_L after connecting L pairs can be read off from the curve below the diagonal. Reflecting this value back to the diagonal line, as indicated by the arrows in Fig. 3, sets the starting value for the purification curve. If F_L lies within the purification interval, then iterated application of the purification map leads back to the initial value F (staircase). Once the initial value F is reobtained, we have N/L^k pairs and can start with the level $k+1$. In summary, each level in the protocol corresponds to one cycle in Fig. 3. Note that if, in the loop, $F_L \leq F_{\min}$, then purification is not possible. Being polynomial in L, the lower curve gets steeper and steeper near $F = 1$ for higher values of L. From this, one sees that for a given starting fidelity F, there is a maximum number of pairs one can connect before purification becomes impossible. For the resources we obtain $M = \prod_m^{m_{\max}} 2/p_{\text{succ}}^{(m)}$, where $p_{\text{succ}}^{(m)}$ is the probability for increasing the fidelity in the m-th purification step. The total number of steps, $m_{\max}$, is the same as in the staircase of Fig. 3.

In Fig. 4(a), M is plotted against the working fidelity F. Due to the discrete nature of the purification process, the fidelity of the repurified pairs need not be exactly the same on each nesting level. The working fidelity is thus defined as the fidelity maintained *on average* when going through different nesting levels. The error parameters for this plot are such that the success probability for the measurements and operations is of the order of 99.5%. One can see that there exists an optimum working fidelity of about 0.94 which requires a minimum number of about 15 resources.

A purification protocol that converges faster and therefore involves fewer parallel channels was proposed by Deutsch *et al.* [9]. We have employed this protocol, using imperfect operations. As is demonstrated in Fig. 4(a), M can be reduced by a factor of the order of 10. Since this number has to be taken to the nth power, this reduces the number of total resources by many orders of magnitude. In Fig. 4(b), M is plotted versus the working fidelity for different error parameters. One can see that for errors in the one-per-cent region, a working fidelity can be maintained with on average 5 L-pairs on each nesting level. We note that the procedure also works for error probabilities up to about 3%, but the number of purification resources gets larger.

This work was supported in part by the Austrian Science Foundation, and by the TMR network ERB-FMRX-CT96-0087.

REFERENCES

1. Einstein, A., Podolsky, B., and Rosen, N., *Phys. Rev.* **47**, 777 (1935).
2. Bell, J., *Physics*, **1**, 195 (1964).
3. Ekert, A. K., *Phys. Rev. Lett.* **67**, 661 (1991).
4. Zbinden, H., Gautier, J. D., Gisin, N., Huttner, B., Muller, A., and Tittel, W., *Electronics Lett.* **7**, 7 (1997).

5. Bennett, C. H., Brassard, G., Crepeau, C., Josza, R., Peres, A., and Wootters, W. K., *Phys. Rev. Lett.* **70**, 1895 (1993).
6. Bouwmeester, D., Pan, J.-W., Mattle, K., Eibl, M., Weinfurter, H., and Zeilinger, A., *Nature* **390**, 575 (1997).
7. Boschi, D., Branca, S., De Martini, F., Hardy, L., Popescu, S., *Phys. Rev. Lett.* **80**, 1121 (1998).
8. Bennett, C. H., Brassard, G., Popescu, S., Schumacher, B., Smolin, J. A., and Wootters, W. K., *Phys. Rev. Lett.* **76**, 722 (1996).
9. Deutsch D., Ekert, A., Josza, R., Macchiavello, C., Popescu, S., and Sanpera, A., *Phys. Rev. Lett.* **77**, 2818 (1996).
10. Tittel, W., Brendel, J., Gisin, B., Herzog, T., Zbinden, H., and Gisin, N., quant-ph/9707042 (1997).
11. Buttler, W. T., Hughes, R. J., Kwiat, P. G., Luther, G. G., Morgan, G. L., Nordholt, J. E., Peterson, C. G., Simmons, C. M., quant-ph/9801006 (1996).
12. Hagley, E., Maitre, X., Nogues, C., Wunderlich, C., Brune, M., Raimond, J. M., and Haroche, S., *Phys. Rev. Lett.* **79**, 1-5 (1997).
13. Monroe, C., Meekhof, D. M., King, B. E., Itano, W. M., and Wineland, D. J., *Phys. Rev. Lett.* **75**, 4714 (1995). Turchette, Q. A., Hood, C. J., Lange, W., Mabuchi, H., and Kimble, H. J., *Phys. Rev. Lett.* **75**, 4710 (1995).
14. Cirac, J. I., Zoller, P., Mabuchi, H., and Kimble, H. J., *Phys. Rev. Lett.* **78**, 3221 (1997).
15. Preskill, J., quant-ph/9712027 (1997).
16. van Enk, S. J., Cirac, J. I., and Zoller, P., *Science* **279**, 205 (1998).
17. Briegel, H.-J., Dür, W., Cirac, J. I., and Zoller, P., unpublished (1998).
18. Knill, E., and Laflamme, R., *Phys. Rev. A* **55**, 900 (1997); Knill, E., and Laflamme, R., quant/ph-9608012 (1996).

Laser Manipulation and Cavity QED with Trapped Ions

H. Walther

*Sektion Physik der Universität München and
Max-Planck-Institut für Quantenoptik
D-85748 Garching*

Abstract. In this paper recent experiments with trapped ions are reviewed. They deal with the spectrum of the resonance fluorescence of a single ion, the ion-trap laser, experiments towards a new frequency standard on the basis of a single In^+ ion, and the investigation of the diffusion of a single ion in a one-dimensional optical lattice.

RESONANCE FLUORESCENCE OF A SINGLE ATOM

The resonance fluorescence of an atom is a basic process in radiation-atom interactions, and has therefore always generated considerable interest. The methods of experimental investigation have changed continuously due to the availability of new experimental tools. A considerable step forward occurred when tunable and narrow-band dye-laser radiation became available. These laser sources are sufficiently intense to easily saturate an atomic transition. In addition, the lasers provide highly monochromatic light with coherence times much longer than typical natural lifetimes of excited atomic states. Excitation spectra with laser light using well-collimated atomic beams lead to widths practically equal to the natural width of the resonance transition; with them it became possible to investigate the frequency spectrum of the fluorescence radiation with high resolution. However, the spectrograph used to analyze the re-emitted radiation was a Fabry-Perot interferometer, the resolution of which did reach the natural width of the atoms, but was insufficient to reach the laser linewidth, see e.g. Hartig et al. [1] and Cresser et al. [2]. Considerable progress in this direction was achieved by investigating the fluorescence spectrum of ultra-cold atoms in an optical lattice in a heterodyne experiment [3]. In these measurements a linewidth of 1 kHz was achieved; however, the quantum aspects of the resonance fluorescence such as antibunched photon statistics cannot be investigated under these conditions since they wash out when more than one atom is involved.

Thus the ideal experiment requires a single atom to be investigated. For some time it has been known that ion traps allow one to study the fluorescence from a single laser-cooled particle practically at rest, thus providing an ideal case for the spectroscopic investigation of resonance fluorescence. The other essential ingredient for achieving high resolution is the measurement of the frequency spectrum by heterodyning the scattered radiation with laser light, as demonstrated with many cold atoms [3]. Such an optimal experiment with a single trapped Mg^+ ion is reviewed in this paper. The measurement of the spectrum of the fluorescent radiation at low excitation intensities is presented. Furthermore, the photon correlation of the fluorescent light has been investigated under practically identical excitation conditions. The comparison of the two results shows a very interesting aspect of complementarity since the heterodyne measurement corresponds to a "wave" detection of the radiation whereas the measurement of the photon correlation is a "particle" detection scheme. It will be shown that under the same excitation conditions the wave detection provides the properties of a classical atom, i.e. a driven oscillator, whereas the particle or photon detection displays the quantum properties of the atom. Whether the atom displays classical or quantum properties thus depends on the method of observation.

The spectrum of the fluorescence radiation is given by the Fourier transform of the first-order correlation function of the field operators, whereas the photon statistics and photon correlation is obtained from the second-order correlation function. The corresponding operators do not commute, and thus the respective observations are complementary. The present theory on the spectra of fluorescent radiation following monochromatic laser excitation can be summarized as follows: fluorescence radiation obtained with low incident intensity is also monochromatic owing to energy conservation. In this case, elastic scattering dominates the spectrum, and thus one should measure a monochromatic line at the same frequency as the driving laser field. The atom stays in the ground state most of the time and absorption and emission must be considered as one process with the atom in principle behaving as a classical oscillator. This case was treated on the basis of a quantized field many years ago, e.g. by Heitler [4]. With increasing intensity upper and lower states become more strongly coupled leading to an inelastic component, which increases with the square of the intensity. At low intensities, the elastic part dominates since it depends linearly on the intensity. As the intensity of the exciting light increases, the atom spends more time in the upper state and the effect of the vacuum fluctuations comes into play through spontaneous emission. The inelastic component is added to the spectrum, and the elastic component goes through a maximum where the Rabi flopping frequency $\Omega = \Gamma/\sqrt{2}$ (Γ is the natural linewidth) and then disappears with growing Ω. The inelastic part of the spectrum gradually broadens as Ω increases and for $\Omega > \Gamma/2$ sidebands begin to appear [2,5].

The experimental study of the problem requires, as mentioned above, Doppler-free observation. In order to measure the frequency distribution, the fluorescent light has to be investigated by means of a high-resolution spectrometer. The first experiments of this type were performed by Schuda et al. [6] and later by Walther et al. [7], Hartig et al. [1] and Ezekiel et al. [8]. In all these experiments, the excitation was performed by single-mode dye-laser radiation, with the scattered radiation from a well-collimated atomic beam observed and analyzed by a Fabry-Perot interferometer. Experiments to investigate the elastic part of the resonance fluorescence giving a resolution better than the natural linewidth have been performed by Gibbs et al. [9] and Cresser et al. [2].

The first experiments that investigated antibunching in resonance fluorescence were also performed by means of laser-excited collimated atomic beams. The initial results obtained by Kimble, Dagenais, and Mandel [10] showed that the second-order correlation function $g^{(2)}(t)$ had a positive slope, which is characteristic of photon antibunching. However, $g^{(2)}(0)$ was larger than $g^{(2)}(t)$ for $t \to \infty$ due to number fluctuations in the atomic beam and to the finite interaction time of the atoms [11,12]. Further refinement of the analysis of the experiment was provided by Dagenais and Mandel [12]. Rateike et al. [13] used a longer interaction time for an experiment in which they measured the photon correlation at very low laser intensities (see Cresser et al. [2] for a review). Later, photon antibunching was measured using a single trapped ion in an experiment that avoids the disadvantages of atom number statistics and finite interaction time between atom and laser field [14].

As pointed out in many papers, photon antibunching is a purely quantum phenomenon (see e.g. Cresser et al. [2] and Walls [15]). The fluorescence of a single ion displays the additional nonclassical property that the variance of the photon number is smaller than its mean value (i.e. it is sub-Poissonian) [14,16].

The trap used for the present experiment was a modified Paul-trap, called an endcap-trap [17]. The trap consists of two solid copper-beryllium cylinders (diameter 0.5 mm) arranged collinearly with a separation of 0.56 mm. These correspond to the cap electrodes of a traditional Paul trap, whereas the ring electrode is replaced by two hollow cylinders, one of which is concentric with each of the cylindrical endcaps. Their inner and outer diameters are 1 and 2 mm, respectively, and they are electrically isolated from the cap electrodes. The fractional anharmonicity of this trap configuration, determined by the deviation of the real potential from the ideal quadrupole field, is below 0.1 % (see Schrama et al. [17]). The trap is driven at a frequency of 24 MHz with typical secular frequencies in the xy-plane of approximately 4 MHz. This required a radio-frequency voltage with an amplitude on the order of 300 V to be applied between the cylinders and the endcaps.

The measurements were performed using the $3\,^2S_{1/2} - 3\,^2P_{3/2}$ transition of the ^{24}Mg$^+$ ion at a wavelength of 280 nm. The heterodyne measurement is performed as follows. The dye laser excites the trapped ion while the fluorescence is observed in a direction of about 54° to the exciting laser beam. However, both the observation direction and the laser beam are in a plane perpendicular to the symmetry axis of the trap. A fraction of the laser radiation is removed with a beamsplitter and then frequency shifted (by 137 MHz with an acousto-optic modulator (AOM)) to serve as the local oscillator. An example of a heterodyne signal is displayed in Fig. 1. The signal is the narrowest optical heterodyne spectrum of resonance fluorescence reported to date. Thus our experiment provides the most compelling confirmation of Weisskopf's prediction of a coherent component in resonance fluo-

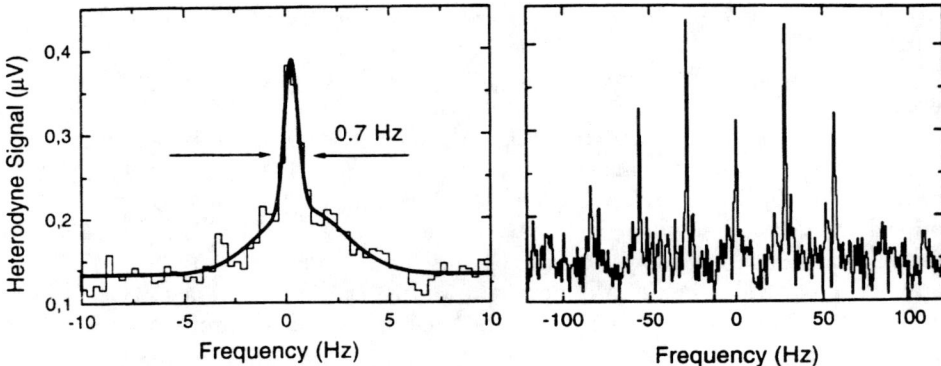

FIGURE 1. Heterodyne spectrum of a single trapped ^{24}Mg$^+$ ion. Left side: Resolution bandwidth 0.5 Hz. The solid line is a Lorentzian fit to the experimental data; the peak appears on top of a small pedestal of 4 Hz width. The latter signal is due to random phase fluctuations in the spatially separated sections of the light paths of the local oscillator and fluorescent light; they are generated by variable air currents in the laboratory. Right side: Heterodyne spectrum of the coherent peak with sidebands generated by mechanical vibrations of the mount holding the trap. The vibrations are due to the operation of a rotary pump in the laboratory. For details see Ref. [19]. Reprinted from [19], with permission from Taylor & Francis.

rescence. The linewidth observed implies that exciting laser and fluorescent light are coherent over a length of 400 000 km. Further details on the experiment are given in Refs. [18] and [19]. The investigation of photon correlations employed the ordinary Hanbury-Brown and Twiss setup. The setup was essentially the same as described by Diedrich and Walther [14]. The results are shown and discussed in Ref. [18] also.

The presented experiment describes the first high-resolution heterodyne measurement of the elastic peak in the resonance fluorescence of a single ion. At identical experimental parameters we have also measured antibunching in the photon correlation of the scattered field. Together, both measurements show that, in the limit of weak excitation, the fluorescence light differs from the excitation radiation in the second-order correlation but not in the first-order correlation. However, the elastic component of resonance fluorescence combines an extremely narrow frequency spectrum with antibunched photon statistics, which means that the fluorescence radiation is not second-order coherent as expected from a classical point of view [20]. The heterodyne and the photon correlation measurement are complementary since they emphasize either the classical wave properties or the quantum properties of resonance fluorescence, respectively.

THE ION-TRAP LASER

There have been several theoretical papers on one-atom lasers in the past [21–25]. This system provides a testing ground for new theoretical concepts and results in the quantum theory of the laser. Examples are atomic coherence effects [26] and dynamic (i.e. self-generated) quantum-noise reduction [27,28,24]. All these aspects are a consequence of a pump process whose complex nature is not accounted for in the standard treatment of the laser. So far there is one experiment where laser action could be demonstrated with one atom at a time in the optical resonator [29]. A weak beam of excited atoms was used to pump this one-atom laser.

A formidable challenge for an experiment is to realize a laser with a trapped ion in the cavity. Mirrors with ultrahigh finesse are required, and a strong atom-field coupling is needed. After the emission of a photon, the ion has to be pumped before the next stimulated emission can occur. Similar to resonance fluorescence experiments that show antibunching [10,14], there is a certain time gap during which the ion is unable to add another photon to the laser field. It has been shown [24] that this time gap plays a significant role in the production of a field with sub-Poissonian photon statistics.

We have investigated the theoretical basis for an experiment of this type. Our analysis takes into account details such as the multi-level structure, the coupling strengths and the parameters of the resonator. It has been a problem to find an ion with an appropriate level scheme. We could show that it is possible to produce a laser field with the parameters of a single Ca^+ ion. This one-atom laser displays several features that are not found in conventional lasers: the development of two thresholds, sub-Poissonian statistics, lasing without inversion, and self-quenching. The details of this work are reported in Ref. [30,31]. In a subsequent paper [32] the center-of-mass motion of the trapped ion is also quantized. This leads to additional features of the ion-trap laser, in particular, a multiple vacuum Rabi splitting has been observed.

The Ca^+ level scheme is sketched in Fig. 2(a). It contains a Λ-type subsystem: the ion is pumped coherently from the ground state to the upper laser level $4P_{1/2}$, and stimulated emission into the resonator mode takes place on the transition to $3D_{3/2}$ at a wavelength of 866 nm. Further pump fields are needed to close the pump cycle and to depopulate the metastable levels.

Although spontaneous relaxation from the upper laser level to the ground state takes place at a relatively large rate of 140 MHz and suppresses the atomic polarization on the laser transition, laser light is generated for realistic experimental parameters due to atomic coherence effects within the Λ subsystem. The occurrence of laser action is demonstrated in Fig. 2 (b) for a resonator with a photon damping rate $A = 1$ MHz and a vacuum Rabi frequency $g = 14.8$ MHz on the laser transition. For the numerical calculation of the realistic scheme, the Zeeman substructure and the polarizations of the fields have to be taken into account. With increasing coherent pump Ω, the mean photon number inside the resonator first increases and then decreases. Both the increase and decease of the intensity are

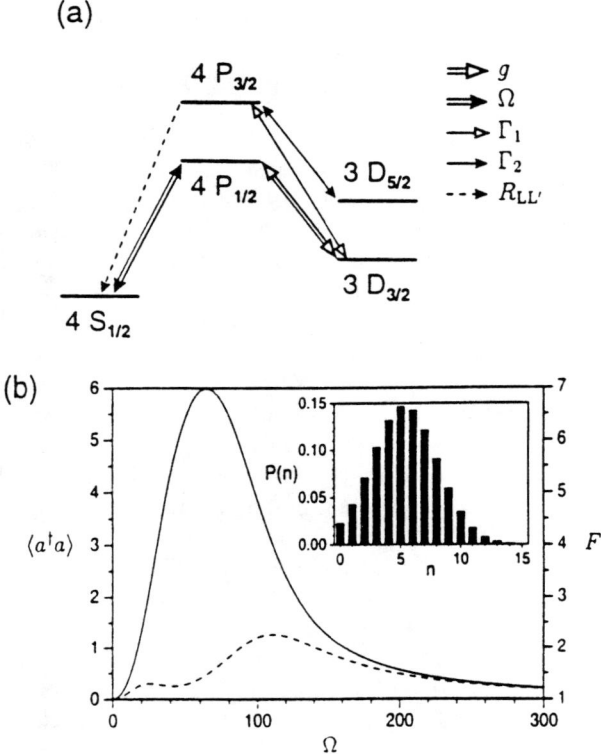

FIGURE 2. (a) Schematic representation of the Ca$^+$ scheme for the ion-trap laser. (b) Mean photon number $\langle a^+a \rangle$ (solid) and Fano factor F (dashed) versus the coherent pump strength Ω. The parameters are $A = 1, g = 14.8, \Gamma_1 = 40$, and $\Gamma_2 = 100$. The inset shows the photon distribution for $\Omega = 50$. All rates are in MHz.

accompanied by maxima in the intensity fluctuations, which can be interpreted as thresholds. Laser action takes place in between these two thresholds. This is confirmed by the Poissonian-like photon distribution given in the inset of Fig. 2 (b). In addition, the linewidth of the output spectrum is up to ten times smaller in the laser region than below the first and beyond the second threshold [31]. Note that for a thermal distribution, the solid and dashed curves in Fig. 2 (b) for the intensity and the intensity fluctuations would coincide.

For a nonvanishing Lamb-Dicke parameter η, higher vibrational states will be excited during the pump and relaxation processes; the amplitude of the atomic motion will increase. Therefore, the ion will in general not remain at an antinode of the resonator mode, and the strength of the atom-field coupling will decrease. However, the atom can be prevented from heating up by detuning the coherent

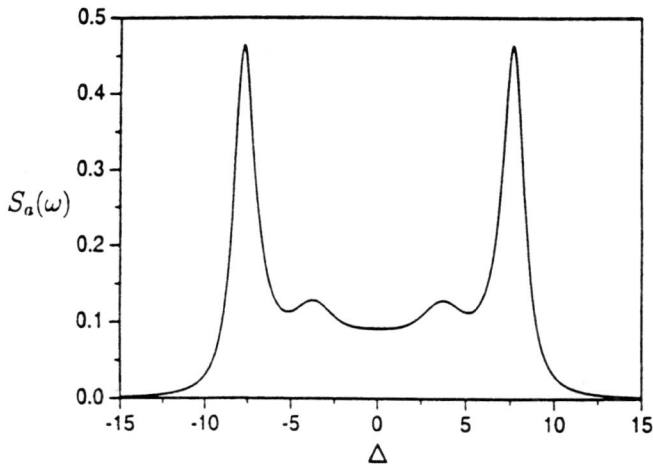

FIGURE 3. Multiple vacuum Rabi splitting in the output spectrum $S_a(\omega)$ for the two-level atom with quantized CM motion. The parameters are $A = 0.1$, $B = 0.05$, $\mu = 0.5$, $R_{AB} = 0.1$, $R_{BA} = 0.001$, and $\eta = 0.7$. All rates are in units of g_0.

pump field. The coupling strength is given by the product of a constant g_0 and a motion-dependent function [32] that is determined by an overlap integral involving the motional wave function of the atom and the mode function of the field.

In a simple two-level laser model with decay rate R_{AB} and pump rate R_{BA}, the cooling process may be incorporated by coupling the atomic motion to a thermal reservoir with cooling rate B and thermal vibron number μ. Already in such a simple model, the discrete nature of the quantized motion shows up below threshold in a multiple vacuum Rabi splitting of the output spectrum [32]. This is illustrated in Fig. 3. The pairs of peaks correspond to different vibrational states with different atom-field couplings.

The cooling mechanism is most transparent in the special case of resolved-sideband cooling. The coherent pump may be detuned to the first lower vibrational sideband so that with each excitation from $4S_{1/2}$ to $4P_{1/2}$ one vibron is annihilated and the CM motion is cooled. Eventually, all the population will collect in the motional ground state of the atomic ground state $4S_{1/2}$ and cannot participate in the lasing process. The coherent pump strength is now given by Ω_0 times a motion-dependent function. In order to maintain laser action in the presence of the cooling, an additional broadband pump field Γ may be applied to the cooling transition. Figure 4 indicates that a field with a mean photon number $|a^+a\rangle = 2.3$ is generated while the mean vibron number is restricted to a value of $|b^+b\rangle = 0.5$. If a larger mean vibron number is acceptable, the pump rate Γ can be increased so that more population takes part in the laser action. This leads to considerably larger mean photon numbers. The calculation shows that it is possible to incorporate a

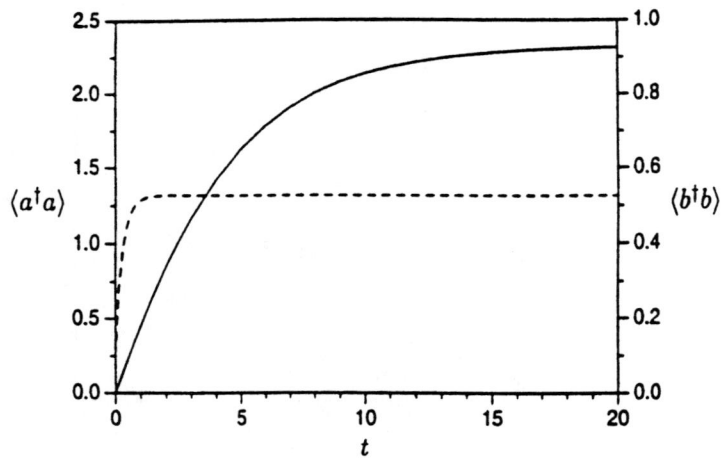

FIGURE 4. Time evolution of the mean photon number (solid) and the mean vibron number (dashed) in the Ca$^+$ ion-trap laser with sideband cooling. The parameters are $A = 0.5$, $g_0 = 14.8$, $\Omega_0 = 100$, $\Gamma = \Gamma_1 = 40$, $\Gamma_2 = 100$, and $\eta = 0.1$ on the laser transition. Initially, the atom is in the ground state and the vibronic distribution is thermal with $\langle b^+ b \rangle = 0.1$. All rates are in MHz.

cooling mechanism in a multilevel one-atom laser scheme and to obtain significant lasing also for nonperfect localization of the atom. Although it is difficult to reach the resolved-sideband limit in an experiment, cooling may still be achieved in the weak-binding regime by detuning a coherent pump field.

WORK TOWARDS A FREQUENCY STANDARD ON THE BASIS OF A SINGLE IN$^+$ ION

We report on progress towards an optical frequency standard in the spirit of the group-III monoion-oscillator proposed by Dehmelt [33], based on the $5s^2\,^1S_0 \to 5s5p\,^3P_0$ transition of the In$^+$ ion at a wavelength of 236.5 nm. This type of forbidden transition between two levels with vanishing electronic angular momentum is favorable for a frequency standard because line shifts due to external stray fields are small. For isotopes with nonzero nuclear spin the hyperfine interaction leads to small admixtures of levels with J = 1 to the $^3P_0$ state leading to a small electronic dipole decay rate of this level. For the case of In$^+$ the natural linewidth of the $^1S_0 \to {}^3P_0$ line becomes 1.1 Hz [34].

Besides the clock transition the ion also has to provide a sufficiently fast optical transition for laser cooling and fluorescence detection. In this respect indium is the most promising candidate among the group-III ions since for this ion the wavelength of the $5s^2\,^1S_0 \to 5s5p\,^3P_1$ transition at 230.6 nm can still be reached by frequency-doubled light of a blue dye laser or by using the fourth harmonic of a high-power near-infrared diode laser. The use of a relatively narrow intercombination line limits in principle the initial cooling power and the single-ion fluorescence signal, but it also results in a very low temperature of the ion. The temperature achieved by sideband cooling is about 100 μK. Another advantage is that the In$^+$ ion can be brought rather easily in the strong binding regime, that means that the oscillation frequencies of the ion in the trap are larger than the linewidth of the cooling transition being 360 kHz [34]. The clock transition at 236.5 nm has the advantage that it coincides with the fourth harmonic of the 946 nm Nd:YAG laser line. Thus, this intrinsically frequency stable solid-state laser can be used as the oscillator for the frequency standard [35].

Among the foreseeable line shifts [33] the most severe one is the Zeeman shift resulting from the small admixtures of $J = 1$ levels into $^3P_0$ which make the g-factor larger than the purely nuclear g-factor of the $^1S_0$ ground state [36]. For the $m_F = 1/2 \rightarrow m_F = 1/2$ component of the clock transition this results in a shift of 240 Hz/G [34].

Experimental Setup

Details of our experimental setup are given in Refs. [34,35] and [37]. The ion is stored in a Paul-Straubel trap [38,17]. The trap consists of just a ring electrode to which the radiofrequency voltage is applied. This design can easily be miniaturized to submillimeter size and provides a large observation angle, because the endcaps are replaced by shield electrodes, which are located far away from the trap center. The ring used for our experiments had a hole diameter of 1 mm and was made of copper-beryllium. The applied AC voltage was 1000 V at 10 MHz and in addition a DC voltage of 30 V was used resulting in a secular frequency of 1 MHz.

The laser radiation at 230.6 nm for cooling of the ion is generated by frequency-doubling blue light from a stilbene-3 ring dye laser with a BBO crystal in an external enhancement cavity. Typically 30 μW of UV light is focused to a spot size of about 50 μm at the position of the ion. The fluorescence signal from the ion is observed using a solar-blind photomultiplier.

The initial search for the forbidden clock transition required a laser source with high output power. Our clock laser system [37] therefore consists of two diode-pumped Nd:YAG lasers: a master laser at 946 nm contains the tuning elements and is frequency locked to a Zerodur cavity with a finesse of 70 000. It produces 30 mW of IR light with a linewidth in the range of 10 Hz. This light is injected into the second diode-pumped Nd:YAG ring laser containing a KNbO$_3$ crystal for intracavity frequency doubling to 473 nm. This light is then doubled again in another external cavity, finally about 0.5 mW of frequency stable UV light are obtained.

To control and measure the absolute wavelengths of both laser systems we use reference lines of the $^{130}Te_2$ molecule in the blue spectral region around 461 nm and 473 nm.

Indium Single-Ion Spectroscopy

In a first series of experiments we measured the wavelength, hyperfine constants, isotope shift and lifetime of the $5s^2\,^1S_0 \to 5s5p\,^3P_1$ cooling transition [34]. Recent results on a double-resonance experiment [39] on the cooling and clock transition are shown in Fig. 5 [40]. Fig. 5(a) shows the cooling transition. For this scan the frequency-doubled dye laser was set to the fourth motional sideband of the transition on the long wavelength side for cooling and a weak sideband of the laser was scanned across the central line. The line observed has a width being slightly larger than the expected value of 360 kHz resulting from a small power broadening. A careful analysis shows that secular sidebands are practically not present so that the ion is predominantly in its vibrational ground state. The measurement with the clock laser gives results of the type shown in Fig. 5(b). The number of induced quantum jumps are plotted versus the laser frequency. The plot shows the result of a run of about 3 min. The clock and cooling laser are chopped and excite the ion alternatively in order to avoid power broadening.

The observed linewidth of the clock transition is at the moment still much broader than the natural linewidth due to the linewidth of the clock laser. It is expected that this deficiency can be eliminated. Presently measurements are also under way to determine the absolute frequency of the clock transition with respect to the Cs clock.

The preliminary results presented here are very promising. There is the hope that the natural linewidth of the clock transition can be achieved quite soon.

DIFFUSION OF A SINGLE ION IN A ONE-DIMENSIONAL OPTICAL LATTICE

Laser cooling of atoms has opened up many new interesting fields, including the possibility of studying random walks [41–46]. The first impressive example was found in the Brownian motion of ultracold atoms in optical molasses [41], where atoms were cooled and confined by photon recoil to form a viscous medium. The cooling and confinement rely on velocity- or position-dependent absorption of laser photons followed by spontaneous emission. The complexity in the absorption process gives diversity to the random walks of laser-cooled atoms; Brownian as well as anomalous diffusion such as Lévy flights [42,47] can be observed, depending on the cooling technique. Lévy flights have generated broad interest over a cross-disciplinary field in recent years [48,49] because of their strange kinetics, self-similarity, and fractal nature. In laser cooling Lévy statistics have been employed to analyze the atomic momentum distribution in velocity selective coherent population trapping [42]; also Lévy walks are predicted to occur in atomic transport in optical lattices [46].

This part of the paper presents a first study of the single-ion dynamics in a one-dimensional optical lattice superimposed on a weak harmonic potential. In the

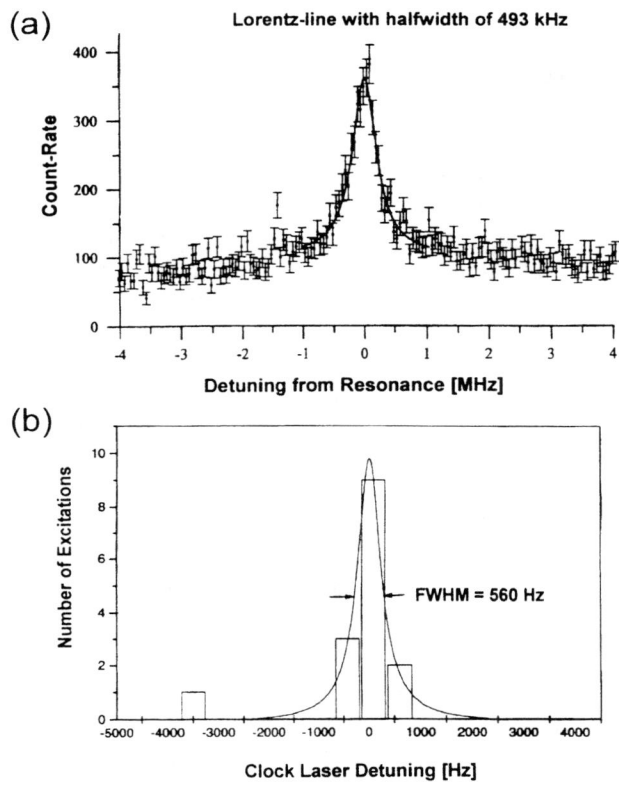

FIGURE 5. Excitation spectra of a single indium ion. Part (a) shows the result for the cooling transition $^1S_0 \to {}^3P_1$ and (b) the one for the clock transition $^1S_0 \to {}^3P_0$ measured in an optical-optical double resonance experiment.

experiment, the trajectory of the ion is measured via its fluorescence photons using a position-resolved, single-photon detection technique. This single-particle observation allows us to see the microscopic processes in the atomic transport, where an anomalous diffusion process such as Lévy walks can best manifest its unique features. The resultant macroscopic effects, the change of the spatial diffusion constant, and the existence of long range correlations, are analyzed by applying statistical techniques to the measured trajectories.

A red-detuned optical lattice with polarization gradients has been successfully used to realize sub-Doppler temperatures with the help of the Sisyphus effect [50]. It can produce atomic samples with kinetic energies less than half of the optical-potential depth, leading to a localization of atoms at the minima of the potential [51,3]. Since the atoms are trapped in the intensity maxima, the atomic residence time in a specific well is limited because energy fluctuations due to spontaneous emission occasionally allow the atom to accumulate enough energy to overcome the potential barrier. This causes random walks of the atoms between lattice sites [43]. To date, these atomic transports have been studied only in relatively deep optical potential wells, where atoms show Brownian (Gaussian) diffusion [45] because the Sisyphus damping force together with the lattice periodicity strongly limit the random walk step size.

For shallower optical potentials, it has been predicted that these energy fluctuations lead to long flights of an atom that can reach over many wavelengths before it is eventually trapped again, since the Sisyphus damping force is negligible for an atom with a large momentum which is far outside the momentum capture range p_c [46,50,52]. The resultant atomic trajectory is reminiscent of random walk phenomena called Lévy walks [48]. Macroscopically this results in a breakdown of the Gaussian diffusion law and causes a divergence of this spatial diffusion constant. Marksteiner et al. [46] discussed the occurrence of this anomalous diffusion in terms of the slowly decaying momentum correlation due to the weak damping force. However, in our work we discussed this phenomenon using energy correlations. The superimposed weak harmonic potential allows us to determine the potential energy of the ion from its spatial distribution measured with an appropriate averaging time. Once the ion acquires enough kinetic energy to significantly exceed the barrier, it becomes insensitive to the Sisyphus damping and oscillates in the superimposed potential. The total energy is then virtually unchanged for some time period, until the ion is captured by one of the optical potentials by means of Sisyphus damping. This results in anomalous fluctuations of the ions kinetic energy with long-range correlations whose origin is attributed to the existence of p_c, as discussed in connection with spatial Lévy walks [46]. In the following, some of our results will be reviewed (for a complete discussion see Ref. [53,54]).

Experimental Setup

An rf quadrupole ring trap was used to confine a single ^{24}Mg$^+$ ion in the radial direction [55]. The trap was operated with an rf frequency of 6.5 MHz, with a resultant secular frequency for the ion's radial motion of 900 kHz. As we used only a small part of the ring with an angle of 200μrad, the trap actually functioned practically as a linear trap. The ion was confined along the free axis by a shallow electric potential with an oscillation frequency of $\omega_{ext}/2\pi = 13$kHz, which enabled us to observe a single ion for long time periods. As mentioned above, a periodic optical potential was produced tangential to the trap axis with a pair of counterpropagating, crossed-linear-polarized laser beams, which were slightly red detuned from the $^2S_{1/2} \rightarrow {}^2P_{3/2}$ atomic transition of ^{24}Mg$^+$ at $\lambda =$280 nm. The natural linewidth of the excited state is $\Gamma/2\pi = 42.7$ MHz. A schematic of this 1-D potential is depicted in Fig. 6.

In the experiment, the position of the ion was measured by its displacement $x(t_i)$ from the minimum of a superimposed external electric potential. The ion position

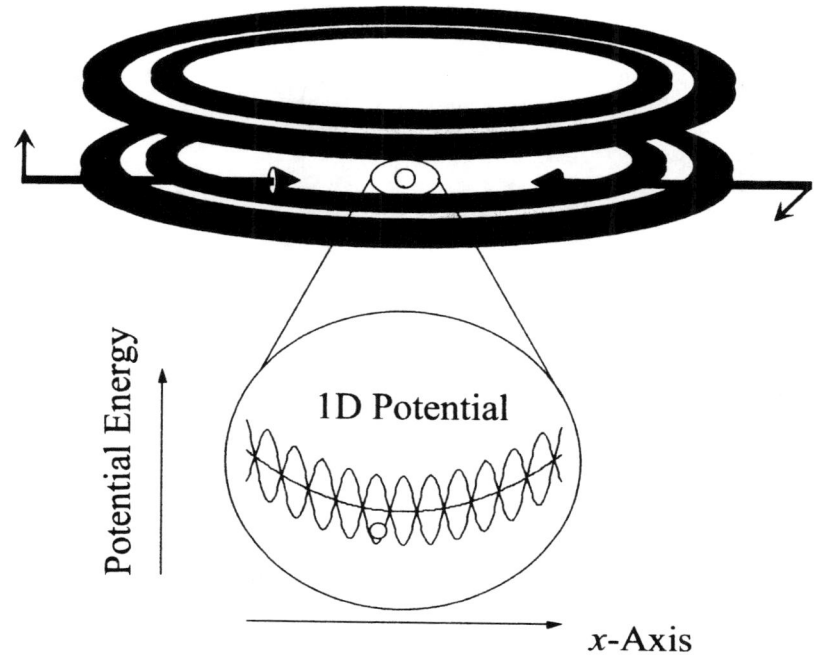

FIGURE 6. Sketch of the ring trap. The shape of the electrostatic and optical potentials is depicted in the inset. The electrostatic potential is formed by two additional electrodes which are not shown in the figure.

was determined by detecting its fluorescent photons that pass through a microscope lens with a numerical aperture (NA) of NA = 0.28. The ion image was projected onto a resistive anode-element-based single-photon counting position analyzer with a time resolution of 10 μs. The effective position resolution of the imaging system was estimated experimentally to be 3 μm, due mainly to the thermal motion of the ion in the focal direction. The total photon detection efficiency was roughly 5×10^{-5}. The signal-to-noise ratio was $10^2 \sim 10^3$, limited mainly by the presence of fluorescence light scattered from the surface of the trap electrodes.

The real-time position information was transferred to a personal computer to record the time history of the single-ion displacement, $\{x(t_i)\}$, $(i = 1, 2, ..., 2^{16})$, which corresponds to an observation time of several seconds to a few minutes depending on the lattice parameters.

Analysis of the Data

In order to extract macroscopic quantities such as spatial diffusion constants and kinetic energies from the measured atomic trajectories $x(t_i)$, we employed an autocorrelation function analysis. The $x(t_i)$ were used to calculate

$$\phi(\tau) = \left\langle [x(t+\tau) - x(t)]^2 \right\rangle = 2 \left\langle x^2 \right\rangle - 2 \left\langle x(0)\, x(\tau) \right\rangle, \tag{1}$$

which leads to the position autocorrelation function $\langle x(0)x(\tau)\rangle$. The angle brackets denote an average over time t. In this analysis the contribution of stray photons and any position uncertainty in the optical imaging result only in a constant offset.

Next, the spatial diffusion coefficient has to be evaluated. For this purpose we start with the Langevin equation for an ion confined to a harmonic potential:

$$m\ddot{x} + \gamma \dot{x} + m\omega_{ext}^2 x = F(t). \tag{2}$$

Due to photon recoils and dipole fluctuations, the ion is subjected to the Langevin force $\langle F(t+\tau)F(t)\rangle = 2\gamma k_B T \delta(\tau)$, where k_B is the Boltzmann constant, γ the friction coefficient, m the mass, and T the temperature of the ion. The friction coefficient γ was introduced to describe spatial diffusion between lattice sites. This phenomenological approach is valid for time scales much longer than the ion's localization time in the optical potential. Inserting Eq. (2) into Eq. (1), we are able to derive in the high-friction limit:

$$\phi(\tau) = \frac{2k_B T}{m\omega_{ext}^2}\left[1 - \exp\left(\frac{-m\omega_{ext}^2 \tau}{\gamma}\right)\right]. \tag{3}$$

This limit applies to deep optical potentials, where the damping time is much shorter than one oscillation period of the ion in the harmonic potential and the atomic transport can be considered as diffusive. The time derivative of Eq. (3) at $\tau = 0$ gives the spatial diffusion coefficient:

$$\phi'(0) = 2k_B T/\gamma = 2D_X. \tag{4}$$

Fig. 7 shows the shape of $\phi(\tau)$ for different optical potential depths $U_0 = \frac{2}{3}\hbar\delta s_0$, where $s_0 = (I/I_{Sat})[1 + (2\delta/\Gamma)^2]^{-1}$, $I_{Sat} = \frac{2}{3}\pi^2 \hbar c \Gamma \lambda^{-3}$, δ, and I are the saturation parameter, the saturation intensity of the transition, the laser detuning, and the intensity, respectively. The optical potential depth is given in units of photon recoil energy $E_R = \frac{\hbar^2 k^2}{2m}$ with $E_R/h = 106$ kHz. The slow rise in the autocorrelation function for the two highest optical potentials indicates that the ion is trapped in the optical potential and therefore moving in the high-friction limit. A spatial diffusion coefficient can be defined.

For the two lowest optical potential depths, a sinusoidal oscillation appears at the frequency ω_{ext} of the external potential. This clearly shows that the atomic transport in the lattice changed from a slow diffusive regime to one where the fast

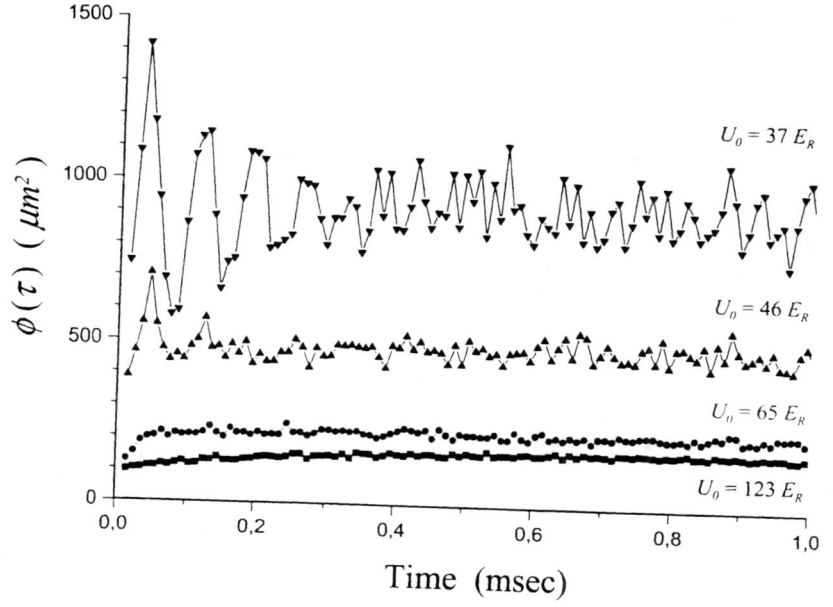

FIGURE 7. Results for the position autocorrelation functions for different optical potential depths. The oscillation at the $U_0 = 37 E_R$ is due to the motion of the ion in the external electric potential. For all the curves the laser intensity is $I = 4 I_{\text{Sat}}$.

oscillation in the external potential is dominant. Sisyphus cooling is unable to cool the ion's kinetic energy below the optical potential depth so that the ion is no longer localized. In this low-friction limit, the result for $\phi(\tau)$ is

$$\phi(\tau) = \frac{2k_B T}{m \omega_{ext}^2} \left[1 - \exp\left(-\frac{\gamma \tau}{2m}\right) \cos(\omega_{ext} \tau) \right], \qquad (5)$$

The ion's secular- and micro-motion are not seen, owing to the time resolution of the detection system of $10 \, \mu s$. The sinusoidal oscillation in the position correlation function also implies an oscillation in the momentum correlation function.

Spatial Diffusion Coefficient

Fig. 8 shows the spatial diffusion coefficient D_x as a function of optical potential depth for a fixed laser intensity of $I = 15 I_{\text{Sat}}$. The optical potential depth was varied by changing the laser detuning δ. The values where obtained by fitting Eq. (3) to the experimental autocorrelation function $\phi(\tau)$ in the high-friction limit and by calculating $\phi'(0)$ [see Eq. (4)].

The experimental results shown in Fig. 8 are compared to theory and discussed in detail in Ref. [54]. The discussion is therefore not repeated here.

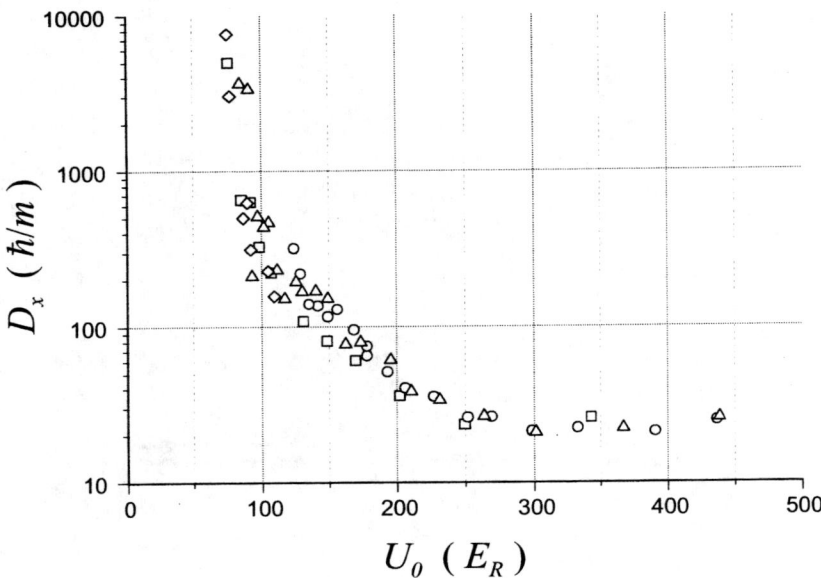

FIGURE 8. The spatial diffusion coefficient D_x as a function of optical potential depth U_0. The symbols mark measurements taken on different days.

In this last part of the paper we describe a new method to investigate the spatial diffusion of a single ion in an optical lattice. The results obtained so far show that the method will be suitable for performing detailed studies of the processes involved. At the moment, there are still some experimental drawbacks such as, e.g., the influence of micro motion being in principle avoidable in future experiments.

So far our study reflects only classical aspects of polarization-gradient cooling. However, it also seems possible to reach the quantum limit if the laser beams used for fluorescence detection and grating formation have different wavelengths. In this way, the lattice distance can be made much larger than the wavelength used for detection. Thus, the optical resolution can be brought into a domain in which jumps between neighboring sites can be resolved. Experiments in this direction are in principle possible and are now in preparation.

REFERENCES

1. Hartig W., Rasmussen W., Schieder R., and Walther, H., *Z. Physik* **A278**, 205 (1976).
2. Cresser, J. D., Häger J., Leuchs, G., Rateike, F. M., and Walther, H., *Topics in Current Physics* **27**, 21 (1982).

3. Jessen, P. S., Gerz, C., Lett, P. D., Phillipps, W. D., Rolston, S. L., Spreuuw, R. J. C., and Westbrook, C. I., *Phys. Rev. Lett.* **69**, 49 (1992).
4. Heitler, W., *The Quantum Theory of Radiation*, Oxford University Press, 1954, Third Edition, pp. 196-204
5. Mollow, B. R., *Phys. Rev.* **188**, 1969 (1969).
6. Schuda, F., Stroud, C., Jr., and Hercher, M., *J. Phys.* **B7**, L198-L202 (1974).
7. Walther, H., *Lecture Notes in Physics* **43**, 358 (1975).
8. Wu, F. Y., Grove, R. E., and Ezekiel, S., *Phys. Rev. Lett.* **35**, 1426 (1975); Grove, R. E., Wu, F. Y., and Ezekiel, S., *Phys. Rev. A* **15**, 227 (1977),
9. Gibbs, H. M., and Venkatesan, T. N. C., *Opt. Comm.* **17**, 87 (1976).
10. Kimble, H. J., Dagenais, M., and Mandel, L., *Phys. Rev. Lett.* **39**, 691 (1977).
11. Jakeman, E., Pike, E. R., Pusey, P. N., and Vaugham, J. M., *J. Phys. A* **10**, L257-L259 (1977).
12. Kimble, H. J., Dagenais, M., and Mandel, L., *Phys. Rev. A* **18**, 201 (1978); Dagenais, M., and Mandel, L., *Phys. Rev. A* **8**, 2217 (1978).
13. Rateike, F. M., Leuchs, G., and Walther, H., results cited in Ref. [2].
14. Diedrich, F., and Walther, H., *Phys. Rev. Lett.* **58**, 203 (1987).
15. Walls, D. F., *Nature* **280** 451 (1979).
16. Short, R., and Mandel, L., *Phys. Rev. Lett.* **51**, 384 (1983).
17. Schrama, C. A., Peik, E., Smith, W. W., and Walther, H., *Opt. Comm.* **101**, 32 (1993).
18. Höffges J. T., Baldauf, H. W., Eichler, T., Helmfrid, S. R., and Walther, H.. *Opt. Com.* **133**, 170 (1997).
19. Höffges J. T., Baldauf, H. W., Lange, W., and Walther, H., *J. Mod. Optics* **44**, 1999 (1997).
20. Loudon, R., *Rep. Progr. Phys.* **43**, 913 (1980).
21. Mu, Y., and Savage, C. M., *Phys. Rev. A* **46**, 5944 (1982).
22. Ginzel, C., Briegel, H. J., Martini, U., Englert, B.-G., and Schenzle, A., *Phys. Rev. A* **48**, 732 (1993).
23. Pellizzari, T., and Ritsch, H. J., *Mod. Opt.* **41**, 609 (1994); *Phys. Rev. Lett.* **72**, 3973 (1994); Horak, P., Gheri, K. M., and Ritsch, H., *Phys. Rev. A* **51**, 3257 (1995).
24. Briegel, H.-J., Meyer, G. M., and Englert, B.-G., *Phys. Rev. A* **53**, 1143 (1996); *Europhys. Lett.* **33**, 515 (1996).
25. Löffler M., Meyer, G. M., and Walther, H., *Phys. Rev. A* **55**, 3923 (1997).
26. For a recent review see Arimondo E., *Progress in Optics*, edited by Wolf, E., Elsevier, Amsterdam 1996, vol. **XXXV**, pp. 257-354.
27. Khazanov, A. M., Koganov, G. A., and Gordov, E. P., *Phys. Rev. A* **42**, 3065 (1990); Ralph, T. C., and Savage, C. M., *Phys. Rev. A* **44**, 7809 (1991); Ritsch, H., Zoller, P., Gardiner, C. W., and Walls, D. F., *Phys. Rev. A* **44**, 3361 (1991).
28. Gheri, K. M., and Walls, D. F., *Phys. Rev. A* **45**, 6675 (1992); Ritsch, H., and Marte, M. A. M., *Phys. Rev. A* **47**, 2354 (1993).
29. An, K., Childs, J. J., Dasari, R. R., and Feld, M. S., *Phys. Rev. Lett.* **73**, 3375 (1994).
30. Meyer, G. M., Briegel, H.-J., and Walther, H., *Europhys. Lett.* **37**, 317 (1997).
31. Meyer, G. M., Löffler M., and Walther, H., *Phys. Rev. A* **56**, R1099 (1997).

32. Löffler M., Meyer, G. M., and Walther, H., *Europhys. Lett.* **40**, 263 (1997).
33. Dehmelt, H., *Proceedings of the Third Symposium on Frequency Standards and Metrology*, J. Physique **42**, C8-299 (1981).
34. Peik, E., Hollemann, G., and Walther, H., *Phys. Rev. A* **49**, 402 (1994).
35. Hollemann, G., Peik, E., and Walther, H., *Opt. Lett.* **19**, 192 (1994).
36. Lahaye, B., and Margerie, J., *J. Physique* **36**, 943 (1975).
37. Hollemann, G., Peik, E., Rusch, A., and Walther, H., *Opt. Lett.* **20**, 1871 (1995).
38. Yu, N., Nagourney, W., and Dehmelt, H., *J. Appl. Phys.* **69**, 3779 (1991).
39. Peik, E., Hollemann, G., and Walther, H., *Proceedings Fifth Symposium on Frequency and Metrology*, edited by J. Berquist, World Scientific, Singapore 1996, pp. 376-379.
40. Peik, E., von Zanthier, J., Becker, T., Abel, J., Fries, M., and Walther, H., to be published.
41. Chu, S. et al., *Phys. Rev. Lett.* **55**, 48 (1985).
42. Bardou, F., et al., *Phys. Rev. Lett.* **72**, 203 (1994).
43. Hemmerich, A., Weidemüller, M., and Hänsch, T., *Europhys. Lett.* **27**, 427 (1994).
44. Hodapp, T. W., et al., *Appl. Phys. B* **60**, 135 (1995).
45. Jurczak, C., et al., *Phys. Rev. Lett.* **77**, 1727 (1996).
46. Marksteiner, S., Ellinger, K., and Zoller, P., *Phys. Rev. A* **53**, 3409 (1996).
47. Bouchaud, J.-P., and Georges, A., *Phys. Rep.* **195**, 125 (1990).
48. Shlesinger, M. F., Zaslavsky, G. M., and Klafter, J., *Nature* **363**, 31 (1993).
49. Viswanathan, G. M., Afanasyev, V., Buldyrev, S. V., Murphy, E. J., Prince, P. A., and Stanley, H. E., *Nature* **381**, 413 (1996).
50. Dalibard, J., and Cohen-Tannoudji, C., *J. Opt. Soc. Am. B* **6**, 2023 (1989), and references therein.
51. Verkerk, P, et al., *Phys. Rev. Lett.* **68**, 3861 (1992).
52. Holland, M., et al., *Phys. Rev. Lett.* **76**, 3683 (1996).
53. Katori, H., Schlipf, S., and Walther, H., *Phys. Rev. Lett.* **79**, 2221 (1997).
54. Schlipf, S., Katori, H., and Walther, H., *Optics Express* **3**, 97 (1998).
55. Waki, I., Kassner, S., Birkl, G., and Walther, H., *Phys. Rev. Lett.* **68**, 2007 (1992).

Dynamics of Autoionizing Rydberg Atoms in an Electric Field

L. D. Noordam and F. Robicheaux

FOM-Institute for Atomic and Molecular Physics
Kruislaan 407,1098 SJ Amsterdam, The Netherlands
Noordam@amolf.nl

Abstract. The dynamics of a Rydberg Stark wave packet above the saddle point of the combined Coulomb-electric field potential is investigated by conventional pump-probe techniques and an atomic streak camera. This new device records when the electron is ejected from the autoionizing system. With some modifications this streak camera can also be used to measure the shape of infrared laser pulses with a picosecond time resolution.

INTRODUCTION

In a Rydberg atom [1,2] the loosely bound electron moves in a large Kepler orbit around the atomic nucleus. The system is very sensitive to external perturbations; for instance, a moderate electric field drastically influences the behaviour of the Rydberg electron. A static field of a few kV/cm is sufficient to change the bound Rydberg atom into a system in which the electron can escape. Within a few picoseconds (10^{-12} s) the atom falls apart. It is an experimental challenge to detect how this decay actually happens. Does the electron come out at once, or are there signatures of the quantum nature of the system which are made manifest by subsequent bursts of probability that the atom emits an electron?

The behaviour of the decaying electron has been studied for more than a decade using spectroscopy, either in the frequency domain [3,4] by measuring the absorption spectrum in the continuum, or in the time domain [5–8] by following the motion of an electronic wave packet created by a short optical pulse. Each of these techniques probe the atom when the excited electron is near the core. If the electron does not escape immediately but returns at least once to the core, this is seen in the frequency domain as a resonance in the absorption spectrum. The same situation can be probed in the time domain by a delayed second pulse that stimulates the electron back to a bound state.

However, spectroscopy does not tell the full story. What if the electron does not leave the atom immediately but also does not returns to the core. A spectroscopist will not observe a recurrence and may conclude that the electron has escaped, which is an incorrect conclusion since the electron is still in the atom but out of the scope of the spectroscopist. An atomic streak camera [9] has been constructed that provides additional information: the camera makes pictures of when the electron leaves the atom [10].

To our surprise, the electron usually makes first a few orbits around the nucleus before escaping. After each orbit there is some probability of escaping, or in more quantum mechanical terms, some fraction of the electron wave function is emitted. The camera directly sees how much electron wave function leaks out after each orbit. In this contribution we discuss in detail the dynamics of electron emission of a highly excited Rydberg state. Moreover, we present some applications of an atom ejecting an electron upon photoexcitation to an autoionizing Rydberg state.

DYNAMICS OF A RYDBERG ELECTRON IN AN ELECTRIC FIELD

The ionization dynamics of Stark states above the saddle point have been studied [7,11] using short optical pulses to excite a coherent superposition of the 'quasi continuum' states. The decay of the wave packet [12] is monitored [13] by measuring the amount of wave function returning to the core region as a function of time. In these experiments a ps optical pulse excites a wave packet above the saddle point energy. After a delay a second pulse is applied to probe the amount of wave function that has returned to the core. In this way the evolution of the wave packet near the core has been observed.

Two types of motion of the autoionizing wave packet above the saddle point of the potential are seen. **Firstly**, the radial motion [14] of the wave packet starting near the core and moving towards the outer turning point and back to the core. This motion is due to beating of eigenstates with different principal quantum number n. Since the energy difference of Rydberg states is given by $dE/dn = 1/n^3$ the beating time is given by $\tau = 2\pi/\Delta E = 2\pi n^3$. This radial oscillation time of the wave packet corresponds to the Kepler motion of an electron around the nucleus. For $n = 20$ this orbit time is 1.2 ps.

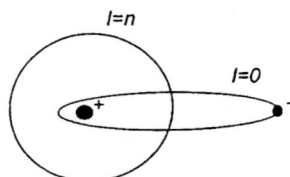

FIGURE 1. The wave packet of a Rydberg electron in an electric field exhibits oscillations in angular momentum.

Secondly, the angular momentum of the electron is not conserved in an electric field F. The wave packet exhibits oscillations in the angular momentum between $l = 0$ and $l = n$ [5], starting from the initial angular momentum state $l \sim 0$ (see Fig. 1).

These angular momentum oscillations are due to the beating of different Stark states k within an n-manifold. Since the energy difference of Stark states is given by $dE/dk = 3Fn$, the beating time is given by $\tau = 2\pi/\Delta E = 2\pi/(3Fn)$. For $n = 20$ and a field of 2 kV/cm, the angular momentum oscillation time corresponds to 6 ps. Recurrences in the optical pump-probe studies are seen whenever the angular momentum and the radial distribution are at the initial values from which the packet was launched by the pump pulse: $r \sim 0$ and $l \sim 0$.

A limitation of using optical techniques to study the ionization dynamics is the fact that the dynamics of the Rydberg electron can only be probed near the core. Using the atomic streak camera, we are able to identify when the electron escapes from the ionic potential (see Fig. 2).

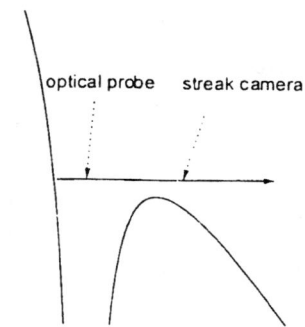

FIGURE 2. The streak camera probes the electron as it escapes over the saddle point, while optical techniques probe recurrences near the core.

ATOMIC STREAK CAMERA

We present a new device, the atomic streak camera [9], which is able to measure the ionization dynamics (i.e., the escape of the electron from the atomic potential) with ps resolution. The design of the atomic streak camera (see Fig. 3) is based on a conventional streak camera. In a conventional streak camera, [15] the optical pulse is transformed into an electron pulse by means of a photocathode. In the atomic streak camera, instead of a photocathode, a low-pressure atomic gas is used to create the electrons. These atoms, which are in a static electric field, are excited by a short laser pulse.

Let us consider the case where the wave packet dynamics are such that the electrons leave the atoms in two bursts. The double electron pulse is accelerated by the electric field and passes the slit in the anode. After the slit the electrons enter the deflection region. The voltage applied to one of the two deflection plates is ramped. Therefore, the electron pulse arriving first between the deflection plates will be deflected less than the second one, resulting in a different impact position on the detector. Thus from the position on the detector, the time-resolved ionization spectrum of the atoms is retrieved.

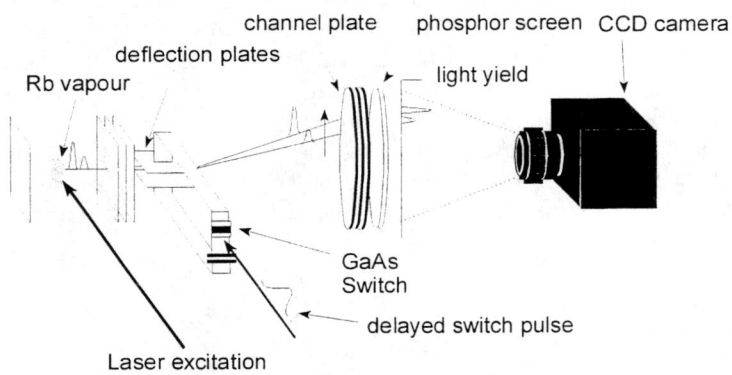

FIGURE 3. The design of the atomic streak camera.

EXPERIMENTAL OBSERVATION OF AN AUTOIONIZING RYDBERG WAVE PACKET

We have monitored the evolution of a Rydberg wave packet in two ways [10]: near the core with a pump-probe technique and near the saddle point with the atomic streak camera. In the experiment rubidium atoms were excited in a static electric field of 2.0 kV/cm by a short optical pulse. The excitation energy chosen was above the saddle-point energy, thus creating an autoionizing wave packet. In Fig. 4 a time-resolved electron emission spectrum of rubidium excited with a 4-ps pulse around 0.87 E_c is shown, where E_c is the saddle point energy.

Inspection of figure 4 shows that instead of observing an exponential decaying type of wave packet the main ionization was surprisingly delayed 12 ps. The polarization of the laser used to excite the wave packet was chosen perpendicular to the static electric field. The wave packet is therefore located perpendicular to the electric field. Since the electron can only escape in the direction of the electric field it is still bound in the directions perpendicular to this axis. The wave packet needs to reorient itself in the direction of the saddle point to escape. This reorientation can occur by scattering from the core electrons. The scattering, in turn, depends on the average value of the angular momentum of the wave packet. For this particular case, the excited wave packet makes an oscillation in angular momentum in

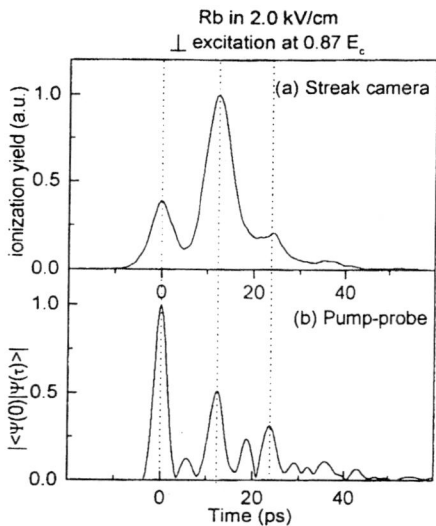

FIGURE 4. In the upper spectrum the electron is probed at the saddle with the atomic streak camera, whereas in the lower spectrum the electron is probed near the core point by a conventional pump-probe technique.

6 ps. Initially excited to a low angular momentum state, the wave packet begins to increase its angular momentum. When the wave packet is in a high angular momentum state, it is located far away from the core electrons ($r_{min} = l(l+1)$) making it impossible to scatter. From Fig. 4(a) we observe that the wave packet ionizes dominantly at the second angular recurrence. In Fig. 4(b) the corresponding recurrence spectrum is plotted. This spectrum is a measure for the amount of wave function that comes back to the $l = 1$ state as a function of time. We see that the amplitude is rather low after the first oscillation in angular momentum at 6 ps. The second angular recurrence gives a high amplitude indicating that a larger fraction of the wave packet returns to low angular momentum states. This causes the scatter event with the core electrons leading to the large ionization at 12 ps observed in fig. 4(a). These spectra have been reproduced by a MQDT calculation [16]. In the calculation one observes the scatter event by the transfer of population from a closed channel (bound states) into an open channel (ionizing states).

TIME-GATED SPECTROSCOPY

We describe a novel spectroscopic tool which provides the time-resolved electron emission as a function of the excitation wavelength: Time-Gated photoionization Spectroscopy (TGS [17]). Using the atomic streak camera, we record the time of electron emission after laser excitation. By setting time gates for the outgoing electron flux and recording the yield in a time gate versus excitation wavelength, we obtain a time-gated spectrum.

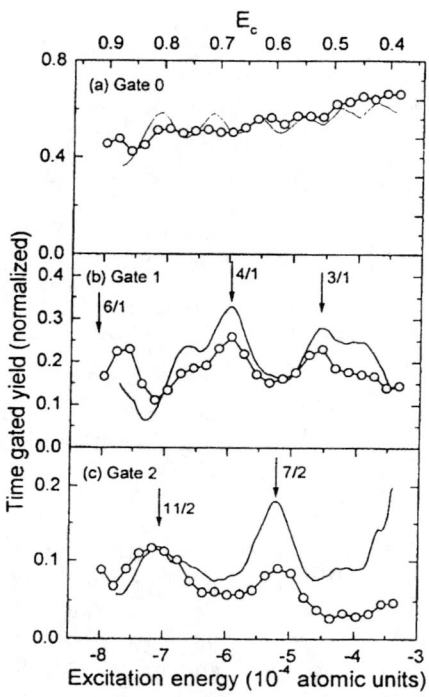

FIGURE 5. Relative ionization yields in time-gated spectroscopy.

In Fig. 5 the relative ionization yields are measured for time gates centered around the prompt peak, the first and the second angular recurrence time of the wave packets. The underlying mechanism responsible for the observed oscillations can be explained in terms of the commensurability between angular and radial periods of the electron dynamics [8]. Within the laser bandwidth, k-states belonging to different n-manifolds are excited. The total wave packet is constructed of two angular wave packets. Depending on the energy of the nk-states with respect to the energy of the $(n+1)k$-states the wave packet will either constructively (same energies) or destructively (interleaved energies) interfere near the nucleus after the first angular oscillation period.

The classical equivalent of this can be explained in terms of angular and radial oscillations of the wave packet. When $E_{nk} = E_{(n+1)k'}$, both oscillation periods are in phase at the first angular recurrence and the wave packet returns to the core. At this time the Rydberg wave packet may scatter off the core since it is the core electrons that break the parabolic symmetry for the valence electron. This scattering event can change the direction of the wave packet and cause it to travel down field giving rise to the observed peaks in the ionization spectrum. When the energies are interleaved, the periods are out of phase at the first angular oscillation time, so that when the wave packet is in a low angular momentum state, it is not located near the core.

The intrinsically large bandwidth of the ps laser pulse does not permit a study of individual resonances, but it does disclose, for a small energy range, the return of the wave packet to the core from which the electron is scattered into an outgoing channel.

CALCULATING THE ELECTRON FLUX FROM AUTOIONIZING RYDBERG ATOMS

We use two different methods to simulate the time-dependent flux of electrons measured by the streak camera [16]. In the first method, we solve for the energy dependent wave function, $\psi_E(\mathbf{r})$, and excitation amplitudes, $A(E)$, using a multichannel WKB approximation. The excitation amplitude is the product of the dipole matrix element connecting the initial state to the final state at energy E above the initial state and the amplitude for the laser pulse to contain a photon at that energy. The time-dependent wave function at any spatial point is given by $\Psi(\mathbf{r}, t) = \int dE\, A(E) \psi_E(\mathbf{r}) e^{-iEt}$. This method allows us to obtain the wave function at distances roughly 1 cm from the nucleus since the $A(E)$ and the asymptotic form of the $\psi_E(\mathbf{r})$ are well known. The flux of electrons into the streak camera may be obtained from Ψ and its spatial derivatives. In the second method, we directly solve the time-dependent Schrödinger equation for the atom in a static electric field and a pulsed laser field using a large basis set in spherical coordinates. We generate the radial part of the basis function by forcing the radial orbital to be zero at a distance that was chosen to be well beyond the saddle point for the experiment being simulated; this distance was typically 3000–5000 Bohr radii. An absorbing potential was placed at distances larger than the saddle point; this potential absorbed the outgoing flux and prevented reflections from the infinite wall that was used to generate the radial basis functions. We generate the time-dependent flux of electrons leaving a sphere of roughly 2000–4000 Bohr radii. Because the electron waves do not disperse in a static electric field, the flux leaving this small sphere very accurately reproduces the flux measured by the streak camera. The results from the two types of calculations were in excellent agreement showing that either method may be used and that the streak camera accurately measures the flux leaving a small region around the atom. While both methods work well when the initial state is compact, the direct time propagation of Schrödinger's equation is the only method that works well when the initial state is already a Rydberg state [20]. It is also easier to use the results from the direct time propagation to plot the wave function at any time; this aids the visualization of the dynamics and is useful for interpretation.

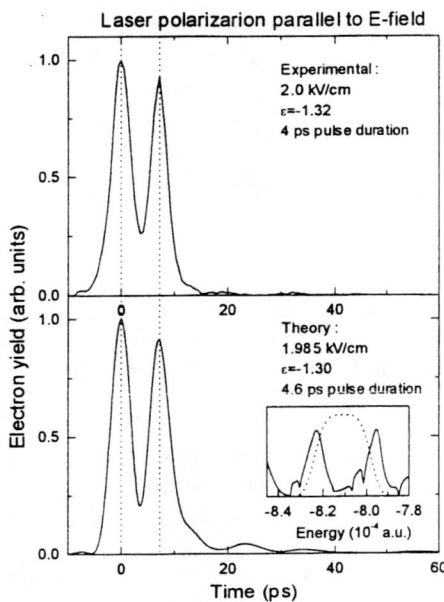

FIGURE 6. The measured and calculated electron flux for Rb in a 2 kV/cm electric field.

As an example, we show a comparison between the measured and calculated electron flux for Rb in a 2 kV/cm electric field in Fig. 6. A pulsed laser linearly polarized in the field direction excites the atom from the ground state to an energy roughly 96 cm^{-1} above the classical ionization threshold in the field, which is 178 cm^{-1} below the zero-field ionization threshold. The ejected electron flux has a simple time dependence because only two resonance states are mainly excited and their energy width is comparable to their energy spacing (see inset). An interesting feature of the time-dependent flux is that the time difference between peaks is roughly 15% longer than expected from the energy spacing of the resonances. This period is longer than expected because the pulsed laser does not fully excite both resonances. By exciting only part of the resonances, the laser pulse effectively moves the resonances closer together and therefore lengthens the period of the electronic motion.

SNAPSHOTS OF THE AUTOIONIZING WAVE PACKET

In Fig. 7 we show two snapshots of the electron probability distribution for the excitation parameters in Fig. 6. These snapshots confirm some expectations of the time dependence of the electron and also demonstrate some surprises. It is clear

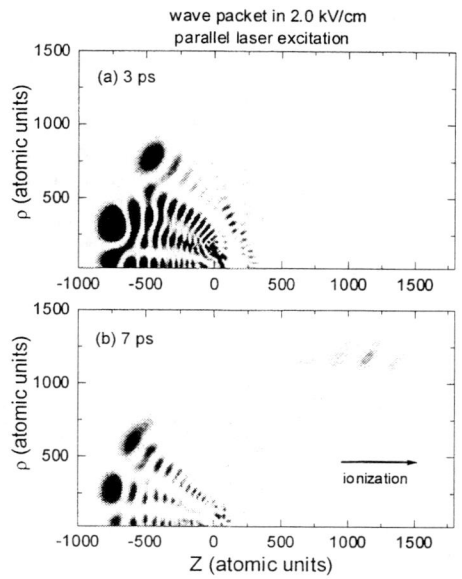

FIGURE 7. Snapshots of the electron probability distribution for the excitation parameters of Fig. 6.

from Fig. 7 that very little electron flux is being ejected at 3 ps while a large amount is being ejected at 7 ps, which is what was expected from Fig. 6. It is also clear that the bound part of the packet ($z < 0$) is performing an oscillation in angle. Although not apparent in Fig. 7, the probability for finding the electron near the nucleus is highly correlated with the amount of ejected flux. Three interesting and unexpected features are apparent in Fig. 7: (1) In the 3 ps snapshot, there is a substantial part of the wave packet that is passing down the potential from the nucleus near a z of 250 Bohr radii. This correlates with the behavior of classical trajectories in a static electric field; trajectories that start at the nucleus and do not have their initial momentum purely in the z-direction will swing by the nucleus at positive z during a portion of the orbit. (2) In the 7 ps snapshot, the majority of the ejected flux is at large ρ. We do not have a simple explanation of this property. (3) In the 7 ps snapshot, the ejected flux is not covering all of the region allowed by energy conservation. For example, a classical electron is energetically allowed at $\rho = 4000$ and $z = 1500$ Bohr radii, but all of the outgoing flux is restricted to $\rho < 1500$ Bohr radii. This again correlates with the behavior of classical trajectories. The only way an electron can be ejected at 7 ps is if it returns close enough to the nucleus to experience the potential due to the core electrons. Classical trajectories that start near the nucleus can not reach all of the regions of space allowed by energy conservation; the outgoing electron flux is restricted to those regions of space that can be reached by a classical electron that starts near the origin.

APPLICATION OF A RYDBERG "PHOTOCATHODE": A FIR STREAK CAMERA

Temporal characterization of infrared radiation is severely limited by the lack of fast IR detectors. Whereas the temporal characterization of IR pulses is cumbersome, visible light pulses are nowadays routinely characterized using a streak camera. The spectral range at which conventional streak cameras can be operated is limited by the spectral response of the photocathode. The work function of the photocathode material sets the long-wavelength limit of a conventional streak camera at 1.6 μm.

Recently, a new type of streak camera has been constructed in our laboratory. Here the photocathode used in conventional streak cameras is replaced by a sample of gas-phase atoms in a Rydberg state. The low binding energy of the electrons in Rydberg atoms and the large photoionization cross section [18] make the Rydberg atom photocathode a very sensitive detector for infrared radiation [19–21] and thus very suitable for use in an IR streak camera. The operation of this streak camera was demonstrated in the infrared at wavelengths ranging from 2.6 μm [22] up to 85 μm [23], well beyond the spectral range of any conventional streak camera.

An essential requirement for proper operation of a streak camera is that the photoelectron emission is prompt. Experiments on the far-infrared ionization behavior [20] of Rydberg atoms in an electric field have shown that states with a negative Stark effect (for which the energy of the states decreases with increasing electric field strength) ionize on a sub-picosecond time scale, independent of the wavelength and polarization of the ionization laser. For states with a positive Stark effect (the energy of the states increases with increasing electric field strength) this is not true. Here the time scale for ionization depends on the wavelength as well as on the polarization of the light and can be as long as 100 ps.

The time resolution of the camera system was tested [23] at a wavelength of 35 μm by using the lowest Stark component of the $n = 16$ manifold in Cs as intermediate level. After absorption of the 35 μm photon the initial kinetic energy of the electrons is only 10 meV. Figure 8 shows the measured temporal profile for a single laser shot. The line is a fit of a Gaussian profile to the data points. The width (FWHM) of the Gaussian profile was found to be 1.4 ps. Note that the oscillation time of the laser field at 35 μm is already 0.12 ps.

CONCLUSIONS

From the comparison of the recurrence spectra, obtained by pump-probe spectroscopy, with the streak spectra, the conclusion can be drawn that the lifetime as measured by an optical technique is not the same as the time it takes the electron to leave the atom. The Rydberg electron can be far away from the core, invisible for optical techniques, but still be captured in the attractive force of the parent ion. When the electron does not pass the core before ionizing, the electron appears to

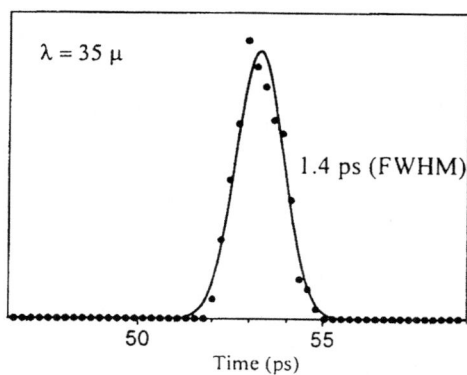

FIGURE 8. A test of the time resolution of the streak camera.

be ionized for an optical technique, but in fact has not yet escaped from the atom. Probing the dynamics of a Rydberg electron in an electric field with half-cycle pulses [24,25] provides additional information on the Rydberg-electron dynamics.

We have described a novel spectroscopic tool called Time-Gated photoionization Spectroscopy (TGS). Information in both the frequency and time domain is obtained simultaneously with this technique. TGS provides additional information compared to conventional absorption spectroscopy. We have experimentally demonstrated the technique on the photoionization of Cs wave packets in a static electric field. It is found that the time-gated spectra reveal information on the commensurability between radial and angular oscillation periods of the excited wave packets. An enhancement in the ionization yield is found whenever the wave packet is in a low angular momentum state and has radially returned to the core, since core scattering is the relevant ionization mechanism. The experimental data is confirmed by MQDT calculations.

We have also demonstrated the operation of an atomic streak camera in the mid to far-infrared (2.6–85 μm), well beyond the spectral range of any conventional streak camera. The temporal resolution of the camera system was found to be in the order of 1 ps over its useful wavelength range.

ACKNOWLEDGMENTS

The work described in this article is performed in collaboration with M. Drabbels, D. I. Duncan, J. H. Hoogenraad, G. M. Lankhuijzen, C. W. Rella, and R. B. Vrijen. It is a pleasure to acknowledge C. W. Rella for critical reading the manuscript. LDN is supported by the "Stichting voor Fundamenteel Onderzoek der Materie (FOM)" which is financially supported by the "Nederlandse Organisatie voor Wetenschappelijk Onderzoek (NWO)."

REFERENCES

1. Gallagher, T. F., *Rydberg Atoms*, Cambridge University Press, Cambridge 1994.
2. Connerade, J.-P., *Highly excited atoms*, Cambridge University Press, Cambridge 1998.
3. Freeman, R. R., Economou, N. P., Bjorklund, G. C., and Lu, K. T., *Phys. Rev. Lett.* **41**, 1463 (1978).
4. Lankhuijzen, G. M., and Noordam, L. D., *Phys. Rev. A* **52**, 2016 (1995).
5. Noordam, L.D., ten Wolde, A., Lagendijk, A., and van Linden van den Heuvell, H.B., *Phys. Rev. A* **40**, 6999 (1989).
6. Broers, B., Christian, J. F., Hoogenruad, J. H., van der Zande, W. J., van Linden van den Heuvell, H. B., and Noordam, L. D., *Phys. Rev. Lett.* **71**, 344 (1993).
7. Christian, J. F., Broers, B., Hoogenraad, J. H., van der Zande, W. J., and Noordam, L. D., *Opt. Commun.* **103**, 79 (1993).
8. Naudeau, M. L., Sukenik, C. I., and Bucksbaum, P. H., *Phys. Rev. A* **56**, 636 (1997).
9. Lankhuijzen, G. M., and Noordam, L. D., *Opt. Commun.* **129**, 361(1996).
10. Lanknuijzen, G. M., and Noordam, L. D., *Phys. Rev. Lett.* **76**, 1784 (1996).
11. Noordam, L. D., Duncan, D. I., and Gallagher, T. F., *Phys. Rev. A* **45**, 4734 (1992).
12. Jones, R. R., and Noordam, L. D., *Advances in Atomic and Molecular Physics* **38**, 1 (1997), Academic Press, San Diego.
13. Noordam, L. D., and Jones, R. R., *J. of Mod. Optics* **44**, 2515 (1997).
14. ten Wolde, A., Noordam, L. D., Lagendijk, A., and van Linden van den Heuvell, H. B., *Phys. Rev. Lett.* **61**, 2099 (1988).
15. Knox, W., and Mourou, G., *Opt. Commun.* **37**, 203 (1981).
16. Robicheaux, F., and Shaw, J., *Phys. Rev. Lett.* **77**, 4154 (1996); *Phys. Rev A* **56**, 278 (1997).
17. Lanknuijzen, G. M., Robicheaux, F., and Noordam, L. D., *Phys. Rev. Lett.* **79**, 2427 (1997).
18. Hoogenraad, J. H., and Noordam, L. D., *Phys. Rev. A* **57**,4533 (1998).
19. Hoogenraad, J. H., Vrijen, R. B., van Amersfoort, P. W., van der Meer, A. F. G., and Noordam, L. D., *Phys. Rev. Lett.* **75**, 4579 (1995).
20. Lanknuijzen,G. M., Drabbels, M., Robicheaux, F., and Noordam, L. D., *Phys. Rev. A* **57**, 440 (1998).
21. Hoogenraad, J. H., Vrijen, R. B., and Noordam, L. D., *Phys. Rev. A* **57**, 4546 (1998).
22. Drabbels, M., and Noordam, L. D., *Optics Lett.* **22**, 1436 (1997).
23. Drabbels, M., Lankhuijzen, G. M. and Noordam, L. D., submitted to IFEE-QE.
24. Jones, R. R., Tielking, N. E., You, D., Raman, C., and Bucksbaum, P. H., *Phys. Rev. A* **51**, 2687 (1995).
25. Vrijen, R. B., Lankhuijzen, G. M., and Noordam, L. D., *Phys. Rev. Lett.* **79**, 617 (1997).

Atoms and Cavities: Explorations of Quantum Entanglement

J. M. Raimond, E. Hagley, X. Maître, G. Nogues, C. Wunderlich, M. Brune and S. Haroche

Laboratoire Kastler Brossel[1], Département de Physique
Ecole normale supérieure, Paris

Abstract. The interaction of circular Rydberg atoms with a high-quality microwave cavity makes it possible to realize complex quantum state manipulations. The state of an atom can be "copied" onto the cavity. Reversing this operation at a later time with a second atom, we realize an elementary "quantum memory" holding an atomic quantum coherence for a while in a cavity mode. We have also generated two-atom entangled states of the Einstein-Podolsky-Rosen type. At variance with previous experiments, this one implies massive particles in a completely controlled process. These entanglement manipulations can be generalized to more complex or to mesoscopic systems and open the way to new tests of fundamental aspects of the quantum world.

INTRODUCTION

One of the most striking features of quantum mechanics is entanglement. This word, used for the first time in this context by Schrödinger [1], depicts a situation where two quantum systems are in a state which cannot be cast as a tensor product of separate wavefunctions. Such a situation occurs, for instance, when the two systems have undergone a nontrivial mutual interaction. Even if they are widely separated, they exhibit strong quantum correlations. Any action, any measurement performed on one of the systems immediately reflects, at a distance, on the state of the other. This is at the heart of the famous Einstein-Podolsky-Rosen (EPR) paradox [2], involving pairs of correlated particles, and of its modern variants with more than two subsystems, like the Greenberger-Horne and Zeilinger (GHZ) [3] triplets of maximally entangled particles.

Entangled states were first used to test predictions of quantum mechanics in unusual situations. The nonlocal nature of quantum mechanics yields a violation of the statistical Bell inequalities [4], verified by any local theory. Experiments,

---

[1] Laboratoire de l'Ecole normale supérieure, associé au CNRS et à l'Université Pierre et Marie Curie

involving mainly pairs of correlated photons emitted by an atomic cascade or emanating from a parametric down-conversion process, have clearly vindicated quantum mechanics as opposed to local hidden-variable theories [5-9]. More and more sophisticated tests are under way to block the last loopholes which could bias the experimental results [10]. The generation of multiparticle entangled states will open the way to even more striking tests of quantum nonlocality [3].

Quantum entanglement may also be used as a tool for practical applications. Quantum cryptography, in its most elaborate versions, uses quantum correlated particle pairs to achieve a secure communication between two stations [11]. Any attempt to eavesdrop on a component of the pair reflects on the state of the other and makes clear that the quantum channel is no longer safe. New versions of quantum communication channels have been proposed recently, using more complex entanglement manipulation schemes [12].

Both fundamental tests and applications require the production of quantum entanglement. Besides the notable exception of the NMR spin manipulation techniques [13], most of the recent schemes belong to quantum optics. Parametric down-conversion is widely used as a source of correlated photons. Teleportation [14] or entanglement swapping [15] are recent examples of the achievements of this approach.

Ions in traps are another extremely promising tool for quantum entanglement. It is now possible to control precisely the internal and vibration states of trapped ions. Entanglement between these degrees of freedom has been demonstrated [16]. More recently, the internal states of two ions in the same trap have been entangled to form an EPR pair [17].

Another route to entanglement uses atoms in cavities [18]. The interaction of atoms with a cavity mode may be used either to prepare a highly nonclassical field state or to generate and manipulate quantum entanglement. We will review briefly experiments performed along these lines in our laboratory with circular Rydberg atoms and superconducting microwave cavities. After a brief description of the experimental setup, we will discuss the atom-cavity resonant interaction. We will show that the Rabi precession of the atom in the empty cavity may be used to perform an atom/cavity state transfer or to create entanglement. We will then describe two experiments. In the first [19], a state transfer is used to realize an elementary quantum memory, holding for a time an atomic coherence in the cavity mode. This operation is essential for the realization of an atom/cavity-based quantum gate [20]. In the second experiment, an EPR pair of atoms is prepared [21]. The entanglement results here from a completely controlled and tailored process. All the phases and weights of this nonlocal quantum state can be freely controlled. We will then conclude and discuss the perspectives opened by these experiments.

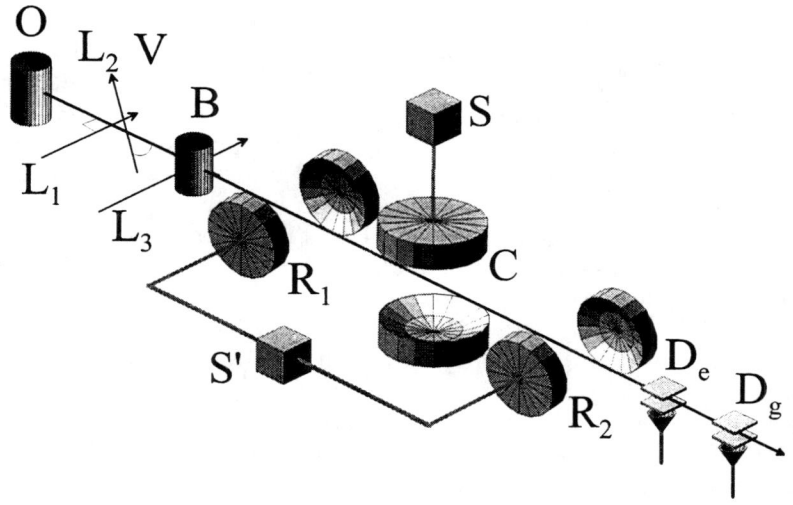

FIGURE 1. Scheme of the atom-cavity experimental setup.

EXPERIMENTAL TECHNIQUES

Circular Rydberg states [22] are very well suited for matter-field coupling experiments. The valence electron of an alkali atom (rubidium in our experiments) is promoted to a level with a large principal quantum number n and maximum orbital and magnetic quantum numbers ℓ and m. Due to the macroscopic size of the orbit, the dipole matrix element for a transition between adjacent states is huge (1250 a.u. for the transition at 51.099 GHz from $n = 51$ (level e in the following) to $n = 50$ (level g). At the same time, the lifetime of these levels (30 ms) is much longer than the duration of the experiment making atomic relaxation negligible. Finally, the circular states may be detected in a sensitive and selective way by the field ionization technique.

In the millimeter-wave domain, superconducting materials like niobium offer the possibility of realizing extremely high-quality cavities. We use Fabry-Perot type cavities, compatible with a small static electric field required to stabilize the circular states. Field-energy lifetimes T_r of the order of 100 to 200 μs are readily achieved (the cavity quality factor Q is of the order of 10^8). This is much longer than the atom-cavity interaction time (20 μs for an atom crossing the centimeter-sized cavity at thermal velocity). Atom-cavity entanglement thus survives the system's separation.

The general scheme of our experiments [19,21,23,24] is depicted in Fig. 1. The ^{85}Rb atoms effuse from an oven O. In zone V, all atoms are optically pumped into the $F = 2$ hyperfine sublevel of the ground state by laser L_1. A short bunch of atoms (2 μs long) in a single velocity class centered around 300 to 400 m/s is then pumped back into the $F = 3$ sublevel by a diode laser L_2 exciting the beam at an angle and resolving the Doppler profile. The width of the selected velocity class varies with L_2 power from 6 to 40 m/s. In zone B, a circular Rydberg-atom sample is subsequently prepared by a combination of lasers L_3 and radiofrequency fields. The preparation process starts from the $F = 3$ sublevel and the L_3 pulse is 2 µs long. The timing, under complete computer control, is set so that L_3 cuts a very narrow slice in the velocity class preselected in zone V. The final velocity of the Rydberg atoms is determined within ±1 m/s. The position of the atoms along their path through the setup is thus known within 1 mm. Precise and independent control of the states of successive atoms is achievable, an essential ingredient in these experiments. In order to operate with single atoms, each preparation pulse produces much less than one atom on the average (about 0.2). When one atom is detected, the probability that a pair was present is below 20%. From the preparation on, the atoms are extremely sensitive to blackbody millimeter-wave fields. The whole atom path is thus contained in an ³He cryostat, cooled down to 0.6K. The mean blackbody photon number per mode is only a few per cent.

The atoms cross the superconducting cavity C. It is made of two spherical niobium mirrors (diameter 50 mm, radius of curvature 40 mm, distance 27 mm). The cavity may be tuned to the atomic line by a mechanical and PZT adjustment of the mirrors' distance. The small electric field applied across the mirrors to stabilize the circular levels may also be used to Stark-shift the line in and out of resonance with the cavity mode at selected times. This allows a precise control of the atom-cavity interaction time. The cavity sustains two orthogonally polarized TEM$_{900}$ modes M_1 and M_2 with comparable decay times (112 and 84 µs respectively). The degeneracy of these modes is lifted by imperfections in the mirrors (giving a frequency difference 70 kHz), and the atoms interact resonantly with only one of them.

The atoms enter finally two ionization detectors D_e and D_g. Atoms in level e are mainly counted in D_e and atoms in g in D_g. About 10% of the atoms are detected in the wrong channel, due to residual spontaneous-emission effects or to imperfections in the ionization field. The overall quantum efficiency of these detectors is about 35%.

Before they enter the superconducting cavity C, the atoms may be prepared in an arbitrary quantum superposition of levels e and g. They interact, in a low-Q cavity R_1, with a classical resonant microwave field producing a rotation of the fictitious spin 1/2 describing the atom (spin up along an arbitrary Oz axis corresponds to e, spin down to g, spin pointing in the equatorial plane of the Bloch sphere represents a mixture of e and g with equal weights). By tuning the amplitude, the phase, and the duration of this field, any rotation can be performed and any atomic state generated.

Another rotation of the fictitious spin can be produced after the interaction with C and before detection in the second low-Q cavity R_2. This rotation, followed by an energy measurement (i.e. by a detection along the Oz axis), is equivalent to a detection along an arbitrary axis. Let us stress that the association of the two

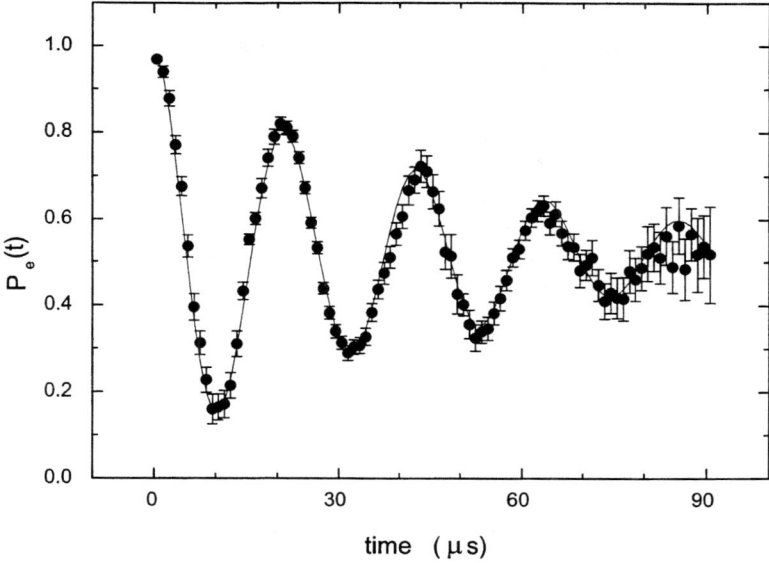

FIGURE 2. Average atomic energy versus effective interaction time. The atom is initially in e and the cavity empty. Four complete vacuum Rabi oscillations are observed.

zones R_1 and R_2 fed by a same microwave source S' is extremely reminiscent of a Ramsey interferometer [25].

An experimental sequence consists in sending one or two atomic samples, separated by a well-defined interval, across the system and detecting them in D_e or D_g. The same sequence is repeated many times, with a repetition period (1.5 ms) longer than T_r, so that C is empty at the beginning of each sequence. Statistics from repeated sequences are then extracted. Samples for joint two-atom probabilities corresponding typically to 15000 events are recorded in about two hours.

RESONANT ATOM-FIELD INTERACTION

The simplest experimental situation corresponds to an atom initially in level e interacting resonantly with an empty cavity mode. One expects a quantum Rabi oscillation between the coupled atom-cavity states $|e, 0\rangle$ and $|g, 1\rangle$. The average atomic energy should be an oscillatory function of time with a frequency $2\Omega/2\pi = 50$ kHz (the Rabi frequency in a single-photon field stored in the cavity), determined by the atomic and field "geometries." This is the oscillatory spontaneous emission, typical of the high-coupling regime of cavity quantum electrodynamics. The photon emitted in the cavity mode remains stored long enough to be reabsorbed by the atom in the lower state and eventually re-emitted.

This Rabi oscillation is easily observed by recording the average atomic energy as a function of the interaction time, determined either by the atomic velocity or by the Stark detuning field inside the cavity [23]. The corresponding signal is shown in Fig. 2 as a function of the effective interaction time, taking into account the spatial structure of the cavity mode. Four complete oscillations of the atomic energy are obtained in the accessible interaction-time range (the lack of very slow atoms in the beam makes it impossible to get interaction times longer than 90 μs). The observed frequency (47 kHz) is extremely close to the expected value. The observation of these oscillations shows that the coherent coupling of the atom with a single photon overwhelms relaxation, a quite unusual situation.

Let us focus on the transformations undergone by the atom-cavity system during this Rabi oscillation. With a proper choice of the relative phases of the atomic levels, after a quarter of a period ($\pi/2$ pulse of oscillatory spontaneous emission) the initial $|e,0\rangle$ state evolves into $(|e,0\rangle + |g,1\rangle)/\sqrt{2}$. The final state is $(-|e,0\rangle + |g,1\rangle)/\sqrt{2}$ when starting from $|g,1\rangle$. In Fig. 2, these states are produced at around 5 and 15 microseconds, respectively. They correspond clearly to an atom-cavity entanglement. Since the preparation time is much shorter than the photon lifetime, the atom moves by a few centimeters outside the cavity before the photon relaxation eventually breaks the nonlocal entanglement.

For a half-period of the Rabi rotation, the transformations are $|e,0\rangle \longrightarrow |g,1\rangle$ and $|g,1\rangle \longrightarrow -|e,0\rangle$ (note that after a full period the state evolves from $|e,0\rangle$ to $-|e,0\rangle$, a well known feature of spin 1/2 rotations). The excitation of the atom and of the cavity are thus exchanged. Assuming an atom entering the empty cavity in a superposition $c_e|e\rangle + c_g|g\rangle$, the atom exits in state $|g\rangle$ and leaves in the cavity a superposition $c_e|0\rangle + c_g|1\rangle$. In other words, the atomic state has been mapped onto the cavity state: a π pulse of spontaneous emission realizes atom/cavity state swapping. Entanglement generation and state swapping are at the heart of two recent experiments [19,21] described in the next sections.

QUANTUM MEMORY

In a first experiment [19], we have used the π spontaneous emission rotation to transfer the state of a first atom onto the cavity field and then back onto a second atom. In other words, we have "written" the state of the atom in the cavity field and then "read" the state of the cavity with a second atom. This is very reminiscent of the operation of a memory, which stores here an elementary quantum information or qubit.

In the simplest version of the experiment, the first atom is prepared in state e. It undergoes a π spontaneous-emission rotation in the cavity and thus leaves a single-photon Fock state in C and exits in g. After a controlled delay T, a second atom is sent in state g through C with an interaction time tuned for a π pulse in a single-photon field. In an ideal experiment, this atom would absorb the photon left by the first one and exit the cavity in state e. In fact, there is a finite probability that

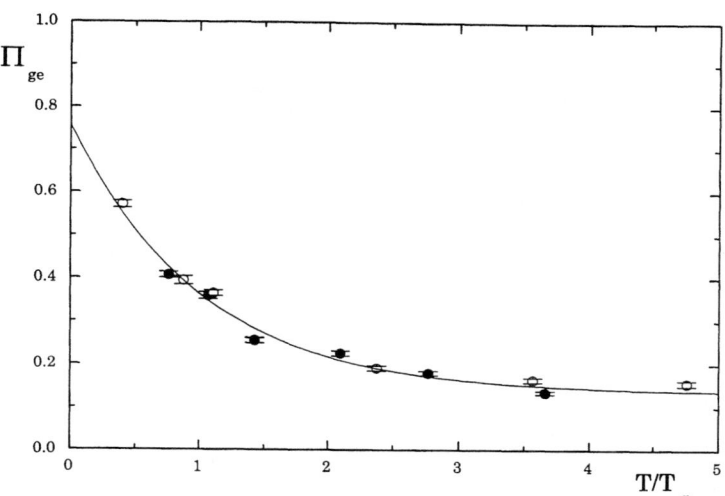

FIGURE 3. Decay of a one-photon Fock state in the cavity: conditional probability $\Pi_{ge}(T)$ versus the delay T between the two atoms expressed in units of cavity mode damping times T_r. Solid and open circles correspond to an experiment performed with one of the two nearly degenerate cavity modes M_1 ($T_r = 112$ μs) or M_2 ($T_r = 84$ μs). The line is an exponential fit with unit time constant and a 13% offset accounting for atomic energy detection errors.

the single photon written by the first atom has decayed away before the second atom enters the cavity. By measuring, as a function of the holding time T, the conditional probability $\Pi_{ge}(T)$ that the second atom exits in e provided the first was detected in g, we measure the lifetime of a single-photon state. The results are displayed in Fig. 3. The maximum probability at $T = 0$ differs from 1 due to the imperfections in the π spontaneous emission rotation. The 13% probability remaining at long times is due to the imperfections of the detectors. As expected, the single-photon lifetime coincides with the lifetime of the classical field energy in the modes, determined independently.

In a second experiment, the first atom enters C in a superposition with equal weights of e and g, prepared by a controlled spin rotation in R_1. With a proper phase choice for the field in R_1, the initial atomic state is $(|e\rangle + |g\rangle)/\sqrt{2}$. The atom undergoes a π spontaneous emission in C. It exits in g and leaves in C the state $(|1\rangle + |0\rangle)/\sqrt{2}$. As a coherent state, this field carries phase information. However, it is nonclassical since it implies a superposition of 0 and 1 photon states.

A variable delay time T after the first atom, a second atom in g is sent in. It absorbs any photon stored in C, which is left empty, and should exit in the state $(-|e\rangle + |g\rangle)/\sqrt{2}$. Besides a π phase shift of the coherence, this is the state of the first atom. To analyze the outgoing state, we use R_2. The detection outcome for the second atom depends upon the relative phase Φ of the atomic coherence in

R_2 with the microwave field. This phase evolves as the product of the detuning between the atomic frequency ω_0 and the frequency ω of the field in R_2, times the total flight time T' of the atomic coherence: $\Phi = \pi + (\omega - \omega_0)T'$. Here, T' includes the time of flight of the first atom from R_1 to C, the holding time T in C, and the time of flight of the second atom from C to R_2 ($T' = T + 216$ μs). We thus expect a modulation of the detection probabilities for the second atom as a function of ω. This modulation is quite reminiscent of the usual Ramsey fringes. Here, however, the pulses are applied to separate atoms, the coherence being transferred through the cavity field.

Figure 4 presents the conditional probability $\Pi_{ge}(\nu)$ to detect the second atom in e provided the first one is detected in g. The observed modulation clearly exhibits the coherence transfer between the two atoms. When the storage time T increases, the fringe spacing reduces as expected, as well as the fringe contrast. This contrast reduction is due to the cavity damping. Figure 5 presents the contrast as a function of T in units of T_r. The solid line is an exponential fit with time constant $2T_r$. Since we consider here a superposition with equal weights of 1 and 0 photon Fock states (lifetimes T_r or infinity, respectively), the lifetime of the coherence is twice the lifetime of a single photon.

This experiment clearly demonstrates that the quantum states of an atom and a cavity can be exchanged at will using spontaneous emission. Let us note that no atom-cavity entanglement is generated here. At any time (but during atom-field interactions) the qubit is carried either by one of the atoms or the cavity and the system's state is factored. This "quantum memory" is an essential ingredient to realize elementary quantum logic operations. An atomic quantum gate [20], for instance, implies an interaction between qubits. There is no direct interaction between circular atoms. Instead, the state of one atom could be copied onto a cavity mode. The interaction with a second atom then takes place, producing the requested entanglement. Finally, a third atom carries away the cavity state. Note that spontaneous emission also allows us to prepare in elementary nonclassical cavity states a simple way.

GENERATION OF AN EPR ATOM PAIR

The atom-cavity interaction produces also atom-cavity or atom-atom entanglement. In this experiment, a pair of entangled atoms in an EPR nonlocal state is generated [21]. A first atom enters C in e and undergoes a $\pi/2$ spontaneous emission. After this interaction, the atom-cavity entangled state is $(|e,0\rangle + |g,1\rangle)/\sqrt{2}$. Since the cavity state cannot be directly analyzed, we copy it onto a second atom before the photon has decayed away. This state-copy operation preserves the entanglement. The second atom, prepared in g, undergoes a π rotation in a single photon field and remains unchanged if the cavity is empty. The cavity is finally empty and the atom pair is in the entangled state $(|e,g\rangle - |g,e\rangle)/\sqrt{2}$.

It is, in terms of spins, the rotationally invariant spin-zero state. An essential

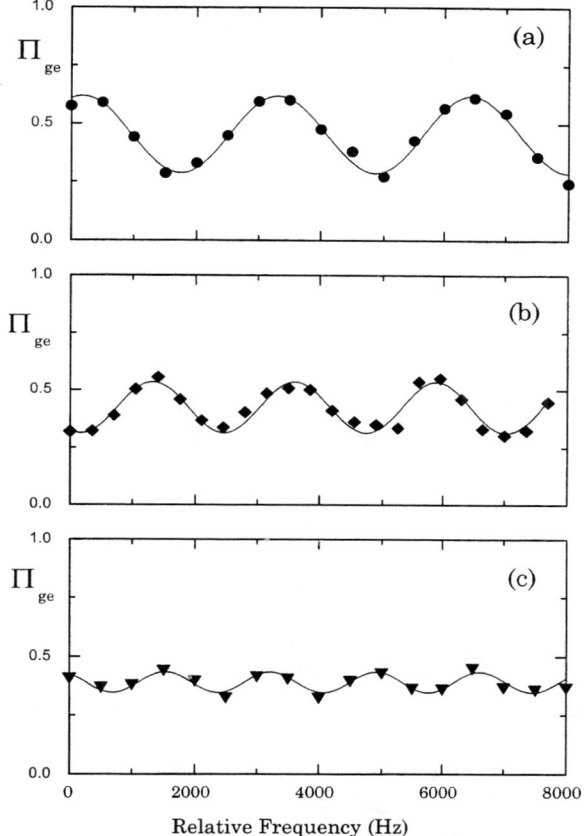

FIGURE 4. Transfer of coherence between two atoms: conditional probability $\Pi_{ge}(\nu)$ versus the frequency $\nu = \omega/2\pi$ of the microwave pulses in R_1 and R_2. The delays $T' = T + 216$ μs between these pulses are 301, 436 and 581 μs respectively from (a) to (c). Cavity mode M_1 has been used.

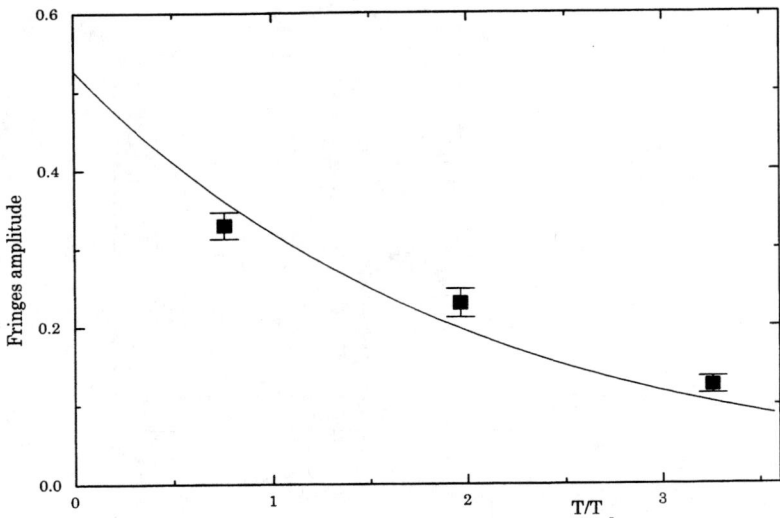

FIGURE 5. Decay of the cavity field coherence: amplitudes of the $\Pi_{ge}(\nu)$ fringes of Fig. 4 versus the delay T expressed in units of $T_r = 112\,\mu s$. Solid line: exponential curve with a time constant 2.

feature is that the outcomes of the detection of the two spins on an *arbitrary* axis are anticorrelated. This anticorrelation exists even if the axis is chosen after the production of the entanglement (delayed choice). Such a basis-independent correlation is the essence of the EPR paradox. Its observation is a clear illustration of the nonlocal nature of quantum states.

We first checked the anticorrelation for an energy measurement (detection axis Oz). The two atoms (velocities 337 and 442 m/s) interact with the cavity at a $T = 26\,\mu s$ time interval. The probabilities of the four possible outcomes should ideally be $P_{eg} = P_{ge} = 1/2$, $P_{ee} = P_{gg} = 0$. We get instead $P_{eg} = 0.44$, $P_{ge} = 0.27$, $P_{gg} = 0.23$, $P_{ee} = 0.06$, with statistical errors of the order of 0.03. The differences are explained by various experimental imperfections. First, the photon left in mode M_1 by the first atom has a finite probability $P_{\text{cav}} = 0.79$ to decay away before the second atom enters C. At the same time, the π rotation for the second atom is not perfect and the probability of absorbing a photon is only $P_{\text{rabi}} = 0.8$ (maximum value in figure 3). Both processes transfer $1 - 0.8 \times 0.79 = 0.47$ of the g, e events into the g, g channel. Adding the detector errors (about 10%), we recover closely the observed probabilities. The imperfections of the "entangler" are well understood. In 63% of the cases, we prepare a genuine EPR pair. In most of the remaining cases, both atoms exit in the ground state.

In order to check the basis-independent anticorrelation, we set R_2 to a $\pi/2$ rotation. The detection of e (g) is thus equivalent to a detection of the spin in the +

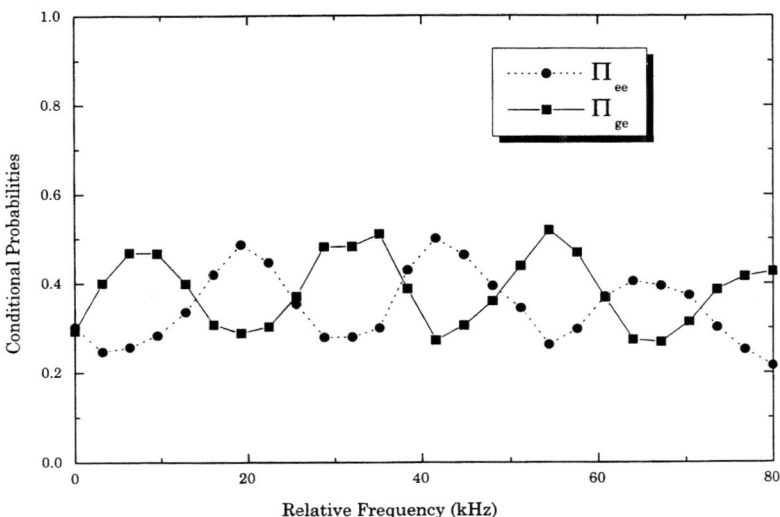

FIGURE 6. Quantum correlation of the EPR atoms pair: conditional probabilities $\Pi_{ee}(\nu)$ (circles) and $\Pi_{ge}(\nu)$ (squares) of measuring the second atom in level e when the first one has been found in e or g respectively, plotted versus the frequency ν of the pulses in R_2. The lines connecting the experimental points have been added for visual convenience.

$(-)$ direction along an axis Ou in the equatorial plane of the Bloch sphere, whose angle ϕ with the Ox axis reflects the phase of the field in R_2. It can be checked easily that the final outcomes of the detection do not depend upon the duration of the atomic path between R_2 and $D_{e,g}$. In the following, we will thus describe the experiment as if the rotation in R_2 and the detection were simultaneous events.

Let us assume, for instance, that the first atom is detected in the $+$ direction along Ou. The second atom is immediately projected onto the $-$ direction along the same axis. Simultaneous detections of the two atoms would provide an exact anticorrelation. In fact, the second atom is 1.5 cm behind the first when this projection occurs (time interval between the two atoms in C: $T = 30$ μs, velocities 413 and 400 m/s). Its state is analyzed in R_2 47 μs after the first atom. During this time interval, T', the relative phase of the atomic coherence and of the field in R_2 changes by $\Phi' = (\omega - \omega_0)T'$. The second atom's state is analyzed along an axis Ou' making an angle Φ' with Ou. When Φ' is an even multiple of π, a perfect anticorrelation is expected. On the other hand, when Φ' is an odd multiple of π, the two detection outcomes should be identical. When scanning ω, a modulation of the detection probabilities for the second atom should be observed.

Figure 6 presents the conditional probabilities Π_{ge} and Π_{ee} to detect the second atom in e provided the first one is detected in g or e, respectively, versus $\nu = \omega/2\pi$. They are modulated with a "period" of $\Delta\nu = 1/T'$. We observe quantum interferences reminiscent of the Ramsey fringes, but the two pulses are here applied to different atoms, after their interaction with C. The coherence is transferred through nonlocal quantum correlations. The contrast of these fringes is limited to 25% by the imperfections of our entangler and, mainly, by the imperfections of the analyzing pulse in R_2. An observation of a violation of the Bell's inequalities [4] would require a substantial increase in contrast. Improvements in our apparatus, now in progress, should allow us to test this violation. More importantly, this experiment shows that two atoms can be entangled in a completely controlled process. Almost any entangled state can be produced by tuning the weights and phases of the superposition. This versatility is extremely important for quantum logic manipulation.

CONCLUSION AND PERSPECTIVES

These two experiments show that circular Rydberg atoms and superconducting cavities are a mature system to study basic quantum properties such as entanglement. Weak relaxation and good control of the atom-field interaction make it possible to prepare a wide variety of quantum states. We have described here experiments based on the resonant atom/field interaction. The index of refraction of a nonresonant atom can also be used to prepare another type of atom/cavity entangled state where the energy of the atom is correlated to the phase of a mesoscopic field. This state implies, as the famous Schrödinger cat [1], a quantum superposition of fields with different macroscopic characteristics. We have been able to monitor the fast relaxation of this state towards a statistical mixture under the influence of decoherence processes [24].

Further studies require some improvements of the setup. The Ramsey interferometer contrast should be increased by better control of the electric fields. The technologically most demanding task is to improve noticeably the cavity quality factor. The mirror preparation is a long process, involving up to ten separate steps. We realized recently a cavity with a photon life time of 1 ms at 51 GHz, 10 ms at 62 GHz. This corresponds to increases of one or two orders of magnitude when compared to the experiments presented here. Further improvements are likely, since we now understand better the main sources of losses.

With such a high-quality factor, a variety of experiments become accessible. The resonant atom-field interaction could be used to prepare a Fock state in the cavity [26]. By cumulative π pulses of spontaneous emission followed by an adiabatic transfer from g to e in an auxiliary classical microwave field, a single atom may be used to pump a few photons into the cavity mode. Up to three-photon Fock states could be prepared and analyzed in the new cavity.

The nonresonant atom-field interaction, in similar experimental conditions, could be used to perform a Quantum Non Demolition (QND) measurement of the cavity field [27]. In the simplest version of this experiment, the photon number is not modified but gets entangled with the state of the atom (g corresponding for instance to one photon—or any odd number—and e to zero photons—or an even number). The atomic detection thus amounts to measuring the photon number (provided no more than one photon is present in the cavity). The repeated measurement of a single-photon field and the observation of quantum jumps due to field relaxation should be possible.

This QND measurement opens the way to a generation of multiparticle maximally entangled states [28]. Let us consider the intermediate state of the EPR pair generation. A first atom interacting with C produced the state $(|e,0\rangle + |g,1\rangle)\sqrt{2}$. A second atom, performing a QND measurement, yields the two-atom–cavity state $(|e,e,0\rangle + |g,g,1\rangle)\sqrt{2}$. A third atom, then copying the state of C, results in an empty cavity and the maximally entangled three atoms state $(|e,e,g\rangle - |g,g,e\rangle)\sqrt{2}$. Besides a trivial modification, this is a GHZ triplet which could demonstrate vividly quantum nonlocality [3]. The method could in principle be generalized to an arbitrary number of intermediate QND atoms, yielding an N-atom complex entangled state. However, the practical realization suffers from low counting rates in multi-atom correlations experiments.

Better cavities open the way also to experiments with larger "Schrödinger cats", deeper in the macroscopic world. A direct determination of the Wigner function of a relaxing Schrödinger cat [29] would be a vivid illustration of the fragility of mesoscopic quantum superpositions. The access to more sophisticated types of entanglement, to states involving larger mesoscopic components, will certainly lead to deeper insights into quantum theory.

REFERENCES

1. Schrödinger, E., *Naturwissenschaften*, **23**, 807, 823, 844 (1935); reprinted in english in Wheeler, J. A., and Zurek, W. H., *Quantum Theory of Measurement*, Princeton Univ. Press, 1983.
2. Einstein, A., Podolsky, B., and Rosen, N., *Phys. Rev.* **47**, 777 (1935).
3. Greenberger, D. M., Horne, M. A., and Zeilinger, A., *Am. J. Phys.* **58**, 1131 (1990).
4. Bell, J. S., *Physics*, **1**, 195 (1964).
5. Kasday, L. R., Ullman, J. D., and Wu, C. S., *Nuovo Cimento* **25**, 633 (1975); Wilson, A. R., Lowe, J., and Butt, D. K., *J. Phys.* **C2**, 613 (1976); Bruno, M., D'Agostino, M., and Maroni, C., *Nuovo Cimento* **B40**, 142 (1977).
6. Lamehi-Rachti, M., and Mittig, W., *Phys. Rev. D* **14**, 2543 (1976).
7. Clauser, J. F., *Phys. Rev. Lett.* **36**, 1223 (1976); Fry, E. S., and Thompson, R. C., *Phys. Rev. Lett.* **37**, 465 (1976).
8. Aspect, A., Grangier, P., and Roger, G., *Phys. Rev. Lett.* **47**, 460 (1981); Aspect, A., Grangier, P., and Roger, G., *Phys. Rev. Lett.* **49**, 91, (1982); Aspect, A., Dalibard, J., and Roger, G., *Phys. Rev. Lett.* **49**, 1804 (1982).
9. Ou, Z. Y., and Mandel, L., *Phys. Rev. Lett.* **61**, 50 (1988); Kwiat, P. G., et al., *ibid.* **75**, 4337 (1995) and references therein.
10. Fry, E. S., Walther, T., and Li, S., *Phys. Rev. A* **52**, 4381 (1995).
11. Bennett, C. H., *Phys. Rev. Lett.* **68**, 3121 (1992).

12. van Enk, S. J., Cirac, J. I., and Zoller, P., *Phys. Rev. Lett.* **78**, 4293 (1997).
13. Gershenfeld, N. A., and Chuang, I. L., *Science*, **275**, 350, (1997).
14. Bouwmeester, D., Pan, J. W., Mattle, K., Eibl, M., Weinfurter, H., and Zeilinger, A., *Nature*, **390**, 575, (1997); Boschi, D., Branca, S., De Martini, F., Hardy, L., and Popescu, S., *Phys. Rev. Lett.* **80**, 1121 (1998).
15. Pan, J. W., Bouwmeester, D., Weinfurter, H., and Zeilinger, A., *Phys. Rev. Lett.* **80**, 3891 (1998).
16. C. Monroe, Meekhof, D. M., King, B. E., and Wineland, D. J., *Science*, **272**, 1131 (1996).
17. Turchette, Q. A., Wood, C. S., King, B. E., Myatt, C. J., Leibfried, D., Itano, W. M., Monroe, C., and Wineland, D. J., "Deterministic entanglement of two ions," (submitted for publication). LANL E-print archive quant-ph/9806012.
18. *Cavity quantum electrodynamics*, Berman, P., editor, Academic Press, 1994, p. 123.
19. Maître, X., Hagley, E., Nogues, G., Wunderlich, C., Goy, P., Brune, M., Raimond, J. M., and Haroche, S., *Phys. Rev. Lett.* **79**, 769 (1997).
20. Domokos, P., Raimond, J. M., Brune, M., and Haroche, S., *Phys. Rev. A* **52**, 3554 (1995).
21. Hagley, E., Maître, X., Nogues, G., Wunderlich, C., Brune, M., Raimond, J. M., and Haroche, S., *Phys. Rev. Lett.* **79**, 1 (1997).
22. Hulet, R. G., and Kleppner, D., *Phys. Rev. Lett.* **51**, 1430 (1983); Nussenzveig, P., Bernardot, F., Brune, M., Hare, J., Raimond, J. M., Haroche, S., and Gawlik, W., *Phys. Rev. A* **48**, 3991 (1993).
23. Brune, M., Schmidt-Kaler, F., Maali, A., Dreyer, J., Hagley, E., Raimond, J. M., Haroche, S., *Phys. Rev. Lett.* **76**, 1800 (1996).
24. Brune, M., Hagley, E., Dreyer, J., Maître, X., Maali, A., Wunderlich, C., Raimond, J. M., Haroche, S., *Phys. Rev. Lett.* **77**, 4887 (1996).
25. Ramsey, N. F., *Molecular Beams*, Oxford Univ. Press, NY, 1985.
26. Domokos, P., Brune, M., Raimond, J. M., and Haroche, S., *European Physical Journal D* **1**, 1 (1998).
27. Brune, M., Haroche, S., Lefèvre, V., Raimond, J. M., Zagury, N., *Phys. Rev. Lett.* **65**, 976 (1990).
28. Haroche, S., in *Fundamental problems in quantum theory*, Greenberger, D., and Zeilinger, A., eds, *Ann. N. Y. Acad. Sci.* **755**, 73 (1995).
29. Lutterbach, L. G., and Davidovich, L., *Phys. Rev. Lett.* **78**, 2547 (1997).

Atom Interferometers and Atom Holography

Fujio Shimizu*, Jun-ichi Fujita[†], Makoto Morinaga**, Tetsuo Kishimoto*,**, and Satoru Mitake*

*Institute for Laser Science and CREST, University of Electro-Communications, Chofu-shi 182-8585, Japan
[†]NEC Fundamental Research laboratories, 34 Miyukigaoka, Tsukuba 305-0841, Japan
**Department of Applied Physics, University of Tokyo, Bunkyo-ku, Tokyo 113, Japan

Abstract. Various techniques of atom manipulation with a binary hologram are discussed and demonstrated experimentally. An atomic beam of metastable neon in the $1s_3$ state and a SiN thin film with holes that expresses the transmission function of the hologram are used to demonstrate this technique. The gray-scale holography of atoms is demonstrated for the first time. Other possibilities of holographic manipulation of atoms are also discussed.

INTRODUCTION

The interferometry of atoms with artificially made structures began in the mid 1980s. In 1986, Gould et al. [1] observed the Fraunhofer interference pattern of a Na atomic beam from a transmission grating. Since then, various interferometric processes that use thin films with holes as diffraction components have been demonstrated. They include Young's double-slit interference [2], grating interferometers with a transmission grating [3], and the focusing of an atomic beam by a Fresnel lens [4].

In previous work, we showed that laser-cooled atoms in a metastable state are able to draw two-dimensional interferometric patterns on a charged-particle detector with spatial resolution. We drew two-dimensional interference patterns from a double slit on a micro-channel plate detector (MCP) by using an ultra-cold metastable neon atomic beam that was released from a magneto-optical trap [5]. A laser that optically pumped $1s_5$ metastable neon atoms to the $1s_3$ state was focused into the magneto-optical trap of the $1s_5$ atoms. The $1s_3$ atoms started to drop, pulled by gravity. The gravitational acceleration reduces the vertical velocity distribution quickly. By placing the atomic source, thin-film structure, and MCP vertically, an interference pattern from a monochromatic atomic de Broglie wave can be obtained relatively easily. When the diffraction angle is small, the interfer-

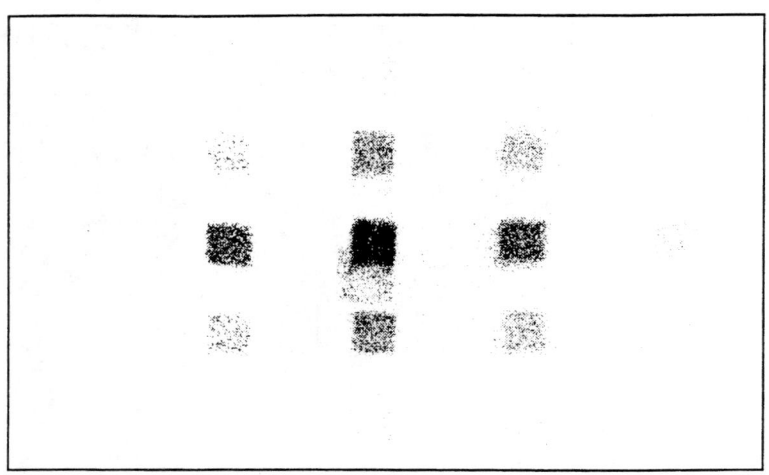

FIGURE 1. Diffraction of atoms from an array of rectangular holes

ence pattern has the same shape as that of an monochromatic optical wave, if we replace the traveling distance of the atoms with their transit time. Figure 1 shows an example of a Fraunhofer interference pattern of atoms that was generated by an array of rectangular holes with $d_x = 250$nm by $d_y = 500$nm. The separation of the holes was 250 nm in both directions. The holes were made on a SiN thin film of 100 nm thickness. The de Broglie wavelength of the atom on the film was 7 nm.

This simple diffraction pattern gives us information that has to be taken into account when using the $1s_3$ metastable atoms for interferometry. The intensity of the first-order diffraction pattern is slightly larger than that expected from the above length ratio. When an atom passes through the hole, it is deflected by the potential that is created by the wall of the hole. If a metastable atom passes through too close to the wall, it decays to the ground state and cannot be detected by the MCP. From the intensity ratio, we estimated a quenching distance of approximately 12 nm. A weak square pattern that overlaps to the zero-order pattern arises from the 74 nm VUV photons. When the $1s_5$ atoms are pumped to the $1s_3$ state, half of the atoms decay to the ground state by emitting a 74 nm VUV photon. The disturbance can be reduced by taking the distance between the source and detector large, because atoms follow parabolic paths, whereas photons move along straight lines. Complete elimination of VUV patterns is possible if we place the thin film and MCP off-axis from the vertical line.

The demonstration of a two-dimensional diffraction pattern shows that it is practically possible to produce an arbitrary pattern of atoms with a thin-film hologram. In this paper we discuss details of atomic de Broglie wave front reconstruction by holographic techniques.

TRANSMISSION FUNCTION OF HOLOGRAM

We consider the configuration shown in Fig. 2. A spherical wave emitted from the atomic source illuminates the hologram. Partial waves that are diffracted by each point of the hologram interfere on the screen and produce a designed pattern of atomic de Broglie wave.

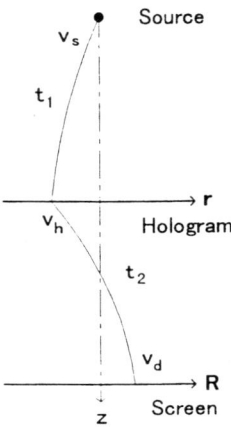

FIGURE 2. The trajectory of an atom from the source to the detector.

The wave amplitude $F(\mathbf{R})$ at a point $\mathbf{R}$ on the screen can be calculated by summing the contribution of all partial waves from the hologram surface,

$$F(\mathbf{R}) = \int f(\mathbf{r}) e^{i\Phi} d\mathbf{r},$$

where $f(\mathbf{r})$ is the transmission function of the hologram and Φ is the integrated phase of the de Broglie wave along the classical path of the atom from the source to hologram and from the source to screen (see Fig. 2). Assuming that the atomic source is at the origin and that the atom is pulled towards the z axis with a constant force, the phase Φ is given by the sum of the two contributions,

$$\Phi_1 = \frac{m}{3\hbar}(v_s^2 + v_s v_h + v_h^2) t_1 + \frac{m}{2\hbar t_1} \mathbf{r}^2 - \frac{m}{8\hbar v_s v_h t_1^3} \mathbf{r}^4 + \cdots$$
$$\Phi_2 = \frac{m}{3\hbar}(v_h^2 + v_h v_d + v_d^2) t_2 + \frac{m}{2\hbar t_2} |\mathbf{r} - \mathbf{R}|^2 - \frac{m}{8\hbar v_h v_d t_2^3} |\mathbf{r} - \mathbf{R}|^4 + \cdots, \quad (1)$$

where t_1 and t_2 are the transit times from the source to hologram and from the hologram to screen, respectively. The velocities v_s, v_h and v_d of the atom are at the source, the hologram, and the screen, respectively.

When **r** and **R** are small, we may take only the quadratic terms to calculate $F(\mathbf{R})$. The result is

$$F(\mathbf{R}) = A\exp(i\frac{m}{2\hbar t_2}\mathbf{R}^2) \int \left[f(\mathbf{r}) \exp\left\{i\frac{m}{2\hbar}(\frac{1}{t_1}+\frac{1}{t_2})\mathbf{r}^2\right\}\right] \exp\left\{-i\frac{m}{\hbar t_2}(\mathbf{rR})\right\} d\mathbf{r}.$$

Since this is a Fourier transform of $f(\mathbf{r})$ multiplied by a spherical phase factor, we may Fourier transform back to obtain the transmission function of the hologram to produce a designed complex amplitude of the atomic de Broglie wave $F(\mathbf{R})$, as

$$f(\mathbf{r}) = A'\exp\left\{-i\frac{m}{2\hbar}(\frac{1}{t_1}+\frac{1}{t_2})\mathbf{r}^2\right\} \int \left[F(\mathbf{R})\exp\left\{-i\frac{m}{2\hbar t_2}\mathbf{R}^2\right\}\right]$$
$$\exp\left\{i\frac{m}{\hbar t_2}(\mathbf{rR})\right\} d\mathbf{R}. \quad (2)$$

Since the Fourier transform can be calculated easily with the FFT algorithm, the quadratic phase approximation is a convenient way to reproduce a fairly complex pattern. The validity of this approximation is estimated from the next term in the expansion (1),

$$\frac{m}{8\hbar v_h v_d t_1^3}|\mathbf{r}-\mathbf{R}|^4 \sim \frac{R^4}{\lambda_{dB} l^3},$$

which should be smaller than unity, where l is the distance from the hologram to screen, and λ_{dB} is the wavelength of the atom. This gives

$$\frac{R}{l} < \left(\frac{\lambda_{dB}}{R}\right)^{1/4}.$$

Laser-cooled atoms typically have de Broglie wavelengths of 10 nm. For an image of the size of 1mm, the quadratic approximation is valid up to the diffraction angle R/l of 0.1.

IMPLEMENTATION OF HOLOGRAM PATTERN

Although it is theoretically possible to generate the arbitrary amplitude function of a wave by a hologram, there are practical limitations. Firstly, since the hologram is a passive component, it is not possible to make a function with extremely large dynamic range. Secondly, an atom cannot transmit coherently through a solid material. The hologram must be composed of holes that represent unity transmission and blocked area representing null transmission. The effect of the restriction is easily understood when the holography is operated near the Fraunhofer-diffraction limit, where the size of the reconstructed image is much bigger than the size of the hologram. In this case, the gross structure of the transmission function of the hologram is the Fourier transform of the reconstructed amplitude function $F(\mathbf{R})$. If $F(\mathbf{R})$ has a uniform phase over a large area, the Fourier transform will have an extremely large peak around the origin, which is practically impossible to implement on the hologram. In such a case the reconstruction is possible only for a very small object that can disperse the large peak at the origin of the Fourier transform. When the patterning of atoms or the intensity distribution of the atomic wave is the object of the operation, the situation is different. We may multiply the amplitude $F(\mathbf{R})$ by a random phase factor, where the correlation length of the random phase function is chosen approximately equal to $l\lambda_{dB}/D$, where D is the size of the hologram. Multiplication by a phase factor does not change the intensity profile of the atomic wave on the screen. However, the amplitude $F(\mathbf{R})$ of each point on the screen is reconstructed by partial waves from a large area of the hologram. Therefore, the sharp peak near the origin disappears.

The technique describe here is a modification for atoms from the binary computer hologram of light [6,7]. To take full advantage of the FFT calculation, we divide a square hologram of the length D by $2^N \times 2^N$ square cells and assign either through hole or opaque area on each cell, representing either unity or zero transmission, respectively. We choose the following method to approximate the complex transmission function of the hologram $f(\mathbf{r})$ by a binary function. We divide first the object figure into $2^N \times 2^N$ meshes. Then, we use for the amplitude F

$$F(n_x, n_y) = \sqrt{I(n_x, n_y)} \exp\{i\phi_{rand}(n_x, n_y)\}, \qquad (3)$$

where $I(n_x, n_y)$ is the density of atoms on the object. We calculate Fourier transform Eq.(2) to obtain the hologram amplitude function,

$$f(n_x, n_y) = A' \exp\left\{-i\frac{\pi}{2^N}(\frac{t_2}{t_1}+1)(n_x^2+n_y^2)\right\}$$
$$\sum_{k_x,k_y=0}^{2^N-1} \left[F(j,k)\exp\left\{-i\frac{\pi}{2^N}(k_x^2+k_y^2)\right\}\right]\exp\left\{i\frac{\pi}{2^{N-1}}(n_xk_x+n_yk_y)\right\}, \qquad (4)$$

where we chose the unit length a of the image on the screen as

$$a = \frac{2\pi}{2^N}\frac{\hbar t_2}{md}. \qquad (5)$$

To approximate the complex function $f(n_x, n_y)$ by a binary function $f_b(n_x, n_y)$, we add the complex-conjugate function $f^\dagger(n_x, n_y)$ to $f(n_x, n_y)$

$$f_r(n_x, n_y) = f(n_x, n_y) + f^\dagger(n_x, n_y), \qquad (6)$$

and then set a threshold f_{th}

$$f_b(n_x, n_y) = \begin{cases} 1 & \text{if } f_r(n_x, n_y) > f_{th} \\ 0 & \text{if } f_r(n_x, n_y) < f_{th} \end{cases}. \qquad (7)$$

We make a hole at the cell (n_x, n_y) with $f_b(n_x, n_y) = 1$.

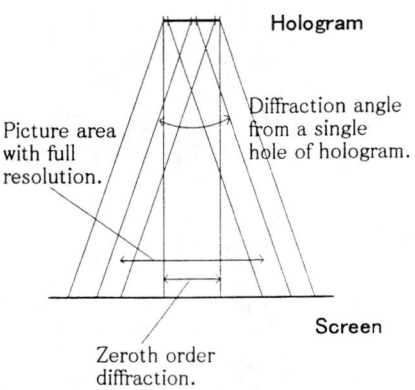

FIGURE 3. The formation of an atomic hologram in the case of near Fraunhofer diffraction.

The addition of the conjugate function $f^\dagger$ and the setting of the threshold produces, in addition to the real image of the object, the conjugate image and the zero-order diffraction spot whose shape is close to the projected pattern of the hologram, respectively. The focal plane of the conjugate image is on the opposite side of the hologram. Therefore, the conjugate image is usually blurred except for the Fraunhofer case. Though the calculation of the hologram pattern is made with discrete variables, the hole has a definite shape. Therefore, the image has various effects arising from the finite size of holes on the hologram. We discuss the case of near Fraunhofer diffraction (see Fig. 3). The atomic de Broglie wave that passes through a single hole produces a Fraunhofer diffraction pattern on the screen. When the hole has a square shape and covers entire cell, the amplitude of the diffracted wave on the screen is a Sinc function with the first zero at the angle $\theta_1 = \lambda_{dB}/d$, which is the direction of the first-order diffraction from the grating with the pitch of d. Diffracted waves from various holes on the hologram overlap completely if the diffraction angle is within $\pm\theta_1$. This is the area where one can draw a pattern of atoms. The zero-order diffraction appears in the center of this area $\theta = 0$. Its size is approximately $(t_1 + t_2)/t_1$ times of the size of the hologram. To avoid overlapping with the zero-order pattern one has to restrict the drawing area on either side of the zero-order pattern and within the first-order diffraction angle from a single hole. One must note that the hologram has the implicit periodicity of the cell size d, although the positions of the holes are random. Therefore, higher-order patterns may appear at various positions on the screen. If the shape of the hole is not equal to the shape of a cell, the higher-order diffraction pattern of the zero-order diffraction appear at angles $\theta = M\theta_1$, where M is an integer. The conjugate image that accompanies to the $M = 1$ diffraction appears on the left side of the angle $\theta = \theta_1$. Therefore, to avoid overlapping of the image with this $M = 1$

conjugate image, the area for the real image has to be confined between $\theta = 0$ and $\theta_1/2$.

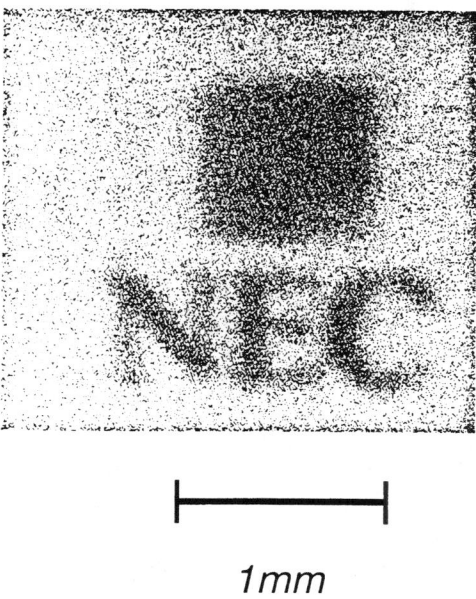

FIGURE 4. The square above the "NEC" is the zeroth-order diffraction (the projection of the hologram). The conjugate image covers an area larger than this picture.

Since the hologram is composed of real penetrating holes, the density of holes is limited to a small number to avoid the dropping of unsupported islands. Unsupported islands can appear even at a very low density of holes depending on the characteristics of the object. It is possible to reduce this risk if one randomizes the hologram pattern $f_r(n_x, n_y)$ by adding a random number before setting threshold. Figure 4 shows an example of holographic image [8] that was generated by the scheme described in this section and with randomization after the Fourier transform. The cell length of the hologram is $500\mu m$, and the number of cells along an axis is 1024 or $N = 10$. The fraction of open cells is approximately 8%. The hologram was placed at 40 cm below the atomic source and the screen at 42 cm further below the hologram. The data was accumulated for 2 hours. The horizontal span of the figure was approximately 2 mm. Since the size of the reconstructed image is of the same order as the size of the hologram, the screen is placed in the Fresnel diffraction region, and the conjugate image spreads over the entire region of the figure. Although the randomization after the Fourier transform reduces the probability of unsupported islands, it also increases the background count of atoms.

TOWARDS A BRIGHTER HOLOGRAM

Unlike the situation with photons, the interaction between atoms is not negligible, and collisions inside the atomic beam destroy the coherence of atomic de Broglie wave. This limits the flux of a coherent atomic beam to a number many orders of magnitude smaller than that of an optical beam. It is important to investigate ways to increase beam intensity. If the beam is given, however, the most practical way is to increase the area of the hologram. As is discussed in the previous section, the unit cell length is determined by the size of the image that we wish to produce. Therefore, to increase the area one has to increase the number of the cell or 2^N. This increases the calculation time of the hologram pattern. To avoid excessively large Fourier transforms, one may divide hologram area into sections and calculate the pattern of each section independently. The resolution of the image is determined by the number of cells in a section, however, the calculation time can be saved by a factor of $(\log_2 M)^2$, where M is the number of sections along an axis. One may further reduce the calculation time by calculating the Fourier transform of a single section. The transmission function of other sections is obtained simply by multiplying by an appropriate phase factor that arises from the shift of the position of the section on the hologram. The position of different section is shifted by an integral multiple of $2^N d$. Because of the choice of the unit length of the image Eq.(5), this shift causes the phase change of an integral multiple of 2π inside the summation of Eq.(4). The information of the shift of the position appears only as a constant phase correction for each cell of the hologram by

$$\exp\left\{-i\frac{t_2}{t_1}2^N \pi (j_x^2 + j_y^2)\right\}.$$

Figure 5 shows an example of the atomic-image reconstruction with a hologram of this configuration. The unit section was composed of 1024 × 1024 cells with the length of 200 nm. Sixteen units were packed in 4 by 4 array making the total area of the hologram 0.8 × 0.8mm. Because of the gain of a factor 16 in the transmission, we could draw a gray scale picture of atoms. Figure (a) is the original object, (b) a part of the hologram pattern, (c) the reconstructed image that was calculated by computer, and (d) the real atomic image. The size of the atomic image of Athena was approximately 3mm by 5mm. The integration time was 30min., and the entire picture is made of approximately 5×10^5 atoms. Figure 5(c) is obtained by inverse Fourier transform of Eq. (4) on the central 2048 × 2048 cells of the hologram. Therefore, the real zero-order diffraction pattern should be twice as large as that of Fig. 5(c).

In this hologram we placed the object on its side. As a result, the hologram pattern was mostly composed of long lines nearly parallel to y axis, which is clearly seen in Fig. 5(b). The fabrication of such hologram was rather difficult because of uneven stress caused by an array of long holes. To reduce this problem we cut a long hole into several pieces so that the longest hole does not exceed three cell lengths. Further a circular hole of 160 nm in diameter was used instead of a 200 × 200 nm

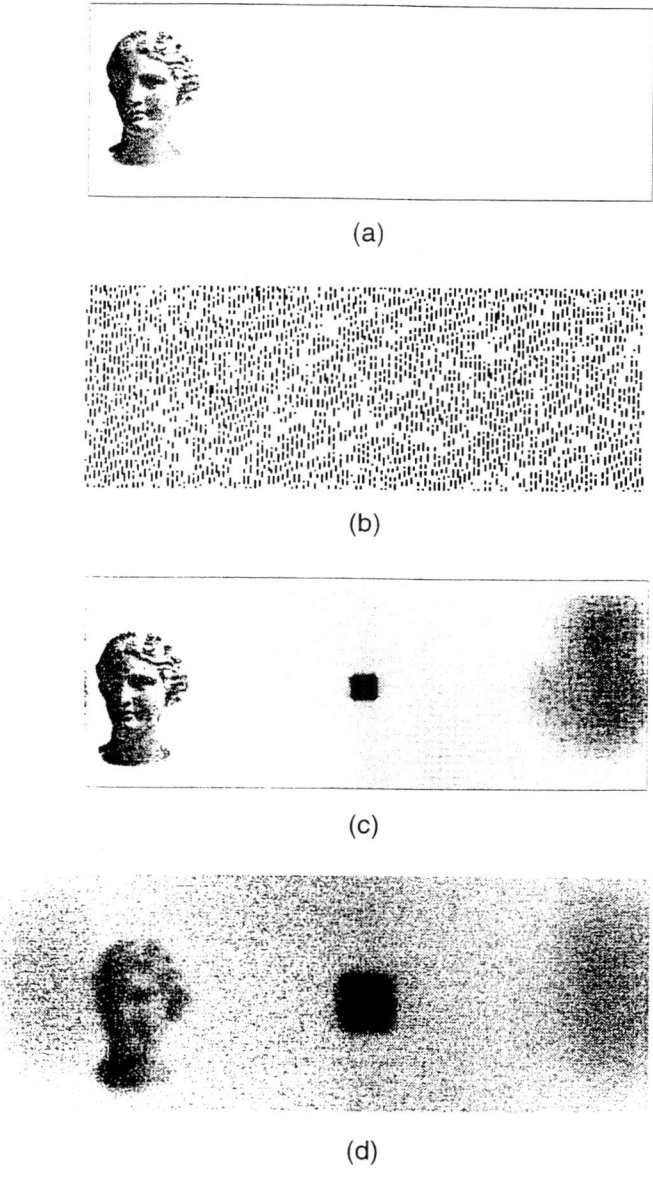

FIGURE 5. A gray-scale hologram. (a) The original picture. (b) Part of the hologram pattern. (c) The image reconstructed by computer. (d) The atomic image.

square hole. This generated a small but clearly observable amount of higher-order images. One of such image is seen on the left side of the image of the Athena statue in Fig. 5(d).

EXPRESSION BY SUBDIVISION OF A CELL

The amplitude transmission function of the hologram can be written more directly by expressing the complex amplitude $f(n_x, n_y)$ with the shape of the hole on each cell (n_x, n_y). We consider the case of ideal Fraunhofer diffraction. The incoming wave is planar, and the image is constructed at infinite distance. Let us take a reference plane that is tilted by $\theta = \lambda_{dB}/d$ along the x axis. The phase difference between the reference plane and the hologram is same for all cells at the boundary of the cell along the x axis except for an integral multiple of 2π. Lines parallel to the y axis and at the distance of 0 and Δx from the boundary have the phase difference of $2\pi\Delta x/d$. Therefore, a thin slit along the y axis of length Δy on Δx from the boundary can express the phase of $\exp(2i\pi\Delta x/d)$ and the relative amplitude of Δy. We have demonstrated this type of Fraunhofer hologram [9], where a cell was divided into 4×4 sub-cells. The four choices of holes along the x direction represent four phases ± 1 and $\pm i$, and four different amplitudes along the y direction.

This technique generates more accurately the real hologram transmission function $f(n_x, n_y)$ than the method described in the previous sections. This will reduce the amount of background. However, this hologram has the basic structure of an intensity hologram with the periodicity of d. It has a periodic pattern of zero-order diffraction with the interval of λ_{dB}/d and is real, and each zero order diffraction is accompanied by a real and conjugate images. Since the width of the slit δx is much smaller than the cell width d, many zero-order diffraction patterns are covered within the diffraction angle from a single cell $\lambda_{dB}/\delta x$.

DISCUSSION

We have shown various techniques of holographic manipulation of atoms with porous thin films. There are many possible holographic techniques to manipulate the atomic de Broglie wave. The intensity hologram described above is accompanied inherently by conjugate and non-diffracted images. To avoid irrelevant images it will be possible to make a phase hologram by using the second-order Stark effect. The Stark potential is $(1/2)\alpha E^2$, where α is the polarizability of the atom, and E is the static electric field. For Ne in the $1s_3$ state, α is roughly 10^{-38}Fm^2. This value of α is sufficiently large to produce a phase shift of π with the electrostatic potential of less than one volt for a micron-size structure. For a one-dimensional pattern an arbitrary pattern can be generated by a grating with electrodes on both sides of each slit that control the phase of the atoms that pass through. It will be difficult to extend this method to two-dimensional patterns. However, we may control a set of discrete phases by using a set of differently shaped holes. In this case we coat both sides of the film with metal, and an electric potential is applied across the film. If the shape of the hole is different, the average phase shift of the atom is also different.

Another possibility is a Bragg hologram by standing waves of light. A Bragg hologram is a three-dimensional structure and is difficult to realize with solid material in atomic holography. A standing wave of near-resonant light that has a wavelength equal to λ_{dB} can be used as a hologram. If the scattering process that generates the holographic image from the hologram is decomposed into Fourier components, each process in k space is equivalent to the degenerate four-wave mixing where two optical waves interact with two atomic waves. Three-dimensional interference patterns of a planar optical wave and the optical wave scattered from an object reconstructs the object of atoms with the same pattern from a planar atomic wave by Bragg diffraction, if both optical and atomic waves have an identical wavelength. Due to gravity, which steadily changes the de Broglie wavelength of the atoms, it will be difficult to realize practical Bragg holography on earth. Simple estimate shows that the maximum interaction length is on the order of several tens of microns. Therefore, only simple patterns can be reconstructed.

REFERENCES

1. Gould, P. L., Ruff, G. A., and Pritchard, D. E., *Phys. Rev. Lett.* **56**, 827 (1986).
2. Carnal, O., and Mlynek, J., *Phys. Rev. Lett.* **66**, 2689 (1991).
3. Keith, D. W., Ekstrom, C. R., Turchette, Q. A., and Pritchard, D. E., *Phys. Rev. Lett.* **66**, 2693 (1991).
4. Carnal, O., Sigel, M., Sleator, T., Takuma, H., and Mlynek, J., *Phys. Rev. Lett.* **67**, 3231 (1991).
5. Shimizu, F., Shimizu, K., and Takuma, H., *Phys. Rev. A* **46**, R17 (1992).
6. Brown, B. R., and Lohman, A. W., *Appl. Opt.* **5**, 967 (1966).
7. Lohmann, A. W., and Paris, D. P., *Appl. Opt.* **6**, 1739 (1967).
8. Morinaga, M., Yasuda, M., Kishimoto, T., Shimizu, F., Fujita, J., and Matsui, S., *Phys. Rev. Lett.* **77**, 802 (1996).
9. Fujita, J., Morinaga, M., Kishimoto, T., Yasuda, M., Matsui, S., and Shimizu, F., *Nature* **380**, 691 (1996).

Alignment and Orientation: Opening Remarks

J. H. Macek

Department of Physics and Astronomy, University of Tennessee, Knoxville, TN 37996-1501
and
Oak Ridge National Laboratory, Post Office Box 2009
Oak Ridge, TN 37831

Abstract. Some brief remarks on the twenty-fifth anniversary of the paper by Fano and Macek on polarization of impact excited fluorescence are presented.

This 25th anniversary symposium is an appropriate time to acknowledge the many contributions made by those who have explored the anisotropy of atomic states over the years. It is their contributions that have sustained the field since the first measurements of the polarization of impact induced fluorescence. Such measurements predate by many years the paper whose anniversary we celebrate [1]. The four talks on our program focus on the current status of the field and thereby show have far the field has come since the early days. This progress would no doubt have occurred without the paper by Fano and myself. The essential content was contained after all in the book on angular-momentum theory by Fano and Racah [2]. The main effect of our paper was to crystallize thinking about practical aspects of the characterization of atomic states. I will confine my remarks to a few general comments about the main points of that paper, and then happily call on my four colleagues to tell you about what is really going on in the field.

The use of density matrix elements to characterize the states of microscopic systems was articulated by Professor Fano in his seminal review/pedagogical paper in the Reviews of Modern Physics [3]. Such use quickly became common, but the review also pointed out something that did not become so widespread, namely, that instead of the density matrix, one could use a complete set of mean values of measurable quantities to characterize a state, thereby bypassing the language of the density matrix. This set of mean values was identified with state multipoles, although that concept still involved the density matrix [4]. Later it occurred to us that the Wigner-Eckart theorem could be used to relate measured quantities to mean values of irreducible tensors constructed from angular-momentum operators. In effect, these mean values were proportional to state multipoles, but density matrix elements need never appear [1].

CP477, *Atomic Physics 16*, edited by W. E. Baylis and G. W. F. Drake
© 1999 The American Institute of Physics 1-56396-752-9/99/$15.00

One aspect of the paper was to show, in the practical example of collision-induced fluorescence, how one could bypass the density matrix and get at the mean values directly. The main purpose of our paper, however, was to show just how such mean values, namely, the alignment and orientation parameters, related to collision-induced fluorescence.

The paper turned out to be timely. Particle-photon coincidence measurements were just starting to have an impact in atomic physics, questions of the time evolution of the polarization of emitted light were coming to the fore, and there was a drive to do complete experiments to test theory on a new level. The alignment and orientation anisotropy parameters turn out to particularly well suited to these studies owing to their relation to angular momentum.

As you are all aware, in any atomic process, but particularly in collisions, energy, momentum and angular momentum are exchanged among the various constituents. All three of these quantities are conserved, and in a classical theory, all three quantities may simultaneously have fixed values. Quantum states, however, cannot be simultaneously eigenstates of linear and angular momentum. If the state of linear momentum is fixed, as it usually is in collision experiments, then the angular momentum is not. Angular momentum is conserved, of course, but we cannot use that information directly. The best we can do is describe the mean value of products of the components of angular momentum. It is just these products that completely specify atomic states. By measuring alignment and orientation one learns about the interchange of angular momentum in atomic collisions. The propensity rules proposed by our first speaker (N. Andersen) are a nice example of this aspect of anisotropy measurements. In these rules, the sharp values of the classical theory are manifest as propensities for the geometrical properties of collision-excited states to follow a classical description.

Disentangling the geometrical from the dynamical aspects of anisotropy is another aspect of alignment and orientation. We know from classical radiation theory that when light is observed, only the components of the source projected onto a plane perpendicular to the direction of observation are imaged. The source must be looked at from different angles to determine its complete electromagnetic configuration. This holds for any measurement of radiation; to see the complete source one must look at it from several angles. Irreducible tensors are selected precisely because they have the simplest possible behavior under changes of direction (rotations). This feature of alignment and orientation was also timely in that it facilitated complete measurements. The second and third speakers (K. Bartschat and A. Crowe, respectively) have been instrumental in testing atomic collision theory via complete determination of collision-excited atomic states. It became clear through their work that continuum atomic states are needed in standard computational methods, even to treat transitions between bound states. Anisotropy measurements thus serve to benchmark atomic collision theory, as noted by our second speaker.

Fragmentation of atoms and molecules into three or more pieces represent final states where the number of anisotropy parameters needed to specify the state becomes infinite. At the present time, we usually use the measured distribution to characterize the state. In this area the appropriate "anisotropy" parameters are still evolving. Our fourth speaker (D. Jaecks) describes the truly novel state consisting of two protons and a negative hydrogen atom. That this state emerges from the breakup of H_3^+ is surprising; one would probably expect to see mostly a single proton and the hydrogen molecule. Perhaps H^- manifests the ionic component ($H^+ : H^-$) of covalent bonds (H:H). In any event, this final state involves three charged particles at relatively low energy, a region where many correlated wave functions have been conjectured. There are probably as many three-particle wave functions as there are theorists working in this field. It may that none of them are particularly relevant; thus analyzing distributions to uncover the pathways to the final state provides an indispensable guide for theory.

This research is sponsored by the Division of Chemical Sciences, Office of Basic Energy Sciences, U.S. Department of Energy, under Contract No. DE-AC05-96OR22464 through a grant to Oak Ridge National Laboratory, which is managed by Lockheed Martin Energy Research Corp. Support by the National Science Foundation under grant no. PHY-9600017 is gratefully acknowledged.

REFERENCES

1. Fano, U., and Macek, J. H., *Rev. Mod. Phys.* **45**, 553 (1973).
2. Fano, U., and Racah, G., *Irreducible Tensorial Sets*, (Academic Press, New York, 1959), ch 19.
3. Fano, U., *Rev. Mod. Phys.* **29**, 76 (1957).
4. Fano, U., *J. Research Natl. Bur. Standards, Rept.* **1214** (1951), unpublished.

Propensity Rules for Orientation and Alignment in Atomic Collisions

Nils Andersen

Niels Bohr Institute
Ørsted Laboratory
Universitetsparken 5
DK-2100 Copenhagen
Denmark

Abstract. A series of propensity rules for orientation and alignment in electron-atom and ion/atom-atom collisions is presented. The current understanding of the origin and the range of validity of these rules is investigated. It is shown how such an analysis may yield deeper insight into the collision dynamics responsible for electronic excitation and transfer in atomic collision processes.

INTRODUCTION

Twenty-five years ago, a paper appeared in the *Review of Modern Physics* that would have a profound influence on the way people subsequently explored the information hidden in the polarisation of light emitted from collisionally excited atoms. The authors were Ugo Fano and Joseph H. Macek, and the title of their paper was "Impact Excitation and Polarization of the Emitted Light" [1]. Several other seminal papers from the same period address aspects of the same question, e.g. [2,3], but what sets this paper apart was that it presented a new formalism that, in the words of its abstract, "disentangles geometrical and dynamical effects and stresses the extraction of data on the alignment and orientation of radiating atoms from observations of the emitted light." Orientation and alignment parameters essentially characterize, respectively, the circulation of the active electron around the atomic core and the shape of the excited electron cloud and its direction in space, thereby allowing new kinds of questions to be addressed. An early example of this is illustrated in Fig. 1, taken from the paper "Orientation by Collision" by Kohmoto and Fano [4]. It serves to illustrate how attractive and repulsive forces lead to collisions with opposite values of $\mathbf{p}_i \times \mathbf{p}_f \cdot \mathbf{J}$. Orientation and alignment parameters thus allow us to go beyond the cross section concept, in favourable cases leading to the so-called "perfect scattering experiment" [5,6], in which the set of quantum-mechanical scattering amplitudes is completely determined.

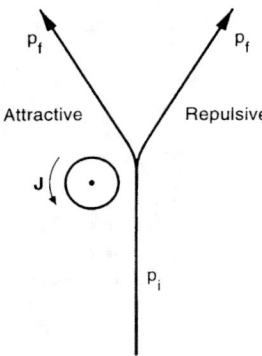

FIGURE 1. Diagram showing how attractive or repulsive forces lead to collisions with opposite values of $\mathbf{p}_i \times \mathbf{p}_f \cdot \mathbf{J}$ [4].

Since 1973, detailed studies of alignment and orientation parameters have, on the one hand, allowed for stringent tests of scattering theories, as illustrated in the subsequent chapters by Bartschat [7] and Crowe [8], and, on the other hand, provided deeper insights into the collision dynamics, leading to the formulation of *propensity rules*, the subject of the present chapter. The concept of a propensity rule has, in the words of Fano, "been introduced with reference to a transition—or to a class of transitions—which is much more likely than alternative but accessible ones. Propensity thus amounts to an attenuated version of a selection rule. Selection rules result from exact symmetries or other properties of a system, "propensities from less clearly identified circumstances" [9]. Below we shall review to which extent collision studies have revealed these circumstances for selected case studies of fundamental excitation processes.

FRAMEWORK

Following the guidelines of Fano and Macek, a convenient framework for systematic analysis of orientation and alignment studies was presented in the review of Andersen, Gallagher, and Hertel [10]. In the simple case of atomic S → P excitation with no change of the quantum state of the collision partner, a perfect scattering experiment may be achieved by measuring the three Stokes parameters (P_1, P_2, P_3) for the light emitted in the direction perpendicular to the scattering plane in coincidence with the scattered particle, see Fig. 2. The Stokes parameters are defined by [11]

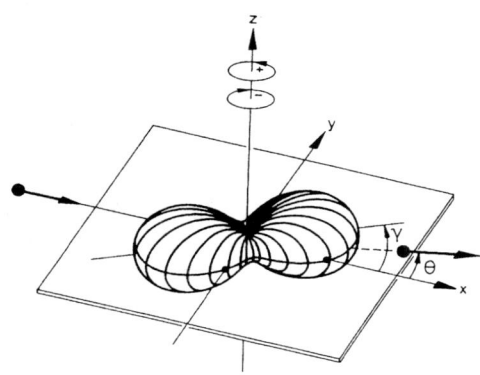

FIGURE 2. A P state may be completely determined in an experimental setup that measures coincidences between particles scattered at an angle Θ and polarization analyzed photons emitted perpendicular to the collision plane.

$$P_1 \equiv \frac{I(0°) - I(90°)}{I(0°) + I(90°)} \tag{1}$$

$$P_2 \equiv \frac{I(45°) - I(135°)}{I(45°) + I(135°)} \tag{2}$$

$$P_3 \equiv \frac{I(RHC) - I(LHC)}{I(RHC) + I(LHC)}. \tag{3}$$

Here $I(\beta)$ is the light intensity transmitted by a linear polarizer with transmission angle β oriented with respect to the incident beam direction, while $I(RHC)$ and $I(LHC)$ are the intensities of right-hand and left-hand circularly polarized light, respectively. In the coordinate frame of Fig. 2, the expectation value of the orbital angular momentum of the excited P state is $\langle \mathbf{L} \rangle = (0, 0, L_\perp)$, and the alignment angle γ is the angle between the incident beam direction and the major axis of the P state charge cloud, as illustrated in Fig. 2. Neglecting for simplicity effects of fine and hyperfine interactions, which may depolarize the radiation, the orientation and alignment parameters are related to the Stokes parameters through the formulas

$$L_\perp = -P_3 \tag{4}$$
$$P_1 + i P_2 = P_\ell e^{2i\gamma}. \tag{5}$$

The linear polarization P_ℓ, which fully characterizes the shape of the orbital, is not an independent parameter, since the Stokes vector $\mathbf{P} = (P_1, P_2, P_3)$ is a unit vector and thus

$$P_\ell^2 + L_\perp^2 = 1. \tag{6}$$

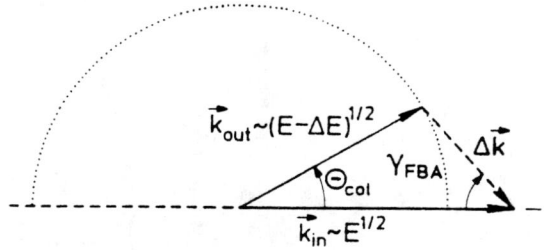

FIGURE 3. In the first Born approximation (FBA) S → P excitation is described as a transfer of linear momentum along the direction of momentum transfer.

The parameter set $(L_\perp, \gamma)$ characterizes the excited P state completely, since the excited state ket vector $|\psi_P\rangle$ can be written as

$$|\psi_P\rangle = \sqrt{\tfrac{1}{2}(1+L_\perp)}\,|+1\rangle + e^{2i\gamma}\sqrt{\tfrac{1}{2}(1-L_\perp)}\,|-1\rangle \ . \qquad (7)$$

We here use the helicity basis with states $|M_L\rangle$ labeled according to the angular momentum projection M_L along a quantization axis perpendicular to the collision plane, see Fig. 2. Note that the P orbital with $M_L = 0$ is not populated in the S → P transition since the atomic reflection symmetry with respect to the scattering plane is a conserved quantity.

CASE STUDIES OF PROPENSITY RULES

Case 1: Orientation for S → P Excitation in Electron-Atom Collisions

We first discuss the simple case of S → P excitation in electron-atom collisions. Within the first Born approximation (FBA), the excitation is described as a transition transferring only linear momentum along the direction of momentum transfer $\Delta\mathbf{k}$, see Fig. 3. Thus

$$L_\perp^{FBA} = 0 \qquad (8)$$

and the alignment angle γ^{FBA} can be calculated by simple trigonometry

$$\tan \gamma^{FBA} = \sin\theta/(\cos\theta - x)\,, \qquad (9)$$

where $x = \sqrt{E/(E - \Delta E)}$. Here θ is the scattering angle, E the impact energy and ΔE the inelastic energy loss in the collision. The classic case of He 2^1P excitation ($\Delta E = 21.2$ eV) at an impact energy of 80 eV is shown in Fig. 4 [12].

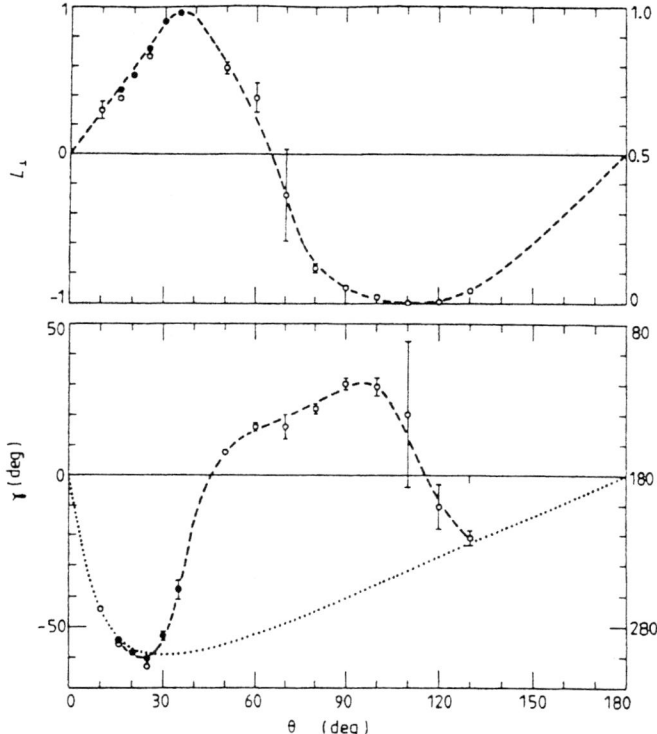

FIGURE 4. The orientation parameter $L_\perp$ and the alignment angle γ as function of scattering angle θ for electron impact excitation of the He $2\,^1P$ state at an impact energy of 80 eV. The dotted curve is the FBA prediction for the alignment angle [12].

Except for forward and backward scattering, the orientation parameter $L_\perp$ deviates dramatically from the FBA prediction, approaching its maximum value of $+1$ near a scattering angle of $35°$ and its minimum value of -1 near a scattering angle of $110°$. At small scattering angles, the alignment angle γ follows the FBA predictions, but beyond an angle of $30°$, dramatic deviations occur.

The pattern shown for $L_\perp$ in Fig. 4, namely that for small scattering angles, the value is positive and increases with angle to a pronounced maximum, has turned out to be a general propensity for electron impact excitation, almost without exceptions. Therefore, considerable efforts have been put into finding a solid theoretical foundation for this empirical propensity rule, the Kohmoto-Fano paper cited in connection with Fig. 1, an early example. A significant step further was the analysis of Madison and Winters [13]. They discussed the orientation parameter in terms of the charge q of the projectile, with $q = \pm 1$ for positron and electron

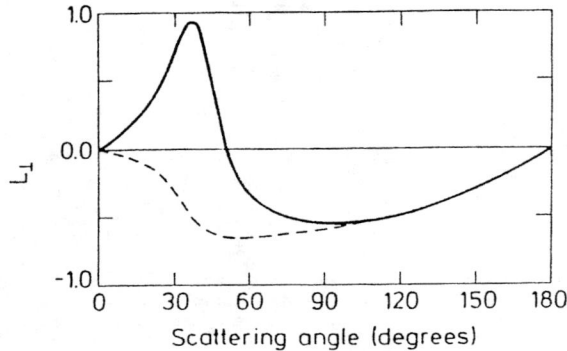

FIGURE 5. The orientation parameter $L_\perp$ for He $2\,^1\mathrm{P}$ excitation at an impact energy of 100 eV for electron (full line) and positron (dashed line) projectiles, respectively [13].

impact, respectively, and expanded the scattering amplitude in a Born series

$$f_{P \leftarrow S} = \langle \Phi_P \mid V \mid \Phi_S \rangle + \langle \Phi_P \mid V G^+ V \mid \Phi_S \rangle + \ldots = q\,t^{(1)} + q^2\,t^{(2)} + \ldots, \quad (10)$$

where V is the interaction potential and G is the free-particle Green's function. This yields the following expression for the transferred angular momentum

$$L_\perp = -2\,q^3\,\mathrm{Im}(t_\sigma^{(1)})\mathrm{Re}(t_\pi^{(2)}) - 2\,q^4\,\mathrm{Im}(t_\sigma^{(2)}t_\pi^{(2)*}) + \ldots . \quad (11)$$

Here, the first term, involving a product of first- and second-order scattering amplitudes, dominates at small scattering angles, and the second one, with a product of two second-order amplitudes, takes over at large angles. Thus, for small scattering angles, the orientation parameter will change sign when switching from electron to positron impact, while at larger angles, the two projectiles will give rise to orientation parameters with the same sign. Figure 5 shows the results of their calculations for the He $2\,^1\mathrm{P}$ state for electron and positron impact, respectively, at an energy of 100 eV. Concerning the actual sign of $L_\perp$ at small scattering angles, however, no definitive, general answer could be derived.

A semi-classical argument for the sign of the orientation parameter, valid for small scattering angles, has been outlined by I. V. Hertel [14] and is summarized in Fig. 6. The starting point is that the electrostatic interaction between an electron and the target atom will always be attractive when the electron begins to penetrate the atomic electron cloud (left column), while the force will be repulsive for positrons (right column). As shown schematically, analysis of the motion of the active electron around the atomic core then yields that $L_\perp > 0$ for electron impact and $L_\perp < 0$ for positron impact.

No decisive improvement has been achieved since then concerning a theoretical, quantum-mechanical foundation of the propensity rule. However, the present satisfactory situation concerning the numerical predictions for electron impact

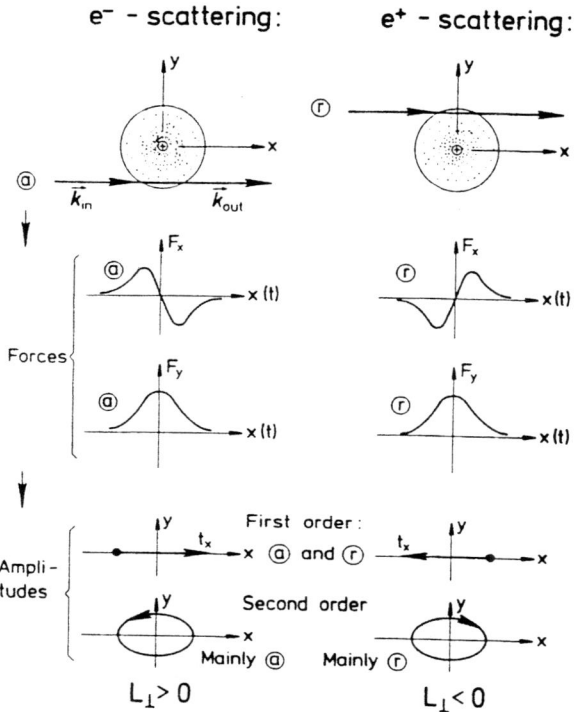

FIGURE 6. A schematic diagram showing the forces acting on the atomic electron charge cloud for small scattering angles of electrons and positrons, together with the resulting excitation amplitudes of the equivalent classical oscillator in a first- and second-order treatment.

excitation of the lighter elements [7,8] might, however, enable further understanding of this problem.

Recent experimental and theoretical progress for spin-polarized electrons has allowed for extraction of spin-dependent orientation and alignment parameters for impact excitation of the Hg $6^1S_0 \to 6^3P_1$ transition [15]. For heavy atoms, explicitly spin-dependent forces, such as the spin-orbit interaction or other relativistic effects, must be taken into account. For this transition, six scattering amplitudes come into play, and a complete scattering experiment becomes necessarily a fairly complicated affair [16]. Here we focus on the spin-resolved orientation parameters only, for which results at 8 eV impact energy display the picture illustrated in Fig. 7, which may be considered a spin-dependent analogue of the Kohmoto-Fano graph, Fig. 1. Note that $L_\perp^{+\uparrow} > 0$ while $L_\perp^{+\downarrow} < 0$ at small scattering angles! While the positive value of $L_\perp^{+\uparrow}$ is in agreement with the well-established propensity rule, the negative value of

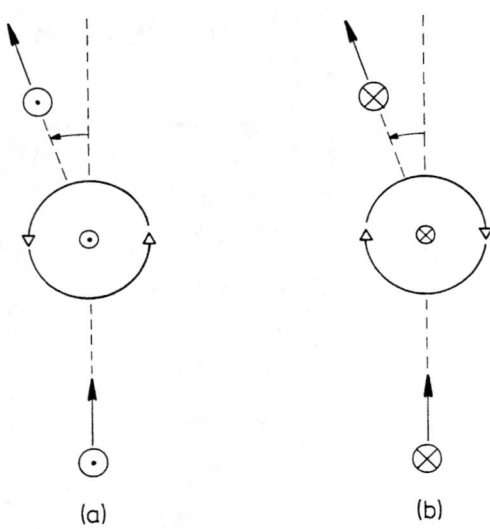

FIGURE 7. Illustration of angular momentum transfer by spin-up (a) and spin-down (b) electrons, as predicted for electron impact excitation of Hg(6s6p)^{3}P$_1$ for small scattering angles at an incident electron energy of 8 eV [15]. Compare with Fig. 1.

$L_\perp^{+\downarrow}$ might seem surprising. Both $L_\perp^{+\uparrow}$ and $L_\perp^{+\downarrow}$ are non-zero for forward scattering, and $L_\perp^{+\uparrow}(0°) = -L_\perp^{+\downarrow}(0°)$ by symmetry requirements.

There is also a large difference between the relative importance of spin-flips in this angular range. Spin-flip is relatively unimportant for spin-up electrons. Spin-flips are very likely for spin-down electrons, but those spin-down electrons, whose spin is not flipped, tend to transfer a *negative* angular momentum to the atom, as indicated in Fig. 7. Presently, the available data for spin-resolved electron impact excitation of heavy atoms are too sparse to reveal if the behaviour shown in Fig. 7 is a singular case or part of a more regular pattern.

Case 2: Orientation for S → P Excitation in Atom-Atom Collisions

As the next case we consider S → P excitation in atom-atom and ion-atom collisions. The main difference when switching from electron impact to ion/atom impact excitation is that for the heavy particle collisions, a trajectory is generally well defined due to the much smaller de Broglie wavelength of the atomic nucleus in the velocity range of interest here. It is thus meaningful to discuss results obtained

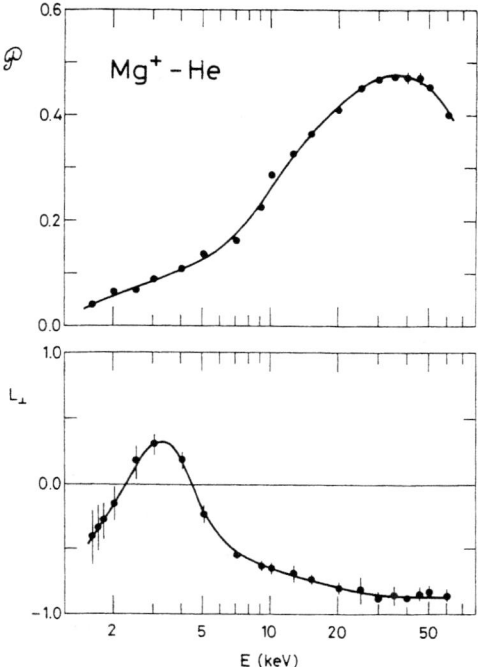

FIGURE 8. Excitation probability and orientation parameter $L_\perp$ versus collision energy for the Mg$^+$–He collision at a fixed impact parameter of 1.0 a_0 ($\tau = 8$ keV·deg) [18].

for, e.g., a fixed impact parameter, corresponding to a fixed value of the reduced scattering angle $\tau \equiv E\theta$, as function of impact velocity or energy.

Complete scattering experiments have been carried out systematically for a series of so-called quasi-one electron systems, i. e., systems in which the active electron is a relatively loosely bound electron outside a closed-shell core, such as an alkali or alkaline earth ion, colliding with a rare gas atom [10,17]. A representative set of results is shown in Fig. 8, which displays the excitation probability and the orientation parameter $L_\perp$ versus impact energy for the Mg$^+$-He collision at a fixed impact parameter of 1.0 a.u. ($\tau = 8$ keV·deg) [18].

The characteristic, general feature gleaned from Fig. 8 is the strong propensity to populate the p_{-1} state near the excitation maximum, where $L_\perp \approx -1$. This is a common feature observed for the direct excitation process in the region of maximum excitation. At energies below the maximum region, this propensity is no longer seen, and the orientation even changes sign, with a $p_{+1} : p_{-1}$ population ratio of 2:1 near 3 keV, and switching once more to predominant p_{-1} population at the lowest energies measured.

A derivation of this propensity rule for the orientation in the region of maximum excitation was actually presented before the experimental verification [19]. Using the minimal, three-state atomic state basis $(|s\rangle, |p_{-1}\rangle, |p_{+1}\rangle)$, from the close-coupling equations one may derive the following first-order expressions for the S → P excitation amplitudes $a_{\pm 1}$ for the two states of opposite helicity, using state amplitudes $(a_s, a_{-1}, a_{+1}) = (1, 0, 0)$ as initial conditions [19] (in atomic units):

$$\begin{cases} a_{-1} = \frac{1}{v} \int_{-\infty}^{\infty} F_{sp} \, e^{i(\frac{\Delta E x}{v} - \phi)} dx \\ a_{+1} = -\frac{1}{v} \int_{-\infty}^{\infty} F_{sp} \, e^{i(\frac{\Delta E x}{v} + \phi)} dx \end{cases} \quad (12)$$

with $F_{sp}(R) = \langle s|V|p_{-1}\rangle = -\langle s|V|p_{+1}\rangle$. Here, V is the interaction between the active electron and the target atom, and $\Delta E = \Delta E(R(x))$ is the S − P energy-level difference at an internuclear distance R at the point x along the trajectory. The angle $\phi = \phi(x)$ is the corresponding rotation angle of the internuclear axis.

The integration extends in principle over the whole trajectory ($x = 0$ corresponds to the distance of closest approach), but in practice only over a finite region a within which the matrix element F_{sp}, typically a smooth, bell-shaped function of internuclear distance R, is different from zero. The angle ϕ rotates by about π during the collision. If one thus tunes the collision velocity v so that the first phase term $\Delta E x/v$ also increases by π during the collision, the two terms will tend to cancel each other during the collision for the a_{-1} amplitude, which thus will acquire its maximum value, while a_{+1} will be very small. The propensity rule thus states that in a velocity range near $v_{\max}$ determined by the "Massey criterion"

$$\frac{\Delta E \, a}{\hbar v_{\max}} = \pi, \quad (13)$$

there will be a strong propensity for populating the p_{-1} state, so that the orientation parameter $L_\perp \simeq -1$. The velocity range for which the above argument applies may be estimated to $\frac{1}{2} < v/v_{max} < 2$, and the most favorable impact parameter by $b \simeq a/\pi$ [19]. From the above formulas, the subsequent oscillations in $L_\perp$ for increasing collision time are also readily understood.

Note that it is important that $\Delta E > 0$. Obviously, if $\Delta E < 0$, in a deexcitation process, then the role of p_{+1} and p_{-1} will be reversed, and $L_\perp \simeq +1$ near the maximum. This prediction still awaits experimental verification. Further interesting consequences are discussed in [20].

The rule is easily generalized to transitions other than S → P excitation. In its general form, the stationary-phase argument above may be stated as

$$\frac{\Delta E \, a}{\hbar v_{\max}} + \Delta m \cdot \pi = 0, \quad (14)$$

with $\Delta m = m_{\text{final}} - m_{\text{initial}}$ the difference between the magnetic quantum numbers of the final and initial states [21].

The detailed collision dynamics responsible for this propensity rule for orientation for direct transitions was further elucidated in a comparative study of hydrogen and helium states excited by proton, antiproton, electron, and positron impact by Lin and collaborators [22]. These studies were soon followed by analysis of the more complicated problem of electron transfer in which the active electron changes center during the collision. Much of the early experimental work was inspired by the pioneering study of Kohring et al. [23], who used the CTMC (classical trajectory Monte Carlo) approach to calculate electron transfer from circular Rydberg states. The quantal description is complicated by the need to use basis states centered on both target and projectile and the need for inclusion of electron translational factors (ETF's) in order to ensure Galilean invariance of the scattering equations. Strong propensities for orientation are found also in this case, again particularly clearly seen when the quantization axis is chosen perpendicular to the scattering plane, see e. g. [24–27].

Case 3: Alignment for P → P Electron Transfer in Ion-Atom Collisions

The third case addresses integral alignment effects in electron-transfer processes. A simple example is the system

$$Li^+ + Na(3p) \rightarrow Li(nl) + Na^+ \qquad (15)$$

with $nl = 2p$ being the most probable final state, constituting a near-symmetric P to P state electron transfer. Collision velocities are comparable to the orbital velocity of the active electron, the so-called "velocity matching region." Here, electron transfer typically takes place at very large internuclear distances, 10-15 a_0, and the corresponding scattering angles are very small, a fraction of a degree [28]. The scattering angle-integrated electron transfer cross section σ, summed over all final states nl, is thus of the order of 100 Å^2, depending on the collision velocity and the optical preparation of the target, i. e., the direction of the Na(3p) orbital in space, conventionally chosen to be parallel ("par") or perpendicular ("perp") to the incoming ion beam direction.

Initial state alignment dependence

The anisotropy with respect to the orbital alignment of the initial P state may be expressed through the dimensionless parameter

$$A = \frac{\sigma_{par} - \sigma_{perp}}{\sigma_{par} + \sigma_{perp}}. \qquad (16)$$

Results for the anisotropy parameter A is shown in Fig. 9, which includes also data for H$^+$ and He$^+$ ions [29–31]. We note a fairly dramatic velocity dependence, with increasingly positive A values beyond the matching velocity. This high velocity dependence might be expected, with electron transfer being favoured for the orbital

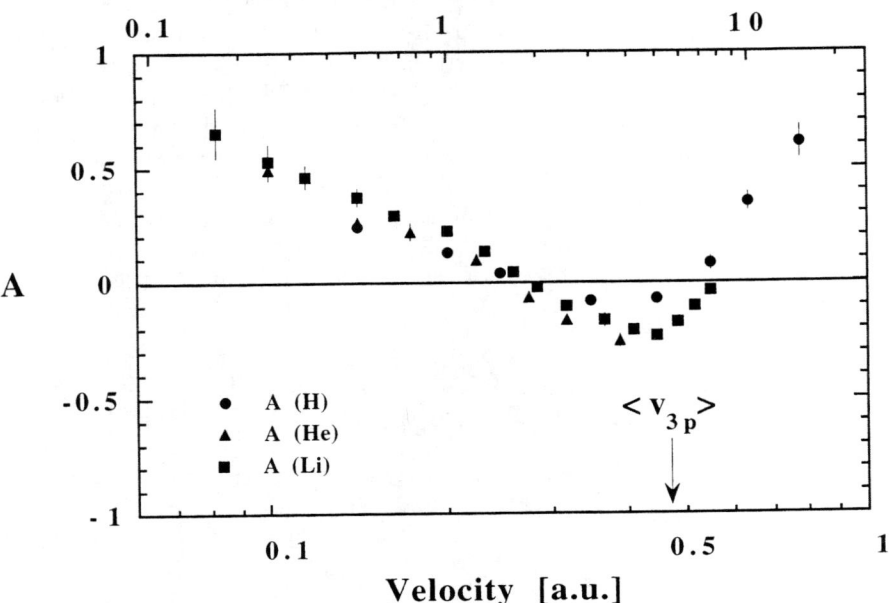

FIGURE 9. The orbital alignment anisotropy parameter A as function of collision velocity for the ions H$^+$(●), He$^+$(▲), and Li$^+$(■) colliding with an optically prepared Na($3p$) target [29-31]. For reference, the mean orbital velocity $\langle v_{3p} \rangle = 0.47$ a.u. of the Na($3p$) electron is indicated

geometry with maximum momentum distribution along the direction of the incident beam. However, the large similarity of the A curves for the three projectiles, including the minimum near $v = 0.4$ a.u., where the perpendicular orbital is the favoured geometry for transfer, is surprising, although in fair agreement with numerical estimates based on close-coupling [28,29,32,33] or CTMC codes [34]. We will return to this feature in the discussion below and here just note that analogous electron transfer studies using He^{2+} [35,36], Ne$^+$ [37], Na$^+$ [38], Ar$^+$ [37], and K$^+$ [39] all yield anisotropy parameters *not* coinciding with the group in Fig. 9. The propensity for perpendicular alignment of the Na($3p$) orbital for electron transfer at a collision velocity near 0.4 a.u. is thus not universal.

Final state alignment dependence

A further step in probing the dynamics is achieved by including also analysis of the final state, focusing on the dominant electron transfer channel

$$\text{Li}^+ + \text{Na}(3p) \rightarrow \text{Li}(2p) + \text{Na}^+ \tag{17}$$

for which further information may be obtained from a determination of the scattering angle-integrated Stokes parameters of the final Li($2p$) state as function

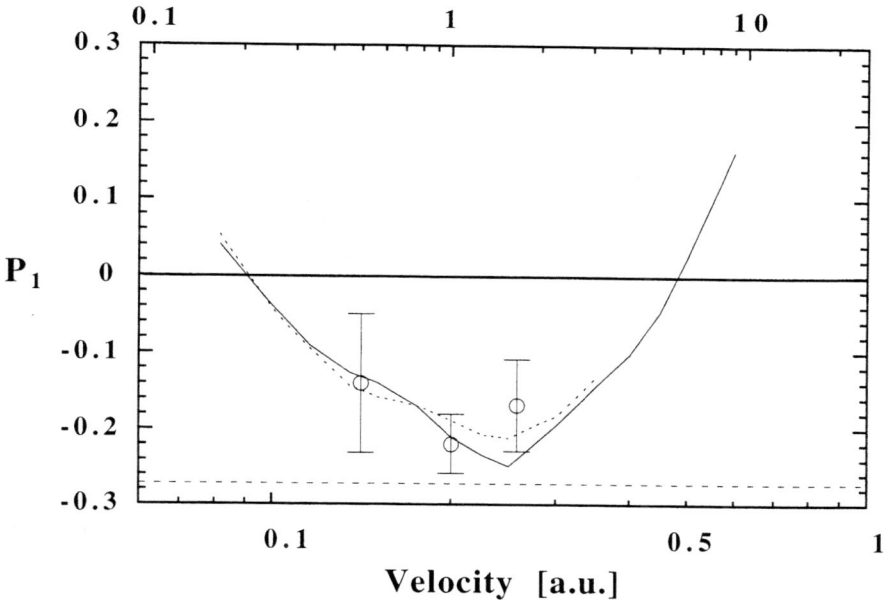

FIGURE 10. The integrated Stokes parameter P_1 for light emitted from the Li(2p) state, measured perpendicular to the ion beam direction, for optical preparation of the Na(3p) target with linearly polarized light along the beam direction. The lower limit, $-3/11$, is shown by a dashed line, and is due to fs and hfs depolarization in initial and final states. Experimental results [41] are shown together with theoretical predictions (full curve [28], dashed curve [33]), including depolarization.

of the initial optical preparation of the Na(3p) target atom. For a P to P state transition such as (17), a complete set of Stokes parameters consists of eight independent components [40].

As a characteristic example, Fig. 10 shows recent results for the integrated Stokes parameter P_1 for light emitted from the Li(2p) state, measured perpendicular to the ion beam direction, for optical preparation of the Na(3p) target with linearly polarized light along the ion beam direction, thus preparing primarily an orbital along the incident beam. Experimental results [41] are shown together with theoretical predictions [28,33]. The lowest possible value, $-3/11$, is shown by a dashed line, and is due to fs and hfs depolarization in initial and final states when using the isotopes ^{6}Li and ^{23}Na and the standard pumping scheme [41]. Without depolarization effects, the range of P_1 would be the full interval $[-1, 1]$. We note that near a collision velocity $v = 0.2$ a.u., P_1 is close to its minimum value (the two companion components P_2 and P_3 being therefore necessarily small), corresponding to a charge cloud density for the excited Li(2p) electron with a very small component along the direction of the incident beam. Remarkably, *this observation is essentially independent of the initial preparation of the target state* [41]. In other words, at this collision velocity, there is thus a universal, strong propensity for the final P state to be aligned perpendicular to the beam direction. Predictions of close-coupling calculations are in agreement with this statement [41], but give no basis for an interpretation of this striking observation.

The Sidky-Simonsen velocity-matching model

The velocity-matching model recently proposed by Sidky and Simonsen [42] offers a convenient framework for understanding of the observed integral alignment propensities for the initial and final P states, respectively, outlined in the two previous paragraphs. Building on earlier work of Brinkmann and Kramers [43] and Schippers *et al.* [44], they present the following picture which holds to a good approximation for large impact parameter collisions at electron velocities near or beyond the matching velocity.

The first step is to transform the scattering problem from configuration space (x, y, z) to a mixed configuration-momentum space, V_{mixed} or (p_x, y, z), by Fourier transforming the coordinate x to the momentum component p_x in the direction of motion of the projectile. In this frame, the transition amplitude $f_{P \leftarrow T}$ for electron transfer from target T to projectile P may be written as (in atomic units)

$$f_{P \leftarrow T} = (factor) \cdot \int_{V_{\text{mixed}}} y \cdot O_{PT} \cdot \delta(p_x - \frac{-\Delta E + \frac{1}{2}v^2}{v}) dp_x \, dy \, dz \qquad (18)$$

where O_{PT} is the overlap of the mixed space wave functions for the projectile final state $\tilde{\psi}_P$ and target initial state $\tilde{\psi}_T$, the former one being shifted in the p_x direction by the collision velocity v and in the y direction by the impact parameter b

$$O_{PT} = \tilde{\psi}_P^*(p_x - v, y - b, z) \cdot \tilde{\psi}_T(p_x, y, z) \,. \qquad (19)$$

The δ-function ensures that energy is conserved in the electron-transfer process. Figure 11 illustrates schematically how the overlap (19) is evaluated. Two situations are of particular interest:

- The δ-function in (18) is centered on the *target* T, or $p_x = 0$. This is seen to be the case if $\Delta E > 0$, i. e., the process is *endoergic*, at a critical velocity $v_c = \sqrt{2\Delta E}$.

- The δ-function in (18) is centered on the *projectile* P, or $p_x = v$. This is seen to be the case if $\Delta E < 0$, i. e. the process is *exoergic*, at a critical velocity $v_c = \sqrt{-2\Delta E}$.

Note that a change of sign of ΔE interchanges the role of target and projectile, as should be the case for the reaction if the direction of time is reversed.

Application of these results to the two situations presented above now explains the observed propensities:

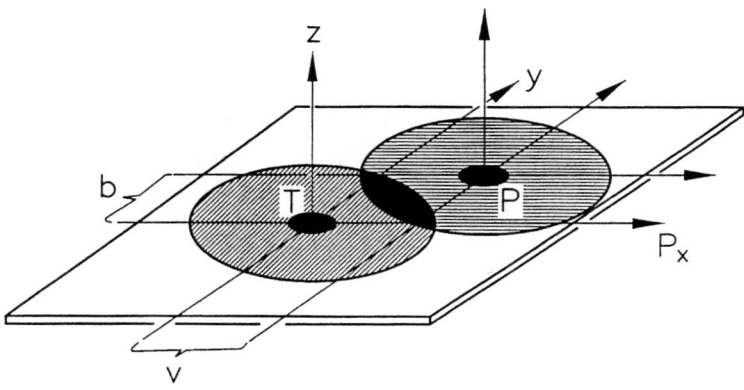

FIGURE 11. In the Sidky-Simonsen velocity-matching model, the electron-transfer scattering amplitude is obtained from an integral in the mixed configuration-momentum space (p_x, y, z) over the overlap of the mixed space wavefunctions of projectile P and target T, taken at the value of p_x fixed by the δ-function in Eq. (18).

- *Initial-state alignment dependence:* For the collisions H$^+$, He$^+$, Li$^+$-Na(3p), $\Delta E > 0$ for the $n = 3$ levels and corresponds to a critical velocity $v_c \approx$ 0.4 a.u. At this velocity, the mixed space wavefunction corresponding to a *target* p orbital aligned along the beam direction will exhibit negative reflection symmetry with respect to the plane defined by the δ-function. The corresponding scattering amplitude (18) is thus zero, giving rise to the observed minimum in the A curve, the negative value corresponding indeed to a propensity for the perpendicular orbital geometry. That this minimum actually is due primarily to the $n = 3$ levels is confirmed by experiment [45] and theoretical calculations [28,29,32–34].

- *Final-state alignment dependence:* For the process (12), $\Delta E < 0$ and corresponds to a critical velocity $v_c \approx 0.2$ a.u. At this velocity, the mixed space wavefunction corresponding to a *projectile* p orbital aligned along the beam direction will exhibit negative reflection symmetry with respect to the plane defined by the δ-function. The corresponding scattering amplitude (18) is thus zero, giving rise to the observed propensity for perpendicular alignment of the charge cloud of the transferred electron.

CONCLUSIONS

Several interesting observations relating to propensities for core orientation [46], inner-shell alignment, alignment in thermal collisions of atoms [47], and recently clusters [48], had to be omitted from this presentation, due to space limitations. Nevertheless, it is my hope that the case studies presented above have served to illustrate how, following the spirit of the Fano-Macek paper, identification and interpretation of propensity rules for orientation and alignment in atomic collisions have served to deepen our understanding of the dynamics of electron-atom and atom-atom collisions.

ACKNOWLEDGMENTS

The author wants to acknowledge many discussions over the years on propensity rules for orientation and alignment with T. Andersen, M. Barat, K. Bartschat, E. E. B. Campbell, C. Courbin, D. Dowek, A. Dubois, U. Fano, S. Grego, J. P. Hansen, I. V. Hertel, J. C. Houver, M. Machholm, D. H. Madison, J. Macek, S. E. Nielsen, R. E. Olson, N. F. Ramsey [49], E. Sidky, H. J. T. Simonsen, and J. W. Thomsen. This work was supported, in part, by the Danish Natural Science Research Council.

REFERENCES

1. Fano, U., and Macek, J. H., *Rev. Mod. Phys.* **45** 553 (1973).
2. Macek, J., and Jaecks, D. H., *Phys. Rev. A* **4** 2288 (1971).
3. Eminyan, M., MacAdam, K. B., Slevin, J., Standage, M. C., and Kleinpoppen, H., *J. Phys. B* **7** 1519 (1974).
4. Kohmoto, M., and Fano, U., *J. Phys. B* **14** L447 (1981).
5. Bederson, B., *Comments At. Mol. Phys.* **1** 41 (1969).
6. Bederson, B., *Comments At. Mol. Phys.* **1** 65 (1969).
7. Bartschat, K., this volume.
8. Crowe, A., this volume.
9. Fano, U., *Phys. Rev. A* **32** 617 (1985).
10. Andersen, N., Gallagher, J. W., and Hertel, I. V., *Phys. Rep.* **165** 1 (1988).
11. Born, M., and Wolf, E., *Principles of Optics*, Oxford: Pergamon Press (1975).
12. Andersen, N., Hertel, I. V., and Kleinpoppen, H., *J. Phys. B* **17** L901 (1985).
13. Madison, D. H., and Winters, K. H., *Phys. Rev. Lett.* **47** 1885 (1981).
14. Andersen, N., and Hertel, I. V., *Comments At. Mol. Phys.* **19** 1 (1986).
15. Andersen, N., Bartschat, K., Broad, J. T., Hanne, G. F., and Uhrig, M., *Phys. Rev. Lett.* **76** 208 (1996).
16. Andersen, N., and Bartschat, K., *Adv. At. Mol. Opt. Phys.* **36** 1 (1996).
17. Andersen, N., Broad, J. T., Campbell, E. E. B., Gallagher, J. W., and Hertel, I. V., *Phys. Rep.* **278** 107 (1997).

18. Panev, G. S., Andersen, N., Andersen, T., and Dalby, P., *Z. Phys. D* **5** 331 (1987).
19. Andersen, N., and Nielsen, S. E., *Europhys. Lett.* **1** 15 (1986).
20. Andersen, N., and Nielsen, S. E., *Z. Phys. D* **5** 309 (1987).
21. Nielsen, S. E., and Andersen, N., *Z. Phys. D* **5** 321 (1987).
22. Lin, C. D., Singhal, R., Jain, A., and Fritsch, W., *Phys. Rev. A* **39** 4455 (1989).
23. Kohring, G. A., Whetmore, A. E., and Olson, R. E., *Phys. Rev. A* **28** 2526 (1983).
24. Campbell, E. E. B., Hertel, I. V., and Nielsen, S. E., *J. Phys. B* **24** 3825 (1991).
25. Toshima, N., and Lin, C. D., *Phys. Rev. A* **47** 4831 (1993).
26. Toshima, N., and Lin, C. D., *Phys. Rev. A* **49** 397 (1994).
27. Salgado, J., Thomsen, J. W., Andersen, N., Dowek, D., Dubois, A., Houver, J. C., Nielsen, S. E., Reiser, I., and Svensson, A., *J. Phys. B* **30** 3059 (1997).
28. Nielsen, S. E., Hansen, J. P., and Dubois, A., *J. Phys. B* **28** 5295 (1995), corrigendum **29** 1419 (1996).
29. Müller, U. *et al.*, *Z. Phys. D* **33** 187 (1995).
30. Thomsen, J. W., *et al.*, *J. Phys. B* **28** L93 (1995).
31. Lauritsen, J. H. V., *et al.*, *J. Phys. B* **29** 1093 (1996).
32. Rod, T. H., and Nielsen, S. E., *J. Phys. B* **28** L607 (1995).
33. Machholm, M., and Courbin, C., *J. Phys. B* **29** 1079 (1996).
34. Lundy, C. J., and Olson, R. E., *J. Phys. B* **29** 1723 (1996).
35. Aumayr, F., Gieler, M., Schweinzer, J., Winter, H., and Hansen, J. P., *Phys. Rev. Lett.* **68** 3277 (1992).
36. Schlatmann, A. R., Hoekstra, R., Morgenstern, R., Olson, R. E., and Pascale, J., *Phys. Rev. Lett.* **71** 513 (1993).
37. Thomsen, J. W., *et al.*, *Z. Phys. D* **37** 133 (1996).
38. Thomsen, J. W., Andersen, N., Campbell, E. E. B., Hertel, I. V., and Nielsen, S. E., *J. Phys. B* **31** 3429 (1998).
39. Thomsen, J. W., *Can. J. Phys.* **74** 950 (1996).
40. Sidky, E., Grego, S., Dowek, D., and Andersen, N., in preparation.
41. Grego, S., private communication.
42. Sidky, E. Y., and Simonsen, H. J. T., *Phys. Rev. A* **54** 1417 (1996), and to be published.
43. Brinkmann, H. C., and Kramers, H. A., *Proc. Acad. Sci. Amsterdam* **33** 973 (1930).
44. Schippers, S., Schlatmann, A. R., and Morgenstern, R., *Phys. Lett. A* **181** 80 (1993).
45. Dowek, D., *et al.*, *Phys. Rev. Lett.* **64** 1713 (1990).
46. Moudry, B. W., Yenen, O., and Jaecks, D. H., *Phys. Rev. Lett.* **71** 991 (1994).
47. Hale, M. O., Hertel, I. V., and Leone, S. R., *Phys. Rev. Lett.* **53** 2296 (1984).
48. Heusler, G., and Campbell, E. E. B., submitted to *Phys. Rev. Lett.*
49. I thank in particular Norman F. Ramsey for kindly drawing my attention to his early, pioneering paper, Ramsey, N. F., *Phys. Rev.* **98** 1853 (1955), demonstrating that the idea of exploring alignment effects in collisions was about 15 years older than I hitherto thought.

Polarization, Alignment, and Orientation in Electron-Atom Collisions: Benchmarks for Atomic Collision Theory

Klaus Bartschat

Drake University
Des Moines, Iowa 50311, USA

Abstract. Experiments with spin-polarized beams of electrons and/or targets are shown to provide very detailed tests of state-of-the-art theory, particularly if the spin-polarization of the reaction partners and/or the polarization of light emitted after impact excitation is measured as well. Although recent advances in the numerical simulation of electron–atom collisions, based upon the "convergent close-coupling" (CCC) and the "R-matrix with pseudo-states" (RMPS) methods, produce impressive results for light quasi-one and quasi-two electron targets, there are still many open questions left regarding the treatment of more complex systems as well as the interpretation of some of the results. Key examples are presented to illustrate the current status of theory and experiment in the field, followed by suggestions for future developments.

INTRODUCTION

Since the fundamental review, *"Impact Excitation and Polarization of the Emitted Light"*, published by Fano and Macek in 1973 [1], scattered-electron–polarized-photon coincidence studies have developed into a key technique that provides data for performing the most detailed tests of atomic collision models to date. Until about a decade ago, most experiments were performed without spin preparation in the initial state and spin analysis in the final state. This work was reviewed by Andersen *et al.* [2] and, with particular emphasis on noble-gas targets, by Becker *et al.* [3].

Except for a few special cases, with the most notable one being $S \to P$ excitation in helium, however, these studies did not represent a "complete experiment" in Bederson's sense [4,5], since averaging over initial spin distributions and failure to observe the spin of the reaction partners after the collision would not allow for an experimental determination of all independent scattering amplitudes. On the other hand, spin-polarization studies without observation of the target had also advanced substantially [6,7].

This situation changed dramatically with the "time-reversed" superelastic e–Na studies of the NIST group [8,9] and the work of the Münster group [10,11] who reported results using spin-polarized incident electrons for excitation of the $(6s^2)^1S_0 \to (6s6p)^3P_1$ transition in mercury. Consequently, the complete experiment became a realistic goal for several collision systems, as demonstrated in several recent reviews [12–14].

In this talk, we present some key examples demonstrating the current status of theory and experiment in the field of polarization, alignment, and orientation studies in electron–atom collisions. After demonstrating the need to perform experiments which test, by investigating independent sets of observables, as many aspects as possible of the theoretical predictions, we review the definitions of polarization, alignment, and orientation and discuss experimental setups for their study. This is followed by a summary of the key ideas behind two currently very promising theoretical approaches to describe electron–atom collisions, namely the "convergent close-coupling" (CCC) and the "R-matrix with pseudo-states" (RMPS) methods. Predictions from these and other methods are then compared with a variety of experimental results to allow for a critical assessment of the present situation and to provide direction for future developments.

DEFINITIONS AND EXPERIMENTAL SETUPS

We begin by reviewing the definitions of polarization, alignment and orientation, as outlined by Kessler [6] and Blum [15]. An ensemble (of electrons) is said to be *spin-polarized* if the (electron) spins have a preferential orientation with respect to a given quantization axis, so that the two possible spin-states ("up" and "down") are not equally populated. The polarization of the ensemble is then defined as

$$P = \frac{N_\uparrow - N_\downarrow}{N_\uparrow + N_\downarrow}. \tag{1}$$

One of many possible consequences of such a spin-polarization is a left-right asymmetry in the differential cross section, an effect called "Mott scattering" [16].

Furthermore, an atomic ensemble is said to be *aligned* if the magnetic sublevels are populated non-isotropically, but the population is invariant against the transformation $M \to -M$. Such alignment is usually related to an axial, and sometimes also to a planar symmetry of the process. In the density-matrix formalism [1,15], it is described in terms of *even-rank* "state multipoles", i.e., irreducible components of the density matrix, and generally leads to a *linear* polarization of the emitted light.

Finally, an atomic ensemble is said to be *oriented* if the magnetic sublevels are populated non-isotropically, but the population is not invariant against the $M \to -M$ transformation. While also related to axial and/or planar symmetry properties of the process, orientation is described by *odd-rank* multipoles and leads to *circular* polarization of the emitted light. Note that systems may be aligned

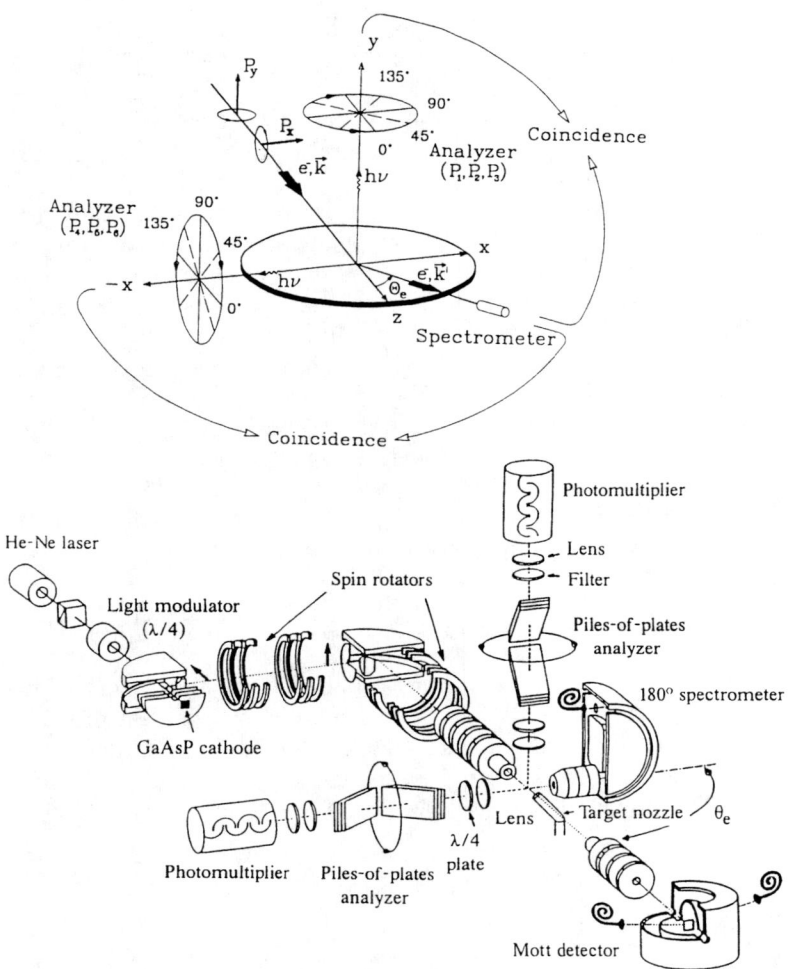

FIGURE 1. Schematic diagram (top) and actual experimental setup (bottom) of the Münster experiment [11] in which spin-polarized electrons are scattered inelastically from atoms at an angle θ. The photons emitted during the decay of the atom in a specific direction are polarization analyzed in coincidence with the scattered electrons without further spin analysis.

and oriented at the same time; often they are simply called *polarized*, but some inconsistencies regarding the nomenclature exist in the literature.

Let us now consider two basic setups, both developed in the Münster group, to perform either scattered-electron–polarized-photon coincidence studies or spin-polarization experiments investigating the projectile spin before and after the collision. The schematic diagram of the first setup and the actual experimental arrangement [11] are shown in Fig. 1. The main difference compared to similar experiments with unpolarized electrons is the GaAs source of spin-polarized electrons which replaces a standard electron gun. After impact excitation of the target, the scattered electrons and the emitted light are observed in coincidence, and the three Stokes parameters,

$$P_1 \equiv \frac{I(0°) - I(90°)}{I(0°) + I(90°)}, \tag{2}$$

$$P_2 \equiv \frac{I(45°) - I(135°)}{I(45°) + I(135°)}, \tag{3}$$

$$P_3 \equiv \frac{I(RHC) - I(LHC)}{I(RHC) + I(LHC)}, \tag{4}$$

are determined. Here $I(\beta)$ is the light intensity transmitted by a linear polarization analyzer oriented at an angle β with respect to a previously defined direction [13,15], while $I(RHC)$ and $I(LHC)$ are the intensities transmitted by filters for right-circularly and left-circularly polarized light, respectively. Note that the result of the polarization measurement depends on the direction of light observation. While various notations have been used to describe the basic cases, the essence of such experiments lies in the determination of the so-called "generalized Stokes parameters" introduced by Andersen and Bartschat [17].

Figure 2 shows the physical interpretation of the results for excitation of an atomic P-state [2]. The alignment angle γ gives the principal axis of the charge cloud relative to the incident beam direction, while the degree of linear polarization P_ℓ indicates the relative length and width; $P_\ell = 0$ thus corresponds to a circular state with the maximum value of the angular momentum transfer $L_\perp$. In some cases, the charge cloud may also exhibit a finite height h that can be measured directly through a $P_1 (\equiv P_4)$ measurement with photon observation *in* the scattering plane.

There are several modifications that can be made to the above apparatus. For example, one might use unpolarized electrons or refrain from a coincidence study and only analyze the light polarization. The latter measurement determines the so-called "integrated Stokes parameters" whose physical importance, especially for polarized incident electron beams, was analyzed by Bartschat and Blum [18]. Alternatively, the same information (and sometimes more) as in electron-photon coincidence experiments can be obtained in "time-reversed" setups that were pioneered by the NIST group [8,9] and involve de-excitation of a laser-prepared target by the incident electron beam.

Figure 3 shows the scheme and an experimental setup where polarized electrons are scattered from unpolarized atoms and the electron polarization after the collision is determined. This type of experiment allows for the determination of the so-called "generalized *STU* parameters" [19] that fully describe the change of an arbitrary initial electron polarization through scattering from unpolarized

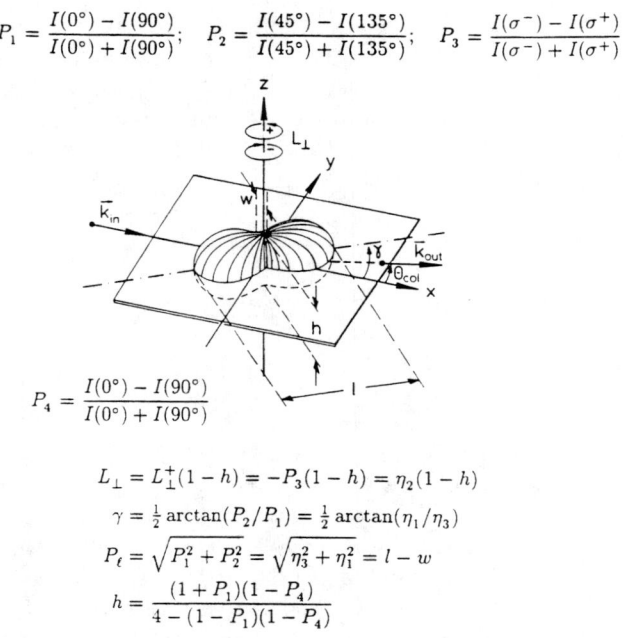

$$P_1 = \frac{I(0°) - I(90°)}{I(0°) + I(90°)}; \quad P_2 = \frac{I(45°) - I(135°)}{I(45°) + I(135°)}; \quad P_3 = \frac{I(\sigma^-) - I(\sigma^+)}{I(\sigma^-) + I(\sigma^+)}$$

$$P_4 = \frac{I(0°) - I(90°)}{I(0°) + I(90°)}$$

$$L_\perp = L_\perp^+(1-h) = -P_3(1-h) = \eta_2(1-h)$$
$$\gamma = \tfrac{1}{2}\arctan(P_2/P_1) = \tfrac{1}{2}\arctan(\eta_1/\eta_3)$$
$$P_\ell = \sqrt{P_1^2 + P_2^2} = \sqrt{\eta_3^2 + \eta_1^2} = l - w$$
$$h = \frac{(1+P_1)(1-P_4)}{4 - (1-P_1)(1-P_4)}$$

FIGURE 2. Physical interpretation of Stokes parameter measurements for excitation of atomic P states. The set $(\eta_3, \eta_1, -\eta_2)$ is a frequently used alternative notation to (P_1, P_2, P_3).

target atoms. Besides the *absolute* differential cross section σ_u for the scattering of unpolarized electrons, there are seven *relative* generalized STU parameters that completely describe the reduced spin density matrix of the scattered projectiles. The polarization function S_P gives the polarization of an initially unpolarized projectile beam after the scattering, while the asymmetry function S_A determines the left-right asymmetry in the differential cross section for scattering of spin-polarized projectiles. Furthermore, the contraction parameters T_x, T_y, T_z describe the change of an initial polarization component along the three Cartesian axes while the parameters U_{xy} and U_{yx} determine the rotation of a polarization component in the scattering plane. Note that the preparation of the projectile and target beams before the collision is identical to the electron–photon coincidence experiment with polarized electrons discussed above, but the photon detector and the coincidence unit for the detection of the final state are replaced by a Mott detector for the spin-polarization analysis of the scattered electrons [20].

As the first example of such experiments and a comparison of their results with theoretical predictions, we show in Fig. 4 the set of Stokes parameters (P_1, P_2, P_3) for electron-impact excitation of the $(5p^5[3/2]6s)$ ("$^3P_1$") state in xenon at an incident energy of 30 eV. (The inverted commas indicate that the LS notation for this state is only an approximation, since it is heavily mixed and must be

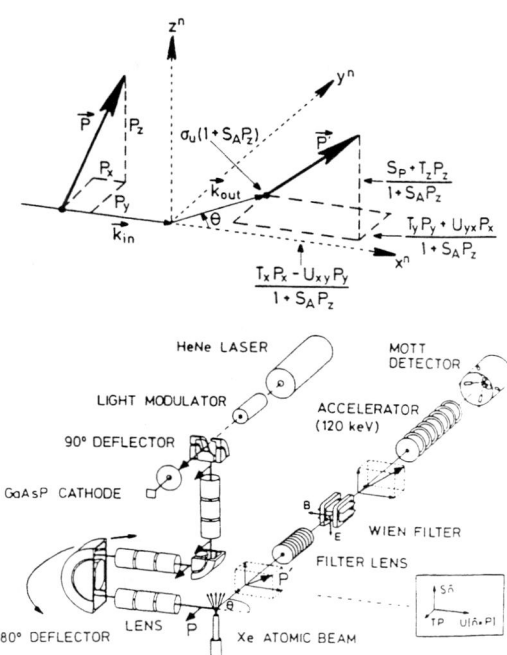

FIGURE 3. Physical meaning of the generalized *STU* parameters (top) for an initial spin polarization **P** which is changed to a final spin polarization **P'** through the scattering process (see text). The bottom part of the figure shows the experimental setup of Berger and Kessler [20].

described by an intermediate coupling scheme. The number in square brackets denotes the angular momentum of the Xe^+ core.) The incident electron beam is unpolarized, and the radiation is observed perpendicular to the scattering plane. After modifying the first-order distorted-wave (DWBA) results of Bartschat and Madison [21] to account for the hyperfine-structure depolarization of the radiation, there is excellent agreement between theory and experiment [22,23] for the set of Stokes parameters. Hence, one might think that the relatively simple, semi-relativistic theoretical approach is sufficient to describe this collision process.

However, this assessment clearly needs to be revised after examining Fig. 5 which displays results for the spin-asymmetry function S_A for impact excitation of the same state as well as the $(5p^5[1/2]6s)$ ("$^1P_1$") state. Investigating an independent observable at the same collision energy apparently reveals major deficiencies in the theoretical model, and the discrepancies between theory and experiment become even more visible due to the wide range of scattering angles for which the experiment was performed. We also note that predictions from a full-relativistic distorted-wave calculation by the Toronto group [24] agrees neither with the semi-relativistic DWBA results nor with experiment over a large range of

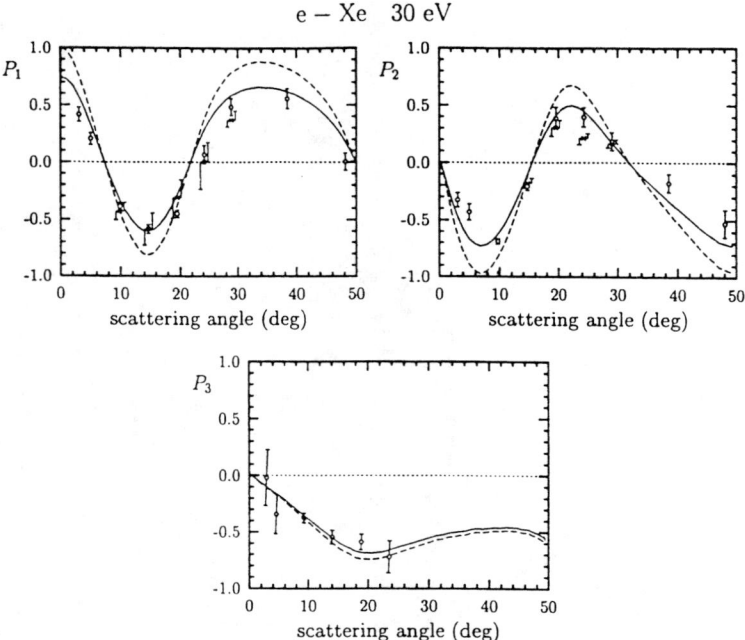

FIGURE 4. Stokes parameters (P_1, P_2, P_3) for electron-impact excitation of the $(5p^5[3/2]6s)$ ("$^3P_1$") state in xenon at an incident energy of 30 eV. The experimental data of Nishimura et al. [22] and of Corr et al. [23] are compared with semi-relativistic DWBA results of Bartschat and Madison [21]. The solid line includes the depolarization effects due to the hyperfine structure.

energies for these and also for other excited states. For more details, we refer to the paper by Dümmler et al. [25].

NUMERICAL METHODS

The above example clearly shows the need for the development of sophisticated theoretical methods, stimulated by numerous such benchmark experiments that were performed over the past 25 years. We therefore summarize the principal ideas behind two recently developed close-coupling methods that account for the effect of the target continuum states in the expansion and keep the coupling between all the physical discrete as well as the bound and continuum pseudo-channels. As will be demonstrated below, the success of these methods for simple target systems such as helium and sodium provide an excellent basis for further improvement. Alternatively, the use of more direct numerical approaches, such as a time-dependent close-coupling approach [26], have become very popular,

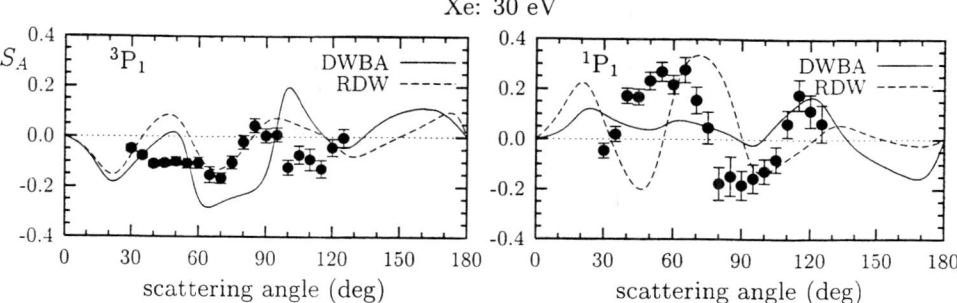

FIGURE 5. Spin asymmetry function S_A for electron-impact excitation of the $(5p^5[3/2]6s)$ ("$^3P_1$") and $(5p^5[1/2]6s)$ ("$^1P_1$") states in xenon at an incident energy of 30 eV. The experimental data of Dümmler et al. [25] are compared with semi-relativistic (DWBA) and full-relativistic (RDW) first-order distorted-wave results of Bartschat and Madison [21] and of Zuo et al. [24].

particularly in light of the recent trends towards massively parallel supercomputers. However, such methods have not yet been used in calculating angle-differential parameters for electron–atom collisions.

The Convergent Close-Coupling Method (CCC)

The details of the CCC theory have been given by Bray and Stelbovics [27]. The method can be thought of as a standard close-coupling approach where, in addition to the discrete target states, the target continuum is treated with the aid of positive-energy, square-integrable pseudo-states. All states are obtained by diagonalising the target Hamiltonian in an orthogonal Laguerre basis. The usage of such a basis ensures that "completeness" is, in principle, approached by increasing the basis size until the results of interest become sufficiently stable.

A key feature of the formalism, as developed by Bray and Stelbovics, lies in the fact that the coupled equations are formulated in momentum space, where they take the form of coupled Lippmann-Schwinger equations for the T-matrix. These are solved separately, upon partial-wave expansion, for each collision energy of interest. As such the method is not ideally suited for the study of detailed scattering behavior as a function of incident energy, as needed, for example, in the vicinity of Rydberg resonance series. However, modern computational resources involving large clusters of workstations allow for the parallel execution of the CCC program at many energies, thereby somewhat reducing this problem. Furthermore, so far the method has only been applied to (quasi-)one-electron and (quasi-)two-electron targets in a non-relativistic framework, although the formalism, as presented by Fursa and Bray [28], is more general.

The R-matrix with Pseudo-States Method (RMPS)

The low-energy R-matrix method (for details, see Burke and Robb [29]) is another method to solve the close-coupling equations, this time in coordinate space. The important difference to the standard formulation is the division of configuration space into two regions, $r \leq a$ and $r > a$, where the R-matrix radius a is chosen in such a way that exchange effects between the projectile and the target electrons can be neglected in the external region. Here the coupled equations (without exchange) are solved for each collision energy and matched, at the boundary $r = a$, to the solution in the inner region. However, instead of solving a set of coupled integro-differential equations in the internal region for each collision energy, the $(N+1)$-electron wavefunction at energy E is expanded in terms of an energy-independent basis set, ψ_k, as

$$\Psi_E = \sum_k A_{Ek} \psi_k. \qquad (5)$$

The basis states ψ_k are constructed as

$$\psi_k = \mathcal{A} \sum_{ij} \Phi_i(1,\ldots,N)\, u_j(N+1)\, a_{ijk} + \sum_i \chi_i(1,\ldots,N+1)\, b_{ik}, \qquad (6)$$

where the Φ_i are N-electron target states, the u_j are members of a complete set of numerical continuum orbitals used to describe the motion of the scattered electron inside the box, and the χ_i are $(N+1)$-electron configurations. The latter are formed from the one-electron orbital basis used to describe the N-electron target. They are included to allow for electron correlation effects when the scattered electron is close to the nucleus, and also to ensure completeness of the total trial wavefunction if the continuum orbitals are constructed orthogonal to the bound orbitals.

Each of the N-electron target states is expanded in terms of a sum of orthonormal configurations

$$\Phi_i = \sum_{j=1}^{m} c_{ij} \phi_j. \qquad (7)$$

The configurations ϕ_j are built up from one-electron orbitals, coupled together to give a function which is completely antisymmetric with respect to the interchange of the space and spin coordinates of any two electrons.

The one-electron orbitals used in constructing the ϕ_j configurations may be of two types: (i) physical (Hartree-Fock) orbitals and (ii) suitably chosen non-physical pseudo-orbitals. In standard R-matrix calculations, performed over many years by the Belfast group and their collaborators worldwide [30], the number of pseudo-orbitals is very small (one or at most two per angular momentum), and generally their sole purpose is to improve the description of the discrete target spectrum.

The R-matrix with pseudo-states (RMPS) method was described in detail by Bartschat *et al.* [31]. The principal idea behind this method is the same as in the

CCC approach, i.e., one includes a large set of pseudo-orbitals which can then be used both (i) to improve the description of the physical target states and (ii) to approximate the effects of the infinite number of discrete and continuum states of the target atom or ion that cannot be included explicitly in the calculation. As in the CCC method, these effects are only represented accurately in a certain region of configuration space (usually the R-matrix box), due to the finite range of the pseudo-orbitals. However, this is expected to be sufficient if one is interested in transitions between discrete states whose range is restricted to within this box. Like the CCC approach, the method also works for ionization if the box is made sufficiently large and some averaging over effective box sizes, determined by different ranges of the pseudo-orbitals, is included [32].

The RMPS method has recently been implemented in updated versions of the Belfast R-matrix codes RMATRX I [33] and RMATRX II [34]. Consequently, it may already be used for complex targets and, in principle, relativistic effects can be accounted for through the inclusion of the one-electron terms in the Breit-Pauli Hamiltonian. Due to the computational demands of such a project, no semi-relativistic RMPS calculation has yet been performed, but work in this direction is currently in progress. Ultimately, an RMPS version of the full-relativistic DARC program [35,36], based upon the Dirac-Breit Hamiltonian, would be highly desirable.

EXAMPLE RESULTS

We now present a few benchmark comparisons between experimental and theoretical results for polarization, alignment, and orientation parameters in electron–atom collisions. We begin with angle-integrated situations where the scattered electrons are not observed. Consequently, the counting statistics are generally much better than in the angle-differential measurements discussed further below.

Angle-Integrated Stokes Parameters

Figure 6 compares recent results for the Stokes parameters (P_1, P_2, P_3) observed by the Perth group [37] after impact excitation of two states in the $(2p^5 3p)$ manifold of neon by a transversely spin-polarized electron beam with polarization P_e as a function of the incident electron energy. The light is observed in a direction perpendicular to the incident beam but parallel to the spin-polarization vector. In this case, P_1 is independent of the electron polarization, whereas P_2 and P_3 are directly proportional to P_e [18].

In the near-threshold regime, there is certainly very satisfactory, though not perfect agreement between the experimental data theoretical predictions from a 31-state (discrete states only) Breit-Pauli R-matrix calculation of Zeman and Bartschat [38]. The very narrow resonance structure predicted by theory was

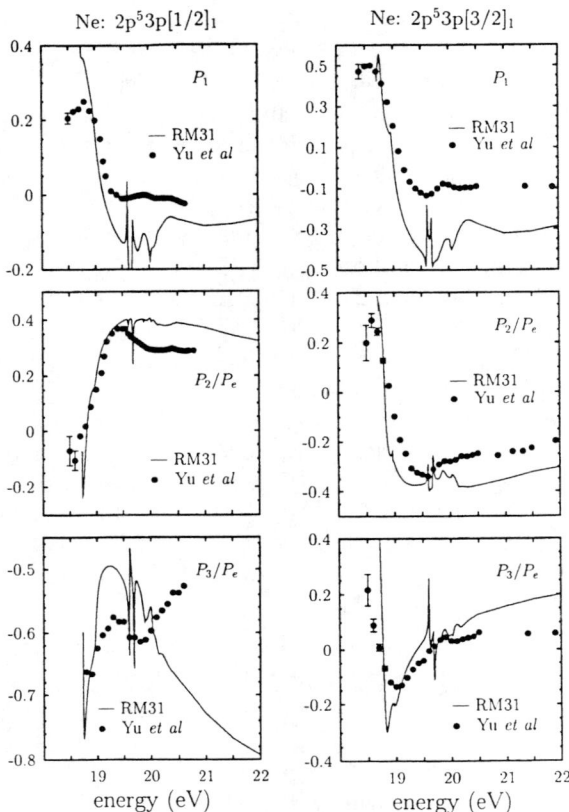

FIGURE 6. Stokes parameters (P_1, P_2, P_3) for electron impact excitation of the $2p^53p[1/2]_1$ (left) and the $2p^53p[3/2]_1$ (right) states in neon and optical decay to $2p^53s[1/2]_1$, plotted as a function of the incident electron energy. The experimental results of Yu et al. [37] (P_2 and P_3 normalized to 100% incident electron polarization) are compared with theoretical predictions based upon a Breit-Pauli R-matrix approach [38].

apparently unresolved in the experiment. We also note that cascade effects, which become substantial for incident energies above approximately 20 eV, generally lead to a depolarization of the emitted radiation and thus explain why theory seems to overestimate the magnitude of the polarization components at higher energies.

Probably the most interesting parameter from a physical point of view is the linear light polarization P_2. While electron exchange, together with some spin-orbit interaction within the target, can lead to non-vanishing values of the circular light polarization P_3, these effects cannot lead to non-zero values of P_2. In other words, spin-polarization of the electron beam can easily be transferred into angular momentum orientation of the excited atomic ensemble, but not into alignment.

In fact, it was shown by Bartschat and Blum [18] that LS-coupling must be violated already *during the collision* for a non-zero P_2 to occur. The two principal mechanisms that have been proposed to cause such a result are:

- The spin-orbit interaction within the target, combined with configuration mixing, makes it impossible to describe the excitation mechanism in LS-coupling. Instead, an intermediate coupling scheme, i.e., combinations of LS-states with different values of L and S but the same value of the total electronic angular momentum J, must be used.

- Coupling of the continuum electron spin to the target angular momenta is so strong that it can effectively rotate the charge cloud (see also Fig. 2). This mechanism represents the optical analogue to Mott scattering, but this time affecting the *target* without even detecting a specific electron scattering angle.

As in the first experimental verification of a non-zero P_2 in an angle-integrated experiment with spin-polarized electrons, carried out by Bartschat *et al.* [39] for electron-impact excitation of the $(6s6p)^3P_1$ state in mercury, the above results can be understood in terms of the intermediate coupling nature of the excited states.

Despite several attempts on different targets [40], however, the optical detection of Mott scattering has not been successful to date. Figure 7 shows the most recent results of Gay's group [41] for the linear polarization P_2 in the $(4p^55p)^3D_3 \rightarrow (4p^55s)^3P_2$ transition in krypton, after impact excitation by spin-polarized electrons. Although krypton is a fairly heavy target ($Z = 36$), the states of interest are well LS-coupled, and hence the intermediate-coupling mechanism is not expected to be relevant. Note that two different ways of determining P_2 did not yield an experimental result that could be interpreted as non-zero in a statistically meaningful way. On the other hand, theoretical predictions from two semi-relativistic 15-state and 31-state Breit-Pauli R-matrix calculations suggest positive values of P_2/P_e in the order of a few percent over an energy range that is free of cascades and several times wider than the energy width of the electron beam. Although the differences between the two theoretical curves clearly indicate a potential lack of convergence with respect to the number of states included in the calculation, the qualitative disagreement between theory and experiment regarding the P_2 parameter is somewhat surprising, since the theoretical model was otherwise very successful in describing near-threshold electron-impact excitation of the noble gases [38,42,44]. The reason for the discrepancy is currently unknown, and hence this problem represents an important open question in the field of polarized electron physics.

We finish this section with one more example related to the P_1 parameter, once again for excitation of the $(2p^53p)$ manifold of states in neon, with subsequent decay to various states of the $(2p^53s)$ manifold. Due to the cylindrical and planar symmetry of the problem, there is only one dynamical parameter, the angle-integrated alignment $\langle t_{20}^+ \rangle$, that fully determines the anisotropy in the emitted dipole radiation. Figure 8 shows results for this parameter as a function of incident

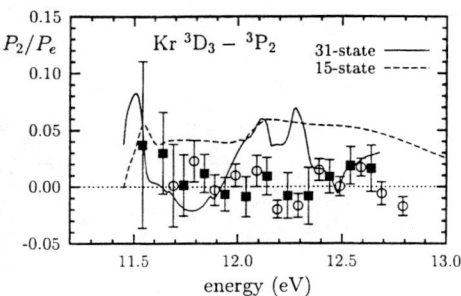

FIGURE 7. Stokes parameter P_2 (normalized to 100% incident electron polarization) for impact excitation of the $(4p^55p)^3D_3 \to (4p^55s)^3P_2$ transition in krypton, plotted as a function of the incident electron energy. The experimental results of Gay's group [41] are compared with theoretical predictions based upon a Breit-Pauli R-matrix approach [38,42].

electron energy for the eight neon states with configuration $2p^53p$ and $J \neq 0$ [42]. (For $J = 0$, $\langle t_{20}^+ \rangle$ is identically zero.) For energies of only a few tenths of an eV above threshold, we note a striking systematic effect: the results cluster by J value of the excited state. Although the agreement between theory and experiment (and between different sets of experimental data) is not perfect, the general trend is clearly confirmed.

This grouping of alignment according to the total electronic angular momentum J of the excited state had not been observed before because of the lack of comprehensive data sets for a given atom, and because much of the previous alignment data was taken with unresolved fine structure. Since the findings can be qualitatively explained using angular momentum coupling rules [42], they are expected to be a general feature for similar excitation processes in Ar, Kr, and Xe, and thus represent an example of a "propensity rule." Such rules are discussed in more detail by Andersen [45].

Angle-Differential Stokes Parameters

Figure 9 shows results for the angular momentum transfer $L_\perp$ and the alignment angle γ or electron impact excitation of the $(1s2p)^1P$ state of helium from the ground state $(1s^2)^1S$ at an incident energy of 50 eV. There is excellent agreement between the CCC [28] and RMPS [46] results, and also with various experimental data. For this optically allowed transition, first-order theories, such as a distorted-wave [51] or a many-body (FOMBT) [52] approach, are generally in reasonably good agreement with experiment as well, especially at small scattering angles and high enough energies. In fact, only the *large-angle measurements* of electron-impact coherence parameters revealed problems with the above first-order theories for this particular collision system—problems that cannot be seen by comparing the predictions for the differential cross section [46].

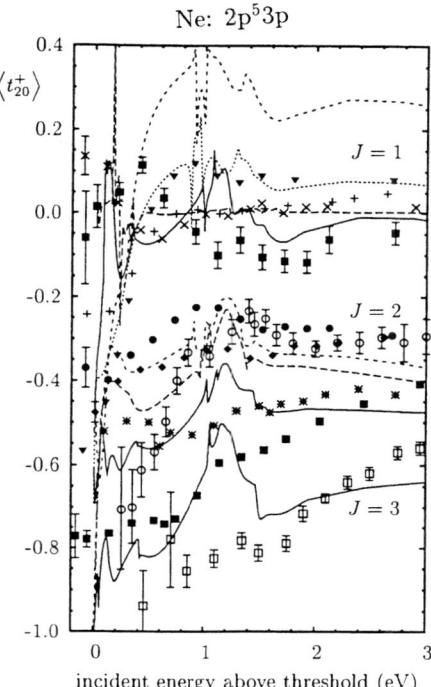

FIGURE 8. Relative alignment parameter $\langle t_{20}^+ \rangle$ after electron-impact excitation of the Ne $2p^5 3p$ manifold. For incident energies more than 0.2 eV above threshold, the top four curves belong to states with $J=1$, the next three to $J=2$, and the bottom one to the $J=3$ state. In detail, the curves and symbols are as follows: $J=1$: solid line, ×, $3p[3/2]_1$; long dashes, squares, $3p[1/2]_1$; short dashes, triangles, $3p'[3/2]_1$; dots, +, $3p'[1/2]_1$. $J=2$: solid line, *, $3p[5/2]_2$; long dashes, diamonds, $3p[3/2]_2$; short dashes, ● and ○, $3p'[3/2]_2$. $J=3$: solid line, open and solid squares, $3p[5/2]_3$. The experimental data are taken from Yu et al. [37,43] and from Zeman et al. [42] (□ for $3p[5/2]_3$ and ○ for $3p'[3/2]_2$).

As shown by Crowe [53,54], however, first-order theories are much less successful in the description of optically forbidden transitions, while CCC and RMPS continue to reproduce the experimental data in a satisfactory way. This is demonstrated in Fig. 10 where the Stokes parameters (P_1, P_2, P_3, P_4) for electron impact excitation of the $(1s3d)^1 D$ state are compared with measurements from the Perth and Newcastle groups.

Figure 11 shows another example of the detailed information that can be obtained in spin-resolved studies of atomic orientation parameters. In the experiment performed by the NIST group [57], spin-polarized electrons were scattered from laser-prepared spin-polarized and angular-momentum-oriented sodium atoms in the $(3p)^2 P$ state. By studying the superelastic signal, i.e., the collision-induced

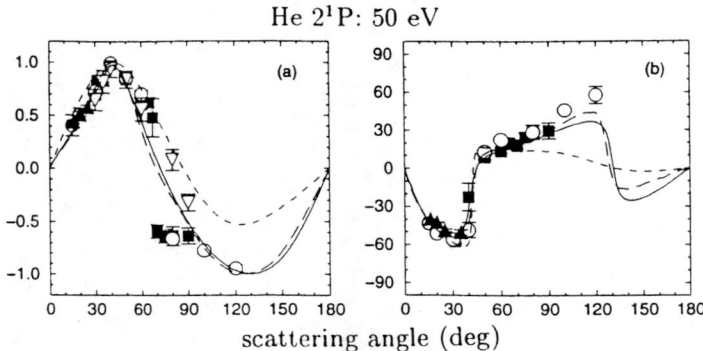

FIGURE 9. Angular momentum transfer $L_\perp$ (a) and alignment angle γ (b) for electron impact excitation of the $(1s2p)^1P$ state of helium at an incident energy of 50 eV. The theoretical predictions from RMPS (solid line), CCC (long-dashed line) and DWBA (short-dashed line) calculations are compared with experimental data of Eminyan et al. [47] (full triangles), McAdams et al. [48] (open circles), Khakoo et al. [49] (open triangles), and Beijers et al. [50] (full squares).

de-excitation to the $(3s)^2S$ state, it was not only possible to determine the corresponding angular momentum transfer $L_\perp$ for the time-reversed inelastic collision, but also the individual contributions from the singlet and triplet spin channels. As seen from Fig. 11, only the CCC method [58] was able to reproduce the experimental results for $L^s_\perp$ (the singlet spin channel) between 30° and 90°. On the other hand, the spin-averaged experimental results are also reproduced in a "close-coupling plus optical potential" (CCO) approach [59], and even by a second-order distorted-wave (DWBA2) approach [60] which one would generally not expect to be suitable at such a low energy of approximately twice the ionization threshold. In fact, the DWBA2 does very well for the triplet spin channel, but it fails completely for the singlet channel where electron-electron correlation effects are generally much more important than in the triplet channel.

As a final example for angle-differential generalized Stokes parameters and the spin-resolved electron-impact coherence parameters derived from their measurement, we show in Fig. 12 results for excitation of the $(6s6p)^3P_1$ state of mercury by a spin-polarized incident electron beam with an energy of 8 eV. For details of this work we refer to the paper by Andersen et al. [61], but it is worthwhile to mention a few important points, namely:

- Based upon their definition, the parameters Q^z_{13}, Q^z_{23}, and Q^z_{33} are identical, but they can be determined experimentally in a variety of ways. Hence, the analysis in terms of generalized Stokes parameters revealed that the original experimental data were not completely consistent.

- The overall agreement with predictions from a 5-state Breit-Pauli R-matrix calculation [62,63,61] is surprisingly good, although one might expect that

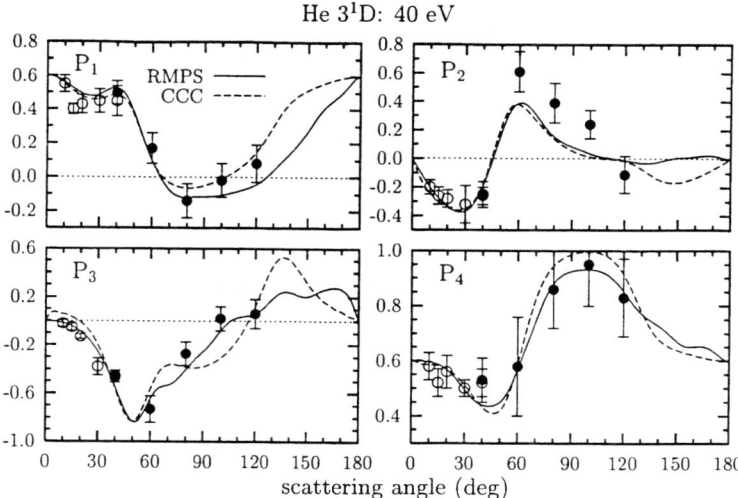

FIGURE 10. Stokes parameters (P_1, P_2, P_3, P_4) for electron impact excitation of the $(1s3d)^1D$ state of helium at an incident energy of 40 eV. The theoretical predictions from RMPS (solid line) and CCC (dashed line) calculations are compared with experimental data from McLaughlin et al. [55] and Mikosza et al. [56].

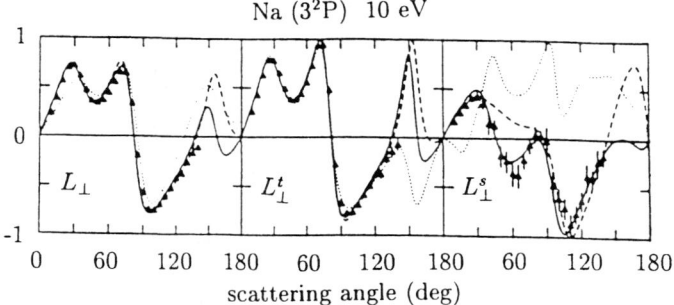

FIGURE 11. Angular momentum transfer $L_\perp$, spin-averaged and separated into contributions from the triplet (t) and singlet (s) spin channels for electron impact excitation of the $(3p)^2P$ state of sodium at an incident energy of 10 eV. The experimental data from the NIST group [57] are compared with theoretical predictions from CCC (solid line) [58], CCO (dashed line) [59], and DWBA2 (dotted line) [60] calculations.

internally consistent experimental data with smaller uncertainties will show the limits of the theoretical model. Such an experiment is currently in preparation [64].

- The negative value of $L_\perp^{+\downarrow}$, i.e., the angular momentum transfer induced by

Hg $(6s6p)^3P_1$: 8 eV

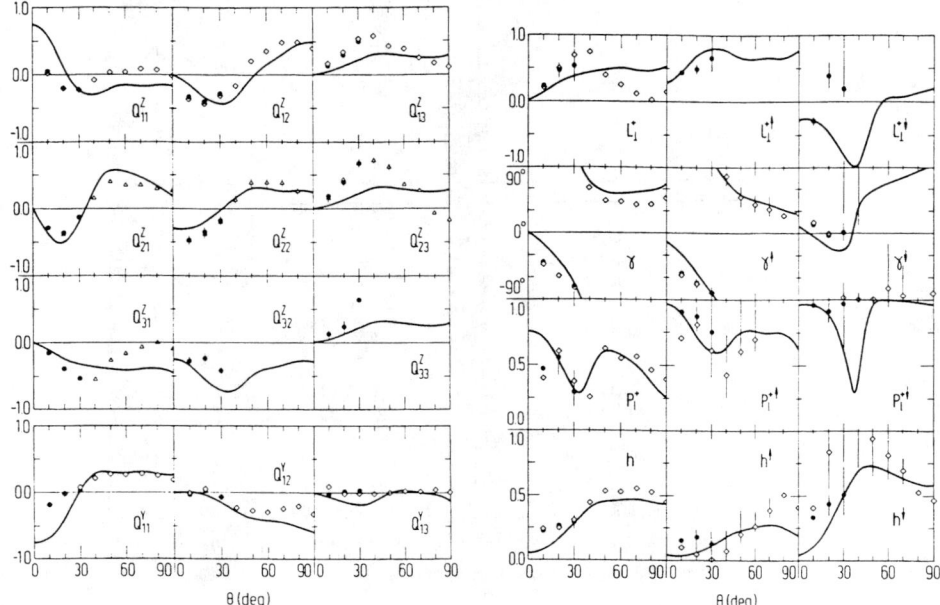

FIGURE 12. Generalized Stokes parameters and the corresponding spin-averaged and spin-resolved electron-impact coherence parameters for electron impact excitation of the $(6s6p)^1P_1$ state of mercury at an incident energy of 8 eV. The theoretical predictions from a 5-state Breit-Pauli R-matrix calculation are compared with experimental data from the Münster group [61].

electrons with spin down relative to the scattering plane, at small scattering angles is a violation of well-established, though empirical, propensity rules found for spin-averaged angular momentum transfers in S → P excitations. For more details, we refer to the discussion by Andersen [45].

- Measurement of the generalized Stokes parameters allows for the determination of all relative sizes and four out of five relative phases between the six (complex) scattering amplitudes that fully determine this excitation process. Together with the differential cross section (for the absolute size) and a generalized *STU* parameter measurement (for the missing phase as well as consistency checks), a *complete experiment* for this complicated process is thus within reach of current technology.

Figure 13 shows an example of the charge cloud characteristics that can be obtained in electron-photon coincidence experiments with spin-polarized incident electrons. If the polarization vector has a component in the scattering plane, the

Hg (6s6p)³P₁: 8 eV

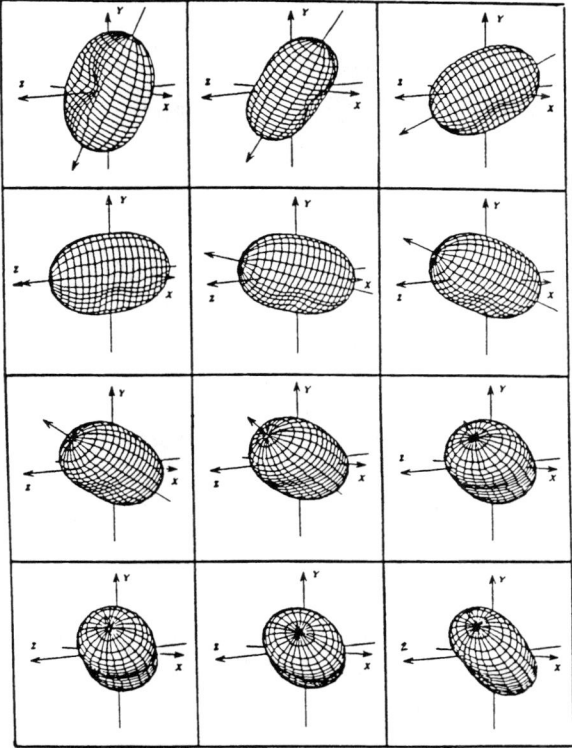

FIGURE 13. Charge cloud representing the excited (6s6p)¹P₁ state of mercury after impact excitation by spin-polarized electrons with polarization vector in the scattering plane at an incident energy of 8 eV [65], according to predictions from a 5-state Breit-Pauli R-matrix calculation. Following the columns from left to right and then the rows from top to bottom, the individual frames correspond to scattering angles of 10°, 30°, 40°, 50°, 70°, 80°, 90°, 100°, 110°, 120°, 140°, and 160°.

planar symmetry of the excitation is broken (since spin polarization corresponds to an *axial* vector). Consequently, the charge cloud cannot only rotate around the normal vector to the scattering plane (resulting in an alignment angle γ) but also twist and tilt out of the reaction plane [65].

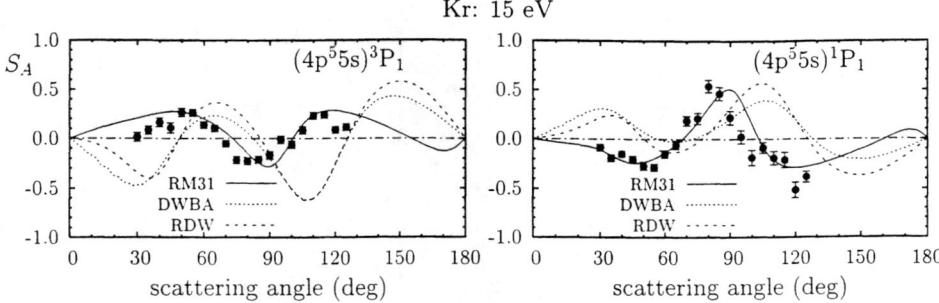

FIGURE 14. Spin asymmetry function S_A for electron-impact excitation of the $(4p^5[3/2]5s)$ ("$^3P_1$") and $(4p^5[1/2]5s)$ ("$^1P_1$") states in krypton at an incident energy of 15 eV. The experimental data of Dümmler et al. [25] are compared with semi-relativistic (DWBA) and full-relativistic (RDW) first-order distorted-wave results of Bartschat and Madison [21] and of Zuo et al. [24], and with predictions from a 31-state Breit-Pauli R-matrix calculation.

Angle-Differential Spin-Polarization Parameters

We finish by presenting two further examples to demonstrate that methods based upon the close-coupling method indeed promise good agreement between theory and experiment even for observables that are very sensitive to the details of the theoretical models. Figure 14 compares results for the spin-asymmetry function S_A (see also Fig. 5) for electron-impact excitation of two krypton states at an incident energy of 15 eV. While there are again major differences between the semi-relativistic (DWBA) and full-relativistic (RDW) predictions and experiment [25], a 31-state Breit-Pauli R-matrix model produces very satisfactory results. However, other comparisons show that RMPS-type calculations would likely be necessary to obtain such good agreement on a routine basis, particularly for slightly higher excitation energies.

Finally, we show results for the left-right spin asymmetry A_1 that occurs in elastic scattering of unpolarized electrons from spin-polarized cesium atoms. It was predicted by Farago [66] that this particular asymmetry should occur if both electron exchange and the spin-orbit interaction between the projectile electron and the target nucleus are important in the process — either mechanism alone is not sufficient. Recently, the Bielefeld group produced benchmark results for elastic e–Cs collisions at an energy of 3 eV for various spin asymmetries [67]. The results for A_1 are shown in Fig. 15 and compared with theoretical predictions from 8-state Breit-Pauli [68] as well as 8-state Dirac-Breit R-matrix [69] calculations. Note that the measured and predicted asymmetries are indeed small for all scattering angles, thereby making the experiment difficult and apparently causing extreme sensitivity of the theoretical results on the details of the model, including the description of the target structure. Since the latter is superior in the Breit-Pauli work, due

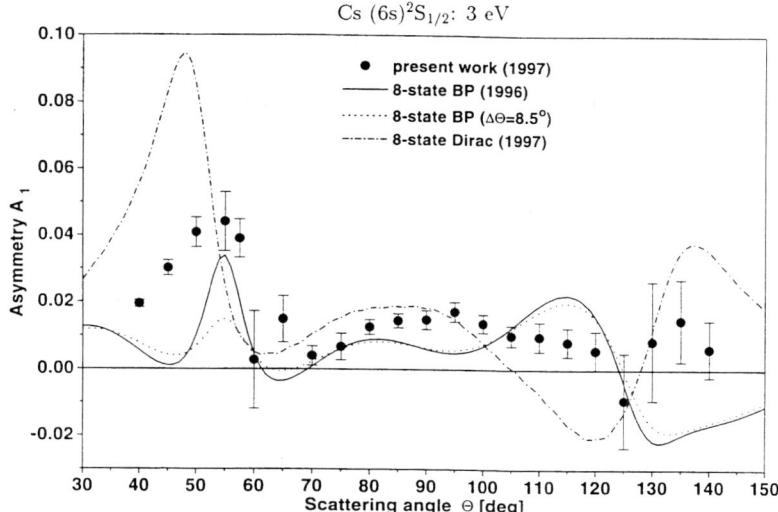

FIGURE 15. Spin asymmetry function A_1 for elastic scattering of unpolarized electrons from spin-polarized cesium atoms at an incident energy of 3 eV. The experimental data of the Bielefeld group [67,70] are compared with predictions from semi-relativistic [68] and full-relativistic [69] 8-state R-matrix calculations. The dotted line was obtained after convoluting the Breit-Pauli results with the experimental angular resolution of 8.5° (FWHM).

to semi-empirical inclusion of core polarization effects, it is not surprising that predictions from the semi-relativistic model are in slightly better, though certainly not satisfactory, agreement with experiment than those from the full-relativistic model. The same conclusion holds for two other spin asymmetries and even the differential cross section, but the differences are much less pronounced in the cross section for unpolarized beams than in the spin-dependent observables [70].

CONCLUSIONS

During the past 25 years, the field of polarization, alignment, and orientation studies, including quantum mechanically complete experiments, has developed to considerable maturity in the area of electron–atom collisions. Data for a few elastic and inelastic scattering processes have provided important benchmarks to test state-of-the art scattering theory.

Advancements in the formulation of the standard time-independent close-coupling method resulting, for example, in the CCC and RMPS approaches, have achieved a break-through in describing electron collisions with simple target systems. However, the methods need to be further developed and tested for more complex systems and also for heavier targets, such as noble gases other than helium, where relativistic effects are likely to become very important. Note that reliable results for many of these systems are of crucial importance for modeling the physics of plasmas, lasers, stars, and planetary atmospheres. In light of the experimental difficulties, particularly with respect to absolute normalization of the cross sections, theoretical CCC and RMPS predictions are increasingly being used to revise the corresponding databases.

Regarding the general formulation of the theory of measurement, it has become evident that all cases can be discussed within a *common framework* by parameterizing the change of the scattered electron polarization by means of generalized *STU* parameters and, for excitation, by the evaluation of generalized Stokes parameters for full characterization of the radiation pattern. Systematic mapping of the various dimensionless parameters provides a much closer insight into the detailed collision dynamics than the differential cross section alone. Being the only *absolute* observable, however, a truly complete experiment cannot be achieved with relative measurements alone, and the need for accurate cross section measurements should not be underestimated.

From an experimental point of view, future progress in the field requires further development of sophisticated coincidence setups and the ability to handle very long data accumulation times under stable conditions. Alternatively, selected *angle-integrated experiments* may provide important benchmarks as well, and more progress is also expected from scattering and (de-)excitation studies involving *optically prepared states*.

Finally, a very promising theoretical alternative lies in the use of time-dependent approaches, particularly in light of the current developments towards massively parallel computing for which such methods are ideally suited. Although it has not yet been done for the electron–atom collision studies reported in this talk, observing the time-development of atomic orientation and alignment in such collisions would undoubtedly provide new insights, perspectives, and further stimulation for this exciting field.

ACKNOWLEDGMENTS

This work would not have been possible without the excellent training that the author received from numerous advisors, particularly K. Blum on the density matrix theory (including the Fano-Macek review), P.G. Burke and N.S. Scott on the R-matrix method, and G.F. Hanne and J. Kessler on experimental aspects. I would also like to thank N. Andersen for his contributions, especially for teaching me the virtues of the natural coordinate frame, and V. Zeman for generating many of the R-matrix results presented for the heavy noble-gas targets. Numerous other colleagues from groups in Adelaide, Belfast, Bielefeld, Boulder, Brisbane, Gaithersburg, Lincoln, Münster, Newcastle, Perth, Rolla, Toronto, and Windsor have also contributed directly or indirectly to this talk, and I thank them for the very fruitful interactions over many years.

This work was supported, in part, by the United States National Science Foundation under grants PHY-9318377 and PHY-9605124, by the Deutsche Forschungsgemeinschaft in SFB-216, and through visiting fellowships at Adelaide, Belfast, Boulder, and Münster.

REFERENCES

1. Fano, U., and Macek, J. H., *Rev. Mod. Phys.* **45**, 555 (1973).
2. Andersen, N., Gallagher, J.W, and Hertel, I. V., *Phys. Rep.* **165**, 1 (1988).
3. Becker, K., Crowe, A., and McConkey, J. W., *J. Phys. B* **25**, 3885 (1992).
4. Bederson, B., *Comments At. Mol. Phys.* **1**, 41 (1969).
5. Bederson, B., *Comments At. Mol. Phys.* **1**, 65 (1969).
6. Kessler, J., *Polarized Electrons* (2nd edition), Berlin: Springer (1985).
7. Kessler, J., *Adv. At. Mol. Opt. Phys.* **27**, 81 (1991).
8. McClelland, J. J., Kelley, M. H., and Celotta, R. J., *Phys. Rev. Lett.* **55**, 688 (1985).
9. McClelland, J. J., Kelley, M. H., and Celotta, R. J., *Phys. Rev. A.* **40**, 2321 (1989).
10. Goeke, J., Hanne, G. F., and Kessler, J., *J. Phys. B* **22**, 1075 (1989).
11. Sohn, M., and Hanne, G. F., *J. Phys. B* **25**, 4627 (1992).
12. Andersen, N., and Bartschat, K., *Adv. At. Mol. Opt. Phys.* **36**, 1 (1996).
13. Andersen, N., Bartschat, K., Broad, J. T., and Hertel, I. V., *Phys. Rep.* **279**, 251 (1997).
14. Andersen, N., and Bartschat, K., *J. Phys. B* **30**, 5071 (1997).
15. Blum, K., *Density Matrix Theory and Applications* (2nd edition), New York: Plenum Press (1996).
16. Mott, N. F., and Massey, H. S. W., *The Theory of Atomic Collisions*, Oxford: Clarendon (1965), ch. IX.
17. Andersen, N., and Bartschat, K., *J. Phys. B* **27**, 3189 (1994); corrigendum: *J. Phys. B* **29**, 1149 (1996);
18. Bartschat, K., and Blum, K., *Z. Phys. A* **304**, 85 (1982).
19. Bartschat, K., *Phys. Rep.* **180**, 1 (1989).
20. Berger, O., and Kessler, J., *J. Phys. B* **19**, 3539 (1986).
21. Bartschat, K., and Madison, D. H., *J. Phys. B* **20**, 5839 (1987).
22. Nishimura, H., Danjo, A., and Takahashi, A., *J. Phys. B* **19**, L167 (1986).
23. Corr, J. J., Plessis, P., and McConkey, J. W., *Phys. Rev. A* **42**, 5240 (1990).
24. Zuo, T., McEachran, R. P., and Stauffer, A. D., *J. Phys. B* **25**, 3393 (1992).
25. Dümmler, M., Hanne, G. F., and Kessler, J., *J. Phys. B* **28**, 2985 (1995).
26. Pindzola, M. S., Robicheaux, F., Badnell, N. R., and Gorczyca, T. W., *Phys. Rev. A* **56**, 1994 (1997).
27. Bray, I., and Stelbovics, A. T., *Adv. Atom. Mol. Phys.* **35**, 209 (1995).
28. Fursa, D. V., and Bray, I., *Phys. Rev. A* **52**, 1279 (1995).
29. Burke, P. G., and Robb, W. D., *Adv. At. Mol. Phys.* **11**, 143 (1975).
30. Burke, P. G., and Berrington, K.A, *Atomic and Molecular Processes – An R-Matrix Approach*, Bristol: Institute of Physics (1993).

31. Bartschat, K., Hudson, E. T., Scott, M. P., Burke, P. G., and Burke, V. M., *J. Phys. B* **29**, 115 (1996).
32. Bartschat, K. and Bray, I., *J. Phys. B* **30**, L571 (1997).
33. Berrington, K. A., Eissner, W., and Norrington, P. H., *Comp. Phys. Commun.* **92**, 290 (1995).
34. Burke, P. G., Burke, V. M., and Dunseath, K. M., *J. Phys. B* **27**, 5341 (1994).
35. Norrington, P. H., and Grant, I. P., *J. Phys. B* **20**, 4869 (1987).
36. Wijesundera, W. P., Parpia, F. A., Grant, I. P., and Norrington, P. H., *J. Phys. B* **24**, 1803 (1991).
37. Yu, D. H., Hayes, P. A., Furst, J. E., and Williams, J. F., *Phys. Rev. Lett.* **78**, 2724 (1997).
38. Zeman, V., and Bartschat, K., *J. Phys. B* **30**, 4609 (1997).
39. Bartschat, K., Hanne, G. F., and Wolcke, A., *Z. Phys. A* **304**, 89 (1982).
40. Furst, J. E., Gay, T. J., Wijayaratna, W. M. K. P., Bartschat, K., Geesmann, H., Khakoo, M. A., and Madison, D. H., *J. Phys. B* **25**, 1089 (1982).
41. Al-Khateeb, H. M., Birdsey, B. G., Bowen, T. C., Johnson, M. L., and Gay, T. J., *Bull. Am. Phys. Soc.* **43**, 1273 (1998).
42. Zeman, V., Bartschat, K., Gay, T. J., and Trantham, K. W., *Phys. Rev. Lett.* **79**, 1825 (1997).
43. Yu, D. H., Hayes, P. A., Williams, J. F., and Furst, J. E., *J. Phys. B* **30**, 1799 (1997).
44. Brunger, M. J., Buckman, S. J., Teubner, P. J. O., Zeman, V., and Bartschat, K., *J. Phys. B* **31**, L387 (1998).
45. Andersen, N., "Orientation and Alignment in Atomic Collisions," in *16th International Conference on Atomic Physics (ICAP'98)* (ed. Drake, G. W.F), New York: American Institute of Physics (1999)
46. Bartschat, K., Hudson, E. T., Scott, M. P., Burke, P. G., and Burke, V. M., *J. Phys. B* **29**, 2875 (1996).
47. Eminyan, M., MacAdam, K. B., Slevin, J., and Kleinpoppen, H., *J. Phys. B* **7**, 1519 (1974).
48. McAdams, R., Hollywood, M. T., Crowe, A., and Williams, J. F., *J. Phys. B* **13**, 3691 (1980).
49. Khakoo, M. A., Becker, K., Forand, J. L., Madison, D. H., and McConkey, J. W., *J. Phys. B* **19**, L209 (1986).
50. Beijers, J. P., Madison, D. H., van Eck, J., and Heideman, H. G. M., *J. Phys. B* **20**, 167 (1987).
51. Madison, D. H., and Shelton, W. N., *Phys. Rev. A* **7**, 499 (1973).
52. Cartwright, D. C., and Csanak, G., *Phys. Rev. A* **38**, 2740 (1988).
53. Crowe, A., "Electron–Helium Correlation Studies," in *Photon and Electron Collisions with Atoms and Molecules* (eds. Burke, P. G., and Joachain, C. J.), New York: Plenum (1997)
54. Crowe, A., "Correlation Studies of Inelastic Electron Scattering from Simple Atoms: Recent Advances," in *16th International Conference on Atomic Physics (ICAP'98)* (ed. Drake, G. W.F), New York: American Institute of Physics (1999)
55. McLaughlin, D. T., Donnelly, B. P., and Crowe, A., *Z. Phys. D* **29**, 259 (1994).
56. Mikosza, A. G., Hippler, R., Wang, J. B., and Williams, J. F., *Z. Phys. D* **30**,

129 (1994).
57. Scholten, R. E., Lorentz, S. R., McClelland, J. J., Kelley, M. H., and Celotta, R. J., *J. Phys. B* **24**, L653 (1991).
58. Bray, I., *Phys. Rev. A* **49**, 1066 (1994).
59. Bray, I., and McCarthy, I. E., *Phys. Rev. A* **47**, 317 (1993).
60. Madison, D. H., Bartschat, K., and McEachran, R. P., *J. Phys. B* **25**, 5199 (1992).
61. Andersen, N., Bartschat, K., Broad, J. T., Hanne, G. F., and Uhrig, M., *Phys. Rev. Lett.* **76**, 208 (1996).
62. Scott, N. S., Burke, P. G., and Bartschat, K., *J. Phys. B* **16**, L361 (1983).
63. Bartschat, K., Scott, N. S., Blum, K., and Burke, P. G., *J. Phys. B* **17**, 269 (1984).
64. Hanne, G. F., *private communication* (1998)
65. Raeker, A., Blum, K., and Bartschat, K., *J. Phys. B* **26**, 1491 (1993).
66. Farago, P. S., *J. Phys. B* **7**, L28 (1974).
67. Baum, G., Raith, W., Roth, B., Tondera, M., *private communication* (1998)
68. Bartschat K., *J. Phys. B* **26**, 3995 (1993).
69. Ait-Tahar, S., Grant, I. P., and Norrington, P. H., *Phys. Rev. Lett.* **79**, 2955 (1997).
70. Baum, G., Raith, W., Roth, B., Tondera, M., Bartschat, K., Bray, I., Ait-Tahar, S., Grant, I. P., and Norrington, P. H., *Phys. Rev. Lett.* **82**, in press (1999).

Correlation Studies of Inelastic Electron Scattering from Simple Atoms: Recent Advances

Albert Crowe

Department of Physics, University of Newcastle
Newcastle upon Tyne, NE1 7RU, United Kingdom

Abstract. Correlation studies of inelastic electron scattering from simple atoms are discussed, largely from an experimental viewpoint with the aim of complementing the contributions of Andersen and Bartschat in these proceedings. After a brief review of experimental advances over the last twenty-five years, major developments over the last four years are highlighted. These include excitation of higher angular-momentum states of helium, excitation of atomic hydrogen, and extension of the electron-photon correlation method to study production of excited-state ions.

INTRODUCTION

The major objective in this article is to discuss recent advances in the study of inelastic electron scattering from small atoms using the scattered electron-decay photon correlation techniques. The discussion will be presented largely from an experimental point of view, relying on the contribution of Andersen in these proceedings for the general background regarding alignment and orientation in atomic collisions and on the contribution of Bartschat for a discussion of the various theoretical models with which experimental data will be compared.

On the 25th anniversary of the first experiment, it is valuable to recall some of the more significant experimental advances in the field.

(i) The pioneering electron-photon angular correlation experiment was carried out by Kleinpoppen and colleagues [1] at the University of Stirling, Scotland for the $2\,^1P$ state of helium at small scattering angles.

(ii) In 1976 Standage and Kleinpoppen [2] reported the first scattered electron-polarized photon correlation study, the polarization analysis being carried out on the 501.6 nm ($3\,^1P$–$2\,^1S$) photon following excitation of the He($3\,^1P$) state.

(iii) In 1978 Dixon et al. [3] reported the first angular correlation measurements for 1s–2p excitation in atomic hydrogen.

(iv) Extension of the He($2\,^1$P) studies to large scattering angles was reported in 1979 [4]. The coincidence signal is proportional to the differential scattering cross-section making these experiments much more difficult. However, they provide a sterner test of scattering theories.

(v) Extension to states of higher angular momentum (for example D states in helium, van Linden van den Heuvell et al. in 1983 [5] and F states, Cvejanović and Crowe in 1998 [6]).

(vi) Development of VUV circular polarizers [7,8] led to the first experimentally complete information on He($2\,^1$P) and H($2\,^2$P) excitation. Previously the sign of a phase had to be assumed, using theory as a guide.

(vii) Study of ionization of helium where the ion is produced in an excited state [9,10]. These very recent electron photon correlation studies for the He$^+$(2p) ion provide more detailed information on ionization than the (e,2e) method [11] which in this case cannot distinguish between the 2s and 2p ion states. The combination of both techniques to perform a complete ionization study remains an intriguing possibility.

(viii) Use of triple coincidence techniques involving the scattered electron, a directly emitted photon and a cascade photon [12] to determine the relative phases of excitation amplitudes uniquely in D state excitation.

Although generally outside the scope of this discussion on light atoms, a number of other important advances should be acknowledged.

(i) Extension to heavier atoms was reported as early as 1975 [13]. The situation for the heavier rare gases has been reviewed by Becker et al. [14].

(ii) Application of stepwise laser excitation to electron-photon correlation studies [15]. This is a particularly useful method for excited metastable states or where the directly emitted photon has an unfavourable wavelength for polarization analysis.

(iii) Excitation of molecules, including the study of an isolated rotational level [16].

(iv) Electron-photon correlation experiments using spin-polarised electrons [17] enabling explicit spin dependent forces such as spin-orbit interactions between the continuum electron and the target to be studied.

Another technique providing detailed information on alignment and orientation in electron-atom collisions is superelastic electron scattering from laser excited states [18]. The equivalence of the correlation and superelastic scattering methods has been discussed in detail by MacGillivray and Standage [19]. Finally an impressive study has been carried out [20] by scattering spin-polarized electrons from spin-polarized laser excited atoms.

ELECTRON-PHOTON CORRELATION TECHNIQUES

Two methods are used. In the electron-photon angular correlation method an angular distribution of the photon is measured in coincidence with the scattered electron from the same collision. In the electron-polarized photon correlation method, the polarization state of the photons is analysed, again in coincidence with the corresponding electrons.

Both methods are illustrated schematically in Fig. 1. The momentum $\mathbf{k_O}$ of the incident electron e_i and $\mathbf{k_S}$ of the scattered electron e_s define a scattering plane, (x, y) in the natural coordinate frame used throughout this discussion. The angular correlations are obtained by rotation of an appropriate photon detector, usually in the (x, y) plane as in the original measurements of Eminyan et al. [1], but where the excited state has a height in the z-direction, then a distribution in another plane is also necessary to describe the excited state [21]. In the polarization correlation method the linear (P_1, P_2) and circular (P_3) Stokes parameters are normally measured for radiation emitted into a small solid angle perpendicular to the scattering plane. For excited states lying in the scattering plane this can be sufficient for a complete experiment, as in the original He($3\,^1$P) excitation experiment of Standage and Kleinpoppen [2]. However, a linear polarization measurement P_4 in the scattering plane is also required to characterise an excited state with a non-zero height, for example, D-state excitation.

There are fundamental differences between these experiments and more traditional electron scattering experiments where only a single particle is detected after the collision. Observation of only photons from a small subset of identically prepared excited states arising from specific momentum transfer $\mathbf{K} = \mathbf{k_O} - \mathbf{k_S}$ collisions is a necessary condition for coherent excitation. The definition of a plane of symmetry as opposed to an axis also has a significant impact on analysis of the experiments.

Figure 2 shows a typical apparatus used in this laboratory for electron-photon polarization correlation measurements. The fixed electron gun produces a horizontal beam of electrons which is intersected at right angles by a beam of atoms. The

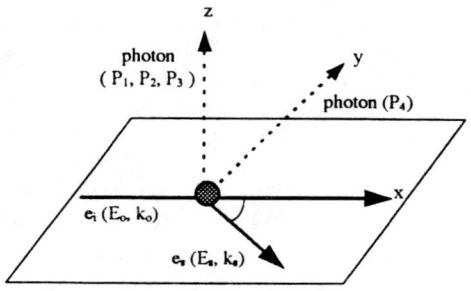

FIGURE 1. Schematic illustration of the electron photon correlation methods.

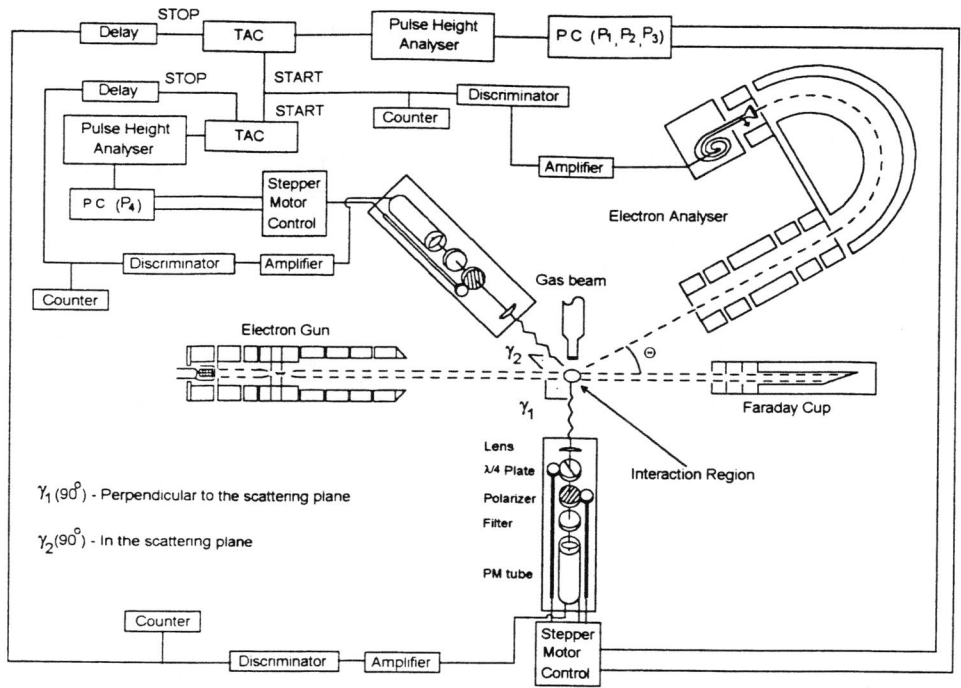

FIGURE 2. An electron photon correlation apparatus used in this laboratory [22]

source of atoms is dependent on the atomic species, but in all cases the atoms reach the interaction region via a long narrow capillary to produce a well collimated beam. The 180° electrostatic electron energy analyser used to select and detect only those electrons which have produced a particular excited state is rotated in the horizontal plane, i.e. the plane of symmetry is horizontal in this case. The photon polarization analysis system shown at the bottom of the diagram is used to determine the linear Stokes parameters P_1, P_2 and the circular Stokes parameter P_3. When necessary, a further linear polarization Stokes parameter P_4 is made using a similar photon polarization analyser in the scattering plane.

By rotating this photon detector (without the polarization components) in the scattering plane this same apparatus can be used to measure electron-photon angular correlations.

For an excited P state both techniques give the shape of the charge cloud in the collision plane. For the polarization correlation method, the intensity transmitted in the z-direction by a linear polarizer simply maps out the shape of the charge cloud since

$$I(\varphi) \propto |\psi^2(\varphi)| \propto 1 + P_\ell \cos 2(\varphi - \gamma), \qquad (1)$$

where the linear polarization is

$$P_\ell = \frac{|\psi|^2_{max} - |\psi|^2_{min}}{|\psi|^2_{max} + |\psi|^2_{min}}, \qquad (2)$$

γ is the direction of the $|\psi|^2_{max}$ axis, and φ is the azimuthal angle, both with respect to the x-axis. In practice the Stokes parameters P_1, P_2 are normally measured to determine the shape.

In the angular correlation method the angular correlation is given by

$$I(\varphi) = 1 - P_\ell \cos 2(\varphi - \gamma), \qquad (3)$$

i.e. the charge cloud rotated through 90° [23].

The relationship between the measured angular correlations and the charge clouds are shown in Fig. 3 for excitation of the $2\,^2P$ state of hydrogen at 54.4 eV and different scattering angles. The rotation through 90° can be clearly seen. In this case the nascent charge cloud excited in the collision is only recovered after taking account of depolarization due to the coupling between the orbital angular momentum and the doublet state spin S=1/2, i.e. the measured angular correlation is related to the so-called reduced linear polarization $\overline{P}_\ell$ by

$$I(\varphi) = 1 - b\,\overline{P}_\ell \cos 2(\varphi - \gamma), \qquad (4)$$

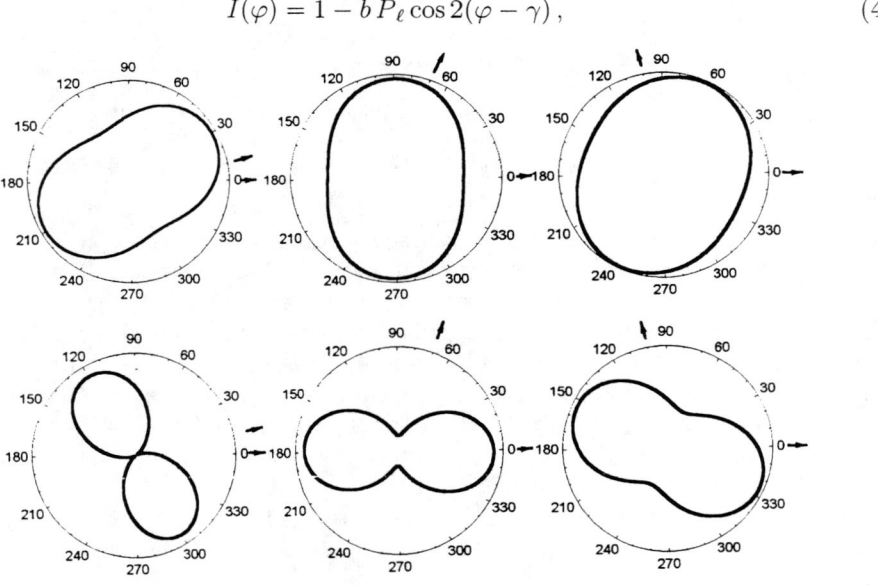

FIGURE 3. The measured angular correlations (top row) and corresponding charge cloud shapes in the scattering plane (bottom row) for excitation of the $2\,^2P$ state of hydrogen excited by 54.4 eV electrons for scattering angles of 10°, 70° and 100° (left to right). The data are those of Yalim et al. [24,25].

where for the ^{2}P state of atomic hydrogen, $b = 3/11$ [23].

A circular polarisation measurement is necessary to determine the angular momentum expectation value. In this respect the angular correlation measurements do not constitute complete experiments.

SPECIFIC EXAMPLES OF RECENT ADVANCES

Helium $1\,^1$S - $2\,^1$P - $1\,^1$S

This system constitutes the simplest system from the point of view of interpretation. Factors leading to this simplicity include lack of both fine and hyperfine structure and the validity of the LS coupling scheme. Moreover for a light two-electron system like helium, no further information can be gained from experiments where the electron spin is determined. Experimentally it is the most studied system but it is worthy of mention because it is only very recently that the large angle data could be completely described by theory.

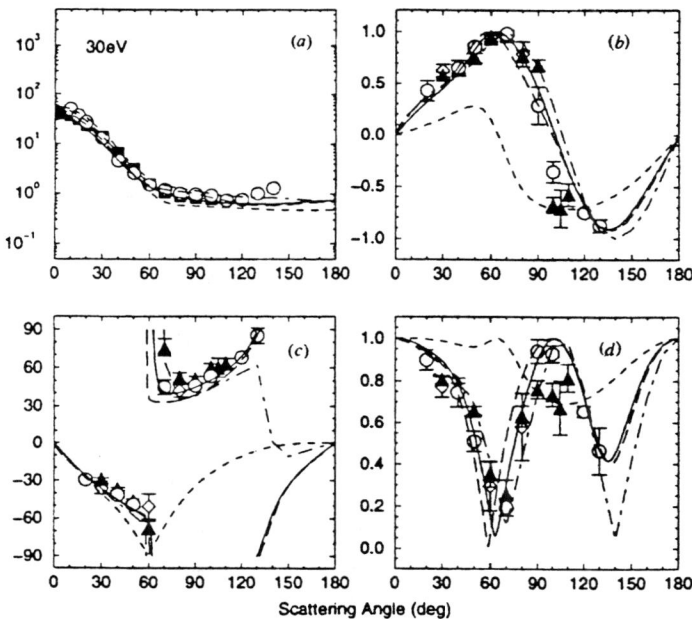

FIGURE 4. Differential cross section (a) and correlation parameters $L_\perp$ (b), γ (c) and P_ℓ (d) for the $2\,^1$P state of helium at an incident electron energy of 30 eV. Experiment: Solid squares, [26]; Open circles (a), [27]; Open circles (b)–(d), [28]; Open diamonds, [29]; Closed triangles, [30]. Theory: Solid line, RMPS, [31]; Dash, CCC, [32]; Short dash, FOMBT, [27]; Chain, 19-state R-matrix [33].

Figure 4 shows a comparison between theory and experiment at an incident electron energy of 30 eV. Only the recent convergent close coupling (CCC) [32] and the R-matrix with pseudo states (RMPS) [31] calculations are in agreement with experiment for the three correlation parameters $L_\perp$, γ and P_ℓ. The value of correlation measurements in evaluating theory is well demonstrated by noting that first-order many-body theory (FOMBT) [27] fails to reproduce the correlation parameters but gives an accurate prediction of the differential cross section.

Helium $1\,^1S$ - $3\,^{1,3}D$ Excitation

Correlation studies for excitation of the D states of helium quickly established that perturbation theories were totally inadequate to describe these processes (see, for example, Crowe and Bray [34]). This situation has been dramatically transformed by development of the CCC and RMPS theories. As a typical example, the measured Stokes parameters P_1–P_4 and the derived charge cloud parameters γ, P_ℓ, $L_\perp$, ρ_{00} and the degree of polarization P are shown in Fig. 5 for the $3\,^3D$ state of

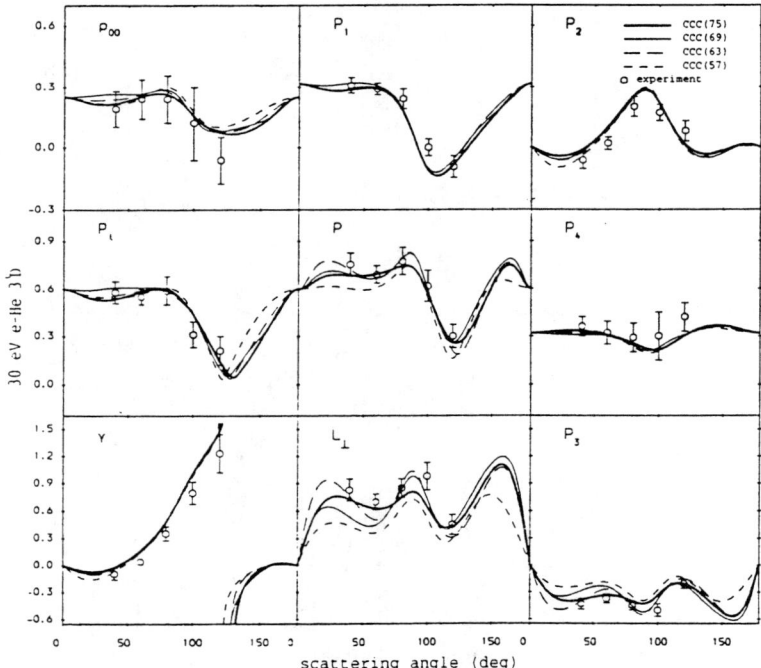

FIGURE 5. Stokes parameters, correlation parameters and degree of polarization P for the $3\,^3D$ state of helium at 30 eV. The experimental results from this laboratory are compared with CCC calculations (from Fursa et al. [22]).

helium at an incident electron energy of 30 eV. The correlation parameters have been derived from the measured Stokes parameters, as discussed by Andersen et al. [23], and after correction for fine structure depolarization as outlined by Crowe et al. [35]. Excellent agreement between the measured parameters and the CCC calculations can be seen over the wide range of scattering angles measured.

A highly significant observation is that the degree of polarization P for the D states never attains a value of unity as expected if both the excitation and decay processes observed are coherent. This loss of coherence can be attributed to an incomplete observation of the D state decay [36] and can be understood by considering the D state decay scheme (Fig. 6) for photons emitted in the z direction. It can be seen that when only the scattered-electron–(D→P) photon coincidences are observed, as for the data in Fig. 5, then the signal corresponds to an incoherent sum of the two decay channels leading to the intermediate $P(M = \pm 1)$ levels, i.e. the experiment does not provide the opportunity to observe interference between the $M = \pm 1$ excitation amplitudes. If we write the excitation amplitudes as

$$a_M = \alpha_M e^{i\beta_M}, \qquad (5)$$

then these experiments leave an ambiguity in the values of the relative phases $\beta_{\pm 2}$ with respect to α_0. This ambiguity can be removed by performing a triple coincidence experiment in which the cascade (P–S) photon is also observed. Such an experiment has recently been performed by Mikosza et al. [12] for the $3\,^1D$ state of helium at 60 eV incident electron energy and a scattering angle of 40°. Unfortunately the triple coincidence signal is around 10^{-3} times that of the two-particle signal, making these experiments extremely difficult. Nevertheless, the measured triple coincidence Stokes parameter $P_2^t = 0.285 \pm 0.371$ was sufficient

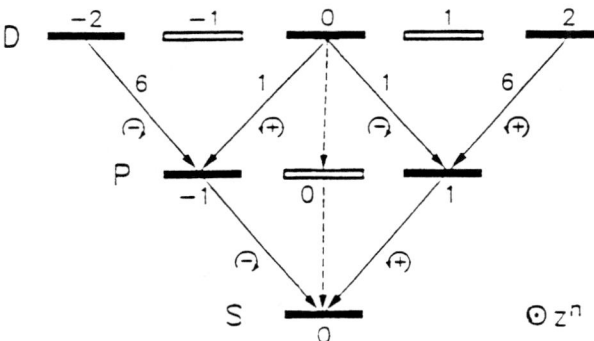

FIGURE 6. Decay scheme (full lines) for a D state through photon emission in the z - direction (natural frame). Note that symmetry arguments permit population of only the $D(M = 0, \pm 1)$ levels.

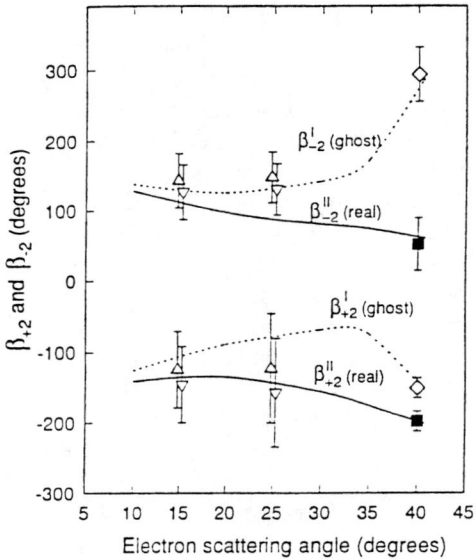

FIGURE 7. The relative phases β_{+2} and β_{-2} of the excitation amplitudes for the $3\,^1$D state of helium at 60 eV. The real set of phases β'' have been determined from an electron-(D-P) photon polarization correlation experiment [12,37] but are indistinguishable from the ghost set β' by this method. The real and ghost solutions at a scattering angle of 40° have been identified from a triple electron–(D–P)–photon–(P–S)–photon coincidence measurement [12]. The solid line is a CCC prediction.

to rule out one of the two pairs of phases implied by the normal electron-photon coincidence experiment. Their results are shown in Fig. 7.

Given the extreme difficulty of triple coincidence experiments and the quality of the CCC and RMPS calculations, the use of theory to resolve this ambiguity from two particle polarization correlation experiments is highly attractive. This is the approach adopted in this laboratory. In our analysis we have used the complete parameter set $(\sigma, \tilde{L}_\perp^\pm, \tilde{\gamma}^\pm)$ introduced by Andersen and Bartschat [36,38], where σ is the absolute differential cross section, and '+' and '−' refer to the helicity of the P–S cascade photons. Here, $\tilde{L}_\perp^\pm$ are readily obtained [36,38] from the incomplete set through

$$\tilde{L}_\perp^\pm = \pm 2\left(1 - \frac{8\rho_{00}}{3(2 \pm L_\perp)}\right). \tag{6}$$

Figure 8 shows $\tilde{L}_\perp^\pm$ for both the $3\,^1$D and $3\,^3$D states at an incident energy of 30 eV. Here, $\tilde{L}_\perp^+$ is close to its maximum value of +2 at all measured scattering angles while the $\tilde{L}_\perp^-$ values are less close to their limiting value of −2.

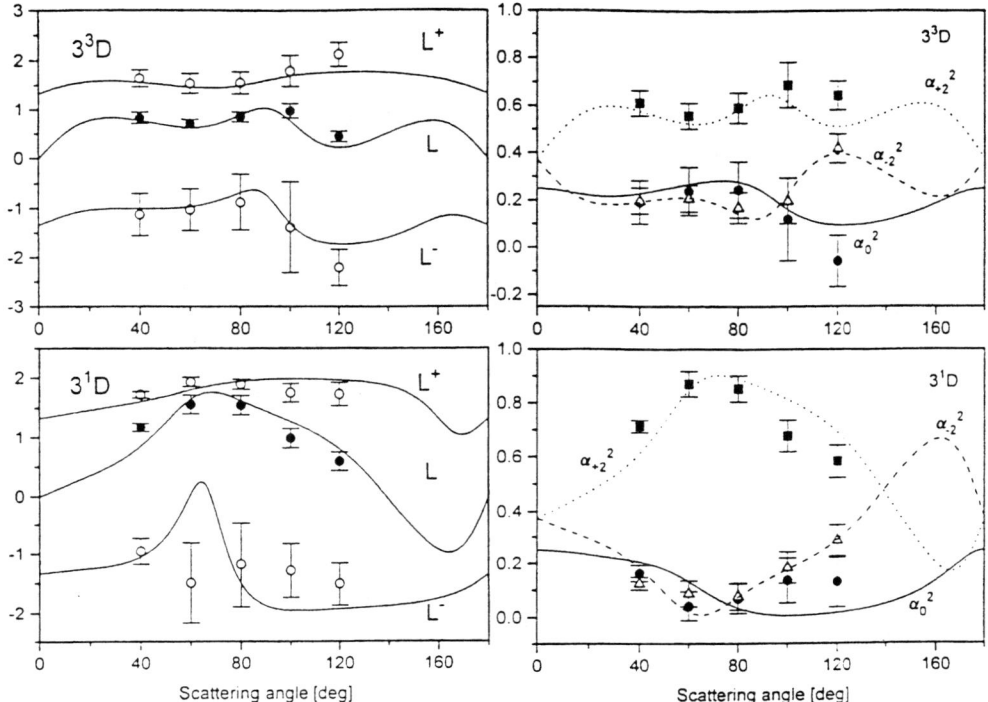

FIGURE 8. Experimental data and CCC theory for $\tilde{L}_\perp^\pm$ (left) and for $\alpha_M{}^2$ (right) for both the $3\,^1D$ and $3\,^3D$ states of helium at an incident energy of 30 eV (from Fursa et al. [22]).

Further insight can be obtained by comparing the relative magnitudes of the squares of the sublevel amplitudes $\alpha_M{}^2$ (Fig. 8). They show the dominance of $\alpha_{+2}{}^2$ over the others, especially for the $3\,^1D$ state over the range of the experimental data. Since the $\tilde{L}_\perp^\pm$ are related to the relative values of $\alpha_{\pm2}{}^2$, this also explains why the $\tilde{L}_\perp^+$ can be determined with much greater certainty than $\tilde{L}_\perp^-$ in Fig. 8.

The angular parameters $\tilde{\gamma}^\pm$ are most directly determined from the incomplete set using the relationship [22]

$$\left(1 - \frac{2}{3}\rho_{00}\right) P_\ell e^{2i\gamma} = \sqrt{\frac{\rho_{00}}{6}} \left(\sqrt{2 + L_\perp - 2\rho_{00}}\, e^{2i\tilde{\gamma}^+} + \sqrt{2 - L_\perp - 2\rho_{00}}\, e^{2i\tilde{\gamma}^-}\right). \quad (7)$$

This relationship can be expressed vectorially as shown in Fig. 9. In this parameterization the ambiguity arising from an electron-(D–P)-polarisation correlation study manifests itself in the indistinguishability of this triangle from one which is a mirror image about **c**, i.e.

$$\tilde{\gamma}^+ = \gamma \mp \frac{\psi}{2}, \quad \tilde{\gamma}^- = \gamma \pm \frac{\chi}{2}. \quad (8)$$

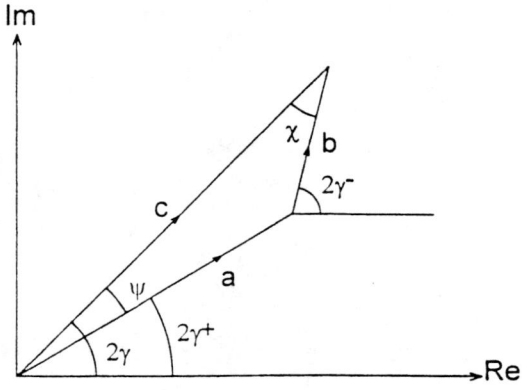

FIGURE 9. Vector diagram corresponding to Eq. (7).

In Fig. 10 we show the real values of $\tilde{\gamma}^{\pm}$ for both D states at 30 eV. We have used CCC theory to remove the ambiguity, although it should be noted that we consistently find

$$\tilde{\gamma}^{+} = \gamma + \frac{\psi}{2}, \quad \tilde{\gamma}^{-} = \gamma - \frac{\chi}{2} \tag{9}$$

for the wide range of kinematics and for both the $3\,^1D$ and $3\,^3D$ states studied in this laboratory [22,35,39,40]. At 30 eV $\tilde{\gamma}^{+} \approx \tilde{\gamma}^{-} \approx \gamma$ out to 100° for the $3\,^3D$ state while for the $3\,^1D$ state $\tilde{\gamma}^{-}$ begins to diverge from $\gamma, \tilde{\gamma}^{+}$ around 10°.

Equation (7) also gives a valuable check on the internal consistency of the Stokes parameters measured in a particular experiment, i.e. the data must satisfy the relationship

$$\mathbf{c} \pm \Delta\mathbf{c} = (\mathbf{a} \pm \Delta\mathbf{a}) + (\mathbf{b} \pm \Delta\mathbf{b}), \tag{10}$$

where $\Delta\mathbf{a}, \Delta\mathbf{b}, \Delta\mathbf{c}$ represent the uncertainties in $\mathbf{a}, \mathbf{b}, \mathbf{c}$ respectively.

Helium $1\,^1S$ - 4F Excitation

Excitation of higher angular momentum states in helium produces new challenges both experimentally and theoretically. Theoretically, even for a light atom, there is substantial breakdown of LS coupling, so that the $4\,^1F$ should be written

$$|^1F_3> = \frac{|^1F_3^0> + \omega|^3F_3^0>}{\sqrt{1+\omega^2}}, \tag{11}$$

where F^0 represents a pure LS coupled state and ω=0.43 [41].

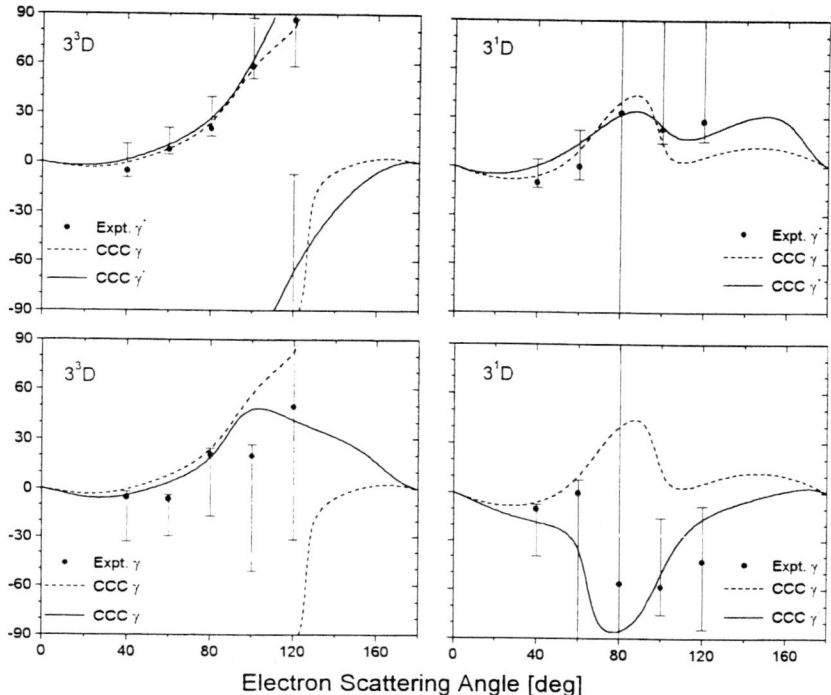

FIGURE 10. Experimental and CCC values of γ, $\tilde{\gamma}^+$, $\tilde{\gamma}^-$ for the $3\,^1$D and $3\,^3$D states of helium excited by 30 eV electrons (adapted from [22]).

For S–F excitation, the $M = \pm 3, \pm 1$ sublevels are populated and so its description requires three relative amplitudes and three relative phases. A brief discussion of this case has been given by Andersen and Bartschat [36].

Experimentally these states become increasingly difficult to isolate by electron spectroscopy and through the photon channel. The ^{1,3}F→^{1,3}D decays yield photons of 1870 nm and 1869 nm respectively. They cannot be isolated nor does the technology exist to observe them as single particles, an essential for a coincidence experiment. We have recently reported a polarization correlation study of the $4\,^1$F state of helium by observation of the $3\,^1$D - $2\,^1$P cascade photon in coincidence with scattered electrons which have excited the $n = 4$ states. Stokes parameters measured at an incident electron energy of 29.6 eV are shown in Fig. 11. As yet no theoretical results are available for comparison.

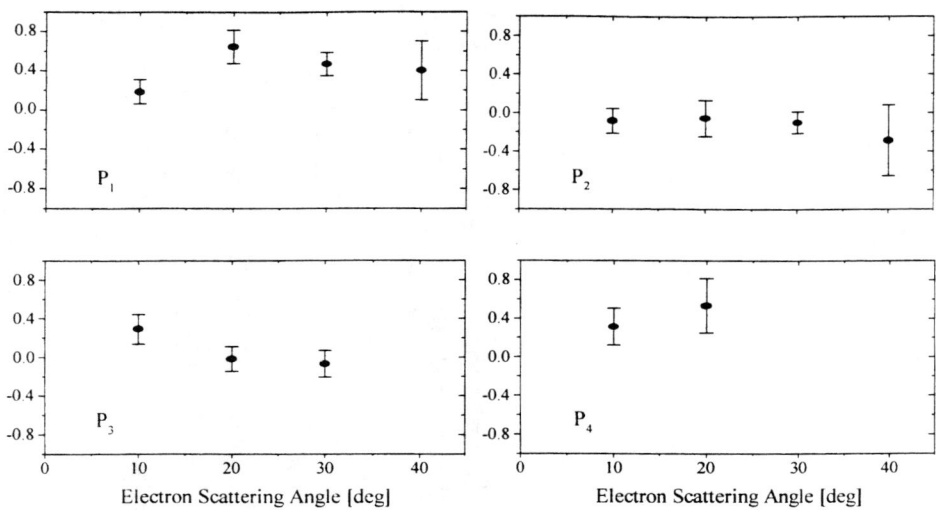

FIGURE 11. Measured Stokes parameters [6] for the $4\,^1$F state of helium at 29.6 eV.

Hydrogen Excitation

Theoretically, electron-hydrogen scattering is normally considered the simplest electron scattering process. However it is less straightforward than helium from the point of view of correlation studies. In particular for a one-electron atom, the excitation is determined by triplet and singlet scattering amplitudes and in principle direct and exchange scattering can be distinguished in spin-resolved studies. The existing experiments using unpolarized beams give an average over singlet and triplet scattering [23].

Until recently a major discrepancy has existed between experiment and theory for excitation of the $2\,^2$P state at 54.4 eV and large scattering angles. In the worst case, the measured and theoretically predicted angular correlations are almost exactly out of phase. This has often been described as the most outstanding problem in fundamental electron-atom scattering [42]. Very recent measurements from this laboratory confirmed the accuracy of the theoretical predictions. Figure 12 shows the present situation for the linear polarization $\overline{P}_\ell$ and alignment angle γ describing the shape of the charge cloud at an incident energy of 54.4 eV.

Although this particular discrepancy is now resolved and despite the excellent agreement seen between CCC theory and experiment for the D states of helium, other major discrepancies exist for hydrogen excitation. Figure 13 illustrates one of these problems where there is complete disagreement between experiments [51,52] and CCC theory [53] for $3\,^2$D excitation, even at small scattering angles.

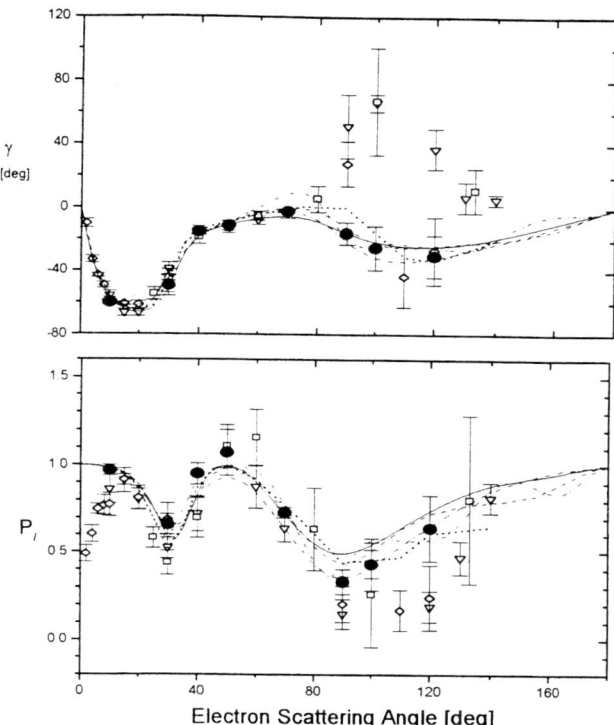

FIGURE 12. The correlation parameters $\overline{P}_\ell$ and γ for the $2\,^2P$ state of hydrogen excited by 54.4 eV electrons. Solid circles [24,25]; diamonds [43]; triangles [44]; squares [45]; solid line [46]; dash-dot line [47]; dash–double-dot line [48]; dotted line [49]; dashed line [50].

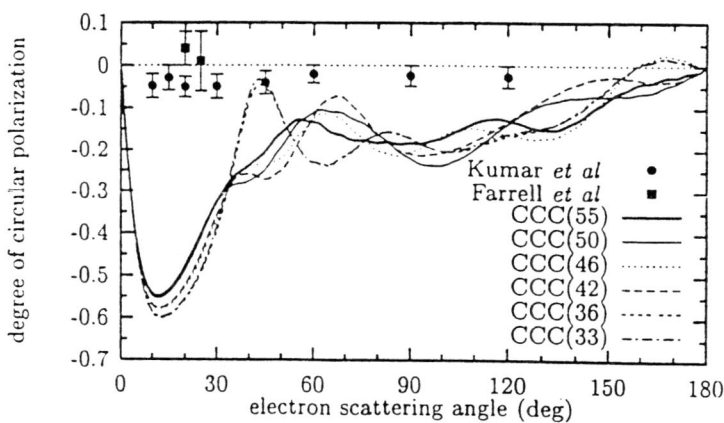

FIGURE 13. Circular polarization, P_3, for the $3\,^2D$ state of hydrogen.

Simultaneous Excitation-Ionization of Helium

Although ionization has been studied in detail using the (e,2e) method for nearly thirty years [54] these measurements yield only triple differential cross sections, i.e. the ejected electron cannot provide the detailed information available through the polarization of the photon in electron-photon polarization experiments. The study of an excited ion using a triple scattered electron-ejected electron-polarized photon technique raises the possibility of a complete ionization experiment. Such an experiment has yet to be performed but some two-particle coincidence experiments have been performed for the He$^+$(2p) ion state. A number of (e,2e) measurements have been performed, see Avaldi et al. [11], but in addition to the above restriction, these experiments cannot resolve the 2s and 2p ion states. Recently two independent electron-photon coincidence studies have been reported [9,10]. Each report two types of measurement. In one the angular correlation between a fast scattered electron and the 30.4 nm, He$^+$(2p) $\rightarrow$ He$^+$(1s), photon is observed for different kinematic conditions. The three alignment parameters describing the excited ion state [55] cannot be uniquely determined from a single angular correlation. Hence, the measured correlations have been parameterized by a function of the form

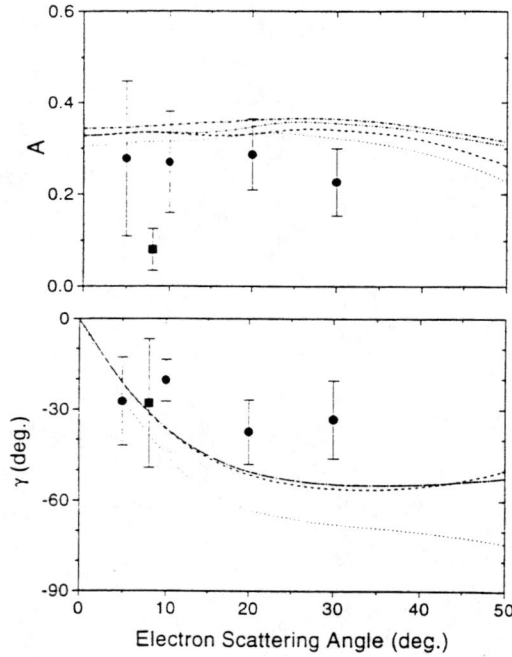

FIGURE 14. The amplitude A and alignment angle γ for the He$^+$(2p) state excited by 200 eV electrons and for an ejected electron energy of 1.2 eV. Circles, experiment of Dogan et al. [10]; squares, Hayes and Williams [9]; lines are calculations of Dogan et al. [10].

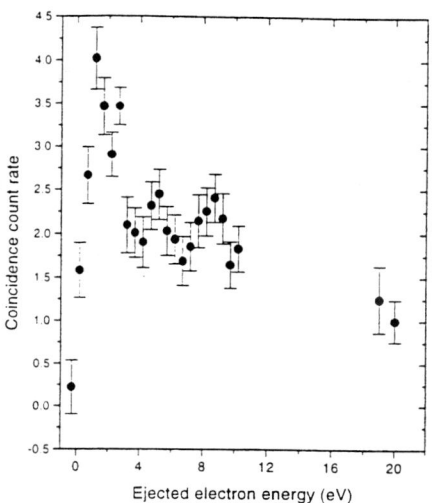

FIGURE 15. Measured double differential cross section for $He^+(2p)$ at an incident electron energy of 200 eV and scattering angle of 5° [10].

$$I(\varphi) \sim 1 - A\cos 2(\varphi - \gamma), \qquad (12)$$

where A is the correlation amplitude and γ is the angle corresponding to its minimum. Figure 14 shows values of A and γ for an incident electron energy of 200 eV and an ejected electron energy of 1.2 eV compared with different plane- and distorted-wave–R-matrix calculations based on the approach of Bartschat [56].

In a second type of coincidence experiment the double differential cross section (DDCS) for $He^+(2p)$ production is measured by varying the energy of the scattered electron detected for a fixed photon angle. The result at an incident energy of 200 eV shows considerable structure. The structure corresponding to ejected electron energies under 5 eV is most likely due to interference between direct $He^+(2p)$ production and its indirect production via $(3\ell, 3\ell')$ autoionization states. Autoionization states could also account for higher energy structure but this is more likely due to cascade from $He^+(3\ell)$ states.

ACKNOWLEDGMENTS

The work in this laboratory has been supported by the U.K. Engineering and Physical Sciences Research Council. The major contributions of the various experimental and theoretical co-authors of cited papers from this laboratory are gratefully acknowledged.

REFERENCES

1. Eminiyan, M., MacAdam, K. B., Slevin, J., and Kleinpoppen, H., *Phys. Rev. Lett.* **31**, 576 (1973).
2. Standage, M. C., and Kleinpoppen, H., *Phys. Rev. Lett.* **36**, 577 (1976).
3. Dixon, A. J., Hood, S. T., and Weigold, E., *Phys. Rev. Lett.* **40**, 1262 (1978).
4. Hollywood, M. T., Crowe, A., and Williams, J. F., *J. Phys. B: At. Mol. Phys.* **12**, 819 (1979).
5. van Linden van den Heuvell, H. B., van Gasteren, E. M., van Eck, J., and Heideman, H. G. M., *J. Phys. B: At. Mol. Phys.* **16**, 1619 (1983).
6. Cvejanović, D., and Crowe, A., *Phys. Rev. Lett.* **80**, 3033 (1998).
7. Khakoo, M. A., Becker, K., Forand, J. L., and McConkey, J. W., *J. Phys. B: At. Mol. Phys.* **19**, L209 (1986).
8. Williams, J. F., *Aust. J. Phys.* **39**, 621 (1986).
9. P. A. Hayes and Williams, J. F., *Phys. Rev. Lett.* **70**, 3098 (1996).
10. Dogan, M., Crowe, A., Bartschat, K., and Marchalant, P. J., *J. Phys. B: At. Mol. Opt. Phys.* **31**, 1611 (1998).
11. Avaldi, L., et al., *J. Phys. B: At. Mol. Opt. Phys.* **31**, 2981 (1998).
12. Mikosza, A. G., Williams, J. F., and Wang, J. B., *Phys. Rev. Lett.* **79**, 3375 (1997).
13. Arriola, H., Teubner, P. J. O., Ugbabe, A., and Weigold, E., *J. Phys. B: At. Mol. Phys.* **8**, 1275 (1975).
14. Becker, K., Crowe, A., and McConkey, J. W., *J. Phys. B: At. Mol. Opt. Phys.* **25**, 3885 (1992).
15. Murray, A. J., MacGillivray, W. R., and Standage, M. C., *Phys. Rev. Lett.* **62**, 411 (1989).
16. McConkey, J. W., Trajmar, S., Nickel, J. C., and Csanak, G. J., *J. Phys. B: At. Mol. Phys.* **19**, 2377 (1986).
17. Sohn, M., and Hanne, G. F., *J. Phys. B: At. Mol. Opt. Phys.* **25**, 4627 (1992).
18. I. V. Hertel and Stoll, W., in *Advances in Atomic and Molecular Physics*, edited by Bates, D. R., and B. Bederson, B., New York: Academic Press, 1977), Vol. 13, pp. 113–228.
19. MacGillivray, W. R., and Standage, M. C., *Phys. Rep.* **168**, 1 (1988).
20. McClelland, J. J., Kelley, M. H., and Celotta, R. J., *Phys. Rev. Lett.* **55**, 688 (1985).
21. Perera, N. W. P. H., and Burns, D. J., *J. Phys. B: At. Mol. Opt. Phys.* **23**, 3007 (1990).
22. D. V. Fursa et al., *J. Phys. B: At. Mol. Opt. Phys.* **30**, 3459 (1997).
23. Andersen, N., Gallagher, J. W., and Hertel, I. V., *Phys. Rep.* **165**, 1 (1988).
24. Yalim, H. A., Cvejanović, D., and Crowe, A., *Phys. Rev. Lett.* **79**, 2951 (1997).
25. Yalim, H. A., Cvejanović, D., and Crowe, A., *J. Phys. B: At. Mol. Opt. Phys.* , (1998), to be published.
26. Brunger, M. J., et al., *J. Phys. B: At. Mol. Opt. Phys.* **23**, 1325 (1990).
27. Cartwright, D. C., Csanak, G., Trajmar, S., and Register, D. F., *Phys. Rev. A* **45**, 1602 (1992).
28. McAdams, R., Hollywood, M. T., Crowe, A., and Williams, J. F., *J. Phys. B: At. Mol. Phys.* **13**, 3691 (1980).

29. Steph, N. C., and Golden, D. E., *Phys. Rev. A* **21**, 1848 (1980).
30. van Linden van den Heuvell, H. B., van Eck, J., and Heideman, H. G. M., *J. Phys. B: At. Mol. Phys.* **15**, 3517 (1982).
31. Bartschat, K., et al., *J. Phys. B: At. Mol. Opt. Phys.* **29**, 2875 (1996).
32. Fursa, D. V., and Bray, I., *Phys. Rev. A* **52**, 1279 (1995).
33. Fon, W. C., Berrington, K. A., and Kingston, A. E., *J. Phys. B: At. Mol. Opt. Phys.* **24**, 2161 (1991).
34. Crowe, A., and Bray, I., in *Selected Topics on Electron Physics*, edited by Campbell, D. M., and Kleinpoppen, H., New York: Plenum Press, 1996, pp. 45–55.
35. A. Crowe et al., *J. Phys. B: At. Mol. Opt. Phys.* **27**, L795 (1994).
36. N. Andersen and Bartschat, K., *J. Phys. B: At. Mol. Opt. Phys.* **30**, 5071 (1997).
37. Mikosza, A. G., Hippler, R., Wang, J. B., and Williams, J. F., *Z. Phys. D* **30**, 129 (1994).
38. Andersen, N., and Bartschat, K., *Adv. At. Mol. Phys.* **36**, 1 (1996).
39. Fursa, D. V. et al., *Phys. Rev. A* **56**, 4606 (1997).
40. McLaughlin, D. T., Donnelly, B. P., and Crowe, A., *Z. Phys. D* **29**, 259 (1994).
41. Paris, R. M., and Mires, R. W., *Phys. Rev. A* **4**, 2145 (1971).
42. Karaganev, V., Bray, I., Teubner, P. J. O., and Farrell, P., *Phys. Rev. A* **54**, R9 (1996).
43. O'Neill, R. W., et al., *Phys. Rev. Lett.* **80**, 1630 (1998).
44. Williams, J. F., *J. Phys. B: At. Mol. Phys.* **14**, 1197 (1981).
45. Weigold, E., Frost, L., and Nygaard, K. J., *Phys. Rev. A* **21**, 1950 (1980).
46. Bray, I., and Stelbovics, A. T., *Phys. Rev. A* **46**, 6995 (1992).
47. Madison, D. H., Bray, I., and I. E. McCarthy, *J. Phys. B: At. Mol. Opt. Phys.* **24**, 3861 (1991).
48. Scholz, T. T., Walters, H. R. J., Burke, P. G., and Scott, M. P., *J. Phys. B: At. Mol. Opt. Phys.* **24**, 2097 (1991).
49. van Wyngaarden, W. L., and Walters, H. R. J., *J. Phys. B: At. Mol. Phys.* **19**, 929 (1986).
50. Wang, Y. D., Callaway, J., and Unnikrishnam, K., *Phys. Rev. A* **49**, 1854 (1994).
51. Farrell, D., Chwirot, S., Srivastava, R., and Slevin, J. A., *J. Phys. B: At. Mol. Opt. Phys.* **23**, 315 (1990).
52. Kumar, M., Stelbovics, A. T., and Williams, J. F., *J. Phys. B: At. Mol. Opt. Phys.* **26**, 2165 (1993).
53. Bray, I., and Stelbovics, A. T., *J. Phys. B: At. Mol. Opt. Phys.* **30**, L493 (1997).
54. Ehrhardt, H., Schulz, M., Tekaat, T., and Willmann, K., *Phy. Rev. Lett.* **22**, 89 (1969).
55. Schwienhorst, R., Raeker, A., Bartschat, K., and Blum, K., *J. Phys. B: At. Mol. Opt. Phys.* **29**, 2305 (1996).
56. Bartschat, K., *Comput. Phys. Commun.* **75**, 219 (1993).

Measured Correlated Motion in the Continuum of Three Coulomb-Interacting Particles, H^+, H^-, H^+

D. H. Jaecks, L. M. Wiese and O. Yenen

Department of Physics and Astronomy
University of Nebraska, Lincoln, NE 68588-0111

Abstract. We have measured the center of mass motion of the three massive Coulomb-interacting particles H^+, H^-, H^+ in the continuum. Starting with excited H_3^+ we have experimentally studied the dissociation of this three-body system by determining the final-state energy sharing configurations and mutual H^+, H^+ correlation angles. Each measured triple dissociation event was mapped onto a Dalitz Plot that elucidated the various dynamic processes. In addition to the direct dissociation of excited H_3^+ into three Coulomb-interacting particles, we have identified intermediate compound states that decay into the observed final-state channel.

INTRODUCTION

The classic "most celebrated of all dynamical problems is known as the Problem of Three Bodies" [1] and has, since the mid-eighteenth century, attracted the interest of many of history's best known thinkers. Whittaker [1] pointed out in his book on analytical dynamics, first published in 1904, that "since 1750 over 800 memoirs, many of them bearing the names of the greatest mathematicians, have been published on the subject"; names we all know, such as Poincaré, Lagrange and Jacobi, to mention but a few. Historically this interest in the "problem of three bodies" resulted from the fact that the motion of three particles acting through attractive forces that vary inversely as the square of the distances between any two of them, and proportional to the product of the masses of the particles, does not lend itself to a closed analytic solution. This fascination and intense interest in the description of systems of three interacting bodies in the continuum continues to this day for atomic systems, where again, there are no closed analytic solutions, even when all the interaction terms of the Hamiltonian are presumably known. One manifestation of this interest is the fact that there were more than 80 papers published in 1997 that were directly related to issues of the three-body problem in quantum systems. We define three-body atomic systems as those containing neutral atoms, atomic ions, bare ions and electrons and, in principle, their antiparticles, and have three identifiable constituents when they separate to large distances. The description of such quantum systems brings additional and diverse, albeit interesting physical aspects to the three-body problem compared to that found in the classical gravitational system. First, in quantum systems the potential energy terms of the Hamiltonian in Schrödinger's equation, used to determine the wave properties of the system, do not contain the masses of the interacting particles; rather they contain other intrinsic physical properties of the particles such as charge and magnetic moment and extrinsic properties such as the magnetic moments due to orbital angular momentum. In the case of molecules or molecular ions containing electrons, where one is interested in the correlated motion of the massive nuclei, although they are surrounded by electrons, the interactions between the massive nuclear constituents can be described by molecular potentials at some level of approximation.

The description of a system of three separated particles in the continuum brings additional challenges because the total energy ε_t of the system may, in principle, be shared an infinite number of ways. The nature of the short and long range interactions given in the Hamiltonian of the three particles, along with the magnitude of the masses, puts constraints on this sharing. In the description of the formation and any subsequent motion of any three-body system, one of the important parameters that characterizes the motion is the manner in which the total energy of the system is shared amongst the three particles. The characteristics of this energy sharing in turn provides information about the transition matrix describing the formation and decay of the three-body complex.

To date, experimental studies of three-body atomic systems in the continuum have centered on primarily those formed in the double photoionization of a neutral atom or a negative ion or in the ionization of an atom by an electron [2], all cases that involve two light electrons and a heavy positive ion that may or may not have added electronic structure. The threshold region, where the two electrons are given just sufficient energy to be put in the continuum of the residual ion, such that they have zero energy at infinite separation, is of particular interest because it is here the effects of the mutual interaction between the three particles is greatest and their relative motion most correlated. In such systems, most of the available energy is carried away by the much lighter electrons. The maximum energy carried away by the residual ion is limited by momentum conservation to a value of $2m_e/M_{ion}$ of the total available energy ε_t, and it is sufficient to measure the emission angles and energies of the two electrons in coincidence to determine the final-state dynamics.

This is not the case if the three particles are of equal or comparable mass, as with the systems such as e^+, e^-, e^- or H^+, H^-, H^+, because in the center-of-mass (c-m) frame the available energy can be more equitably shared among the three particles. For a system of three equal masses the maximum available energy from momentum conservation is 2/3 of the total energy available. In an effort to gain insights into the fundamental three-body dynamics of systems of nearly equal masses, where at the same time the long-range interactions are of a Coulomb nature, we have carried out an investigation of the formation and breakup of excited H_3^+ into the final-state

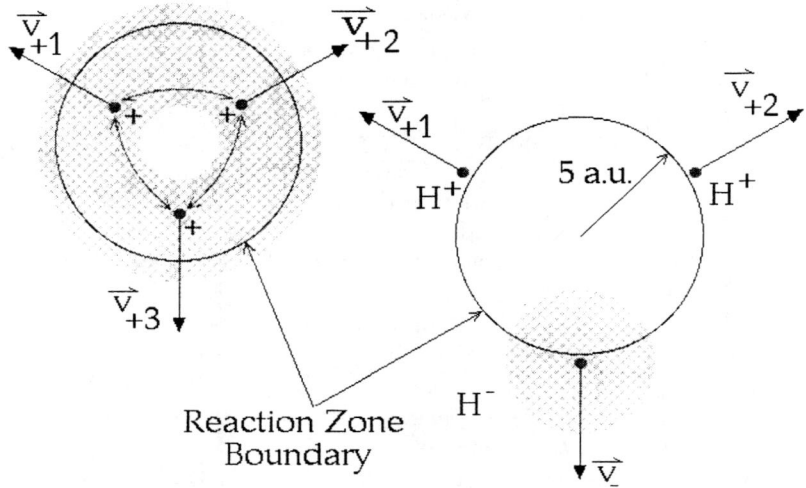

FIGURE 1. Schematic illustration of reaction and Coulomb zone in the dissociation of H_3^+ to H^+, H^-, H^+.

channel of H^+, H^-, H^+. A more ideal three-body system would be one in which the negative ion, H^-, is replaced by an antiproton, but this presents severe additional experimental challenges.

The dissociation of excited H_3^+ into the final-state channel can be considered to take place in several steps. The initial unexcited H_3^+ is considered to be in its lowest mode of vibration, which is the symmetric breathing ode. The dissociative states are formed by collisions with the He target, which for our purposes, is an "excitation hammer." The subsequent three-body breakup consists of three distinct physical regions. The first is often called the reaction zone and for this case the two electrons are shared by the three protons, but shared in such a way that the net force on the three protons is repulsive since we are in the continuum. By a given radius (see Fig. 1) the covalent state(s) of the system coalesces to an ionic, Coulomb-interacting state consisting of two protons and an H^- ion. In this outer Coulomb-interacting zone, the three particles interact over the entire trajectories of their motion as they move to very large separations, leading to the possibility of significant correlated motion. A similar type of process occurs in the dissociation of excited H_2 molecules which transform from a covalent to an ionic state, $H^+ + H^-$, over a small range of internuclear separations [3], with the subsequent interaction being purely Coulomb. At sufficiently large separations the two electrons remain on one proton and do not tunnel back to the other proton.

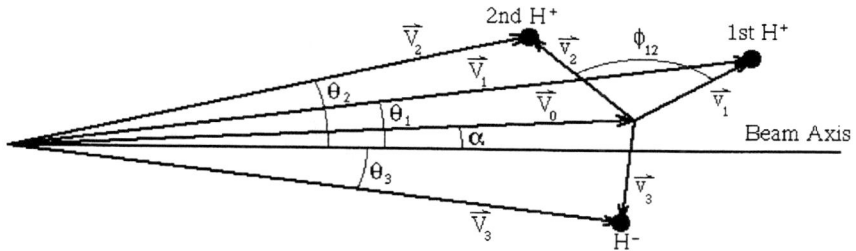

FIGURE 2. Newton diagram of 4 keV H_3^+ dissociating into H^+, H^-, H^+.

The Experiment

Measurements of the c-m characteristics of the three-body dissociation of H_3^{+*} into the final Coulomb interacting system, H^+, H^-, H^+, were carried out by first forming a beam of 4 keV H_3^+ ions and exciting the ions to anti-bonding states by colliding them with a static gas target of He. The subsequent decay in flight of the excited state(s) of H_3^{+*} is represented by the Newton diagram of Fig. 2, where $\mathbf{V}_0, \mathbf{V}_i$ and $\mathbf{v}_i$ represent the laboratory velocity of excited H_3^+ before decay, the laboratory velocities of the three ions, and the c-m velocities of the three ions. Of particular interest in this study is the c-m velocity or c-m energy distributions of the three particles and how the available c-m energy is shared amongst the three particles. The rate R of formation of a particular final state of the system, expressed in the center of mass, can be expressed in terms of the well-known equation

$$R = 2\pi |\mathbf{M}_{fi}|^2 \rho_f, \qquad (1)$$

where ρ_f is the density of final states in the c-m system and $\mathbf{M}_{fi}$ is the transition matrix that contains all the dynamical information about the excitation and decay of the three-body system. $\mathbf{M}_{fi}$ will in general be a function of the c-m energies of the three particles, the nature of their interaction, as well as the nature of the excitation process. For example, the character of the prepared excited H_3^+ state that is formed before decay may depend on the orientation of the triatomic molecule relative to the beam axis The nature of $\mathbf{M}_{fi}$ can be determined from the measured c-m energies of all three particles for a given available total c-m energy ε_t and displayed on a Dalitz plot [4].

Before discussing some of our results of the three-body dissociation, we briefly outline our experimental technique. Referring to Fig. 2, the basic principle of the experiment is to determine the laboratory energies and angles of all three particles coming from the same dissociation event in a triple coincidence measurement. From the Newton diagram and conservation of energy and momentum, one can readily express the c-m energy of each particle ε_i in terms of the measured laboratory energies E_i.

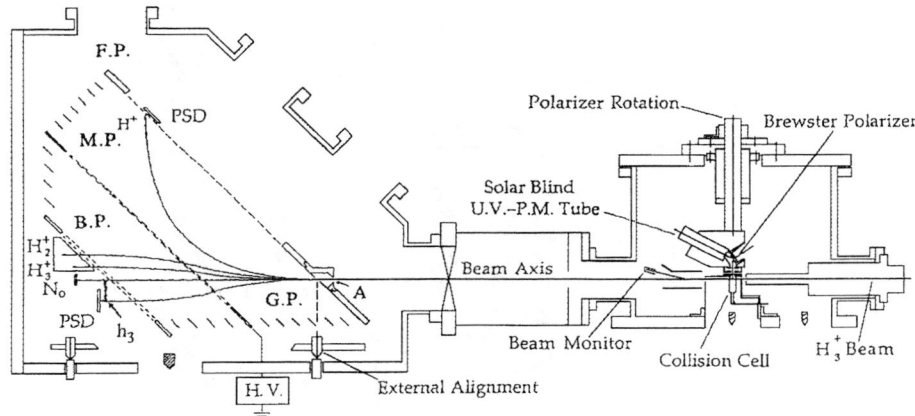

FIGURE 3. Side view of the apparatus for measuring the laboratory energies and angles of all three particles in coincidence.

$$\varepsilon_i = (1/9)[(4E_i + E_j + E_k) - 4(E_i E_j)^{1/2} \cos(\theta_i - \theta_j)$$
$$-4(E_i E_k)^{1/2} \cos(\theta_i - \theta_k) + 2(E_j E_k)^{1/2} \cos(\theta_j - \theta_k)] \qquad (2)$$

A physical quantity of significance is the correlation angle ϕ_{12} between the two protons in the c-m system, and it can be calculated once the c-m energies have been determined.

$$2\cos\phi_{12} = [\varepsilon_3/(\varepsilon_1\varepsilon_2)^{1/2} - (\varepsilon_1/\varepsilon_2)^{1/2} - (\varepsilon_2/\varepsilon_1)^{1/2}] \qquad (3)$$

The above equations hold for a geometry in which the plane of the molecule is the same as the plane defined by the collision center and the long, narrow slit of an electrostatic analyzer that is 0.845 meters from the collision center. The electrostatic analyzer system used to measure the laboratory energies has been described in great detail in an earlier publication [5]. One advantage of measuring the c-m energies using the present beam method is that in the Galilean transformation from the laboratory to the c-m system, the absolute errors in the energies undergo a compression resulting in small errors in the c-m system. The error in the total c-m energy ε_t given by $\Sigma\varepsilon_i$ is 3%, with typical errors in the individual c-m energies being 100 meV. Total c-m energies were on the order of 6 eV. Another advantage of the present beam method is that the ions are moving with laboratory energies on the order of 1.3 keV and therefore are detected with reasonably efficiency by position-sensitive microchannel plate detectors.

As shown in a side view of the apparatus in Fig. 3, the distance from the point of formation of the excited H_3^+ to the analyzer is about 0.845 meters, while the resolution of the position-sensitive detectors is 250 microns. Since all three particles are coming from the same position (event), the results are insensitive to the physical width of the ion beam (0.75mm), or its energy spread. It should be noted that the determined c-m energies ε_i given in the above equation do not depend on the value of the primary beam energy E_0 of the H_3^+ ion. However, a determination of the inelastic energy loss during the collision requires knowledge of E_0.

Experimental Results

The nature of the interaction and the correlated motion of the three particles in the continuum, as manifested in the experimentally determined values of ε_i and sharing of the total available internal energy ε_t, for each measured triple event, can be best seen using a Dalitz plot, utilized first in high energy physics to analyze the three-body decay [4] $K^\pm \to \pi^\pm + \pi^+ + \pi^-$. The Dalitz plot utilizes the property of an equilateral triangle that for any point inside the triangle, the sum of the three perpendicular distances from that point to the three sides of the triangle is a constant and is equal to the height of the triangle. This is shown in Fig. 4, where the measured total c-m energy of the three-body system is set equal to the height of the triangle and the three c-m energies ε_i are the perpendicular distances to the three sides. Thus any correlated three-body event of equal masses, where the measured c-m energies of the three equal-mass particles are ε_1, ε_2 and ε_3, with $\varepsilon_1 + \varepsilon_2 + \varepsilon_3 = \varepsilon_t$, can be represented by a single point within the triangle, with the height of the triangle given by ε_t. The position of that event point completely describes the final state c-m dynamics of the system. Momentum conservation limits the event points to lie within a circle that is tangent to the three sides at the mid-point of each side. Loci of points with constant correlation angle ϕ_{12}, as shown in Fig. 4, are given by ellipses that are always tangent to the two sides at their mid-point, with $\phi_{12} = 180°$ and $0°$ representing the circle limit. This also represents the collinear configuration of the three particles. The distribution of event points within the Dalitz plot provides a measure of the transition matrix $|\mathbf{M}_{fi}|^2$ from the initial state of H_3^+ system to the final three-body state of H^+, H^-, H^+.

Fig. 4 shows the results of a Dalitz plot construction of one set of measurements in which the three particle detectors were placed within the energy analyzer such that they sampled a distribution of final states where the three particles were in nearly a collinear configuration. [6] The 40 mm active area of the detectors did not allow us to sample the entire available energy space of the Dalitz plot simultaneously, with equal efficiency. However, the detectors were movable without breaking vacuum so that the entire region of the energy distribution can be sampled. The 59 real triple-coincidence events shown in Fig. 4 are for those in which the total energy available to the three particles lies between 6 and 7 eV. One observes from the Dalitz plot that the correlation angle varied from 180° (collinear) to 155°. The data shown is part of a sample of 250 measured triple-coincidence events in which the internal energy available varied from 4 to 14 eV. The data were taken during one data run that represented 30 days of continuous running of the ion beam. The count rate of real events was about 10^{-4} Hz. One experimental difficulty with triple coincidence

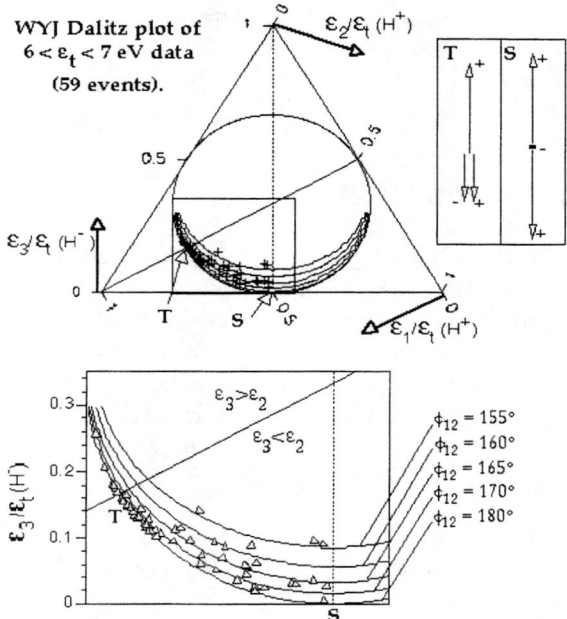

FIGURE 4. Dalitz Plot of sample measured triple coincidence events.

measurements is that the accidental triple coincidence rate increases as the cube of the He target density and the ion beam current while the real coincidence rate increases linearly, therefore one had to take great care in determining the optimum operating conditions of target pressure and beam current.

One can begin to understand some of the interesting dynamics of the three-body breakup from the distribution of event points. In Fig. 4, the point labeled **T** represents a final-state configuration in which the internal energy, ε_3 of H^-, and the internal energy ε_2 of one proton have identical energies. The c-m energy of the third proton is near it's maximum as allowed by conservation of energy and momentum. Points at **T** and along the 180° circle represent a collinear configuration of the particles in the c-m frame. We observe from Fig. 4 that this collinear configuration, as represented by $\phi_{12} = 180°$, has a propensity for the c-m energy of H^- being slightly less than the c-m energy of the proton labeled "2." This indicates that the H^- has a propensity to lie between the two protons in the collinear configuration. These results can result from a process whereby a compound state of H_2^{**} is formed and then subsequently decays, in the presence of the first proton, to H^+ and H^-. Event points resulting from the decay of compound states have specific, recognizable signatures on the Dalitz plot as indicated by the band of events about **T**.

$$H_3^{+**} \rightarrow H_2^{**} + H^+ \rightarrow H^+ + H^- + H^+ \quad (4)$$

We observe from the Dalitz plot that the energies of the H^+ and H^- in the c-m frame of the H_2^{**} compound state are nearly zero and are at most on the order of meV. The long range Coulomb interaction decreases the c-m energy of the H^- and slightly increases the c-m energy of the proton. Since the compound state H_2^{**} is in the presence of the first proton, it would be represented by some linear combination of unperturbed states of the free, excited H_2, thus allowing the H_2 system to couple to the final-state Coulomb-interacting channel that leads to H^+, H^-.

Once the H^- c-m energy ε_3 falls below 0.1 of the total available energy, event points fall off the collinear, $\phi_{12} = 180°$, configuration and a new physical dissociation mechanism manifests itself. These events result from a direct breakup of the H_3^{+**} into the three Coulomb-interacting particles. The mechanism is schematically shown in Fig. 1, with the details depending on the symmetry of the reaction-zone states that are excited and that subsequently decay to the three Coulomb states. Until the system has cleared the reaction zone and three identifiable Coulomb-interacting particles are formed, the dissociation dynamics are determined by the symmetries of the internal electron distributions. The extent to which the long-range Coulomb interaction modifies the velocity directions and magnitudes will depend upon the velocities at the reaction-zone boundary, with the Coulomb interaction dominating as $\varepsilon_t \to 0$.

ACKNOWLEDGEMENTS

This work was supported by the National Science Foundation under Grant No. PHY-9419505.

REFERENCES

1. Whittaker, E. T., *Analytical Dynamics*, New York: Dover Publications, 1944, p. 339.
2. Jones, S. and Madison, D. H., "Ionization of Atoms as a Three-Body Process," in *The Physics of Electronic and Atomic Collisions, AIP Conf. Proc. No. 360*, 1995, pp. 341-345.
3. Yenen, O., Calabrese, D., and Jaecks, D. H., *Nucl. Instr. and Meth.* **B79**, 103-105 (1993)
4. Hagedorn, R., *Relativistic Mechanics*, New York : Benjamin, 1963, p. 101.
5. Calabrese, D., Yenen, O., Wiese, L. M., Jaecks, D. H., *Rev. Sci. Instrum.* **65**, 116-122 (1994).
6. Wiese, L. M., Yenen, O., Jaecks, D. H., *Phys. Rev. Lett.* **79**, 4982-4985 (1997).

Precision Measurements of Atomic Polarizabilities

William Arie van Wijngaarden

*Physics and Astronomy Dept.
4700 Keele St., Petrie Bldg., York University
Toronto, Ontario, Canada, M3J 1P3
wvw@yorku.ca*

Abstract. Substantial progress has occurred during the last decade to develop improved methods for determining atomic polarizabilities. We review work that has been done on ground states as well as the study of Stark shifts of transitions to excited states that have been made with a precision of parts in 10^4. Measurements done on alkali atoms have yielded significantly more accurate data than previously available, stringently testing atomic theory.

INTRODUCTION

Polarizabilities describe the response of an atom to a static external electric field and are essential to determine a wide variety of quantities including charge exchange cross sections, van der Waals constants, dielectric constants, indices of refraction, electric fields in plasmas, etc. [1,2]. It is surprising that the polarizabilities of most elements have not been measured and few have been precisely determined. Several review articles describing various techniques to measure polarizabilities have been written including ones by Miller and Bederson [3,4] that describe the study of atoms and molecules and a text by Bonin and Kresin that discusses work done on clusters [5].

During the last decade, several groups have developed techniques that have determined atomic polarizabilities of alkali atoms with accuracies as high as parts in 10^4 [6-10]. This represents an improvement in accuracy of more than two orders of magnitude. Hence, measured polarizabilities provide a stringent test of atomic theory. Work has been done on ground states as well as on excited states that have been studied by precisely measuring Stark shifts of transition frequencies. Alkali atoms are of particular interest to theorists as they possess only a single valence electron and can therefore be relatively easily modeled. Applications of alkali atoms include atomic clocks [11], Bose-Einstein condensation [12] and parity-violation experiments [13,14].

This article briefly reviews some of the newly developed techniques to measure polarizabilities and discuss their results. This paper is organized as follows. First, the definitions of polarizabilities are stated and work studying the ground state is reviewed. This is followed by a presentation of optical techniques to accurately measure Stark shifts of alkali D lines as well as transitions to Rydberg states. Finally, conclusions are discussed.

DEFINITION OF POLARIZABILITIES

An external electric field $\mathbf{E}$ polarizes the electron cloud of an atom shifting the energy of an atomic state ψ. The first order shift is given by $\langle\psi|e\mathbf{E}\cdot\mathbf{r}|\psi\rangle$ where e is the electron charge and $\mathbf{r}$ is the electron position vector. This matrix element vanishes for atomic states of alkali atoms which have a definite parity. Hence, in general the Stark shift is second order in the electric field and is given by the following Hamiltonian [15].

$$H_{Stark} = -\left\{\alpha_0 + \alpha_2\frac{3J_z^2 - \mathbf{J}^2}{J(2J-1)}\right\}\frac{E^2}{2} \tag{1}$$

Here, $\mathbf{J}$ is the total electronic angular momentum of the state and J_z is the operator along the quantized direction z specified by the electric field direction. The second term is absent when $J = 0, 1/2$. The scalar α_0 and tensor α_2 polarizabilities are given by

$$\alpha_0 = \frac{r_o}{4\pi^2}\sum_{J'}\lambda_{JJ'}^2 f_{JJ'} \tag{2}$$

$$\alpha_2 = \frac{r_o}{8\pi^2}\frac{1}{(2J+3)(J+1)}\sum_{J'}\lambda_{JJ'}^2 f_{JJ'}[8J(J+1) - 3X(X+1)] \tag{3}$$

where $X = J'(J'+1) - J(J+1) - 2$, r_o is the classical electron radius, $\lambda_{JJ'}$ is the wavelength for a transition between states J and J', and $f_{JJ'}$ is the oscillator strength. The eigenstates of the Hamiltonian are $|Jm_J\rangle$ where m_J is the azimuthal quantum number and the corresponding eigenenergy is given by

$$E_{Jm_J} = -\left\{\alpha_0 + \alpha_2\frac{3m_J^2 - J(J+1)}{J(2J-1)}\right\}\frac{E^2}{2}. \tag{4}$$

The scalar and tensor polarizabilities have units of length cubed. Convenient atomic units are $a_o^3 = 1.4818\times 10^{-25}$ cm^3 where a_o is the Bohr radius. Alternatively, polarizabilities can be expressed in terms of the frequency shift produced by an electric field, i. e. 1 kHz/(kV/cm)$^2 = 4.0189 a_o^3/h$ where h is Planck's constant. Polarizabilities strongly depend on the electron position and are proportional to n^7 where n is the principal quantum number [16]. Hence, excited states have much larger polarizabilities than do ground states.

GROUND STATE POLARIZABILITIES

The earliest work studied ground states by passing an atomic beam through an inhomogeneous electric field [17]. The atoms experienced a force $\mathbf{F} = \alpha_0 \nabla(E^2/2)$ in the direction transverse to their velocity. The deflection of the atoms was measured using a hot wire detector which ionized the atoms and measured the resulting current. This method determined the alkali ground state polarizabilities to about 10% accuracy. The uncertainty arises principally from the determination of the electric field gradient which is calculated from the electrode geometry. Another complication is the spread of atomic velocities which broadens the spatial profile of the deflected beam. Later versions of the experiment used a velocity selector and obtained a result of $\alpha_0 = 41.0 \pm 2.9$ kHz/(kV/cm)^2 for the sodium ground state [18].

The so called E-H gradient balance, developed by Bederson et al. [19], was much less sensitive to the distribution of atomic velocities. The electrodes used to generate the electric field gradient also served as the pole pieces of a Stern Gerlach inhomogeneous magnetic field dH/dz. Atoms were not deflected if the electric and magnetic forces canceled i.e. $\alpha(m_J)EdE/dz = \mu(m_J)dH/dz$ where μ_J is the magnetic moment. This force cancellation is independent of the atom's velocity. Hence, this method had a larger signal to noise ratio than the earlier technique. The accuracy was limited by the determination of the electric field gradient which was found using a beam of metastable helium atoms whose polarizability was known. A value of 39.6 ± 0.8 kHz/(kV/cm)^2 was obtained for the sodium ground state polarizability.

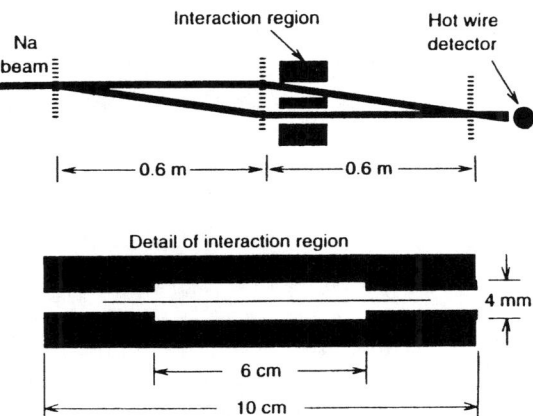

FIGURE 1. Apparatus used to determine ground state polarizability of sodium using atom interferometry. (Reprinted with permission from D. Pritchard [20].)

Recently, a significantly improved value for the ground state sodium polarizability was obtained using atom interferometry [20]. The apparatus consisted of a three grating Mach-Zender interferometer illustrated in Fig. 1. The transmission gratings had a 200 nm period and generated two atomic beams separated by 55 μm. One beam passed through an electric field created by applying a voltage across two metal foils which generated a relative phase shift between the two atomic beams given by

$$\Delta\phi = \frac{1}{\hbar v}\int_0^L \frac{\alpha_0 E^2}{2} dx. \quad (5)$$

Here v is the velocity of the atoms, α_0 is the ground state polarizability, E is the electric field and L is the length of the electric field region. The two beams were recombined and the resulting interference pattern was studied using a movable hot wire detector. Phase shifts of up to 60 rad were observed using fields of several kV/cm and α_0 was determined to be 40.56 ± 0.14 kHz/(kV/cm)^2. The uncertainty is due to statistical and systematic effects. The latter is dominated by geometrical effects such as fringing electric fields that affect the interaction length L. Another complication was modeling the velocity distribution of the atoms in the atomic beam to estimate the average velocity.

POLARIZABILITIES OF FIRST EXCITED STATES

The methods used to determine polarizabilities of ground states are not applicable to study excited states that have short radiative lifetimes. Several techniques exist to determine excited state polarizabilities including optical double resonance [21] and quantum beat spectroscopy [22]. We shall describe how more accurate data has been obtained by measuring Stark shifts.

An electric field shifts a transition frequency by an amount

$$\Delta\nu = \frac{K}{2}E^2. \quad (6)$$

The Stark shift rate, K, depends on the polarizabilities of the upper and lower states of the transition. For a $S_{1/2} \rightarrow P_{3/2}$ transition,

$$K = \alpha_0(S_{1/2}) - \alpha_0(P_{3/2}) - \alpha_2(P_{3/2})(m_J{}^2 - 5/4). \quad (7)$$

A typical experiment is illustrated in Fig. 2. A laser excites an atomic beam as it passes through a field free region and in an electric field. The laser frequency ν is scanned across the resonance while two detectors (Det 1 & 2) monitor fluorescence produced by the radiative decay of the excited atoms. The change in laser frequency is found using a Fabry-Perot interferometer that transmits light whenever the laser frequency changes by a free spectral range $\nu_{\text{FSR}} = c/2nL$, where c is the speed of light, n is the index of refraction of the air occupying the cavity and L is the etalon length. This method has been used to study a number of transitions,

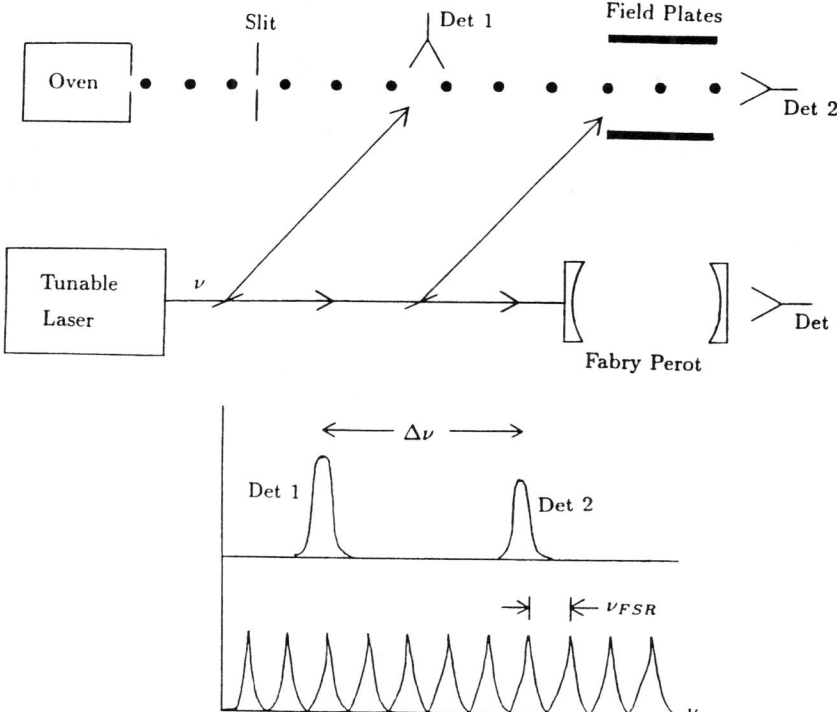

FIGURE 2. Typical apparatus used to measure Stark shifts using a Fabry-Perot interferometer to calibrate frequency.

including the sodium D lines [7]. The frequency calibration is limited by vibrations of the cavity and pressure/temperature fluctuations that perturb the refractive index n. These problems are reduced by using a HeNe laser that is locked to an iodine transition to stabilize the etalon length and enclosing the interferometer in a vacuum chamber [10]. This paper focuses on techniques that determine frequency shifts more accurately and conveniently.

Tanner and Wieman studied the cesium $6S_{1/2} \rightarrow 6P_{3/2}$ transition using the apparatus illustrated in Fig. 3 [6]. A diode laser at 852 nm was locked to the transition observed using saturation spectroscopy in a cell while an acousto-optic modulator (AO) frequency shifted part of the laser beam into resonance with atoms experiencing an electric field. This permitted the determination of $\alpha_0(6P_{3/2}) - \alpha_0(6S_{1/2}) = 308.6 \pm 0.6$ and $\alpha_2(6P_{3/2}) = -65.3 \pm 0.4$ kHz/(kV/cm)^2. These data agree with values of 308.0 and -65.1 kHz/(kV/cm)^2 computed by Zhou and Norcross using a semiempirical potential composed of a Thomas-Fermi potential plus a term describing the polarization of the inner electron core [23].

Another method to measure Stark shifts is the laser heterodyne technique illustrated in Fig. 4. It was used to study the cesium $6S_{1/2} \rightarrow 6P_{1/2}$ transition [8]. One

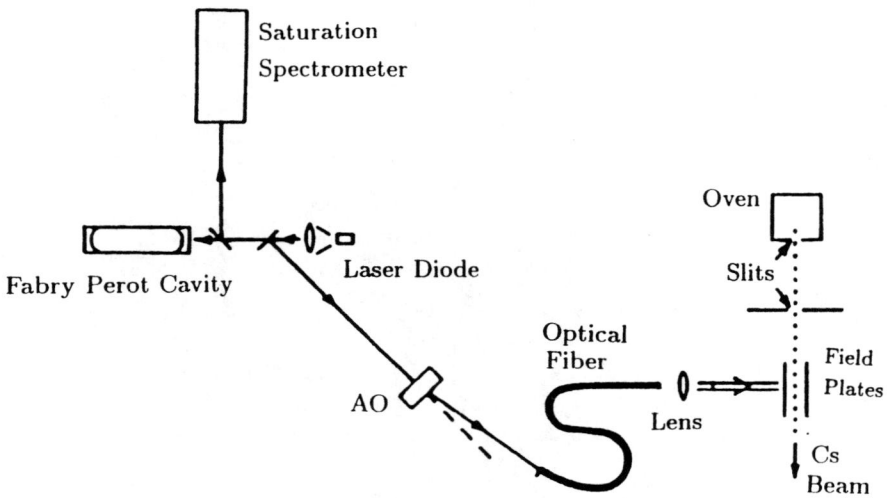

FIGURE 3. Apparatus used to study the cesium $6S_{1/2} \to 6P_{3/2}$ transition. (Reprinted with permission from C. Tanner [6].)

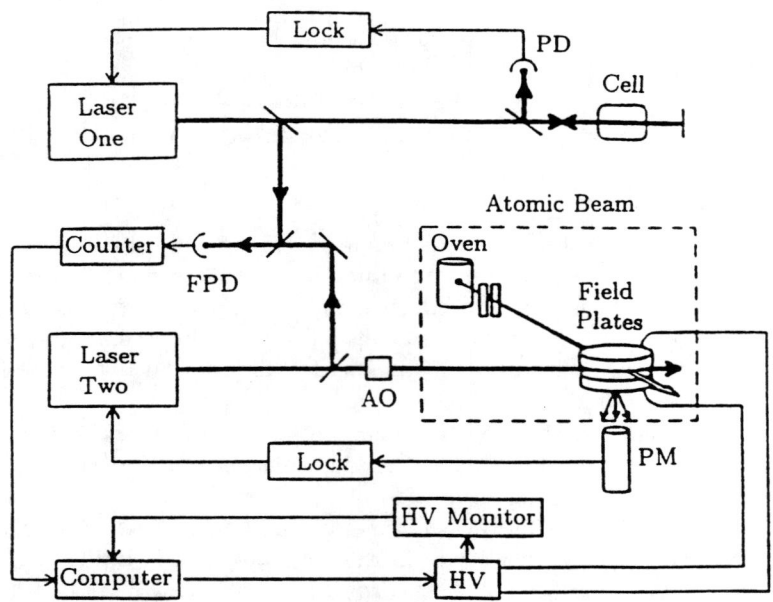

FIGURE 4. Apparatus used to determine Stark shifts of the alkali D lines.

Reprinted from Optics Communications 94, L. R. Hunter et al., Precise measurement of the Stark shift of the cesium D1 line, 210, © 1992, with permission from Elsevier Science.

TABLE 1. Accuracy of the electric-field determination.

Effect	Contribution
Plate Spacing	0.025
Voltage Divider	0.015
Voltmeter	<0.005
Voltage Stability	<0.01
Fringing Field Effects	<0.01
Total Uncertainty	0.029

diode laser was locked to a transition observed in a cesium cell by a photodiode (PD) while a second laser was locked to a Stark shifted transition observed by a photomultiplier (PM). The two laser beams were superimposed onto a fast photodiode (FPD) and a frequency counter measured the beat frequency as a function of the high voltage (HV) applied across the field plates. The latter consisted of two aluminum coated glass plates that formed an etalon. The plate spacing was found using a laser and a wavemeter to measure the etalon's free spectral range. The result of $\alpha_0(6P_{1/2}) - \alpha_0(6S_{1/2}) = 230.44 \pm 0.03$ kHz/(kV/cm)^2 compares to a value of 230.5 kHz/(kV/cm)^2 computed by Zhou and Norcross [23]. The experimental accuracy is limited primarily by uncertainty in the determination of the voltage applied across the field plates.

POLARIZABILITIES OF RYDBERG STATES

Rydberg states can be populated via stepwise excitation using several lasers as is illustrated in Fig. 5. In our experiments, the cesium $6P_{3/2}$ state was excited using a diode laser (SDL 5712-H1) at 852 nm. A ring dye laser (Coherent 699) then excites either a S or D state whose transition wavelength is in the range of 535 to 605 nm, using Pyromethene 556 or Rhodamine 6G laser dyes.

The apparatus is illustrated in Fig. 6. An atomic beam was generated by an oven enclosed in a vacuum chamber that was pumped to a pressure of 2×10^{-7} torr by two diffusion pumps and a liquid nitrogen trap. The atoms were collimated using several slits to produce a beam having a divergence of 1 milliradian. The atomic beam was intersected by laser beams in a region free of electric fields as well as between the two field plates. The latter were made of stainless steel and have a diameter of 12.70 cm. The plate spacing was measured using machinists calibration blocks to be 2.5395 ± 0.0006 cm. Voltages of up to 50 kV can be applied across the plates. The voltage was determined using a voltage divider having an accuracy of 0.015% and a precision voltmeter (HP 34401A). Other effects such as voltage stability of the power supply and fringing fields which were numerically modeled, were negligible as is shown in Table 1.

The diode laser produced about 100 mW of single mode light. Part of this laser beam was passed through a pyrex cell that had been evacuated and loaded with cesium metal. The cesium $6P_{3/2} \rightarrow 6S_{1/2}$ transition excited in a cell has a Doppler broadened width of about 0.5 GHz as compared to a natural linewidth of 5 MHz when observed using an atomic beam. Hence, the cesium cell was useful to find the transition. Two infrared cameras monitored the fluorescence produced by the radiative decay of the $6P_{3/2}$ state.

The ring dye laser was electronically stabilized and has a manufacturer specified linewidth of 0.5 MHz. The laser beam passed through either an acousto-optic or electro-optic modulator. The modulation signal was supplied by a signal synthesizer (Hewlett Packard 8647A) with an accuracy of one part in 10^6 and amplified to a power of 4 W (Amplifier Research 4W1000). An acousto-optic modulator was used to study the $(10-13)S_{1/2}$ states [9] while an electro-optic modulator was used for the $(10-13)D_{3/2,5/2}$ states [24,25]. In general, acousto-optic modulators operate at lower modulation frequencies. We used an acousto-optic modulator (Brimrose TEF 27-10) that frequency shifted over 50% of the incoming laser light for modulation frequencies in the range of 220–320 MHz. The electro-optic (EO) modulator (ν Focus 4421) shifted over half of the incoming light for modulation frequencies between 995 and 1005 MHz. Electro-optic modulators have the advantage of not spatially deflecting the frequency shifted laser components unlike acousto-optic modulators. This simplifies the alignment of the laser and atomic beams. Precision measurements with acousto-optic modulators have been reviewed [10], and the remainder

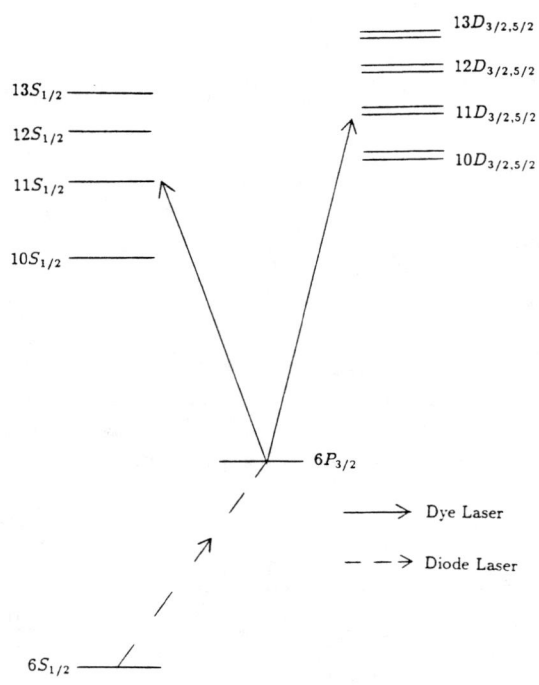

FIGURE 5. Laser excitation of cesium Rydberg states.

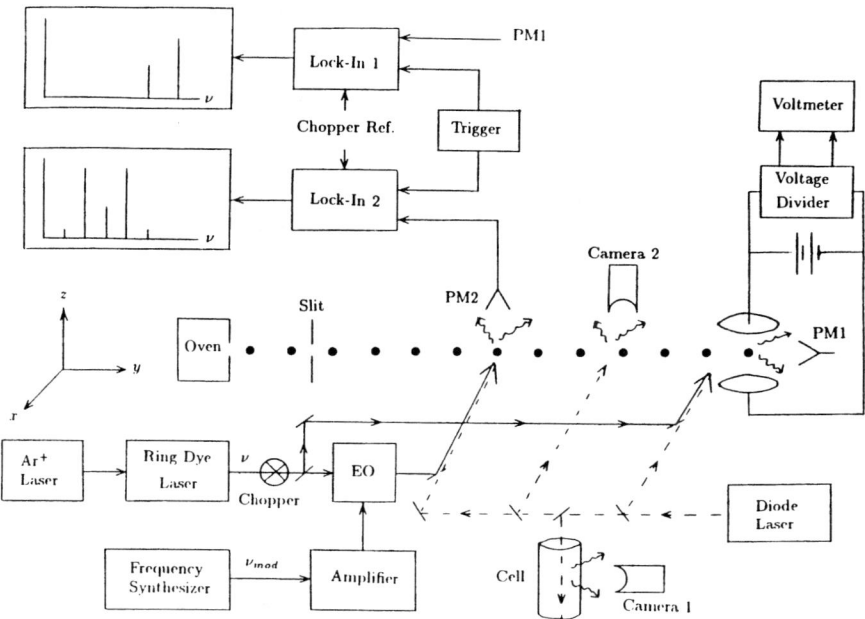

FIGURE 6. Apparatus used to study Stark shifts of cesium $6P_{3/2} \rightarrow (10-13)D_{3/2,5/2}$ transitions.

of this article therefore discusses work done using electro-optic modulators.

The diode and dye laser beams were superimposed and intersected the atomic beam orthogonally to eliminate the first order Doppler shift. Fluorescence, produced by the radiative decay of the excited state, was detected from the field free region and from between the electric field plates by two photomultipliers (Hamamatsu R928). Their signals were processed by two independent lock-in amplifiers (SRS 850) where the reference signal was supplied by a chopper that modulated the dye laser beam at a frequency of 2 kHz. The lock-in amplifiers digitized the demodulated signal when externally triggered by a signal generator at a rate of 256 Hz.

Sample signals are shown in Fig. 7. This scan took about 2 minutes and consisted of approximately 30,000 points. Five peaks are shown in Fig. 7a due to excitation of the $6P_{3/2} \rightarrow 11D_{3/2}$ transition by the laser frequencies ν, $\nu \pm \nu_{\text{mod}}$, $\nu \pm 2\nu_{\text{mod}}$ where the modulation frequency $\nu_{\text{mod}} = 1000.000$ MHz. The number of points separating the various 1 GHz intervals was found using several hundred different scans and were found to be consistent with each other, showing that the frequency scan is linear to better than one part in 1000 [10]. Each point corresponded to an average value of 0.1898 ± 0.0003 MHz.

Data were first taken without any voltage applied to the electric field plates to check whether the atoms in the two regions where the laser intersects the atomic

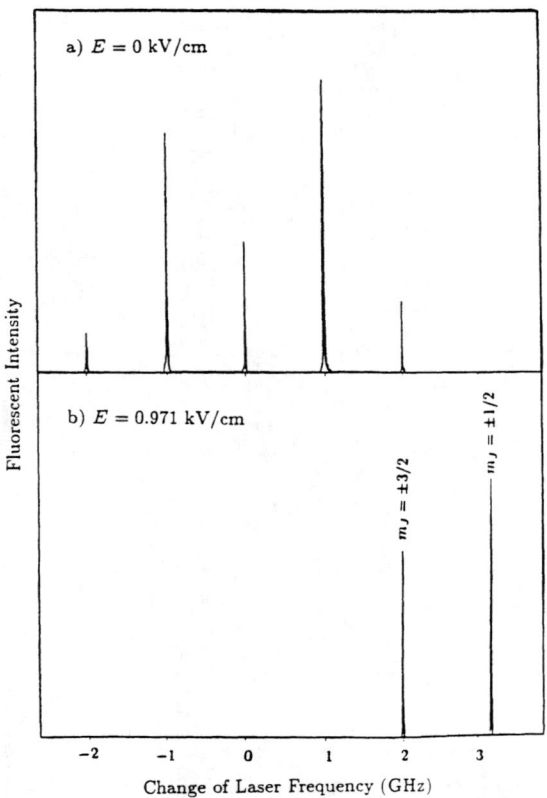

FIGURE 7. Fluorescence versus change of laser frequency. Fig. 7a shows the signal generated when the frequency modulated laser excited the atomic beam in a field-free region, while Fig. 7b shows the signal produced when atoms in an electric field are excited by the unmodulated laser beam.

beam were excited by the same laser frequency. A frequency offset ν_{off} occurs if there is a slightly different alignment of the laser beams in the two regions with the atomic beam creating a different residual first order Doppler shift. ν_{off} was measured to have an average value of 7.97 ± 0.45 MHz.

Data were next taken by applying an electric field across the plates. The dye laser was linearly polarized parallel to the electric field and therefore excited the atoms from the $6P_{3/2}$ state to the $|m_J| = 1/2, 3/2$ levels of the $D_{3/2,5/2}$ states. Hence, the fluorescence signal shown in Fig. 7b shows two peaks. The lock-in amplifier located the peak centers by fitting a Gaussian function to the data. The results were found to be independent of voltage polarity and the dye laser power, which was varied by a factor of 100 using neutral-density filters. The lineshape was broadened at the higher powers but the peak was not frequency shifted. Each dye laser beam was attenuated to a power of less than 1 mW to reduce power broadening while data such as shown in Fig. 7, were collected.

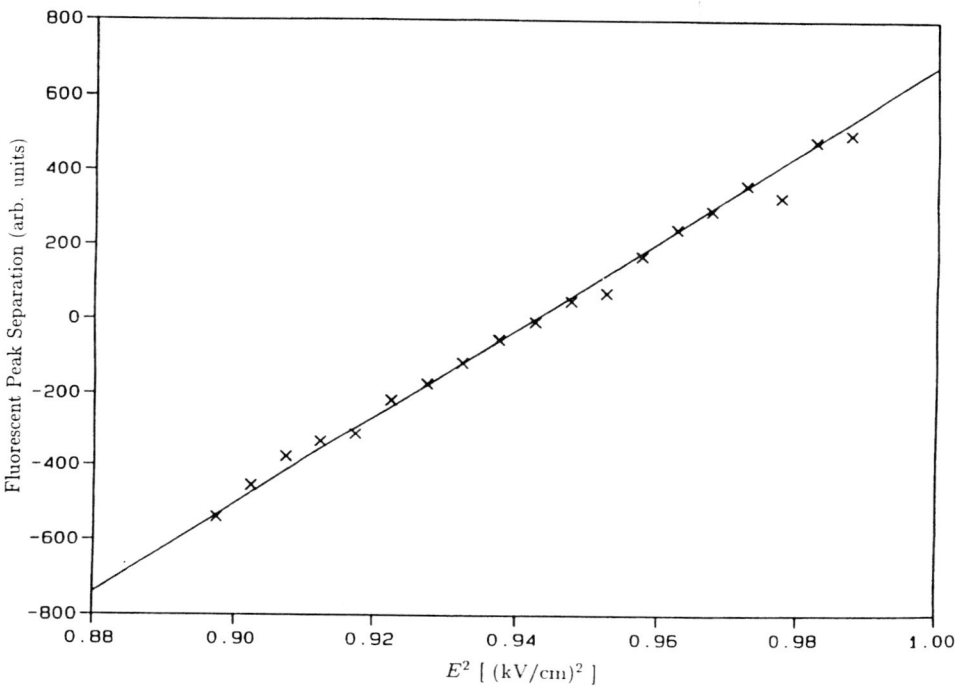

FIGURE 8. Fluorescent peak separation versus electric field squared for the $11D_{5/2}, |m_J| = 3/2$ level.

The Stark Shift rate K was found by measuring the electric field such that the peak observed in the electric field region overlapped with the peak generated by the second laser sideband. The exact field strength was found by plotting the frequency separation of the two peaks versus the square of the electric field as is shown in Fig. 8. The data was fitted to a straight line and a Stark shift of 2000 MHz was found to occur at an electric field of 970.82 ± 0.42 V/cm.

Figure 9 shows data taken when the dye laser excited the $6P_{3/2} \rightarrow 11D_{5/2}$ transition. The frequency shift does not depend quadratically on the electric field due to Stark mixing of the fine structure states which can be seen by diagonalizing the Hamiltonian

$$H = a\mathbf{L} \cdot \mathbf{S} - \left\{\alpha_0 + \alpha_2 \frac{3L_z^2 - \mathbf{L}^2}{L(2L-1)}\right\} \frac{E^2}{2}. \tag{8}$$

The first term is the spin orbit interaction, a is the coupling constant, $\mathbf{L}$ is the orbital electronic angular momentum and $\mathbf{S}$ is the electronic spin. The eigenenergies are plotted in Fig. 10. For weak fields, the Stark shift rate K is

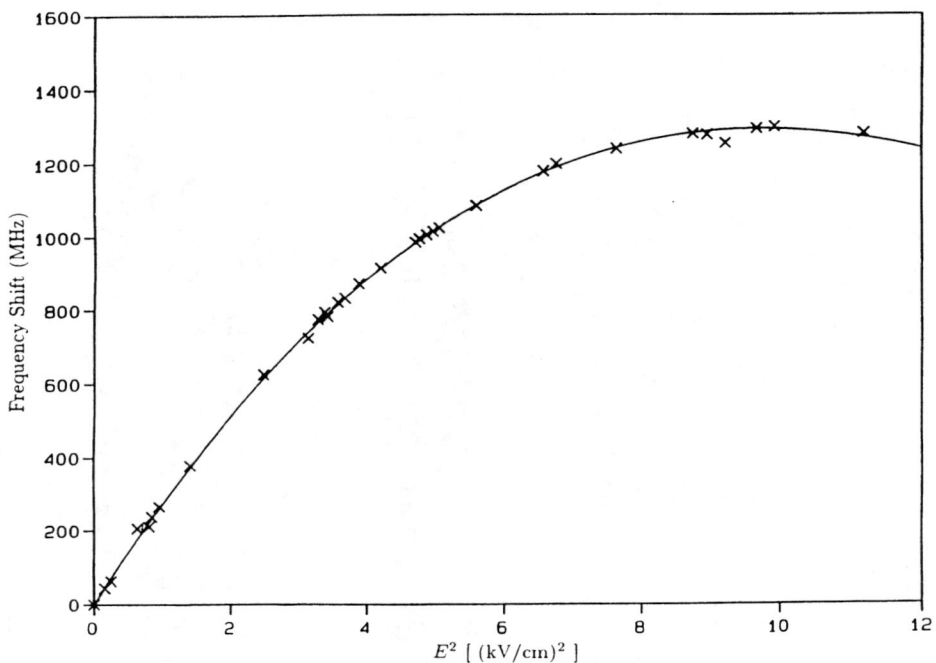

FIGURE 9. Frequency shift versus electric field squared for the $11D_{3/2}, |m_J| = 3/2$ level. The fitted function is given by $y = 293.55E^2 - 19.4E^4 + 0.29E^6$ where y is the frequency shift (MHz) and E is the electric field measured in kV/cm.

$$\begin{aligned} K(D_{5/2}, |m_J| = 1/2) &= -\alpha_0(D_{5/2}) + \frac{4}{5}\alpha_2(D_{5/2}) + \frac{3}{250}\frac{\alpha_2(D_{5/2})^2}{a}E^2 \\ K(D_{5/2}, |m_J| = 3/2) &= -\alpha_0(D_{5/2}) + \frac{1}{5}\alpha_2(D_{5/2}) + \frac{9}{125}\frac{\alpha_2(D_{5/2})^2}{a}E^2 \\ K(D_{5/2}, |m_J| = 5/2) &= -\alpha_0(D_{5/2}) - \alpha_2(D_{5/2}) \\ K(D_{3/2}, |m_J| = 1/2) &= -\alpha_0(D_{3/2}) + \alpha_2(D_{3/2}) - \frac{6}{245}\frac{\alpha_2(D_{3/2})^2}{a}E^2 \\ K(D_{3/2}, |m_J| = 3/2) &= -\alpha_0(D_{3/2}) - \alpha_2(D_{3/2}) - \frac{36}{245}\frac{\alpha_2(D_{3/2})^2}{a}E^2. \end{aligned} \quad (9)$$

For the $|m_J| = 3/2$ levels of the $(10-13)D_{3/2}$ state, data was fit to the function $y = a_1 E^2 + a_2 E^4 + a_3 E^6$. Equation (9) shows a_1 can be identified as $-\alpha_0 - \alpha_2$. The uncertainty of this quantity was conservatively estimated by considering the effect of deleting the term proportional to E^6 in the fitting function. For the $|m_J| = 1/2$ $(10-13)D_{3/2}$ states and the $|m_J| = 1/2, 3/2$ levels of the $(10-13)D_{5/2}$ states, a straight line was fit to determine an initial value of the Stark shift rate K_0. Next,

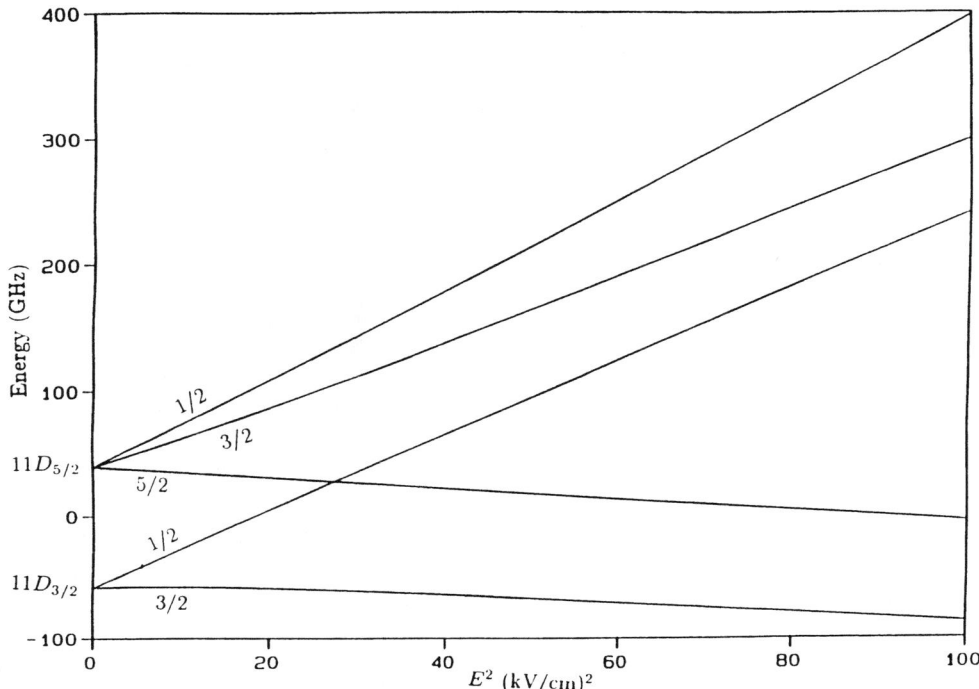

FIGURE 10. Eigenenergies computed for the 11D states for the Hamiltonian given in equation 8 using $a = 39.0$ GHz. To first order, the eigenenergies depend quadratically on the electric field. The departure from this quadratic dependence is most noticeable for the $|m_J| = 3/2$ level of the $11D_{3/2}$ state.

the perturbative correction in K quadratic in the field was evaluated. The Stark shift rate was then recalculated and the polarizabilities were obtained as shown in Tables 2 and 3.

Table 3 also lists results obtained by Fredriksson and Svanberg [26]. They used a rf lamp to excite the $6P_{3/2}$ state and a dye laser having a linewidth of 75 MHz to populate the S and D states. Their fluorescent signals had a significantly lower signal to noise ratio than the present work. The change in laser frequency was also determined using an interferometer as illustrated in Fig. 2 while the electric field was determined to an accuracy of a few percent.

The results were also compared to data computed using a so called Coulomb approximation [27,28]. This is a non *ab initio* theoretical model that inputs measured energies into the Schrödinger equation which is then solved using a Coulomb potential to model the interaction between the alkali valence electron with the potential generated by the nucleus and inner electron core. This potential does not include relativistic effects nor the spin orbit interaction which for the $10D_{3/2,5/2}$

TABLE 2. Determination of the Stark shift rate. Units are MHz/(kV/cm)^2.

State	$|m_J|$	K_0	Correction	K
$10D_{3/2}$	1/2	1886.5 ± 1.4	0.66 ± 0.01	1887.5 ± 1.4
	3/2	195.0 ± 1.0		
$10D_{5/2}$	1/2	2677.0 ± 2.6	−0.90 ± 0.01	2676.1 ± 2.6
	3/2	1667.4 ± 1.4	−8.71 ± 0.09	1658.7 ± 1.4
$11D_{3/2}$	1/2	4799.7 ± 4.6	+2.33 ± 0.02	4802.1 ± 4.6
	3/2	587.1 ± 5.8		
$11D_{5/2}$	1/2	6777.4 ± 5.7	−3.21 ± 0.03	6774.2 ± 5.7
	3/2	4259.6 ± 3.8	−30.7 ± 0.3	4229.0 ± 3.8
$12D_{3/2}$	1/2	10864.7 ± 6.4	7.10 ± 0.07	10871.8 ± 6.4
	3/2	1488.6 ± 7.6		
$13D_{5/2}$	1/2	31553 ± 24	−27.0 ± 0.3	31526 ± 24
	3/2	20137 ± 17	−254.1 ± 2.5	19882 ± 18

states is 140 GHz. It is therefore surprising that the calculated polarizabilities are within 1% of the experimental results. Similar agreement between experimental and theoretically determined polarizabilities has been found for the other alkali atoms [29-31]. Clearly, more elaborate theoretical models such as many body perturbation theory [32,33] that take into account effects of polarization of the inner electron core, electron correlation, relativistic corrections can be expected to yield more accurate polarizability values.

CONCLUSIONS

Techniques that determine scalar and tensor polarizabilities with accuracies of one part in a thousand or higher have been demonstrated in the last decade. Atom interferometry is especially well suited to study ground states as was demonstrated for sodium. Polarizabilities of excited states can be found by measuring Stark shifts of optical transitions. The laser heterodyne technique is attractive to study first excited states that can be populated using relatively inexpensive diode lasers. Hunter et al used this method to investigate Stark shifts of D lines of several alkali atoms [10]. In general, transitions whose wavelengths are inaccessible to diode lasers, including ones to Rydberg states, can be studied using acousto/electro optical modulators to precisely determine Stark shifts.

The measurements challenge our theoretical understanding of many electron atoms. The polarizabilities computed using the simple Coulomb approximation were found to be within 1% of the experimental data. It is however essential to take into account Stark induced mixing of the fine structure states. Improved agreement between theory and experiment is likely to result from more elaborate calculations as discussed previously. Existing theoretical work has been largely

TABLE 3. Polarizabilities of some cesium Rydberg states. Units are MHz/(kV/cm)^2.

State	α_0	α_2	Ref.
$10S_{1/2}$	123 ± 6		26
	119.06 ± 0.28		9
	118		27
$11S_{1/2}$	322 ± 16		26
	309.70 ± 0.26		9
	309		27
$12S_{1/2}$	720 ± 45		26
	713.48 ± 0.58		9
	709		27
$13S_{1/2}$	1650 ± 170		26
	1491.2 ± 1.2		9
	1490		27
$10D_{3/2}$	−1150 ± 170	840 ± 40	26
	−1041.3 ± 0.9	846.3 ± 0.9	25
	−1050	848	27
$11D_{3/2}$	−2694.6 ± 2.7	2107.5 ± 2.7	24
	−2712	2120	27
$12D_{3/2}$	−6180.0 ± 4.9	4691.4 ± 4.9	25
	−6245	4753	27
$13D_{3/2}$	−12935 ± 18	9620 ± 18	25
	−12989	9679	27
$10D_{5/2}$	−1340 ± 130	1770 ± 90	26
	−1319.5 ± 2.1	1695.7 ± 4.9	25
	−1319	1704	27
$11D_{5/2}$	−3790 ± 350	4010 ± 400	26
	−3379.5 ± 5.4	4242 ± 11	24
	−3384	4255	27
$12D_{5/2}$	−7660 ± 15	9501 ± 39	25
	−7738	9530	27
$13D_{5/2}$		19000 ± 1000	26
	−16001 ± 25	19406 ± 49	25
	−16099	19533	27

confined to the cesium D lines which are of importance for tests of parity violation.

The work discussed in this paper is readily applicable to other atomic systems. It should be relatively straight forward to apply the atom interferometric method to study ground states of other alkali atoms. Similarly, Stark shifts have only been investigated for a small number of transitions in alkali and alkaline-earth atoms [10]. These experimental results complement information about of atomic wavefunctions obtained by measurements of radiative lifetimes [34] and oscillator strengths [35]. Polarizabilities can however be determined with much higher precision. Hence, precision measurements of polarizabilities will be of continued future interest for investigating atomic structure.

ACKNOWLEDGMENTS

This work was supported by the Natural Sciences and Engineering Research Council of Canada and York University. We would like to thank J. Clarke for proof reading the manuscript.

REFERENCES

1. Kadar-Kallen, M. A., and Bonin, K. D., *Phys. Rev. Lett.* **72**, 828 (1994).
2. Lawler, J. E., and Doughty, D. A., *Adv. At. Mol. Phys.* **34**, 171 (1995).
3. Miller, T. M., and Bederson, B., *Adv. At. Mol. Opt. Phys.* **13**, 1 (1977).
4. Miller, T. M., and Bederson, B., *Adv. At. Mol. Opt. Phys.* **25**, 37 (1988).
5. Bonin, K. D., and Kresin, V., *Electric Dipole Polarizabilities of Atoms, Molecules and Clusters*, Singapore: World Scientific Press, 1997.
6. Tanner, C. E., and Wieman, C., *Phys. Rev. A* **38**, 162 (1988).
7. Windholz, L., Musso, M., *Phys. Rev. A* **39**, 2472 (1989).
8. Hunter, L. R., Krause, D., Miller, K. E., Berkeland, D. J. and Boshier, M. G., *Opt. Commun.* **94**, 210 (1992).
9. van Wijngaarden, W. A., Hessels, E. A., Li, J., and Rothery, N. E., *Phys. Rev. A* **49**, R2220 (1994).
10. van Wijngaarden, W. A., *Adv. At. Mol. Opt. Phys.* **36**, 141 (1996).
11. Itano, W. M., Lewis, L. L., and Wineland, D. J., *J. de Physique* **42**, 283 (1981).
12. Anderson, M. H., Ensher, J. R., Matthews, M. R., Wieman, C. E., and Cornell, E. A., *Science*, **269**, 198 (1995).
13. Wood, C. S., Bennett, S. C., Cho, D., Masterson, B. P., Roberts, J. L., Tanner, C. E., and Wieman, C. E., *Science* **275**, 1759 (1997).
14. Dzuba, V. A., Flambaum, V. V., and Sushkov, O. P., *Phys. Rev. A* **51**, 3454 (1995).
15. Khadjavi, A., Lurio, A., and Happer, W., *Phys. Rev.* **167**, 128 (1968).
16. Bethe, H. A., and Salpeter, E. E., *Quantum Mechanics of One and Two Electron Atoms*, New York: Plenum Press, 1977.
17. Chamberlain, G. E., and Zorn, J. C., *Phys. Rev.* **129**, 677 (1963).
18. Hall, W. D., and Zorn, J. C., *Phys. Rev. A* **10**, 1141 (1974).

19. Molof, R. W., Schwartz, H. L., Miller, T. M., and Bederson, B., *Phys. Rev. A* **10**, 1131 (1974).
20. Ekstrom, C. R., Schmiedmayer, J., Chapman, M. S., Hammond, T.D., and Pritchard, D. E., *Phys. Rev. A* **51**, 3883 (1995).
21. Rinkleff, R. H., *Z. Phys. A* **296**, 101 (1980).
22. Kulina, P., and Rinkleff, R. H. *Z. Phys. A* **304**, 371 (1982).
23. Zhou, H. L, and Norcross, D. W., *Phys. Rev. A* **40**, 5048 (1989).
24. van Wijngaarden, W. A., and Li, J., *Phys. Rev. A* **55**, 2711 (1997).
25. Xia, J., Clarke, J., Li, J., and van Wijngaarden, W. A., *Phys. Rev. A* **56**, 5176 (1997).
26. Fredriksson, K., and Svanberg, S., *Z. Phys. A* **281**, 189 (1977).
27. van Wijngaarden, W. A., and Li, J., *J. Quant. Spectrosc. Radiat. Transfer* **52**, 555 (1994).
28. Bates, D. R., and Damgaard, A., *Philos. Trans. R. Soc. London* **242**, 101 (1949).
29. Gruzdev, P. F., Soloveva, G. W., and Sherstyuk, A. I., *Opt. Spectrosc.* **71**, 513 (1991).
30. van Wijngaarden, W. A., *J. Quant. Spectrosc. Radiat. Transfer* **57**, 275 (1997).
31. van Wijngaarden, W. A., and Xia, J., *J. Quant. Spectrosc. Radiat. Transfer* in press (1998).
32. Dzuba, V. A., Flambaum, V. V., Silvestrov, P. G., and Sushkov, O. P., *Phys. Lett. A* **140**, 493 (1989).
33. Blundell, S. A., Johnson, W. R., and Sapirstein, J., *Phys. Rev. A* **43**, 3407 (1991).
34. Rafac, R. J., Tanner, C. E., Livingtston, A. E., Kukla, K. W., Berry, H. G. and Kurtz, C. A., *Phys. Rev. A* **50**, R1976 (1994).
35. van Wijngaarden, W. A., Bonin, K. D., Happer, W., Miron, E., Schreiber, D., and Arisawa, T., *Phys. Rev. Lett.* **56**, 2024 (1986).

Mass Measurements far from Stability: Modern Approaches

Georg Bollen

EP-ISOLDE, CERN
CH-1211 Geneva 23
Switzerland

Abstract. Today a variety of powerful techniques exist for direct mass measurements of unstable isotopes. This paper gives an overview of the most modern approaches, which enable us to investigate with high accuracy nuclear binding energies in regions very far from the valley of β-stability.

INTRODUCTION

By means of a simple mass spectrograph, Thompson [1] and Aston [2] discovered that many elements have several isotopes, each with integer atomic weight. In subsequent systematic measurements on different isotopes, Aston noticed tiny deviations from this integer number rule. He improved his mass spectrograph and investigated these deviations for a number of isotopes. In 1927, over seventy years ago, he published his famous 'packing fraction' curve [3], showing the average binding energy per nucleon. From the observation, that the 'packing fraction' is practically constant it was concluded that the forces that bind the nucleons to form a nucleus have a saturation character. This led to the development of the liquid-drop model [4–6].

Today a number of models for the description of the atomic nucleus have been developed [7–9]. They reach from microscopic theories based on more or less realistic nucleon-nucleon interactions to phenomenological models that allow the whole surface of masses to be reproduced with relatively small computational effort. The development of these models was only possible because of a systematic experimental study of the nuclear binding energy (and of course also of other nuclear properties) as a function of mass number and isospin. Not single measurements but the observation of trends in the nuclear binding energy and the departure from a smooth behaviour give us information about nuclear-structure effects that have to be explained by theory.

Since Aston, it has been possible to continually raise the accuracy of the determination of atomic masses and to extend the investigation to more and more exotic nuclei. An important step was the production of artificial unstable isotopes which became possible with the construction of nuclear reactors, accelerators and on-line mass separators. Today more than 3000 isotopes have been identified, but only for about 2000 of them have mass values been determined [10,11]. Close to stability, the atomic masses are known with a typical accuracy of several keV and in a few cases even better. However, the uncertainties of experimental mass values increase steadily when leaving the valley of stability and reach values of $\delta m/m \approx 10^{-4}-10^{-6}$ at the border line of nuclides with known mass values.

For a long time the direct determination of masses with mass spectrometers was only possible for stable or very long-lived isotopes. Q-values of nuclear decays and reactions were employed for the determination of masses of short-lived isotopes. Indeed, the majority of mass values of radioactive isotopes have to date been obtained from such measurements. Q-value mass determinations require a knowledge of the decay and nuclear level scheme of the nuclei involved. Far from stability, such information is unfortunately often not known. Furthermore, the large number of links required to connect a very exotic nucleus to a nucleus with well-known mass can lead to large uncertainties and, if systematic errors occur, to wrong mass values. Therefore, the demand for direct mass measurements far from stability has always been strong [12].

Measurements on short-lived isotopes by means of mass spectrometers started in the seventies with experiments carried out at CERN at the proton-synchrotron PS [13] and later at the on-line mass separator ISOLDE [14]. This pioneering work was carried out by conventional mass spectrometry based on magnetic-field and voltage measurements. In the following years a trend away from the measurement of voltages to the measurement of times of flight and frequencies was observed that today culminates in the availability of various powerful techniques for mass measurements far from stability. This paper gives an overview of these modern approaches.

TIME-OF-FLIGHT MASS MEASUREMENTS

Recoil Spectrometers

A new generation of mass experiments started with extended investigations on neutron-rich light isotopes performed by two groups at LAMPF/Los Alamos and GANIL/Caen. The measurements were carried out with special spectrometers using recoil nuclei. This allowed very short-lived isotopes to be investigated. The Los Alamos experiment was performed with the time-of-flight spectrometer TOFI [15,16], at GANIL the spectrometer SPEG [17] was used. In both cases the motivation was the detailed investigation of a region of enhanced neutron binding around the magic neutron number $N = 20$, first discovered in the early CERN experiments.

In the case of TOFI, exotic nuclei have been produced by an 800 MeV proton beam via target fragmentation and via fission reactions in a Th target. The recoil ions are sent through the TOFI spectrometer, which consists of four dipole magnets. The configuration of magnets is designed such that both a temporal and spatial focusing is achieved; ions with higher energies perform larger trajectories and reach the detector simultaneously with slower ions. The charge-to-mass ratio q/m of the ions can be directly obtained from the time of flight $T \propto \sqrt{q/m}$ through the spectrometer which is determined with appropriate fast detectors. The additional measurement of stopping power and energy of the ions together with their velocity allows the particles to be identified. The resolving power of the spectrometer is on the order of 10^4. With TOFI, mass measurements were performed on neutron-rich isotopes of the elements lithium through nickel [18–22].

In the case of SPEG, the mass determination is performed via a combined measurement of the velocity of the ions and their magnetic rigidity. The exotic nuclei are produced by projectile fragmentation of primary 40-60 keV argon, calcium or krypton beams shot onto a tantalum or nickel target. After a first momentum analysis, the velocity v of the ions is determined via a time-of-flight measurement in an approximately 100-m long beam line section. The magnetic rigidity $B \cdot \rho$ of the ions in a following magnetic dipole sector field with strength B is determined via a reconstruction of each single ion trajectory with position-sensitive detectors in front and behind the magnet. From the time of flight of the ions and a measurement of their stopping power their charge state q can be determined. The mass is then determined via $m = b \cdot \rho \cdot q/v$. The resolving power of SPEG is comparable to that of TOFI. With SPEG, mass measurements were performed on isotopes of the elements boron through phosphor [17,23,24]. The most recent measurements have been performed on proton-rich isotopes up to $A \approx 70$ [25].

With a few exceptions there is good agreement between the results of both the TOFI and SPEG experiments. Larger discrepancies exist with earlier measurements [13] (see above) performed at PS/CERN for exactly those isotopes that caused the interest in this region. The results show that ^{31}Na is 2 MeV and ^{32}Na is 4 MeV less bound than expected before. In the new experiments another region of enhanced neutron binding was found in the vicinity of ^{53}Sc.

Cyclotrons

At GANIL a very different approach has been employed for the determination of masses in the vicinity of the doubly magic nucleus ^{100}Sn. It makes use of the very extended flight path of ions in the second of the two large GANIL cyclotrons, CSS1 and CSS2. [26,27]. Neutron-rich nuclei with Z≈50 are produced by fusion evaporation reactions. A ^{50}Cr^{9+} beam is accelerated to an energy of 265 MeV by the first cyclotron CSS1 and sent to a rotating and cooled ^{58}Ni target. The reaction products are injected into the second cyclotron CSS2 for further acceleration and a time-of-flight measurement. Inside CSS2, for additional particle identification, the accelerated ions are detected by a silicon-detector telescope that can be moved radially into the cyclotron. The time of flight, or equivalently the phase of the particles with respect to the cyclotron radio frequency, is measured. The relative mass difference $\delta m/m$ between two ions is obtained via the observation of the time separation

$$\delta t/t = \delta m/m, \tag{1}$$

where t is the total time of flight in the cyclotron. With this technique, a resolving power of about 1 million has been achieved. In the measurements on secondary ions close to $A = 100$ [27], it was possible to obtain a mass value for the doubly magic isotope ^{100}Sn, shortly after its discovery [28,29].

As at GANIL, cyclotron mass measurements on $A = 80$ isobars have been carried out at ISN/Grenoble with the second cyclotron of the SARA complex. The main difference in the technique compared to GANIL is the use of a start detector in front of SARA and the location of the stop detector outside the cyclotron.

Storage Rings

Another approach is the measurement of the time of flight of a relativistic ion beam circulating in a heavy-ion storage ring [30,31]. First successful tests of such a scheme have recently been carried out at the experimental storage ring ESR at GSI/Darmstadt, aiming for the investigation of very short-lived isotopes delivered by the fragment separator FSR. For the time-of-flight measurements the ESR is operated in an isochronous mode similar to the TOFI experiment. Compared to TOFI, higher resolution and a higher accuracy is achieved due to the several hundred circulations of the beam in the ESR after injection. This mode of operation of the storage ring is expected to provide a powerful tool for the study of very short-lived isotopes.

FREQUENCY MEASUREMENTS

Penning traps

Penning traps have proven to be the most accurate mass spectrometers. A large variety of mass measurements at high accuracy have been performed on stable, mostly light particles [32]. A common property of all mass experiments with Penning-trap spectrometers is the determination of the cyclotron frequency

$$\omega_c = (q/m) \cdot B \tag{2}$$

of ions with a mass-to-charge ratio m/q confined in a strong magnetic field B. The value of the magnetic field B required for the mass determination can be determined from the cyclotron frequency ω_c of any ion with well-known mass.

For the determination of masses of unstable isotopes, ISOLTRAP (Fig. 1) is up to now the only operational Penning-trap mass spectrometer [33]. It is installed at the on-line mass separator ISOLDE at CERN, which provides low-energy radioactive ion beams for a large number of elements. The ISOLTRAP spectrometer consists of three ion traps, a radiofrequency quadrupole (RFQ) trap and two Penning traps. The RFQ trap system [34–37] has the task of stopping the continuous 60-keV ISOLDE ion beam, accumulating them in a buffer gas, and delivering the accumulated ions as short ion bunches of low energy to the first Penning trap. This trap serves for the purification and cooling of the ions and for the production of ion bunches [38]. A mass-selective buffer-gas cooling technique [39,40] is employed, which is based on the simultaneous application of a buffer gas and an rf-excitation of

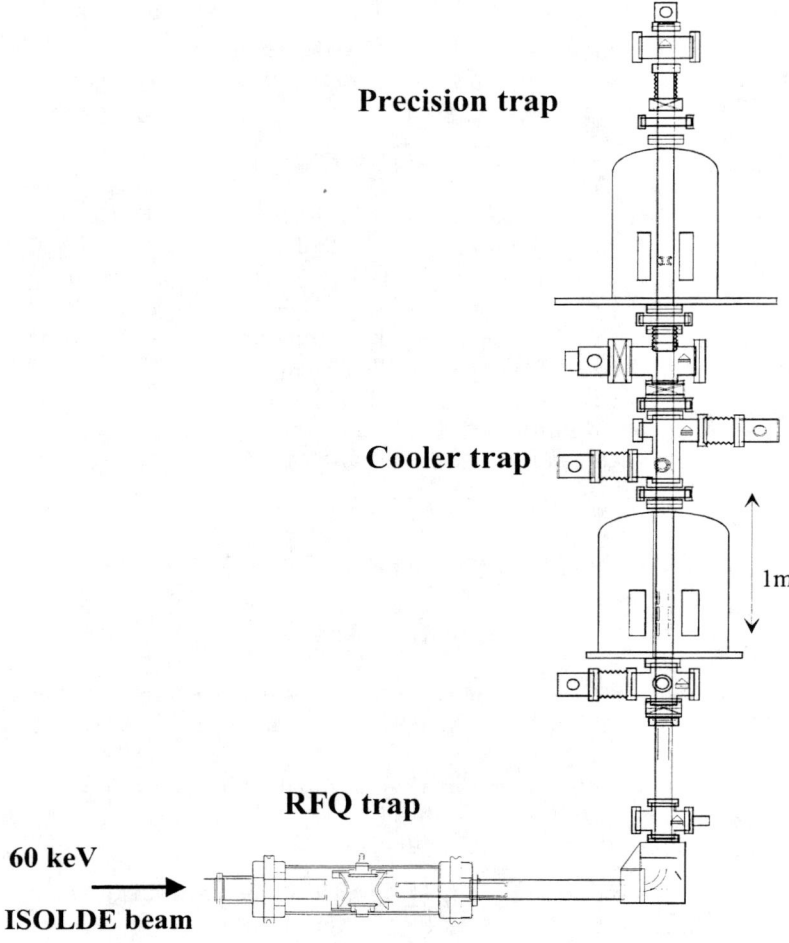

FIGURE 1. Experimental set-up of the ISOLTRAP Tandem Penning-trap spectrometer installed at the on-line mass separator ISOLDE at CERN.

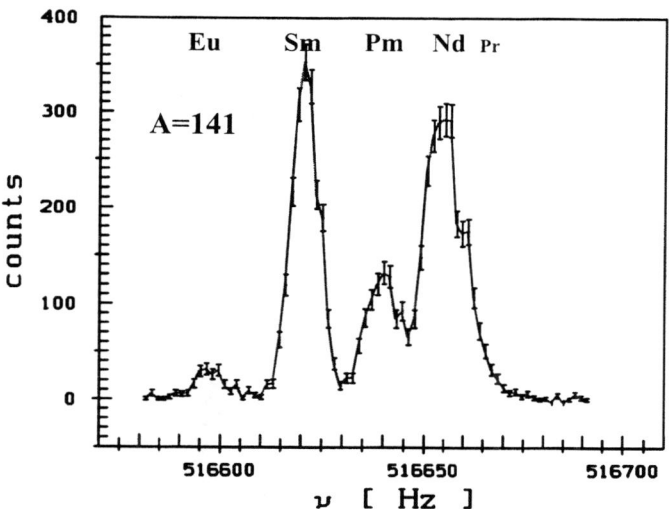

FIGURE 2. 'Mass scan' performed with the cooler trap. Shown is the number of ions that can be extracted from the cooler trap as a function of the applied radiofrequency.

the motion of the stored ions. This technique allows one to operate the first trap as an isobar separator with a resolving power of $R \approx 10^5$ for $A = 100$. Figure 2 shows a 'mass scan' performed with the cooler trap for ions of mass number $A = 141$ delivered by ISOLDE. The isobars of several rare earth elements are clearly resolved. The second trap is a high-precision trap [33,41] for a high-precision determination of the cyclotron frequency [40] of the isobar-separated ions delivered by the cooler Penning trap. The ion motion of the captured ion is driven by an rf-field, subsequently the ions are ejected as a short bunch from the trap and their time of flight to an ion detector is measured. The signature of having hit the cyclotron resonance is a reduction in this time of flight. With the ISOLTRAP spectrometer, a mass resolving power of up to $R = 10^7$ can be achieved, which is sufficient to separate isomeric and ground states even for excitation energies much less than 200 keV [46,42,43]. As an example, Figure 3 shows a cyclotron resonance curve of ^{141}Sm. The resonances of the ground and isomeric states are resolved. The figure illustrates that high resolution can be mandatory for an unambiguous determination of ground-state masses.

The first series of measurements with ISOLTRAP, starting at the ISOLDE 2 separator, were carried out on long isotopic chains of alkali and alkali earth isotopes [44,46,45,47–49]. Further series of mass measurements have been performed on rare-earth isotopes in the vicinity of the semi doubly-magic nucleus ^{146}Gd, discussed in detail in [43]. The most recent measurements concentrated on neutron-rich isotopes of mercury and heavier isotopes [50]. The typical accuracy achieved in all ISOLTRAP measurements is $\delta m/m \approx 1 \times 10^{-7}$.

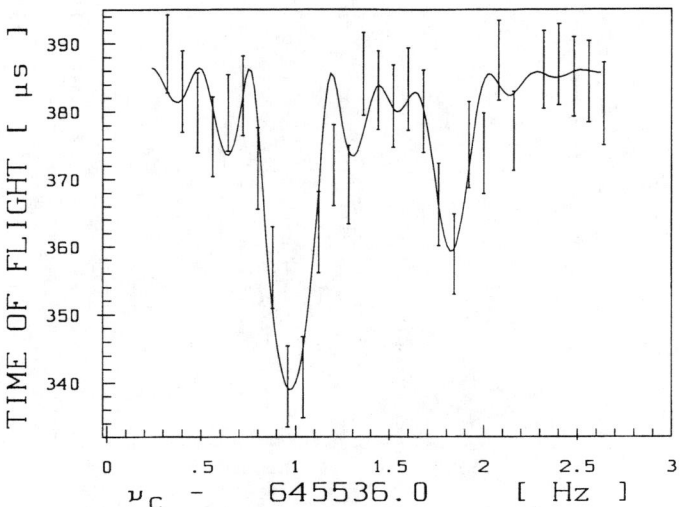

FIGURE 3. Cyclotron resonance curve of ^{141}Sm obtained with a resolving power of $R = 3 \times 10^6$. Both isomer ($E = 175\,\text{keV}$) (left resonance) and ground state (right resonance) are resolved.

The construction of another Penning-trap mass spectrometer [51] initially undertaken at Chalk River is nearly finished. The system is now installed at Argonne and will come into full operation soon, using radioactive recoil ions from a gas-filled recoil separator.

Radiofrequency Transmission Spectrometers

At ISOLDE, another project for direct mass measurements has started [52,53], namely MISTRAL, which uses a radiofrequency transmission spectrometer. Here the mass measurement is also carried out via the determination of the cyclotron frequency of an ion: 60-keV ISOLDE ions are injected through a narrow slit into the strong field of a large magnet in which they perform two turns on a helicoidal path before they pass through a narrow exit slit, are extracted out of the magnetic field and finally detected. After a half turn, the ions pass through an rf cavity where they experience a modulation and after the next turn a demodulation, of their energy by an rf field. Only ions for which the modulation is zero are transmitted through the exit slit of the spectrometer and can finally be detected. This is only achieved if the frequency of the rf field is a half-integer multiple of the cyclotron frequency of the ions in the magnetic field:

$$\omega_{rf} = (n + 1/2)\omega_c . \tag{3}$$

Compared to ISOLTRAP, the resolution of the spectrometer and the accuracy that can be reached is expected to be about one order of magnitude lower. However,

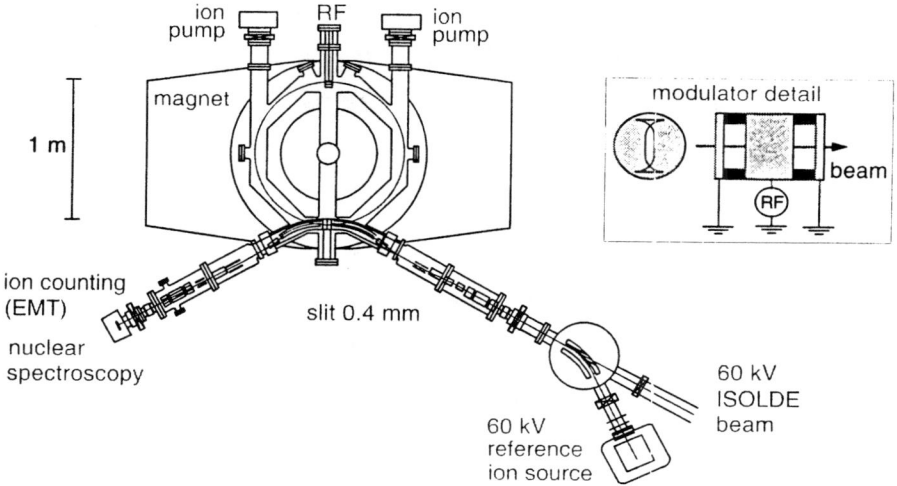

FIGURE 4. Experimental setup of the transmission rf spectrometer MISTRAL at ISOLDE.

MISTRAL will allow very short-lived isotopes to be studied since it is a transmission spectrometer. Very recently a first series of measurements has been performed successfully on neutron-rich sodium isotopes with half-lives down to 50 ms.

Storage Rings

In the last few years it has been proven that a storage ring [54] is a powerful high-resolution mass spectrometer for radioactive isotopes [55–57]. At GSI, Au and Bi fragments produced and separated by the fragment separator FRS have been injected at relativistic energies into the experimental storage ring (ESR) (see Fig. 5). After electron cooling the image currents of the circulating ions are recorded and Fourier-transformed into the frequency domain. The technique shows a very high sensitivity; even the signal of a single stored heavy ion shows up in the rich, so-called Schottky spectrum, which includes different ions in various charge states. Since the velocities $\beta \cdot c = C \cdot \nu_r$ of the ions circulating in the storage ring with the circumference C are identical within $\leq 10^{-6}$, the difference in their revolution frequencies $\Delta\nu_r$ is a measure of their difference in the mass-over-charge ratio given by

$$\Delta\nu_r/\nu_r = -\alpha_p \cdot \Delta(m/q)/(m/q), \qquad (4)$$

with the momentum compaction factor α_p, which is a characteristic factor determined by the magnetic rigidity of the ring and the ion trajectory.

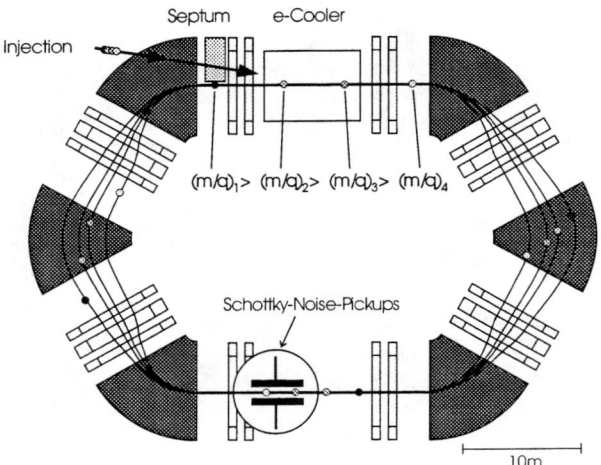

FIGURE 5. Principle of Schottky mass spectrometry as performed at the Experimental Storage Ring (ESR) at GSI.

The measurements at ESR have shown that Schottky mass spectrometry in a storage ring is a very general technique. Addressing the several heavier and medium heavy regions of the nuclide chart by projectile fragmentation of relativistic gold and bismuth primary beams at GSI, more than 300 isotopes have been investigated [56,57] with a resolving power of up to $R \approx 10^6$ and an accuracy of better 10^{-6}.

CONCLUDING REMARKS

In the past years, a number of new techniques for mass measurements of radioactive isotopes have been developed. For a comparison of the most recent approaches based on time-of-flight and frequency measurements, one has to take into account the different schemes for the production of the radioactive isotopes, the half-life of the isotopes to be investigated, and the achievable accuracy.

Time-of-flight spectrometers and storage rings are able to make direct use of secondary beams produced by fragmentation or fusion evaporation reactions at relativistic energies. ISOL beams of low energy fit very well the requirements for mass measurements with Penning traps and transmission rf-spectrometers.

Figure 6 shows the range of operation for the different approaches with respect to the accuracy that has (or can) be reached and the half-life of the isotopes that can be studied. For the investigation of isotopes with very short half-lives, time-of-flight techniques offer clear advantages but cannot provide mass data as accurate as that available from Schottky mass measurements in storage rings or mass measurements in Penning traps.

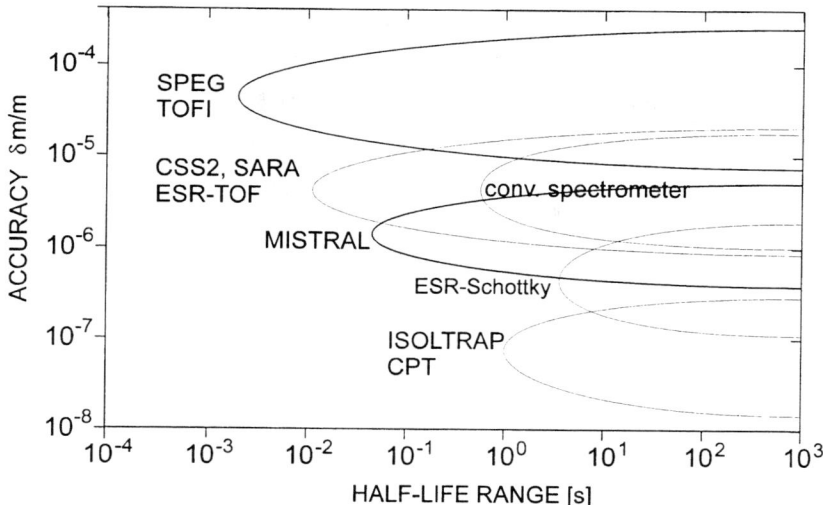

FIGURE 6. Comparison of various approaches for mass measurements on radioactive isotopes. Shown is the typical accuracy versus the half-life of the isotopes that can be studied.

There will be no large benefit from accelerating ISOL beams to high energies for subsequent mass measurements in storage rings. However, a deceleration of relativistic radioactive ions for Penning-trap mass measurements with highest accuracy is interesting. A project in this direction, SHIPTRAP, has recently been launched at GSI for the investigation of isotopes of heavy and super-heavy elements after the velocity filter SHIP.

In conclusion, with direct mass measurements very high accuracies can be achieved today and very short-lived isotopes are accessible. The different approaches are well adapted to the different production schemes for radioactive isotopes. The new techniques and their further development will help to obtain new and reliable information about nuclear binding far from stability.

REFERENCES

1. Thomson, J. J., *Philos. Mag.* **VI 24**, 209, 668 (1897).
2. Aston, F. W., *Philos. Mag.* **45**, 934 (1923).
3. Aston, F. W., *Proc. R. Soc. London, Series A* **115**, 487 (1927).
4. Gamow, G., *Proc. R. Soc. London, Series A*, bf 126, 632 (1930).
5. von Weizsäcker, C. F., *Z. Phys.* **86**, 431 (1935).
6. Bethe, H. A., and Bacher, R. F., *Rev. Mod. Phys.* **8**, 82 (1936).
7. Bohr, A., and Mottelson, B., *Nuclear Structure*, Vol. 1, New York: W.A. Benjamin Inc., 1969, p. 262.

8. Ring, P., and Schuck, P., *The Nuclear Many-Body Problem*, New York: Springer Verlag, 1980.
9. Haustein, P. E., (ed.), *At. Data Nucl. Data Tables* **39**(2) (1988).
10. Audi, G., et al., *Nucl. Phys.* **A565**, 193 (1993).
11. Audi, G., and Wapstra, A. H., *Nucl. Phys.* **A595**, 409 (1995).
12. Wapstra, A. H., Proc. AMCO 6 and NFFS 9, Bernkastel-Kues, Germany, 1992, *Inst. Phys. Conf. Ser.* **132**, 979 (1993).
13. Thibault, C., et al., *Phys. Rev. C* **12**, 644 (1975).
14. Audi, G., et al., *Nucl. Phys.* **A378**, 443 (1982).
15. Vaziri, K., et al., *Nucl. Instrum. Methods* **B26** 280 (1987).
16. Wouters, J. M., et al., *Nucl. Instrum. Methods* **B26** 286 (1987).
17. Gillibert, A., et al., *Phys. Lett.* bf B 176, 317 (1986).
18. Tu, X.L., et al., *Z. Phys. A* **337**, 361 (1990).
19. Zhou, X. G., et al., *Phys. Lett. B* **260**, 285 (1991).
20. Viera, D. J., et al., *Phys. Rev. Lett.* **57**, 3253 (1986).
21. Wouters, J. M., et al., *Z. Phys. A* **331**, 229 (1988).
22. Vieira, D.J., et al., Proc. ENAM 95, Arles, France, 1995, Editions Frontieres, Gif-sur Yvette Cedex, p. 59.
23. Orr, N. A., et al., *Phys. Lett. B* **258**, 29 (1991).
24. Gillibert, A., et al., *Phys. Lett. B* **192**, 39 (1987).
25. Chartier, M., et al., Proc. ENAM 95, Arles, France, 1995, Editions Frontieres, Gif-sur Yvette Cedex, p 59.
26. Auger, G., et al., *J. Phys. G* **17**, 463 (1991).
27. Chartier, M., et al., *Phys. Rev. Lett.* **77** 2400 (1996).
28. Lewitowicz, M., et al., *Phys. Lett.* **B332** 20 (1994).
29. Schneider, R., et al., *Z. Phys.* **A348**, 241 (1994).
30. Balog, K., et al., *Nucl. Instrum. Methods* **B70**, 459 (1992).
31. Wollnik, H., et al., *Nucl. Phys.* **A626**, 327c (1997).
32. For an overview see: Proc. Nobel Symposium 91 on "Trapped Charged Particles and Related Fundamental Physics," Lysekil, Sweden, 1994, *Physica Scripta* **T59**, (1995).
33. Bollen, G., et al., *Nucl. Instrum. Methods* **A 368,** 675 (1996).
34. Moore, R. B., Rouleau, G., *J. Mod. Optics* **39**, 361 (1992).
35. Moore, R. B., et al., *Physica Scripta* **46**, 569 (1992).
36. Moore, R. B., et al., *Physica Scripta* **T59**, 93 (1995).
37. Bollen, G., et al., "A Radio Frequency Quadrupole Ion Beam Buncher for ISOLTRAP," in *Proc. Int. Conf. on Exotic Nuclei and Atomic Masses ENAM 98*, June 1998, Bellaire, MI, USA, *AIP Conf. Proc.* (in print)
38. Raimbault-Hartmann, H., et al., *Nucl. Instrum. Methods* **B126**, 374 (1997).
39. Savard, G., et al., *Phys. Lett. A* **158**, 247 (1991).
40. König, M., et al., *Int. J. Mass Spectrom. Ion Proc.* **142**, 95 (1995).
41. Becker, St., et al., *Int. J. Mass Spectrom. Ion Proc.* **99**, 53 (1990).
42. Beck, D., et al., *Nucl. Instrum. Methods* **B126**, 378 (1997).
43. Beck, D., et al., *Nucl. Phys.* **A626**, 343c (1997).
44. Stolzenberg, H., et al., *Phys. Rev. Lett.* **65**, 3104 (1990).
45. Bollen, G., et al., *Hyperfine Int.* **38**, 793 (1987).

46. Bollen, G., et al., *Phys. Rev. C* **46**, R2140 (1992).
47. Kluge, H.-J., Physica Scripta **T22**, 85 (1988).
48. Bollen, G., et al., *J. Mod. Opt.* **39**, 257 (1992).
49. Otto, T., et al., *Nucl. Phys.* **A567**, 281 (1994).
50. Bollen, G., et al., "Mass measurements with a Penning trap mass spectrometer at ISOLDE," in *Proc. Int. Conf. Exotic on Nuclei and Atomic Masses ENAM 98*, June 1998, Bellaire, MI, USA, *AIP Conf. Proc.* (in press).
51. Sharma, K. S., et al., *Hyperfine Interact.* **81**, 217 (1993).
52. de Saint Simon, M., et al., *Physica Scripta* **T59**, 406 (1995).
53. Lunney, M. D., et al., *Hyperfine Interact.* **99**, 105 (1996).
54. Franzke, B., *Nucl. Instrum. Methods* B24, 18 (1987)
55. Franzke, B., et al., *Physica Scripta* **T59**, 176 (1995).
56. Schlitt, B., et al., *Nucl. Phys.* **A626**, 315c (1997).
57. Radon, T., *Phys. Rev. Lett.* **78**, 23 (1997).

The Muon $g-2$ Experiment at Brookhaven

F. J. M. Farley*
for the $(g-2)$ collaboration[1]

Department of Physics, Yale University

Abstract. The new muon storage ring at Brookhaven is intended to measure the anomalous magnetic moment a_μ of the muon to sub ppm accuracy. A preliminary run in 1997 gave $a_\mu = 1\,165\,925\,(15) \times 10^{-9}$ in agreement with previous CERN measurements. The principles of the experiment and numerous detailed improvements are described.

INTRODUCTION

A new measurement of the g-factor of the muon is underway at Brookhaven. This experiment is intended to test the standard electro-weak theory and to look for new physics. What is new physics? It stands for any speculation that has not yet been disproved by experiment.

Why the muon? Well the muon is a structureless, apparently point particle and that is what you need for testing fundamental theory. How do we know that it is

---

[1] H. N. Brown[b], G. Bunce[b], R. M. Carey[a], P. Cushman[h], G. T. Danby[b], P. T. Debevec[g], H. Deng[l], W. Deninger[g], S. K. Dhawan[l], V. P. Druzhinin[i], L. Duong[h], W. Earle[a], E. Efstathiadis[a], G. V. Fedotovich[i], S. Giron[h], F. Gray[g], M. Grosse Perdekamp[l], A. Grossmann[e], U. Haeberlen[f], M. Hare[a], E. S. Hazen[a], D. W. Hertzog[g], V. W. Hughes[l], M. Iwasaki[k], K. Jungmann[e], D. Kawall[l], M. Kawamura[k], B.I. Khazin[i], J. Kindem[h], F. Krienen[a], I. Kronkvist[h], R. Larsen[b], Y. Y. Lee[b], W. Liu[l], I. Logashenko[i], R. McNabb[h], W. Meng[b], J.-L. Mi[b], J. P. Miller[a], W. M. Morse[b], C. J. G. Onderwater[g], Y. Orlov[c], C. Pai[b], C. Polly[g], J. Pretz[l], R. Prigl[b], G. zu Putlitz[e], S. I. Redin[l], O. Rind[a], B. L. Roberts[a], N. Ryskulov[i], R. Sanders[b], S. Sedykh[g], Y. K. Semertzidis[b], S. Serednyakov[i], Yu. M. Shatunov[i], E. Solodov[i], M. Sossong[g], A. Steinmetz[l], L. R. Sulak[a], C. Timmermans[h], A. Trofimov[a], D. Urner[c], D. Warburton[b], D. Winn[d], Q. Xu[b], A. Yamamoto[j], D. Zimmerman[h]; ([a]Department of Physics, Boston University, [b]Brookhaven National Laboratory, [c]Newman Laboratory, Cornell University, [d]Physics Department, Fairfield University, [e]Physikalisches Institut der Universität Heidelberg, Germany, [f]MPI für Med. Forschung, Heidelberg, Germany, [g]Department of Physics, University of Illinois at Urbana-Champaign, [h]Department of Physics, University of Minnesota, [i]Budker Institute of Nuclear Physics, Novosibirsk, Russia, [j]KEK, Japan, [k]Tokyo Institute of Technology, Tokyo, Japan, [l]Department of Physics, Yale University)

a structureless point particle? Because up to now it conforms to the theory! It would also be interesting to find some structure of the muon, or a new interaction that might explain the $\mu - e$ mass difference. One used to speculate that it might somehow be coupled to the nucleon but that is not evident. Now the favoured coupling is to the Higgs.

Also because the muon is more massive than the electron, its g-factor is 40,000 times more sensitive to new physics. The anomalous moment of the electron is known to amazing accuracy, 4 parts per billion (ppb), 2000 times more accurate than the present muon data; but this can only test QED to 44 ppb, the present accuracy in the fine structure constant α. Therefore the present muon data is 200 times more sensitive to typical departures from the standard model. The new BNL experiment aims to reduce the error in a_μ by a further factor of 20.

The g-factor of the muon is $g_\mu = 2(1 + a_\mu)$ where a_μ, called the "anomalous magnetic moment," is a correction due to quantum fluctuations in the electric field around the particle. It is a small effect but fortunately a_μ can be measured directly using the so-called $(g-2)$ principle.

The orbital angular frequency of muons in flight in a magnetic field B is

$$\omega_c = \frac{eB}{mc}, \tag{1}$$

while the spin precession frequency is

$$\omega_s = g \frac{e}{2mc} B. \tag{2}$$

If $g = 2$ these frequencies are exactly equal, so longitudinally polarised muons will stay that way. But if $g > 2$ then the spin turns faster than the momentum vector at a relative angular frequency

$$\omega_a = \omega_s - \omega_c = a_\mu (e/mc) B, \tag{3}$$

where $a_\mu = (g-2)/2$. By measuring the change in spin direction relative to the momentum vector one can measure a_μ directly.

The magnetic field is calibrated by the proton NMR frequency ω_p and

$$\lambda = \omega_s/\omega_p = \mu_\mu/\mu_p \tag{4}$$

is obtained from the very accurate measurements of muon precession at rest [1], again in fields calibrated by proton NMR, while e/mc is given by the muonium hyperfine structure [2]. One then finds

$$a_\mu = \frac{R}{\lambda - R} \tag{5}$$

where $R = \omega_a/\omega_p$ is the ratio measured in the $(g-2)$ experiment.

FIGURE 1. The Brookhaven Muon Storage Ring

The $(g-2)$ precession formula (3) was derived above in non-relativistic approximation, but it turns out to be correct at high energies also [3]. While the muons live longer in the laboratory, ω_a is not slowed down by time dilation. Therefore, by using high-energy muons, one can measure more precession cycles and thus get better accuracy.

MUON STORAGE RING

To make the measurement one ideally needs a large ring magnet, continuous in azimuth with no straight sections and a uniform magnetic field calibrated by proton NMR. A uniform field provides horizontal focusing but no vertical focusing, so the muons would soon be lost. If a weak magnetic gradient is added to focus vertically [4], then ω_a becomes a function of radius limiting the accuracy to order 100 ppm. In the last CERN experiment [5] the magnetic field was flat and the vertical focusing was provided by electric quadrupoles, all of the same sign, distributed around the ring. One can show that this combination gives weak focusing with an effective gradient $n_{\text{eff}} = 2V\rho/a^2\beta B$ where V is the voltage applied to quadrupole plates distance a from the centerline and ρ is the radius of the ring.

The vertical component of the electric field does the focusing, but is accompanied by a horizontal component, varying with radius, which in general alters the $(g-2)$ frequency so one is no better off:

$$\omega_a = (e/mc)\left[a_\mu B - (a_\mu - 1/(\gamma^2 - 1))\beta \times E\right] . \tag{6}$$

At CERN we chose $\gamma = \sqrt{1 + 1/a_\mu} = 29.3$ so the second term in (6) is zero and the electric field then has no effect on the spin precession. To satisfy this condition the stored muons must have the so-called "magic" momentum of 3.094 GeV/c.

Polarised muons were produced inside the storage volume by injecting pions of about 3.1 GeV/c. During the first turn, 20% decay to muons, and some of these, from forward decay, fall onto permanently stored orbits. About 10^4 π were needed to give one stored μ.

Electrons from decay of muons in orbit have energies varying from 0 to 3.1 GeV, so they are bent more sharply by the magnetic field and emerge on the inside of the ring where they are detected in energy-sensitive calorimeters. By requiring high energy in the lab one selects forward-emitted electrons in the muon rest frame, so as the spin turns the number of electrons is modulated (see Fig. 1), and one can read off the precession frequency. At CERN this gave a_μ to 7 ppm.

IMPROVEMENTS

In the new BNL experiment the following improvements have been introduced:
• A larger aperture ring magnet with more uniform magnetic field. The field is 14.5 T; the orbit radius, 711 cm.
• Superconducting coils, implying more stable current and smaller temperature effects in the iron.
• Special shimming devices, including pole-face windings, to correct lower-order multipoles.
• A storage aperture of circular cross section (diameter 9 cm). The average field seen by the stored muons is consequently much less sensitive to higher field multipoles.
• An NMR trolley that travels round the ring inside the vacuum chamber to map the field without turning the magnet on and off.
• 24 special windows to allow the decay electrons to exit with minimum showering and so give a larger-amplitude ($g - 2$) modulation.
• Polarised muons are produced outside the ring in a $\pi - \mu$ decay channel and injected with the aid of a fast full-aperture kicker. This gives a factor of 10 improvement in the stored intensity and much less background.
• Higher beam intensity due to accelerator improvements at the AGS since 1970.
• 24 calorimeters with separate timing circuits to reduce signal overlap.
• Electronics to record individual electron hits for off-line processing and algorithms for dealing with overlapping signals.
• Improved electron calorimeters with good stability in gain and timing.

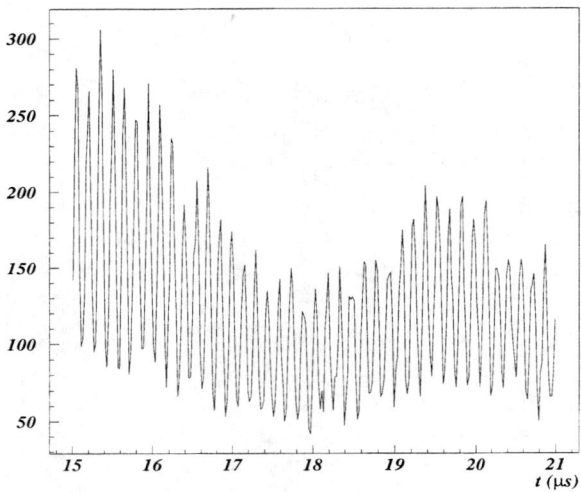

FIGURE 2. Decay electron counts at early times showing muon orbit period (149 ns) plus slow $g-2$ modulation

- Use of the whole AGS beam in 8 separate RF buckets at 33 ms intervals. This limits the peak counting rates; in effect the experiment is made 8 times per AGS cycle of 2.3 seconds.
- Measurement of the muon distribution inside the storage volume by
    (a) trace-back of the decay electron tracks using drift chambers
    (b) pick-up electrodes.
- Position-sensitive detectors in front of the calorimeters.
- A direct-current superconducting inflector [6] to cancel the magnet field along the path of the incoming beam.

PRELIMINARY TESTS

A preliminary engineering run was made in early 1997 to test the system. At this stage the fast kicker for direct muon injection was not available, so we used pion injection as in CERN and experienced the anticipated severe background in the detectors. As a result the fit to the $(g-2)$ precession was not started until 80 μs after injection. Nevertheless data taken over 8 days gave a_μ to 13 ppm.

A Fourier fit to the data at early times (see Fig. 2), when the muons are still bunched, gives the distribution of rotation frequencies shown in Fig. 3 compared

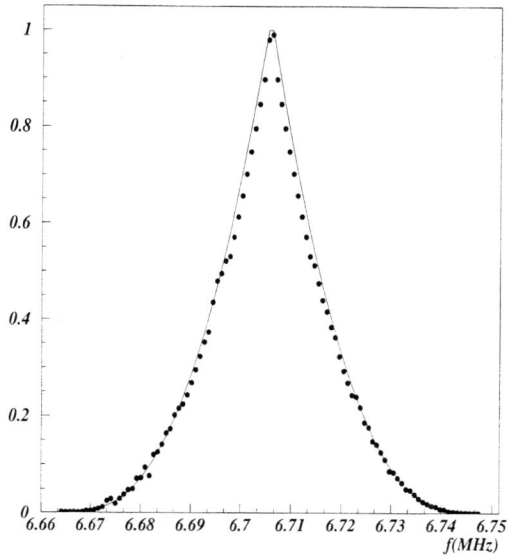

FIGURE 3. Distribution of rotation frequencies calculated from Fig. 2 (•), and phase-space prediction for a circular aperture (solid line)

with the prediction for a uniformly populated circular aperture; the agreement is excellent.

Figure 4 shows the time distribution of the decay electrons at later times with the $(g-2)$ modulation. Figure 5 gives the result of our measurement with the CERN results for comparison. Agreement is good. Combining these numbers for μ^+ and μ^- and using $\lambda = 3.18334547(47)$ [7], the current weighted mean value for the anomalous moment is

$$a_\mu = 1\,165\,923.4\,(7.2) \times 10^{-9}\quad(6.2\,\mathrm{ppm}). \tag{7}$$

COMPARISON WITH THEORY

According to the Standard Model

$$a_\mu(theory) = a_\mu(QED) + a_\mu(weak) + a_\mu(hadron)\,. \tag{8}$$

The QED contribution consists of a series expansion in (α/π) that has been evaluated over a period of 40 years with ever increasing accuracy [8]. At present,

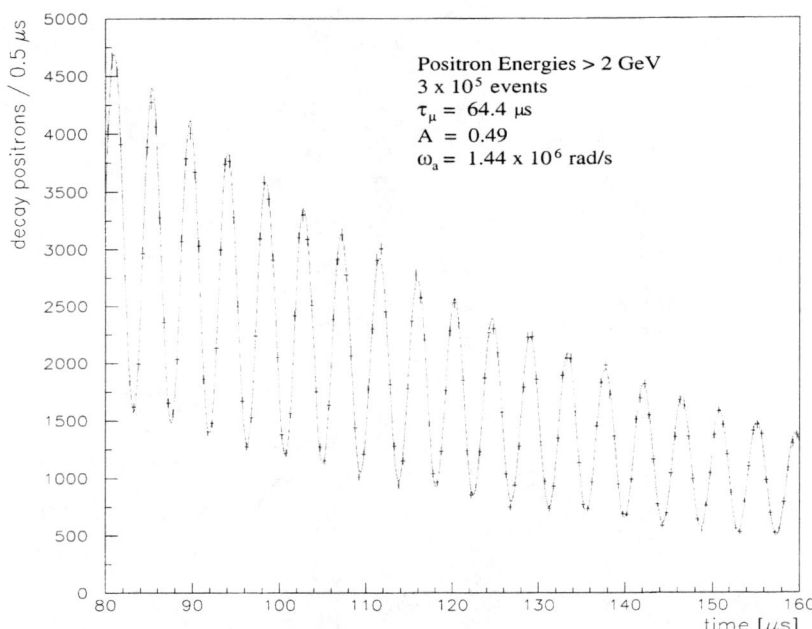

FIGURE 4. Decay electron distribution at late times showing $g-2$ modulation. A fit to this data gives a_μ.

$$a_\mu(QED) = 116\,584\,706\,(2) \times 10^{-11} \quad (16\,\text{ppb}). \tag{9}$$

The weak effect comes from two first-order diagrams (Fig. 5) [9] plus second-order loops:

$$a_\mu(W) = \frac{G_F m_\mu^2}{8\pi^2\sqrt{2}}\left(\frac{10}{3}\right) = 389 \times 10^{-11} \tag{10}$$

and

$$a_\mu(Z) = \frac{G_F m_\mu^2}{8\pi^2\sqrt{2}}\left(\frac{1}{3}\right)\left[(3 - 4\cos^2\theta_W)^2 - 5\right] = -194 \times 10^{-11}, \tag{11}$$

giving first-order weak terms $a_\mu(W) + a_\mu(Z) = 195 \times 10^{-11}$ (1.6 ppm). But the second-order weak contribution is [10] -44×10^{-11} giving in total

$$a_\mu(weak) = 151 \times 10^{-11} \quad (1.3\,\text{ppm}). \tag{12}$$

The theorists are very interested in having an experimental check on $a_\mu(weak)$ because this would be the first test of loop diagrams involving Z and W.

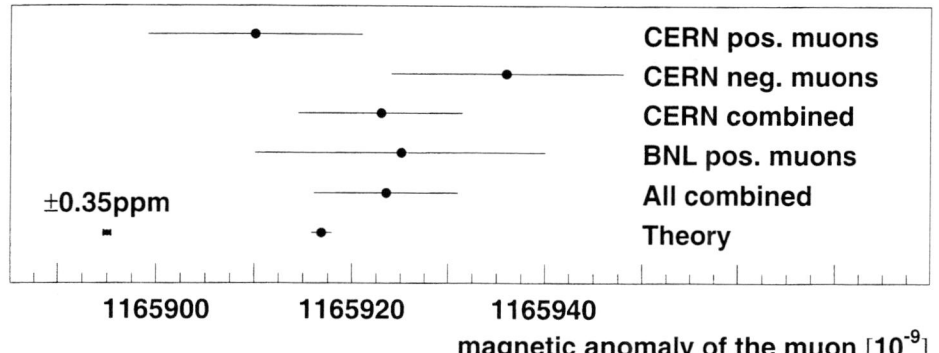

FIGURE 5. Results for $a_\mu \times 10^9$ compared with theory.

However, the water is clouded by $a_\mu(hadron)$, which arises from virtual photons producing virtual hadron loops (see Fig. 7). The contribution to a_μ can be related via a dispersion relation to the total cross section for hadron production in e^+e^- collisions:

$$a_\mu(had1) = \left(\frac{\alpha m_\mu}{3\pi}\right)^2 \int_0^\infty K(s)R(s)ds, \tag{13}$$

where $K(s)$ is an algebraic expression in s and

$$R(s) = \frac{\sigma_{tot}(e^+e^- \Longrightarrow hadrons)}{\sigma_{tot}(e^+e^- \Longrightarrow \mu^+\mu^-)}. \tag{14}$$

A long program of careful experiments at CMD2 and VEPP [11] is devoted to the measurement of $R(s)$.

One obtains another handle on $R(s)$ from hadronic τ-decay via a virtual $W^\pm$. The virtual W and the virtual photon are related through a conserved vector current [12] (see Fig. 8). In addition one must include second-order hadronic diagrams [13], -0.86 ppm, and hadronic light by light scattering [14], -0.68 ppm. The overall current value is

$$a_\mu(hadron) = 6771\,(77) \times 10^{-11}\ \ (58.6 \pm 0.82\,\text{ppm}). \tag{15}$$

Thus the error in $a_\mu(hadron)$ is about half the total weak effect that we hope to see. Further work still in progress will gradually improve the situation.

Adding all these terms together the current theoretical prediction is

$$a_\mu(theory) = 116\,591\,628\,(77) \times 10^{-11}. \tag{16}$$

Therefore,

$$a_\mu(experiment) - a_\mu(theory) = 6.1 \pm 6.2 \text{ ppm}. \tag{17}$$

Thus, there is an offset of about one standard deviation, which may mean nothing. The next run at BNL with muon injection, just starting, should push the error down to about 1 ppm, with further improvements expected to follow.

If there turns out to be a difference, the most likely explanation with "new physics" comes from supersymmetry

$$a_\mu(SUSY) = 1.2 \text{ ppm} \times \tan\beta \left(\frac{100 \text{ GeV}}{\tilde{m}}\right)^2, \tag{18}$$

where $\tilde{m}$ is the typical supersymmetric mass (perhaps 100–200 GeV) and the favored values of $\tan\beta$ are either about 1.5 or in the region of 40. So $a_\mu(SUSY)$ could be large and easily detectable.

Stay tuned to find whether the standard model is confirmed, or if on the other hand there are indications in favour of supersymmetry or possibly some structure of the muon.

REFERENCES

1. Camani, M., Gygax, F. B., Klempt, F., Ruegg, W., Schenck, A., Schilling, H., Schulze, R., and Wolf, H., *Phys. Lett.* **77B**, 326 (1978).
2. Mariam, F. G., Beer, W., Bolton, P. R., Egan, P. O., Gardiner, C. J., Hughes, V. W., Lu, D. C., Sonder, P. A., Orth, H., Vetter, J., Moser, U., and zu Putlitz, G., *Phys. Rev. Lett.* **49**, 993 (1982).
3. Mendlowitz, H., and Case, K. M., *Phys. Rev.* **97**, 33 (1955); Bargmann, V., Michel, L., and Teledgi, V. L., *Phys. Rev. Lett.* **2**, 435 (1959).
4. Bailey, J., Bartl, W., von Bochmann, G., Brown, R. C. A., Giesch, M., Jöstlein, H., van der Meer, S., Picasso, E., and Williams, R. W., *Nuovo Cim.* **A9**, 369 (1972).
5. Bailey, J., Borer, K., Combley, F., Drumm, H., Eck, C., Farley, F. J. M., Field, J. H., Flegel, W., Hattersley, P. M., Krienen, F., Lange, F., Lebee, G., McMillan, E., Petrucci, G., Picasso, E., Rúnolfsson, O., von Rüden, W., Williams, R. W., and Woicicki, S., *Nucl. Phys.* **B150**, 1 (1975).
6. Krienen, F., Loomba, D., and Meng, W., *Nucl. Instrum. Methods* **A283**, 5 (1989).
7. Particle Data Group, *Phys. Rev. D* **54**, 1 (1996)
8. Kinoshita, T., *Phys. Rev. D* **47**, 5013 (1993); Kinoshita, T., *Rep. Prog. Phys.* **59**, 1459 (1996); Hayakawa, M., and Kinoshita, T., *Phys. Rev. D* **57**, 465 (1998); for a review, see Kinoshita, T., *Quantum Electrodynamics*, World Scientific (1990).
9. Calmet, J., Narison, S., Perrottet, M., and de Rafael, E., *Rev. Mod. Phys.* **49**, 21 (1977).
10. Czarnecki, A., Krause, B., and Marciano, W. J., *Phys. Rev. D* **52**, R2619 (1995), *Phys. Rev. Lett.* **76**, 3267 (1998).

11. Akhmetshin, R. R., et al., *Phys. Lett.* **B398**, 423 (1997); *Phys. Lett.* **B415**, 445 and 452 (1997); Eidelmann, S., and Jegerlehner, F., *Z. Phys.* **C67**, 585 (1985); Brown, D. H., and Worstell, W. A., *Phys. Rev. D* **54**, 3237, (1996).
12. Alemany, R., Davier, M., and Höcker, A., *Eur. Phys. J. C* **2**, 123 (1998); Davier, M., and Höcker, A., *Phys. Lett.* **B419**, 419 (1998).
13. Kinoshita, T., Nizić, B., and Okamoto, Y., *Phys. Rev. D* **31**, 2108 (1985).
14. Hayakawa, M., Kinoshita, T., and Sanda, A. I., *Phys. Rev. D* **53**, 3137 (1996);Bijnens, J., Pallante, E., and Prades, J., *Nucl. Phys.* **B474**, 379 (1996).

Rhenium-187 and the Age of the Galaxy

Fritz Bosch

GSI, Planckstr.1, 64291 Darmstadt, Germany

Abstract. The amount of ^{187}Os in meteorites resulting from beta decay of the long-lived ^{187}Re nuclide has been used as a measure of the time span of nucleosynthesis in our galaxy. During the galactic history, however, the rhenium atoms can be "astrated" several times into newly forming stars, where they are stripped of most or all of their electrons. An experiment conducted at the ion storage ring ESR at Darmstadt showed that for bare ^{187}Re ions, the lifetime is shortened by more than *nine orders of magnitude*. This observation strongly suggests that the *effective* lifetime of ^{187}Re during the galactic evolution might differ significantly from the lifetime of neutral ^{187}Re. Furthermore, it enables a recalibration of the rhenium aeon clock in the framework of chemical-evolution models of our galaxy. Based on this new calibration, a preliminary lower limit of 12×10^9 yr for the age of our galaxy has been derived, which is, a fortiori, also a lower limit for the age of the universe. In combination with the Hubble constant, this limit provides narrow constraints for actual cosmological models.

AEON CLOCKS

Does the world have a beginning and an end, or has it always existed and will it last forever? Homo sapiens sapiens has pondered this since the dawn of history, when he left the rain forest and made his first steps into the African savannah. Those thoughts are reflected in a thousand ways in the myths, religions and traditions of all peoples and at all times.

Until well into this century, most astronomers favoured a stationary, that is—viewed on a large scale—an unchanging universe. In the late twenties, Edwin Hubble made the epochal discovery at the Mount Wilson Observatory of a nonstationary, expanding universe when he noticed that, the further spiral nebulae are away, the more rapidly they move away from us. From the first preliminary data of a universal recession velocity per distance—later on called the "Hubble constant H_0"—of 500 km/s per megaparsec (1 megaparsec (Mpc) = 3.26 million light years), Hubble deduced an "origin" of the universe two billion years ago by *assuming* a constant expansion. Now, however, H_0 is believed to be much smaller and to lie somewhere between 50 and 100 km/s per Mpc, which for a constant expansion corresponds to an age T_U of the universe between 10 and 20 billion years.

In reality, however, the expansion velocity will be slowed-down by the mutual

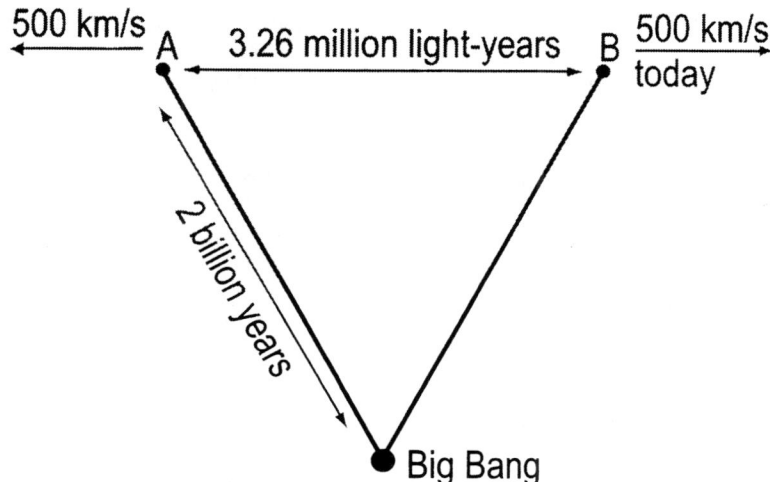

FIGURE 1. Hubble's first estimate of the age of the universe. From his measurements of the redshift of the light from distant galaxies, Hubble deduced an overall expansion velocity of 500 km/s per megaparsec, which corresponds to an elapsed time of two billion years since the Big Bang if one assumes a constant expansion.

gravitational attraction of all the masses contained in the universe. Therefore, the Hubble "constant" as observed today, H_0, can serve as a "clock" for the age of the universe only in association with a known (or assumed) mass density. For the limiting case of an empty universe (constant expansion), the age T_U of the universe is just the inverse of the Hubble constant, $(H_0)^{-1}$, the so-called "Hubble time." On the other hand, if the expansion is approaching a standstill an infinite time from now, which is what happens at the "critical density," the time elapsed since the origin ("Big Bang") equals two-thirds of the Hubble time.

Therefore, already the day after Hubble's exciting discovery of a universe finite with respect to both space and time, astrophysicists searched for "clocks" independent of the Hubble constant that could measure the age of the universe or at least provide reliable constraints for it. Perhaps the most famous of those clocks is hidden in the Hertzsprung-Russell diagram, where the absolute luminosity of stars is depicted versus their colour, i.e., their temperature. From the point on this plot where the oldest stars, the "globular clusters," leave the main sequence, their age can be estimated: according to the most recent, comprehensive review of D. A. VandenBerg and co-workers [1], 15 billion years, with a safe lower limit of 12 billion years.

Besides the globular clusters and other *astronomical* objects like white dwarfs, there are, so to speak, autochronous chronometers, the aeon clocks, which show both the endurance of cosmic marathon runners and the precision of Swiss watches. Nature has created them, radiochemists have detected them, and geologists and astrophysicists have learned to understand and interpret their message. They are long-lived radioactive mother nuclei, N_m, and their *direct* decay daughters, N_d. From the abundance ratio $R(t)$ of them as measured at time t, and from the known decay constant λ of the mother nucleus, the elapsed time can be deduced, irrespective of the amount of the mother nuclei, $N_m(0)$, at the starting point of the decay—provided only that no daughter atom pre-existed at this time.

Probably the most striking achievement of these aeon clocks was the determination of the age of our solar system, T_s. The incredibly precise number for it, $T_s = (4.56 \pm 0.05) \times 10^9$ years, yielded from today's relative abundance of ^{238}U and its (final) decay product ^{206}Pb, as found in meteorites formed just at the decoupling of the solar system from our galaxy.

This method cannot be immediately extended to the time before decoupling, when matter that would become later the sun, earth and living beings was still scattered and mixed in the stars and in the interstellar space of our galaxy. For an estimate of the duration T_N of pre-solar nucleosynthesis, one had to rely on the abundance ratio of ^{232}Th and ^{238}U. Both are long-lived radionuclides but *not* belonging to a common decay chain. To get reliable information on the elapsed time, one has to know in addition the production probabilities during nucleosynthesis of both, ^{238}U and ^{232}Th. They depend, however, strongly on details not yet understood of the "r-process" of nucleosynthesis and are thus uncertain to a large extent. Therefore, the literature values for T_N, based on the uranium/thorium clock, are spread over a factor of two.

THE AEON CLOCK RHENIUM-187

D. D. Clayton pointed out as early as 1964 [2] that, by a clever choice of a couple of *connected* radionuclides, the congenital defect of the uranium/thorium clock, namely the basically unknown relative production probabilities of uranium and thorium, respectively, could be very elegantly avoided. He proposed as "the best of all those couples" the pair ^{187}Re/^{187}Os. For this mother-daughter relation, any (unknown) details of the production of the mother nucleus ^{187}Re drop out. ^{187}Re undergoes β^--decay to the ground state of ^{187}Os with a half-life of 42×10^9 years. That is long enough, on the one hand, to cover the full time-range of about 10×10^9 years, to be expected for the duration of the pre-solar galaxy, but also short enough, on the other hand, to produce still detectable amounts of ^{187}Os.

A short time after Clayton had proposed the rhenium/osmium clock, the astrophysicists E. M. D. Symbalisty, D. N. Schramm, and independently, the cosmochemist G. J. Wasserburg, developed a mean-age approach [3]. They showed that the abundance ratio of a directly connected radioactive pair, taken at decoupling, renders both, an upper $\left(T_N^{MAX}\right)$ and a lower $\left(T_N^{MIN}\right)$ limit of T_N, the duration of the pre-solar galaxy, for the extreme cases of a sudden (one-spike) and a steady nucleosynthesis. These limits differ by a factor of only two and read, for the couple rhenium/osmium and for $\lambda T_N \ll 1$ as:

$$T_N^{MIN,MAX} = \frac{1(2)}{\lambda} \left(\frac{^{187}Os^\beta}{^{187}Re}\right)_{dec.}, \qquad (1)$$

where $1/\lambda = \tau$ is the mean lifetime of ^{187}Re, 61×10^9 yr, and where the $^{187}Os^\beta/^{187}$Re ratio has to be taken at decoupling, i.e. 4.6×10^9 years ago. This simple expression left only one problem to be solved: In equation (1) only the "β-part," $^{187}Os^\beta$, stemming from β-decay of ^{187}Re has to be taken into account. There is, however, another contribution to the total ^{187}Os abundance, due to slow neutron capture of ^{186}Os during the "s-process" (nucleosynthesis via the interplay of "slow" beta decay and neutron capture).

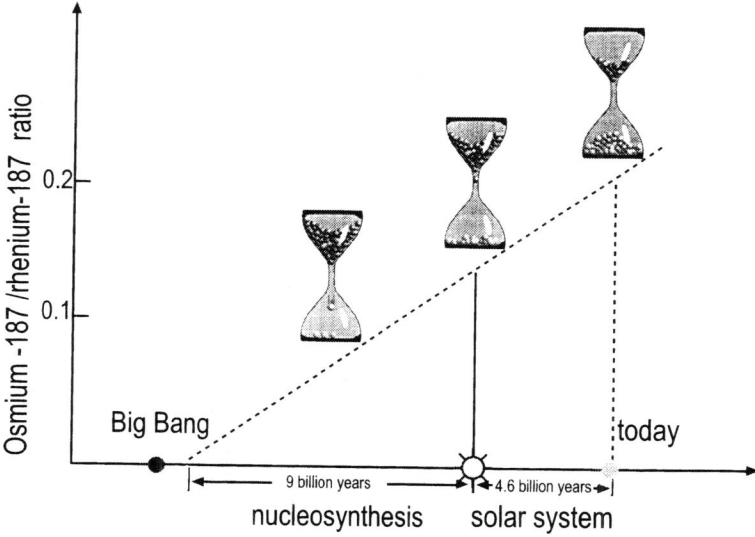

FIGURE 2. A minimum period of about 9 billion years for the pre-solar nucleosynthesis can be derived, in the framework of the Symbalisty-Schramm-Wasserburg model, from the abundance ratio (sketched by the hour-glasses) of ^{187}Os/^{187}Re at decoupling of the solar system and from the half-life of 42 billion years for neutral ^{187}Re. This yields, together with the solar age of 4.6 billion years, a minimum age of about 14 billion years for our galaxy and, a fortiori, for the universe.

Thus, in previous decades, the effort in rhenium cosmochronology focused on determining this s-fraction as reliably as possible. The main problem was to estimate the fraction of neutron capture into the first excited state of ^{187}Os at about 10 keV—facing that this level will be considerably populated at s-process temperatures of typically 30 keV. Nowadays this battle seems to be over and it is commonly believed [4] that the s-contribution (about 40 % of the total ^{187}Os abundance [4]) can be settled with an accuracy of better than 10 %. The halflife of ^{187}Re has meanwhile been precisely determined—after a long period of contradictory results—to $(42.3 \pm 1.3) \times 10^9$ yr [5]. However, the relative abundances of both, ^{187}Os and ^{187}Re are still uncertain by about 6 % [6]. From all these data one extracts, within the "mean age approach" of Eq. (1), a minimum time-span T_N^{MIN} of roughly 9 billion years for the pre-solar nucleosynthesis. By adding the well-known solar age T_s of 4.6 billion years to this number, one gets as a lower limit T_G^{MIN} for the age of our galaxy, with

$$T_G^{MIN} = T_N^{MIN} + T_s, \qquad (2)$$

about 14×10^9 years. This number can serve, obviously, also as a lower limit for the age T_U of the universe. The value of 14 billion years fits fairly well with the 12 billion years established as a safe lower limit for T_U from globular clusters [1], keeping in mind that the dating methods differ widely.

HOW ABOUT HIGHLY IONIZED RHENIUM-187?

The Symbalisty-Schramm-Wasserburg model contains some tacit simplifications that probably do not correspond to the real scenario of nucleosynthesis. In particular, the ability of the rhenium clock to provide a lower limit for the duration of the pre-solar nucleosynthesis relies on the assumption that there is *one* and only one decay constant λ, namely the one determined for *neutral* ^{187}Re with a full cloud of 75 electrons. In 1983, the astrophysicists K. Takahashi, K. Yokoi and M. Arnould aired well-founded reservations about the correctness of this assumption [7]. They suggested that ^{187}Re—like any other nuclide—could have been partially or even completely ionized in the course of galactic history and that this might have dramatically changed its lifetime. Like all atomic nuclei created by rapid neutron capture ("r-process") during nucleosynthesis, ^{187}Re is formed within a matter of seconds, probably during the explosive phase of a supernova, and is immediately flung into interstellar space. At a later stage the resultant galactic gas probably compresses again, with both ^{187}Re and the small portion of ^{187}Os produced in the meantime, and contributes to the formation of a new star. Some of the atoms go through this process of "astration" several times. Depending on the location in the star and the temperatures prevailing there, the atoms are more or less strongly ionized.

In this state, a special type of beta decay, the so-called "bound beta decay" (β_b) becomes possible: the electron created in the nucleus does not leave the atom but remains in an unoccupied (inner) electron shell (see Fig. 3).

In neutral atoms where the inner atomic orbits are occupied ("Pauli-blocked") β_b decay is almost negligible as compared to continuum decay. When going to highly-ionized, heavy atoms (such as occur abundantly in a stellar plasma), where the strongly bound inner shells (K, L,...) are completely or partially empty, the situation can change totally: most of the beta-decay probabilities are enhanced

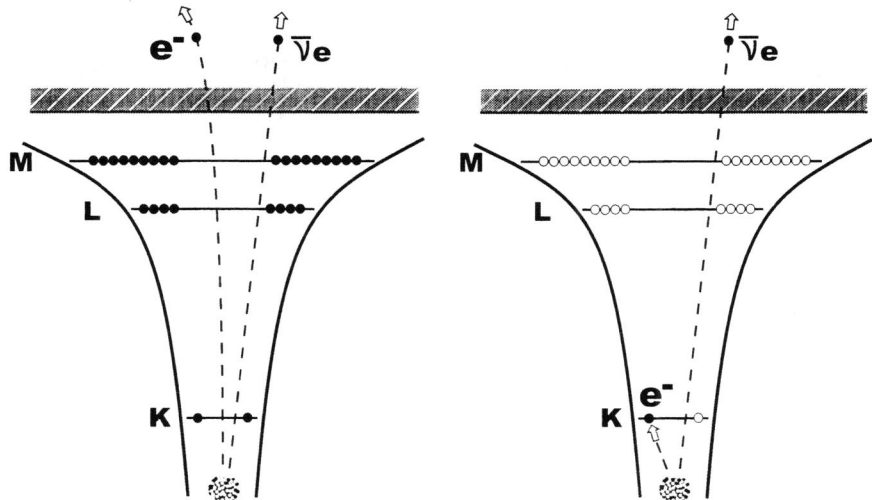

FIGURE 3. Sketch of the well-known "continuum" beta decay for a neutral atom (left) and of bound beta decay for a bare nucleus (the difference of the corresponding potentials is not drawn). In both decay modes a neutron of the nucleus is transformed to a proton, an electron and an antineutrino. Bound beta decay enforces a well-defined energy of the created bound electron. Therefore, a monochromatic antineutrino emerges in that case.

considerably by β_b transitions to deeply bound states, as no work is required for the electron to escape the attraction of the positively charged atomic nucleus. In a rough approximation, the beta-decay energy (Q value) is enlarged by the binding energy of the final electron state. In special cases, even nuclei that are stable when embedded in their full electron cloud can become unstable with respect to β_b decay if they are highly ionized.

It is not surprising in this context that β_b decay—predicted as the time-mirrored form of the well-known orbital-electron capture already in the forties By R. Daudel, M. Jean and M. Lecoin [8]—was only observed once heavy-ion storage rings became operational. There for the first time highly-charged ions could be provided, stored and preserved for many hours [9]. The possible enhancement of decay in q-times ionized ^{187}Re atoms had long been recognized theoretically. The higher the atomic charge state q becomes, the more β_b decay into empty states of the $n = 4, 5...$ electron shells competes with continuum beta decay, keeping in mind the extremely small Q-value of the latter of 2.66 keV (taken for neutral ^{187}Re). Starting from a charge state q of about 40, continuum decay becomes impossible and is fully replaced by β_b decay.

What Takahashi and co-workers [10] noted for the first time, however, was that—at least for bare and hydrogen-like ^{187}Re ions—also the first *excited* nuclear state of ^{187}Os at 9.75 keV can be fed by β_b decay. Because here a change of the nuclear spin

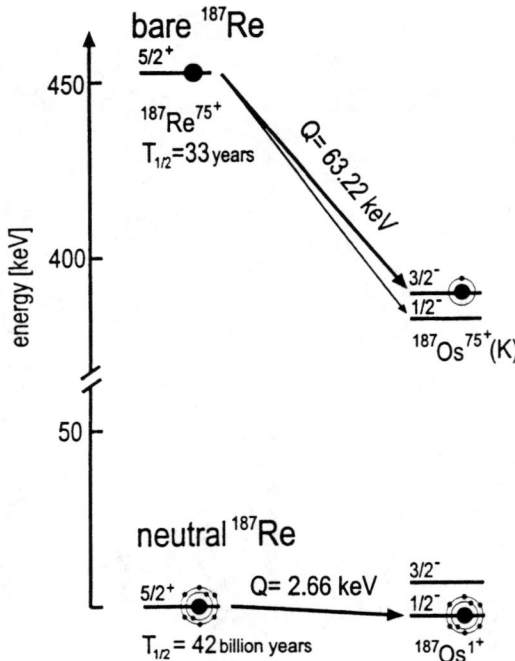

FIGURE 4. Decay schemes for neutral ^{187}Re (bottom) and bare ^{187}Re. Neutral ^{187}Re has a beta half-life of 42 billion years due to both, a small Q-value (2.66 keV) and a large difference of angular momenta of parent ($5/2^+$) and daughter ($1/2^-$) nucleus. If completely ionized, it decays significantly faster as β_b decay to the first excited state of ^{187}Os ($3/2^-$) at 9.75 keV becomes allowed. The experiment conducted at the ESR established a half-life of only 33 years for that decay.

I of only one unit ($\Delta I = 1$) is involved, this transition should be much stronger than the corresponding one to the ^{187}Os ground state, where the nuclear spin has to change by two units. A detailed inspection of the energy balance [11] shows that for the example of bare ^{187}Re^{75+} and for the created electron being bound in the K shell of ^{187}Os^{75+}, a $Q^K_{\beta_b}$ value of 63.22 keV arises for this transition to the first excited state (see figure 4). The nuclear matrix element for this decay *cannot* be provided from the time-mirrored electron capture (EC) process, because the excited 9.75 keV state does not undergo any EC decay during its short lifetime. Only a first guess might be deduced from known matrix elements of neighbouring nuclear states of similar structure. From this comparison Takahashi estimated, for bare ^{187}Re , a half-life of 14 years [10] for β_b transition to the 9.75 keV state (almost identical with the total half-life of bare ^{187}Re). This would be a spectacular change by more than 9 orders of magnitude with respect to the 42×10^9 yr half-life of neutral ^{187}Re.

To set this prediction on safe ground, it was mandatory to measure [11] the half-life of bare ^{187}Re, especially since the heavy-ion storage ring ESR at Darmstadt, Germany, had already proved to be tailor-made for this kind of experiment. The challenge was that a 100 to 1000 times smaller decay rate had to be expected as compared to the pioneering experiment with bare dysprosium-163, where β_b decay was observed for the very first time [9]. It was an open question whether such long β_b lifetimes (on the order of several ten years) could be addressed at all with the technique designed for a 50-day half-life.

MEASURING THE DECAY OF BARE RHENIUM-187

Like the stable neutral atom ^{163}Dy, the quasi-stable ^{187}Re (63 % natural abundance, 42×10^9 year half-life) can be introduced in an ion source, moderately ionized there, and subsequently accelerated to 370 MeV/u in the linear accelerator UNILAC and the heavy-ion synchrotron SIS installed at GSI. After the ions leave the synchrotron, their remaining electrons are stripped off in a foil; the naked rhenium is transported to the ion storage ring ESR where it is cooled to a sharp energy with the aid of an "electron cooler." After that it can be stored for many hours in the ESR without substantial losses, due to an ultra-high vacuum of 10^{-11} millibar pressure.

The electron cooler is an ingenious tool invented in the sixties by the Russian physicist G. Budker. A monoenergetic, i.e. cold, electron beam travels parallel to the ion beam along a two-meter section of the storage ring (see Fig. 5). At each turn, i.e. every 500 nanoseconds, the ions undergo collisions with the electrons, thus steadily exchanging momenta with them. In contrast to the ions, the electrons are extracted from the interaction zone after one passage and steadily replaced by new ones. Due to this extremely clever trick, the large original velocity spread and angular divergence of the ions is narrowed by many orders of magnitude until the transverse and longitudinal temperatures become equal for both the electrons and the ions. In particular for heavy, highly-charged ions, the cooling is efficient as well as fast; for typical electron currents of about 100 milliamperes, cooling times of less than a second can be obtained.

Electron cooling not only generates brilliant, monoenergetic beams, it also provides a sharp *common velocity* for all circulating ions over the whole period of storage. Therefore, the revolution frequencies of electron-cooled ions depend solely on their *mass-to-charge ratio*. By transforming these short pulses, induced onto pick-up plates by every circulating ion ("Schottky-noise"), into frequencies (by Fast Fourier Transform), cooler rings may serve as superb mass spectrometers. This has been convincingly demonstrated at the ESR within the last few years [12]. During the storage period, some of the naked ^{187}Re-ions decay through β_b decay, becoming hydrogen-like ^{187}Os-ions with one bound electron, mostly in the K shell. If the storage time is small with respect to the β_b half-life, the number of osmium ions increases proportional to the storage time, the β_b decay-constant, and to the number of primary bare ^{187}Re-ions (for the exact treatment, including all the various

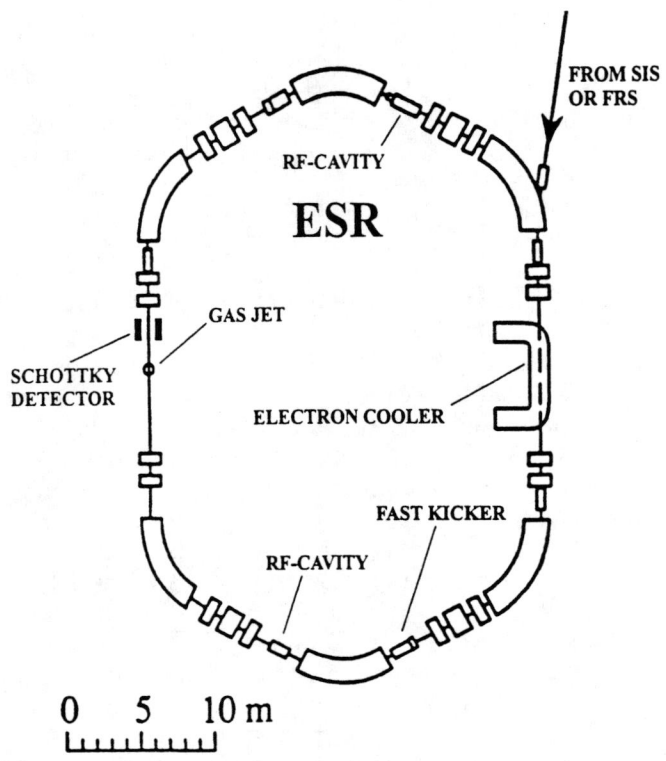

FIGURE 5. Sketch of the experimental storage ring ESR at Darmstadt together with its main installations: the electron cooler, the gas jet, and the Schottky pick-up plates, where the revolution frequencies and hence the mass-to-charge ratio of the stored and cooled ions can be determined.

corrections needed, see [9,11].

At first the hydrogen-like ^{187}Os daughter ions follow almost the *same* trajectory as their parent nuclei and with nearly the same revolution frequency, since both have almost exactly the same mass-to-charge ratio. The mass difference between mother- and daughter is only 63 keV/c^2, which corresponds to a relative mass difference $\Delta m/m$ of about 4×10^{-7}. The best mass resolution, achieved in Schottky mass-spectrometry at the ESR for some thousand stored ions, is about 2×10^{-6} as shown in Fig. 6. Thus, a direct separation of the ^{187}Os-daughters from the ^{187}Re mother-nuclei is *not* far beyond the capability of the ESR. However, to generate a few tens of decay products within several hours of storage—for an expected half-life of some ten years—it was necessary to store at least 10^7 primary ions. For such a large number, however, the presently achievable mass resolution $\Delta m/m$ is not much better than 10^{-5}. Therefore, the ^{187}Os-daughters can*not*, unfortunately, leave their own fingerprints in the Schottky spectrum of revolution frequencies.

Until this direct observation of β_b decay can be achieved at *small* Q-values and for

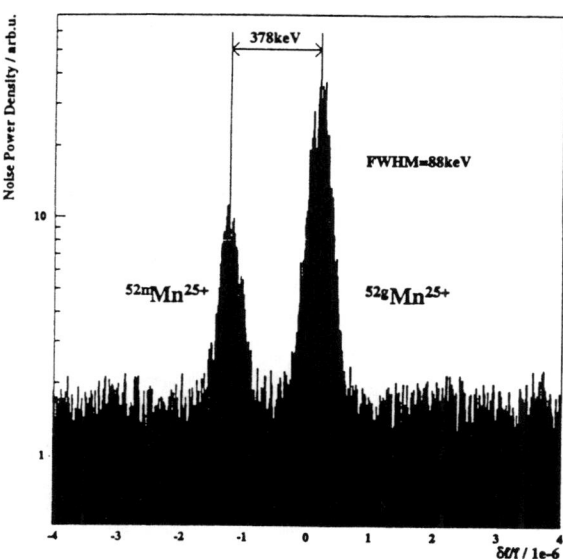

FIGURE 6. Schottky spectrum of mass-resolved groundstate (r.h.s.) and isomeric state at 378 keV of bare, cooled ^{52}Mn ions, vs. the relative difference of revolution frequency f, $\delta f/f$, which corresponds—for cooled ions—to a well-defined change in the mass-to-charge ratio m/q. The relative mass resolution $\Delta m/m$, achieved for the about 2000 stored and cooled ^{52}Mn ions, amounts to 2×10^{-6}.

a large number of primary ions, one has to be content with a less direct detection technique: after a variable storage time, the electron created by β_b decay and bound in the daughter atom is stripped off in the internal gas jet of the ESR (see Fig. 5), which is turned on for a few minutes. The resulting change of atomic charge state by one unit modifies both the trajectory and the revolution frequency of the daughter atom, which is thus easily detected either by Schottky spectroscopy or by a particle detector intercepting the new orbit. The procedure is simplest for *bare* mother nuclei, where the β_b daughters eventually become bare ions, too.

However, even in this case an auxiliary experiment is needed to determine the fraction of ionization in the gas jet (about 72% for $Z = 75$ and an energy of 370 MeV/u) with respect to recombination. Our detection method, somewhat indirect for reasons discussed above, leads—with corrections for electron capture in the gas jet—to another complication: ^{187}Os ions might also be generated by nuclear charge exchange (n,p reaction) of the ^{187}Re ions with the argon atoms of the gas jet. This, however, does *not* depend on the storage time (in sharp contrast to the number of β_b daughters) but only on the amount of primary bare ^{187}Re ions and on the time-integrated intensity of the gas jet. Therefore, this nuclear background can be determined in an experiment with zero storage time.

Finally, a plot of the number of ^{187}Os nuclei created (normalized to the amount of ^{187}Re parent nuclei) versus the storage time directly provides the decay rate and hence the half-life of *bare* ^{187}Re. The number of ^{187}Os ions was determined—as discussed above—in two ways, either by counting them directly in a detector introduced into the ring aperture, or by measuring the signal strength of (bare) ^{187}Os in the Schottky spectrum recorded after the operation of the gas jet. Both methods have specific pros and cons: the particle counter has a 100% detection efficiency for all generated ^{187}Os nuclei but does not allow a complete suppression of nuclei with different nuclear charge; the Schottky signal, on the other hand, providing a clear-cut identification of bare ^{187}Os, has to be calibrated in terms of an *absolute* particle number—not a simple task. Very helpful in this respect was the astonishing sensitivity of Schottky spectroscopy at the ESR: even *one single* stored, highly-charged ion can be detected and unambiguously identified [13].

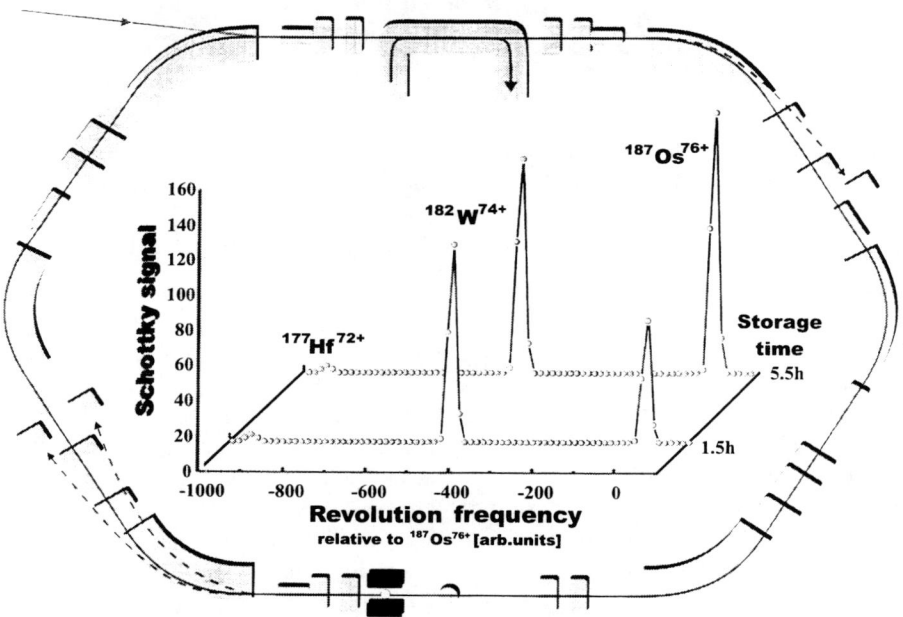

FIGURE 7. Observation at the ESR of ^{187}Os from β_b decay of bare ^{187}Re ions. The figure shows corresponding Schottky signals for storage times of 1.5h and 5.5h, respectively. The spectra were normalized onto the same number of primary bare ^{187}Re ions. The daughter atoms ^{187}Os are clearly identified after the single electron of ^{187}Os has been removed by switching on an argon gas jet. The number of ^{187}Os ions increases significantly with storage time, in contrast to the intensities of nuclear reaction products like tungsten (W) and hafnium(Hf).

From these two independent experiments, which were carried out as a collaboration between GSI and the Technical University of Munich, a half-life of (33 ± 2) years for bare ^{187}Re was obtained, where both detection methods agreed within the experimental error margins. This amazing result shows that indeed the lifetime of ^{187}Re is more than a billion time shorter if all of its bound electrons are stripped away. From this half-life of the *bare* ion, the nuclear matrix element for the β_b decay to the first *excited* state of ^{187}Os, and hence the corresponding log ft-value, can be derived. From this, the half-life of ^{187}Re at any charge state q (at any temperature) can be reliably calculated, too.

RECALIBRATING THE RHENIUM CLOCK

Does the dramatic difference in lifetime between neutral and fully ionized ^{187}Re render the rhenium clock useless? How can we possibly know all the details of the history of both ^{187}Re and ^{187}Os, namely the location in the star after each astration, the temperature there (i.e. the distribution of atomic charge states q), or the duration of stay? All that would be needed, however, to get a new *effective* decay constant λ_{eff} is the mean of the q-dependent $\lambda(q)$. The values of $\lambda(q)$ can now be calculated precisely, based on the result of our experiment, but they still have to be weighted according to their relative occurrence during the whole period of nucleosynthesis!

Equipped with this new knowledge of the lifetime of any charge state, K. Takahashi is calculating the probable "life story" of the couple ^{187}Re/^{187}Os. Although this is a mammoth task, it is not as hopeless as it might seem at first sight: astrophysicists have created reliable models to figure out the chemical evolution of our galaxy. They are now able to quantitatively explain the observed abundances of almost all the nuclides as found in the solar system. Moreover, some circumstances might decisively facilitate Takahashi's work: the decay to the first excited state and, hence, the most dramatic change in lifetime is possible only for *bare and hydrogen-like* ^{187}Re ions, since already for lithium-like ^{187}Re ions the corresponding β_b decay (the electron has to populate the L shell of osmium in that case) has a (slightly) negative Q-value.

To produce substantial amounts of bare or hydrogen-like rhenium ions, temperatures of about 9×10^8 K are needed, and these prevail for very short periods and only in the innermost zones of massive stars. In this extremely hot and dense stellar plasma, a large number of highly energetic, free electrons surround the ions. Therefore, the newly created ^{187}Os ions can also *decay back* to ^{187}Re, just by capturing from the excited 9.75 keV state (which is strongly populated at those temperatures) one of these electrons, provided its energy is above 63 keV. (This process of "free electron capture" is best known from the decay of bare ^{7}Be in the interior of the sun). As the probability of this back-decay depends on the *same* matrix element as the β_b decay to the 9.75 keV state, it can be easily calculated, given only density and temperature of the electron plasma.

K. Takahashi deduced, very preliminarily, from this recalibration of the rhenium clock an age of the galaxy of $\left(15^{+2}_{-3}\right) \times 10^9$ years [14]. The lower limit of 12 billion years, being a fortiori a lower limit for the age of the universe, is significantly, albeit not dramatically, smaller than the 14 billion years as derived within the much simpler "mean age approach" for neutral rhenium. This outcome is perhaps surprising. Obviously, the back-decay by capture of *free* electrons from the excited ^{187}Os state acts to counterbalance β_b-decay to a large extent, and it is even more effective at higher temperatures and charge states. Takahashis result, if confirmed by the final evaluation, would exactly coincide with the minimum age of our galaxy as determined from globular clusters.

PROBING THE STANDARD COSMOLOGICAL MODEL

Carefully evaluated lower limits for both the Hubble constant H_0 and the age of the universe T_U provide serious constraints on current cosmological models and, thereby, also on the future fate of the universe: the product of the Hubble constant H_0, on the one hand, and the age of the universe, T_U, on the other, is a function of the energy-matter density Ω in the universe. Usually, Ω is given as fraction of the "critical" density $\Omega_c = 3H_0^2/(8\pi G)$, with G the gravitational constant; Ω_c exactly defines the borderline between an infinitely expanding universe and a universe for which, somewhere in the future, the expansion will turn back to a contraction (closed universe).

Presently, the preferred cosmological model is the Einstein-de Sitter universe. In this cosmos the metric is Euclidean, i.e. the curvature is zero, and the density is exactly the critical one ($\Omega = \Omega_c$). This choice is also most often required in models of the inflationary universe that just a few years ago got new and strong support from the COBE sky map of the $3K$ microwave radiation, left over from the Big Bang [15].

In the Einstein-de Sitter universe, the so-called cosmological constant, Λ, is taken to be zero. The constant Λ was introduced by Einstein into the field equations of his general theory of relativity to counterbalance the gravitational attraction of the masses by a kind of a repulsive energy density of the vacuum, because he sought solutions providing a *stationary* universe. Later on, Einstein called the postulate of Λ "the biggest blunder of my life." Perhaps Λ will experience a revival very soon! A direct consequence of the postulated critical density for the universe is that more than 90 % of matter must be "dark," because only about 5 % of the critical density is visible.

If the cosmological constant Λ is taken to be zero, a simple relation between the Hubble constant H_0, the age of the universe T_U, and the density Ω holds:

$$H_0 \cdot T_U \to 978, \quad \Omega \to 0$$
$$H_0 \cdot T_U = 652, \quad \Omega = \Omega_c \tag{3}$$

when H_0 is measured in terms of recession velocity (km/s) per megaparsec (Mpc), and T_U in billions of years.

If $H_o \cdot T_U > 978$ then the cosmological constant Λ has, *necessarily*, a positive value and the universe would expand forever.

From equation (3) we may immediately learn how to sensitively probe the Einstein-de Sitter universe: searching for *independent* lower limits for both H_0 and T_U, and looking thereafter whether their *product* fits to the underlying assumption of both a critical density and a vanishing cosmological constant. That is exactly what makes safe lower limits for both H_0 and T_U, such important.

Very recently the detection of δ-Cepheids in the M100 galaxy of the Virgo cluster brought a breakthrough for determining the *true* distances of very far galaxies, the biggest obstacle so far to settle H_0. δ-Cepheids are variable stars with a periodically changing luminosity, where a well-known relation exists between period and absolute luminosity. They serve, therefore, as calibration marks for cosmic distances. After the δCepheids in the M100 galaxy had been found, the distance of the Virgo cluster could, for the first time, be reliably deduced to 15 megaparsec and, thereafter, the Hubble constant H_0 to about 80 km/s per Mpc with a lower limit of 63 km/s per Mpc (taking into account the recalibration of the distances of the nearest δ-Cepheids by the Hipparcos satellite). This limit, although consistently derived with both the Hubble Space Telescope and the Canada-France-Hawaii Telescope [16,17], is nonetheless still the object of extremely violent struggles in the community of Hubble-experts.

At the time being, almost all of the clocks in use point to a lower limit of about 12 billion years for the age of the universe. That is true not only for the globular clusters and the recalibrated rhenium clock, but also for other astronomical clocks like the luminosity distribution of white dwarfs, or for the aeon clock ^{238}U/^{232}Th (with a large error margin, however). Combining this value of 12 billion years and the new lower limit of 63 km/s per Mpc for H_0, one gets about 760 as the product of both (in the practical units of equation (3)), which lies significantly above the value of 652 postulated by the Einstein-de Sitter model (Fig. 8).

If these values are corroborated by future observations, the standard cosmological model will fail in the sense that the density Ω of the universe has to be less than the critical one and/or that the cosmological constant Λ has to be different from zero. However, current limits for H_0 and T_U are hardly water-tight, keeping in mind all the intricate and peculiar methods needed to derive them. It is of utmost importance to make further checks on the reliability and consistency of these clocks and, if possible, also to develop new ones.

Each individual new result, either for H_0 or for T_U, is like a grain of sand in a thousand-piece puzzle. It seems, nonetheless, that the time is not too far off when we may get a first impression of the image hidden in this puzzle. No matter what it turns out to be, it will reveal the future fate of the universe.

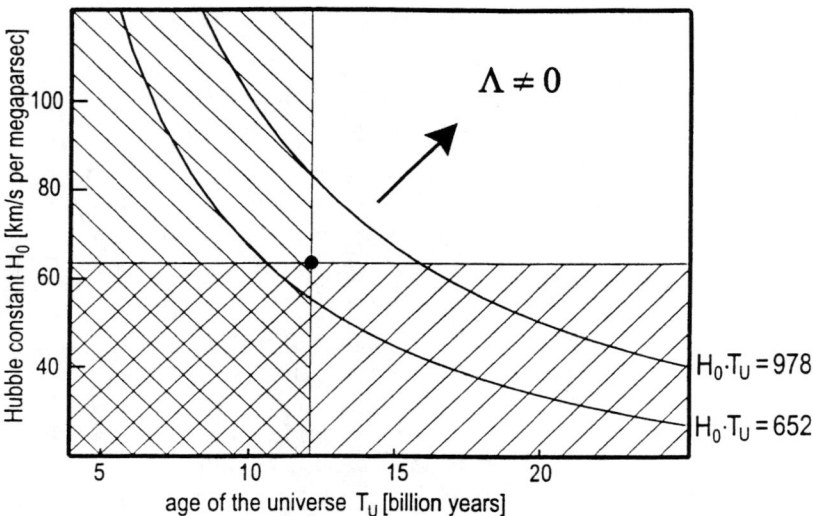

FIGURE 8. Plot of the Hubble constant H_0 vs. the age of the universe T_U for a cosmological constant $\Lambda=0$. The upper hyperbola corresponds to an empty universe, the lower one to a universe with critical density. The numerical values result if H_0 is taken in km/s per Mpc and T_U in billion years. The presently favoured lower limits of 63 km/s per Mpc for H_0 and of 12 billion years for T_U (from both globular clusters and the recalibrated rhenium clock) signal a density less than the critical one and/or a nonvanishing cosmological constant Λ.

SYNOPSIS

The rhenium-187 aeon clock is an example that brings to light—in a rather spectacular manner—the influence of the atomic charge state on nuclear and astrophysical properties. It has been recognized for a long time that the number and configuration of electrons bound in the atom can significantly alter the lifetime of beta decay. Prominent examples are the Fermi function that denotes the correction to the wave function of the emitted charged lepton as a function of the atomic charge state, or the obvious impact of the number of bound electrons on the probability of orbital electron capture. However, those effects could be not investigated until recently, because only neutral atoms were available in the laboratories. That was the more regrettable the more the fundamental role of beta decay in nucleosynthesis was realized. The synthesis always occurs in hot stellar plasmas, where even the heaviest atoms are considerably ionized.

This disappointing situation only changed with the advent of ion storage-cooler rings and ion-traps a few years ago, in which ions in a well defined charge state can be accumulated, cooled and stored for many hours. Only then did it become possible to study weak decays under conditions similar to those prevailing in hot stellar plasmas during nucleosynthesis.

In trying to get to the root of the rhenium clock, we may realize that the enormous impact of the number of bound electrons is by no means accidental. A perfect aeon clock has to satisfy two conditions: first,it should consist of a mother/daughter couple (to avoid such problems as inherent in the uranium/ thorium clock, for instance) but with a stable daughter nucleus. Secondly, the lifetime of the mother has to be—at least—on the order of the age of the universe. The latter condition, restricting candidates to a couple connected by beta decay, can only be satisfied by either a small Q-value or a high angular momentum of the beta transition, or both. It is exactly at a high atomic charge state that both the Q-value and the angular momentum of the transition can change significantly once the door is opened to bound-state beta decay into strongly bound electron shells.

Interesting aspects of beta decay of highly charged ions will be investigated in the near future at the Darmstadt ion cooler ring ESR. With the help of a "fragment separator" (FRS, see Fig. 5) beta-unstable, highly charged ions of any kind can be injected into the ESR, cooled, and analyzed there. In particular the powerful technique of Schottky spectroscopy, together with an almost 100% detection efficiency for beta decay, will allow us to address a wealth of still unexplored problems of astrophysics. There is no question that ion-traps, with their unrivaled precision, can and will also be used in the near future for key experiments on the beta decay of highly charged ions.

REFERENCES

1. VandenBerg, D. A., Bolte, and M., Stetson, P. B., *Annual Rev. Astron. Astrophys.* **34**, 461 (1996).
2. Clayton, D. D., *Astrophys. J.* **139**, 637 (1964).
3. Symbalisty, E. M. D., and Schramm, D. N., *Rep. Prog. Phys.* **44**, 293 (1981)
4. Winters, R. R., Carlton, R. F., Harvey, J. A., and Hill, N. W., *Phys. Rev. C* **34**, 840 (1986).
5. Lindner, M., *Geochim. Cosmochim. Acta* **53**, 197 (1989).
6. Anders, E., and Grevesse, N., *Geochim. Cosmochim. Acta* **53**, 197 (1989).
7. Yokoi, K., Takahashi, K., and Arnould, M., *Astron. Astrophys. J.* **117**, 65 (1983).
8. Daudel, R., Jean, M., and Lecoin, M., *J. Phys. Radium* **8**, 238 (1947).
9. Jung, M., Bosch, F., Beckert, K., Eickhoff, H., Folger, F., Franzke, B., Gruber, A., Kienle, P., Klepper, O., Koenig, W., Kozhuharov, C., Mann, R., Moshammer, R., Nolden, F., Schaaf, U., Soff, G., Spädtke, P.,Steck, M., Stöhlker, Th., and Sümmerer, K., *Phys. Rev. Lett.* **69**, 2164 (1992).
10. Takahashi, K., Boyd, R. N., Mathews, G. J., and Yokoi, K., *Phys. Rev. C* **36**, 1522 (1987).
11. Bosch, F., Faestermann, T., Friese, J., Heine, F., Kienle, P., Wefers, E., Zeitelhack , K., Beckert, K., Franzke, B., Klepper, O., Kozhuharov, C., Menzel, G., Moshammer, R., Nolden, F., Reich, H., Schlitt, B., Steck, M., Stöhlker, Th., Winkler, Th., and Takahashi, K., *Phys. Rev. Lett.* **77**, 5190 (1996).

12. Schlitt, B., Beckert, K., Bosch, F., Eickhoff, H., Franzke, B., Fujita, Y., Geissel, H., Hausmann, M., Irnich, H., Klepper, O., Kluge, J., Kozhuharov, C., Kraus, G., Münzenberg, G., Nickel, F., Nolden, F., Patyk, Z., Radon, T., Reich, H., Scheidenberger, C., Schwab, W., Steck, M., Sümmerer, K., Winkler, Th., Beha, T., Falch, M., Kerscher, Th., Löbner, K.E.G., Jung, H.C., Wollnik, H., and Novikov, Yu., *Nucl. Phys.* **A626**, 591c (1997).
13. Bosch, F., and Schlitt, B., *Phys. Blätter* **53**, 27 (1997).
14. Takahashi, K., private communication and to be published.
15. Wilkinson, D., Science Spectra **4**, 52 (1996).
16. Freedman, W. L., Madore, B. F., Mould, J. R., Hill, R., Ferrarese, L., Kennicutt, R. C., Saka, A., Stetson, P. B., Graham, J. A., Ford, H., Hoessel, J. G., Huchra, J., Hughes, S. M., and Illingworth, G. D., *Nature* **371**, 757 (1994).
17. Pierce, M. J., Douglas, L. W., McClure, R. D., van den Bergh, S., Racine, R., Stetson, P. B., *Nature* **371**, 385 (1994).

X-Ray Emission from Comets

Konrad Dennerl

Max-Planck-Institut für extraterrestrische Physik
Giessenbachstraße, D-85748 Garching, Germany

Abstract. When comet Hyakutake (C/1996 B2) encountered Earth in March 1996 at a minimum distance of only 15 million kilometers (40 times the distance of the moon), x-ray and extreme ultraviolet emission was discovered for the first time from a comet. The observations were performed with the astronomy satellites ROSAT and EUVE. A systematic search for x-rays from comets in archival data, obtained during the ROSAT all-sky survey in 1990/91, resulted in the discovery of x-ray emission from four additional comets. They were detected at seven occasions in total, when they were optically 300 to 30 000 times fainter than Hyakutake. These findings indicated that comets represent a new class of celestial x-ray sources. Subsequent detections of x-ray emission from additional comets with the satellites ROSAT, EUVE, and BeppoSAX confirmed this conclusion. The x-ray observations have obviously revealed the presence of a process in comets which had escaped attention until recently.
This process is most likely charge exchange between highly charged heavy ions in the solar wind and cometary neutrals. The solar wind, a stream of particles continuously emitted from the sun with ≈ 400 km s^{-1}, consists predominantly of protons, electrons, and alpha particles, but contains also a small fraction ($\approx 0.1\%$) of highly charged heavier ions, such as C^{6+}, O^{6+}, Ne^{8+}, Si^{9+}, Fe^{11+}. When these ions capture electrons from the cometary gas, they attain highly excited states and radiate a large fraction of their excitation energy in the extreme ultraviolet and x-ray part of the spectrum. Charge exchange reproduces the intensity, the morphology and the spectrum of the observed x-ray emission from comets very well.

INTRODUCTION

Comets are the remains of the formation of the solar system. As kilometer-sized aggregates of ices and dust, they usually orbit the sun at distances of several billion kilometers. Due to the lack of external heating and the very low surface gravity, the primordial matter is well conserved. Sometimes a comet approaches the sun. Then the ices sublimate and drag off dust from the surface, creating a 10^4 km sized coma of dust and an even larger coma of neutral gas around the cometary nucleus. The dust particles are slowly deflected by the repulsive force of solar radiation pressure into a broad dust tail. The gas molecules eventually become ionized by, e.g., the solar ultraviolet radiation. Then they are quickly swept away by the solar wind, a continuous stream of charged particles emitted from the sun with a typical velocity of 400 km/s, and form a narrow plasma tail pointing almost directly away from the sun.

With tail lengths of 10^8 km and more, comets can become spectacular sights. Their sudden, dramatic appearance and their peculiar apparent paths in the sky have attracted fear and excitement since historical times. In this century, comets have become subject to increasingly advanced astrophysical investigations, which so far culminated in detailed observations of comet 1P/Halley, including *in-situ* measurements, during its apparition in 1986. Nevertheless, one aspect of cometary activity seems to have escaped attention until recently, when surprisingly bright x-ray emission from comets was discovered in 1996 [1–3].

Comets are cool objects, and the processes which had been previously observed to take place in a comet operate at moderate temperatures. Strong x-ray emission was not a natural consequence of the current picture of what was thought to be going on in a cometary coma of gas and dust. One possibility for the creation of x-rays was already suggested in 1990 [4]: dust grains in the cometary coma might collide with interplanetary dust particles. Due to the high relative velocities of $\gtrsim 50$ km/s, the particles would immediately evaporate and create a plasma clot of $T \approx 10^6$ K, which would briefly radiate in the soft x-ray range. Another possibility was that electrons might be accelerated by an interaction of the comet with the solar wind, similar to a terrestrial aurora [5], and this idea motivated the first observation of a comet in x-rays, which was performed with the satellite Einstein in 1980. No x-ray emission, however, was found [5], and this result seemed to fit well into the general view that comets do not emit a significant amount of x-rays.

THE DISCOVERY OF COMETARY X-RAY EMISSION

The first record of cometary x-ray emission dates back to July 1990. Comet 45P / Honda-Mrkos-Pajdušáková happened to cross the field of view of the x-ray astronomy satellite ROSAT [6], which was scanning the sky in order to produce the first map of the whole sky to be obtained with an imaging x-ray telescope [7]. The comet, however, was quite faint, and was moving between the individual telescope scans. In the uncorrected data, it did not leave a conspicuous signal among the $\sim 80\,000$ x-ray sources that were detected during the all-sky survey. Other comets shared the same fate.

X-rays from a comet were discovered for the first time in March 1996, when comet C/1996 B2 (Hyakutake) encountered Earth at a minimum distance of only 15×10^6 km. Observations with ROSAT [1] and the Extreme Ultraviolet Explorer EUVE [3] resulted in the unexpected discovery of bright x-ray and EUV emission from this comet. Unlike the optical appearance, the emission came from a crescent-shaped region that was symmetric about the sun-comet line and offset sunward by $\approx 18\,000$ km (Fig. 1). Also in contrast to the optical behaviour, the ROSAT

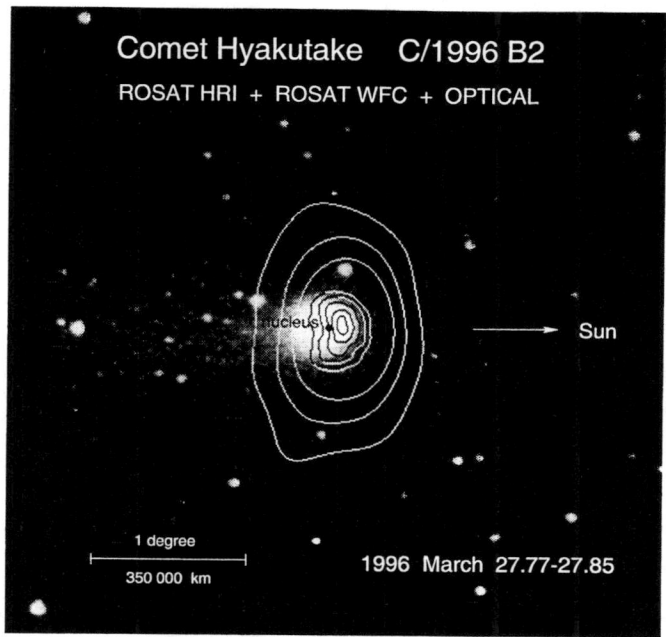

FIGURE 1. Contours of the x-ray and extreme ultraviolet emission of comet C/1996 B2 (Hyakutake) superimposed on an optical image. The inner five contour lines show the intensity distribution of the x-ray emission, observed with the ROSAT High Resolution Imager (HRI), while the outer three contours refer to the extreme ultraviolet emission, measured simultaneously with the ROSAT Wide Field Camera (WFC). The optical image was taken by the author during the ROSAT observations. It is evident that the x-ray emission is coming mainly from the sunward side of the coma and not from the cometary nucleus (marked).

Reprinted with permission from C. M. Lisse et al., *Discovery of X-Ray and Extreme Ultraviolet Emission from Comet C/Hyakutake 1996 B2*, SCIENCE, **274**, pp 205-209, ©1996 American Association for the Advancement of Science.

observations showed that the x-ray emission was highly variable on time scales of hours. The total x-ray luminosity varied between 4×10^{15} ergs s^{-1} and 16×10^{15} ergs s^{-1}. A comparison of the x-ray and extreme ultraviolet flux indicated a soft x-ray spectrum, with a typical energy $E < 1$ keV [1].

Early Models

The challenge of understanding the origin the x-ray emission motivated many activities on the theoretical side. Attempts to explain it by scattering of solar x-rays in gas or normal-sized dust grains failed to reach the required intensity. While very small dust particles (≈ 30 Å size) are more efficient in scattering soft x-rays [8], the temporal variations of the Hyakutake x-ray flux were considerably different from the 2–12 keV solar x-ray flux, monitored simultaneously with the

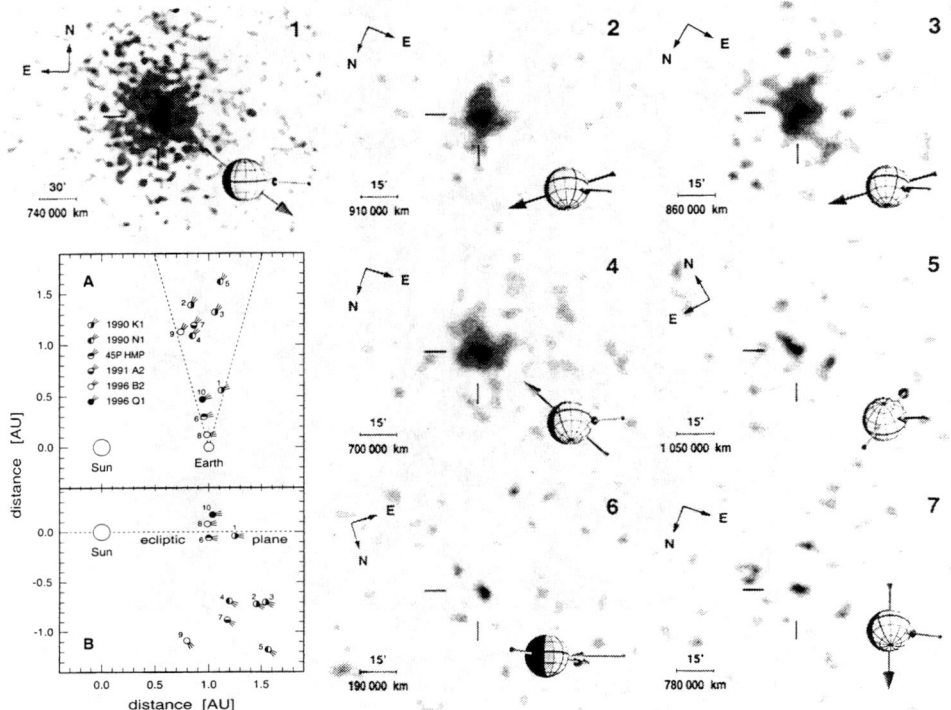

FIGURE 2. X-ray images of the comets detected in the ROSAT all-sky survey 1990/91 in the energy range 0.1 to 0.4 keV. The images were accumulated in the rest frame of each comet (the position of the nucleus according to the ephemeris is marked) and aligned with respect to the direction of the sun, which is to the right. They were corrected for exposure variations across the field and smoothed with a gaussian filter with $\sigma = 2'$. The directions toward North and East are marked in the upper left corner; linear and angular scales are given at lower left. The illumination of the sphere at lower right indicates the phase angle (angle sun-comet-observer). An arrow penetrating the sphere shows how the comet was moving with respect to the solar system barycenter; the direction of the solar wind as seen from the moving comet is indicated by an arrow touching the sphere. The diagram at left shows the position of the comets with respect to the sun and Earth (**A**), and to the ecliptic plane (**B**) during the observations. The dashed lines in (**A**) enclose the ROSAT observing window. Different symbols indicate different comets, identified in (**A**). Numbers at the comet symbols correspond to the numbers in the comet images. Additional numbers indicate the position of comet C/1996 B2 (Hyakutake: 8, 9) and C/1996 Q1 (Tabur: 10) when x-ray emission was detected in pointed ROSAT observations.

Reprinted with permission from K. Dennerl et al., *X-Ray Emissions from Comets Detected In the Röntgen X-Ray Satellite All-Sky Survey* , SCIENCE, **277**, pp 1625-1630, ©1997 American Association for the Advancement of Science.

GOES-8 satellite. Furthermore, the observed morphology could be matched only by postulating a very specific spatial distribution of such particles. The suggestion that the X-ray emission was caused by collisions between cometary and interplanetary dust particles [4] failed to reproduce the observed intensity and morphology. An interaction of the comet with the solar wind, similar to a terrestrial aurora [5], was unlikely, because no reconnection events in the tail were observed. Other interactions with the solar wind, either with the protons, heavier ions, energetic electrons, or its magnetic field, were considered as an alternative cause for the x-ray emission, but no mechanism was found at this early stage that explained the observed properties in a straightforward way. In view of these difficulties there was also the possibility that Hyakutake had been observed in an unusual "x-ray bright" state (of unknown origin), and that such intense x-ray emission would not be a general property of comets.

Comets: a New Class of X-Ray Sources

On the observational side, the fact that one comet was now known to be capable of emitting x-rays, triggered a systematic search in ROSAT archival data: in 1990/91 ROSAT had performed the first survey of the whole sky with an imaging x-ray telescope. There was the possibility that comets might have crossed the field of view and left traces in the data. For this search, all of the comets that were known to have passed their perihelion between 1985 and 1997 were selected [9]. Of these 214 comets, 25 turned out to have been in the ROSAT field of view when they were at heliocentric distances of less than three astronomical units (AU). To obtain the maximum detection efficiency, the search was performed in the rest frame of each comet. This was possible since the arrival time and the detector coordinates of each detected x-ray photon were recorded. Thus, the apparent proper motion of the comet as seen from the satellite could be tracked, given the satellite position together with the comet ephemeris.

The search resulted in the discovery of x-ray emission from four comets, seen on seven occasions at different states [2]. These comets are C/1990 K1 (Levy), C/1990 N1 (Tsuchiya-Kiuchi), 45P/Honda-Mrkos-Pajdušáková and C/1991 A2 (Arai). They were optically 300 to 30 000 times fainter than Hyakutake. In the case of comet C/1991 A2 (Arai), the x-ray observations took place even six weeks before the comet was (optically) discovered.

These findings establish comets as a new class of celestial x-ray sources and allow, for the first time, class properties to be studied over a wide optical brightness range. All of the comets that were observed at heliocentric distances of less 2 AU and with total visual magnitude $m_1 < 12$ mag were detected in x-rays (Fig. 2). X-ray emission was revealed as a general product of cometary activity.

Observational Clues to the Origin of the X-Ray Emission

The ROSAT all-sky survey was performed with a Position Sensitive Proportional Counter (PSPC [10]) in the focal plane of the x-ray telescope. The unlimited field of view provided by the scanning mode together with the low instrumental background of the PSPC and its high sensitivity to soft x-rays makes it well suited for studying the morphology of faint extended objects with soft spectra.

Most of the x-ray images show emissions that are roughly spherically symmetric or elongated perpendicular to the sun–comet line, with the brightest peak usually being offset sunward by some 10^4 km, reminiscent of a bow shock structure. In the case of comet Levy, it is possible to trace the x-ray emission out to a radial distance of at least 2×10^6 km, while in none of the images evidence for x-ray emission from the plasma or dust tail is found. There is no clear dependence of the x-ray morphology on the velocity vector of the comet.

All the information on the x-ray spectrum of comet Hyakutake had to be derived from a comparison of its flux in two different spectral ranges [1]. The comets found in the ROSAT all-sky survey, however, were observed with a proportional counter. With its energy resolution of $\Delta E/E = 0.43 \cdot \sqrt{0.93/E}$ [keV] over a bandpass of 0.1 – 2.4 keV, the ROSAT PSPC allows considerably improved spectroscopic studies of the cometary x-ray emission. It turned out that in all the seven observations the spectrum was similarly soft: about 95% of the detected photons were at $E < 0.4$ keV. This corresponds to 83% of the energy flux in the total PSPC bandpass. While certain spectral models can be definitively ruled out, for example a line spectrum consisting only of oxygen and carbon K_α fluorescence lines, the softness of the x-ray emission reduces the usable spectral bandwidth and does not allow to determine the incident spectrum in a unique way. For comet Levy, where the largest number of photons were recorded, acceptable fits can be obtained for a thermal bremsstrahlung spectrum ($kT = 0.23 \pm 0.06$ keV) and emission from a hot optically thin plasma in thermal equilibrium (Raymond and Smith [11]) with solar abundance ($kT = 0.12 \pm 0.02$ keV). Formally a power law (photon index $\alpha = -3.0 \pm 0.3$) also fits the data, but the residuals indicate systematic deviations.

Spectral analysis with a thermal bremsstrahlung model, which provides a simple, straightforward method for estimating the characteristic temperature, shows that the temperatures of all comets are consistent with a mean value of $kT = 0.23 \pm 0.04$ keV, independent of heliocentric distance, ecliptic latitude, or relative velocity between the comet and the solar wind. While a physically more appropriate model will be described in the next section, the fact that a similar spectral shape is observed among comets that differ in their x-ray luminosity by more than two orders of magnitude, hints to a common, stable emission mechanism.

Another clue to the origin of the x-ray emission comes from a comparison of the x-ray luminosity L_x with the optical luminosity L_opt (Fig. 3). L_x is about proportional to L_opt, with $L_\mathrm{x} \approx 10^{-5} \cdot L_\mathrm{opt}$, although there are differences in the $L_\mathrm{x}/L_\mathrm{opt}$ ratio

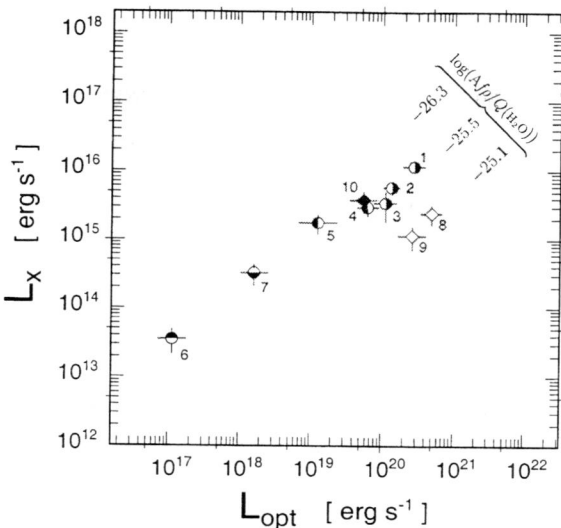

FIGURE 3. Comparison between the optical and x-ray luminosity. The symbols and numbers are the same as in Fig. 2. Shaded regions illustrate the slope for constant L_x/L_{opt} ratios, for different dust-to-gas ratios (expressed by the quantity $\log(Af\rho/Q(H_2O))$). The relative gas content increases from lower right to upper left.

Reprinted with permission from K. Dennerl et al., *X-Ray Emissions from Comets Detected In the Röntgen X-Ray Satellite All-Sky Survey*, SCIENCE, **277**, pp 1625-1630, ©1997 American Association for the Advancement of Science.

between different comets. These differences are not related to the solar activity cycle: comets in the all-sky survey were observed at the solar maximum, while the recent comet observations took place during solar minimum. The differences may, however, be related to the ratio between the dust and gas in the coma. This ratio is usually different from comet to comet, and can also exhibit temporal variations for one specific comet. It depends on how much dust is embedded in the sublimating ice layer on the cometary nucleus. The optical luminosity of a comet is mainly determined by the amount of dust in the coma and tail, which scatters sunlight and appears almost white. The gas is only visible due to its fluorescent emission, which makes the coma appear green and the plasma tail blue. As Fig. 3 shows, there is a tendency for L_x/L_{opt} to increase with the relative gas content. Since L_{opt} is controlled mainly by cometary dust, this suggests that L_x is determined by cometary gas.

The positive correlation of the L_x/L_{opt} ratio with the gas/dust ratio between different comets favors the gas coma as the target for the interaction that is responsible for the x-ray emission. This is further supported by the fact that the x-ray emission can be traced over several 10^6 km, because only cometary gas, in particular hydrogen, is known to extend over such distances. Since no correlation between the x-ray intensity of a comet and the solar x-ray flux is observed, it is unlikely that solar x-ray radiation is primarily responsible for the x-ray emission. Another argument against solar x-rays is the fact that no x-ray emission is observed from the tail. All observational evidence favors the idea that the x-ray emission is caused by an interaction between the solar wind and the gas in the coma. Similar morphologies and spectral shapes indicate that this interaction is basically the same in all cases, and that it is independent of the distance and velocity of the comet.

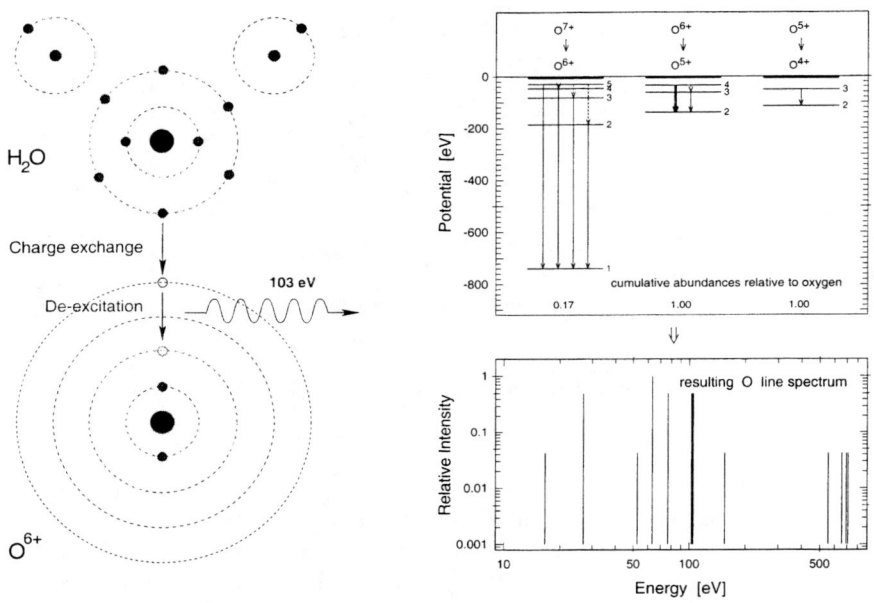

FIGURE 4. Example of the charge exchange process in a comet: In the diagram at left, an O^{6+} ion of the solar wind encounters a neutral water molecule of the comet and captures one of its outer electrons. The captured electron populates predominantly the $n = 4$ state, thereby producing an excited O^{5+} ion. De-excitation can take place directly from $4 \to 2$ (as shown here; see also the thick transition arrow and the corresponding thick emission line in the diagrams at right), or by two transitions, from $4 \to 3$ and $3 \to 2$, and release photons of 103.5 eV, or 26.8 eV and 76.7 eV, respectively. Together with the other O^{5+} ions in the solar wind, the resulting O^{5+} ion is again able to capture an electron and emit a photon of 63.3 eV via the $3 \to 2$ transition. Due to the cumulative abundance effect this line is the strongest of all oxygen charge exchange lines in the extreme UV and soft x-ray range. Additional lines arise from charge exchange with solar wind O^{7+} ions.

CHARGE EXCHANGE

The most straightforward process for this interaction is charge exchange between highly charged heavy ions in the solar wind and cometary neutrals [12]. Among the protons, electrons and alpha particles, the solar wind also contains a small fraction, about 0.1%, of heavier particles in highly charged states, such as C^{6+}, N^{6+}, O^{6+}, Ne^{8+}, Si^{9+}, Fe^{11+}. They obtain this high degree of ionization in the hot solar corona, which has a temperature of several million degrees, before they leave the sun at some hundred kilometers per second. On their trip through the solar system in the tenuous solar wind, these ions usually have no chance to capture the missing electrons. They remain in the highly ionized state until they hit sufficiently dense matter, e.g. a comet, where electrons are available in large numbers, mostly bound in neutral water molecules or oxygen atoms. When solar wind ions capture such electrons, they attain highly excited states and radiate a large fraction of the excitation energy in the extreme ultraviolet and x-ray range (Fig. 4). The ionized cometary molecule or atom may subsequently capture a free electron from the solar wind and return to its neutral state.

Thus, the comet acts like a catalyst: it allows the positive and negative charges in the solar wind to recombine, without being significantly altered in this process. An attractive feature of this model is that the heavy ions are capable of storing the high energies available in the solar corona and of transporting them over large distances. It offers a straightforward explanation for the unexpectedly high radiation temperatures of several million degrees that we observe in comets: what we see can be considered as part of the solar corona, frozen in the charged ions and transported by them to the comet.

The charge-exchange model makes detailed quantitative predictions of the properties of the x-ray emission, and these can be tested.

X-Ray Spectrum

Charge exchange produces a characteristic line spectrum. Its basic properties can be calculated in the following way [13]: After electrons have been transferred from cometary atoms or molecules to solar wind ions, they populate predominantly states with a certain principal quantum number n, which depends on the charge q of the solar wind ion and the ionization potential I_p of the cometary target [14]:

$$n \leq q \left(2|I_p| \left(1 + \frac{q-1}{2\sqrt{q}+1} \right) \right)^{-1/2}$$

where I_p is measured in atomic units (1 au = 27.2 eV). Since the ionization potentials of, e.g., H_2O (12.6 eV), O (13.6 eV), OH (13.8 eV) and H (13.6 eV) are very similar, the chemistry of the comet does not influence the x-ray spectrum significantly. An approximate de-excitation spectrum can be calculated by considering

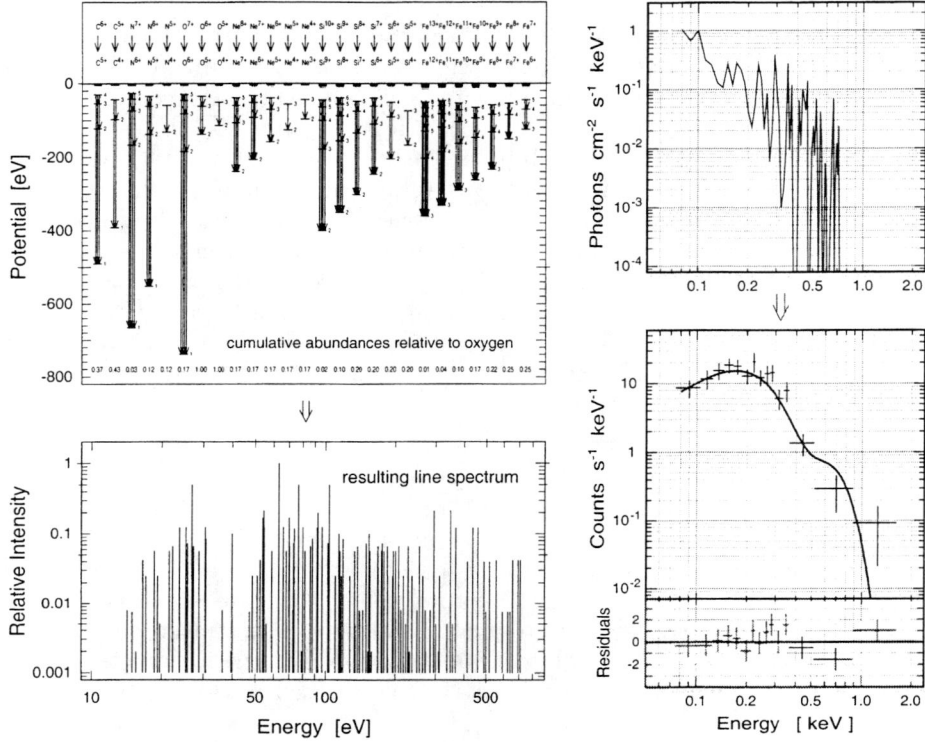

FIGURE 5. Charge exchange spectrum. The diagram at upper left summarizes the individual transitions which are expected to dominate the charge exchange process. By weighting these transitions with the (cumulative) abundances of the solar wind ions, the corresponding emission line spectrum at lower left is obtained. In the diagrams at right, this spectrum is compared with the ROSAT PSPC x-ray spectrum of comet Levy. All the relative line intensities were fixed according to the theoretical prediction, and only a global intensity normalization was allowed as a free parameter in the fit. The photon spectrum for the best fit, with a resolution of 10 eV, is shown at top. In the diagram below, the thick line is the pulse height distribution which would result from this photon spectrum after convolution with the ROSAT PSPC detector response. The crosses (with 1σ error bars) represent the observed pulse height distribution, which is well described by the charge exchange spectrum. This can also be seen in the residual deviations of the observed data (crosses with 1σ error bars) from the model, plotted at bottom. The deviations are statistically not significant, as the fit yields $\chi^2 = 11.5$ for 14 degrees of freedom. Thus, the observed pulse height distribution is consistent with that expected from charge exchange with a probability of 65%.

only the quantum transitions from n to the ground state n_0 in a single step, and, with the same probability, in two steps via an intermediate state (Fig. 4). Then, neglecting all fine structure, the line energies result from the difference between the approximate energy levels Z_{eff}^2/n^2, where the effective nuclear charge Z_{eff} is calculated from the ionization potential of the ground state n_0. The complete charge exchange spectrum (Fig. 5) is obtained by adding these individual de-excitation spectra, weighted according to the solar wind composition [15] and ionization state [16], and taking the cumulative abundance effect (Fig. 4) into account.

This spectrum contains so many lines that it cannot be resolved with the spectroscopic instruments which are currently available for astrophysical investigations. It is, however, possible to simulate what would be observed from such an incident spectrum with the spectral resolution provided by the ROSAT PSPC. As the diagram at lower right of Fig. 5 shows, the measured ROSAT spectrum of comet Levy agrees well with that expected from charge exchange, and this agreement was obtained by adjusting only the total intensity. In view of the fact that the theoretical charge exchange spectrum was derived as an approximation, by just applying general concepts of atomic physics and taking data on the solar wind from the literature, this is a remarkable result.

Peak X-Ray Surface Flux

Charge exchange in a cometary coma has an interesting property: due to the large cross section ($\sigma_{\text{cxe}} > 2 \times 10^{-15}$ cm^2), all ions lose their highly charged state when they penetrate the inner coma of an active comet. This implies that the peak x-ray surface flux is limited by the supply of heavy ions, and independent of specific properties of the comet. Although the coma is collisionally thick with respect to charge transfer, it is still tenuous enough that self-absorption of x-rays is negligible. Thus, the peak x-ray surface flux should scale directly with the heavy ion flux of the solar wind.

The peak surface x-ray flux $\hat{s}_x$, as seen from the direction of the sun, can be estimated in the following way [2,12]:

$$\hat{s}_x \approx v_{\text{sw}} \, n_{\text{sw}} \, f_h \, f_y \, E_{\text{ion}}$$

with
$$\begin{cases} v_{\text{sw}} \approx 400 \text{ km s}^{-1} & \text{velocity of the solar wind} \\ n_{\text{sw}} \approx 10 \text{ cm}^{-3} & \text{density of solar wind ions (at } r_h = 1 \text{ AU)} \\ f_h \approx 0.001 & \text{fraction of heavy ions} \\ f_y \approx 0.1 & \text{x-ray yield} \\ E_{\text{ion}} \approx 200 \text{ eV} & \text{energy released per heavy ion} \end{cases}$$

The resultant value, $\hat{s}_x \approx 1.3 \times 10^{-5}$ erg cm^{-2} s^{-1} = 8.6×10^{-14} erg cm^{-2} s^{-1} arcmin^{-2}, agrees with that derived from the ROSAT observations (Fig. 6).

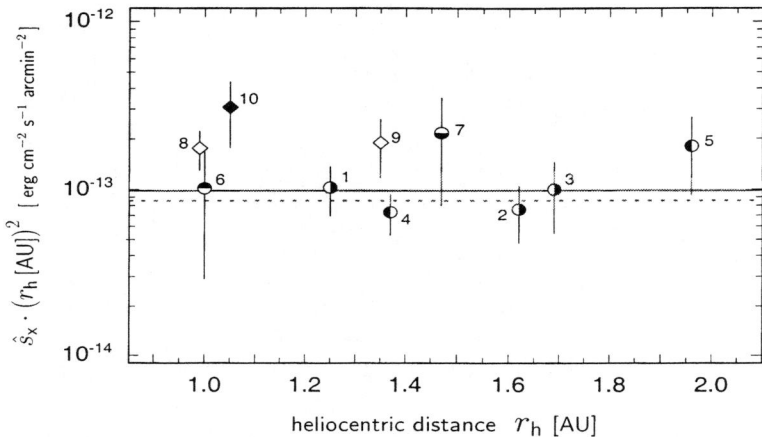

FIGURE 6. Peak surface x-ray flux, normalized to a heliocentric distance $r_h = 1$ AU and a phase angle (angle sun-comet-observer) of zero, for the 10 comet observations identified in Fig. 2. The data are consistent with a mean value $(1.0 \pm 0.1) \times 10^{-13}$ erg cm^{-2} s^{-1} arcmin^{-2} (shaded region) with a confidence of 31%. The dashed line, at 8.6×10^{-14} erg cm^{-2} s^{-1} arcmin^{-2}, is the value resulting from the parameters given in [12]. It is only one standard deviation below the observed mean value.

Reprinted with permission from K. Dennerl et al., *X-Ray Emissions from Comets Detected In the Röntgen X-Ray Satellite All-Sky Survey*, SCIENCE, **277**, pp 1625-1630, ©1997 American Association for the Advancement of Science.

X-Ray Morphology

The x-ray morphology which would result from the charge exchange process has been calculated for comet Hyakutake with several methods [2,12,13,17]. All agree with each other and with the observation (Fig. 7). The crescent shape is the result of the large cross section with respect to charge transfer, which causes the energy stored in the ions to be consumed before they reach the innermost part of the coma. Such an exhaustion was observed with Giotto near comet 1P/Halley [18,19]. Since most of the x-rays are emitted before the solar wind flow is significantly perturbed by the comet, a simple model, which ignores any hydrodynamic effects, provides already a remarkably good morphological description. The fact that the ions are discharged after having passed the coma also explains why no x-rays are observed from the (antisolar) cometary tail.

Temporal Variability

A characteristic feature of the x-ray emission from comets is pronounced variability on time scales of hours and less. It is not unlikely that this is caused by fluctuations in the heavy ion content of the solar wind. The helium content of the solar wind is known to be highly variable [20], and the oxygen flux is strongly correlated with the helium flux [21]. No such observations, however, have so far

FIGURE 7. Comparison of the x-ray morphology of comet Hyakutake, observed with the ROSAT HRI [1] (left), with that expected from the charge-exchange process. The frame in the middle shows the result of a magnetohydrodynamic simulation [17], which is very similar to that of a hydrodynamic treatment [13]. The image at right was obtained by simply irradiating the r^{-2} density distribution in the cometary coma with a parallel beam of solar wind ions, ignoring any hydrodynamic effects [2]. Even such a simplified treatment of the charge exchange process reproduces the general properties of the observed x-ray image remarkably well.

been made sufficiently near to a comet, and so this final test of the charge-exchange model will have to await *in-situ* measurements of the heavy ion flux at a comet, accompanied by simultaneous x-ray observations.

OTHER MODELS

Charge exchange is not the only process which has been suggested to explain the origin of the cometary x-ray emission. Competing theories state that they can reproduce the observed properties as well, by scattering of solar x-rays on cometary dust [8], or by bremsstrahlung x-rays and characteristic x-rays induced by nonthermal energetic electrons [22–24], which are created either by magnetic field reconnection [25] or by lower hybrid waves [26,27]. Also collisions between cometary and interstellar dust grains [4] or between very small dust grains in the inner coma [28] may contribute to the x-ray flux.

From the observational point of view, the ROSAT data indicate that charge exchange is the dominant emission mechanism. EUVE observations of the comets C/1996 B2 (Hyakutake), C/1995 O1 (Hale-Bopp), and 6P/d'Arrest support this conclusion [3,29]. Only the BeppoSAX observation of C/1995 O1 (Hale-Bopp) is interpreted in a different way: by scattering of solar x-rays on cometary dust [30].

It seems that none of the many models which have been suggested can be definitively ruled out. On the contrary, it is even likely that these processes do occur and do contribute to the cometary x-ray emission. The question, however, is to which extent. An answer to this question is very difficult, because many of the suggested processes depend on parameters which are not or only poorly known. Consequently, estimates on the efficiency in producing x-rays are quite uncertain, and allow only global statements to be made. Clear predictions on observable x-ray properties, such as the intensity, spectrum, morphology, or variability, however, are necessary in order to verify or falsify a model. In this context, the charge-exchange process has the advantage that the parameters are known well enough to make such predictions. It has passed all the tests, which have been applied already, so successfully that, from an observational point of view, there is no strong need for additional mechanisms.

CONCLUSIONS

X-ray emission from comets, first discovered in C/1996 B2 (Hyakutake) [1], is now well established as a general feature of these objects [2]. There is a lot of evidence that this emission is a direct consequence of charge exchange between highly charged heavy ions in the solar wind and cometary neutrals [12,13]. Although it is not yet clear to what extent other processes contribute to the total x-ray output of comets, there is no doubt that charge exchange is an important process for the generation of x-rays in comets.

The discovery of cometary x-ray emission came as a surprise to many scientists, and those who had thought about this possibility earlier did not consider charge exchange as a possible mechanism. Even after the discovery, the potential of charge exchange was overlooked for some time. Obviously the importance of this effect for the generation of soft x-rays was underestimated. It may therefore be worthwhile to consider the presence of this effect also in other astrophysical environments, where highly charged heavy ions are created and interact with extended clouds of relatively cool matter.

One such example may occur in our immediate environment: the solar wind may interact with the extended upper atmosphere of Earth in a similar way as it does with the coma of a comet, causing the sunward side of the geocorona to glow in x-rays. From the low ROSAT orbit, this glow would appear as a temporally variable soft component of the diffuse x-ray emission. This might explain the occasional increases in the overall soft x-ray background with a typical duration of up to eight hours, which were observed during the ROSAT all-sky survey. It might also provide an explanation for the soft x-rays which were observed with ROSAT from the direction to the dark side of the moon, at about one percent of the flux of the bright side [31].

In comets, the charge-exchange process would lead to interesting applications. Since the peak x-ray surface flux scales directly with the highly charged heavy ion content in the solar wind, comets with a sufficiently dense coma could be used as "standard candles" for probing this specific property of the solar wind. By high resolution spectroscopy, it would even be possible to determine the chemical composition and ionization structure in the solar wind, since each ion leaves its characteristic signature in the charge-exchange spectrum. Observations with high spatial resolution would allow a two-dimensional monitoring of solar wind density fluctuations. Comets could thus be used as natural space probes for sampling properties of the solar wind at various positions in the inner solar system, which would otherwise be only accessible by *in-situ* measurements.

REFERENCES

1. Lisse, C. M., Dennerl, K., Englhauser, J., Harden, M., Marshall, F. E., Mumma, M. J., Petre, R., Pye, J. P., Ricketts, M. J., Schmitt, J., Trümper, J., and West, R. G., *Science* **274**, 205–209 (1996).
2. Dennerl, K., Englhauser, J., and Trümper, J., *Science* **277**, 1625–1630 (1997).
3. Mumma, M. J., Krasnopolsky V. A., and Abbott, M. J., *Astrophys. J.* **491**, L 125–128 (1007).
4. Ibadov, S., *Icarus* **86**, 283–288 (1990).
5. Hudson, H. S., Ip, W.-H., and Mendis, D. A., *Planet. Space Sci.* **29.12**, 1373–1376 (1981).
6. Trümper, J., *Adv. Space Res.* **2.4**, 241 (1983).
7. Voges, W., Proc. of Satellite Symp. 3: Space Sciences with particular emphasis on High-Energy Astrophysics, from the 'International Space Year' Conference held in Munich, Germany, 30 March – 4 April 1992 (ESA ISY-3, July 1992, p. 9) (1992).
8. Wickramasinghe, N. C. and Hoyle, F., *Astrophysics and Space Science* **239**, 121–123 (1996).
9. Catalogue of Cometary Orbits, 11th Edition 1996, Central Bureau for Astronomical Telegrams / Minor Planet Center (1996).
10. Pfeffermann, E., Briel, U. G., Hippmann, H., Kettenring, G., Metzner, G., Predehl, P., Reger, G., Stephan, K.-H., Zombeck, M. V., Chappell, J., and Murray, S. S., *Proc. SPIE* **733**, 519 (1986).
11. J. C. Raymond and B. W. Smith, *Astroph. J. Suppl. Ser.* **35**, 419 (1977).
12. Cravens, T. E., *Geophysical Research Letters* **24.1**, 105–108 (1997).
13. Wegmann, R., Schmidt, H. U., Lisse, C. M., Dennerl, K., and Englhauser, J., *Planet. Space Sci.* **46/5**, 603–612 (1998).
14. Mann, R., Folkmann, F., and Beyer, H. F., *J. Phys. B: At. Mol. Phys.* **14**, 1161–1181 (1981).
15. Bochsler, P., *Physica Scripta* **T18**, 55–60 (1987).
16. Bame, S. J., *Solar Wind*, NASA SP-308, 1972, p. 535.
17. Häberli, R. M., Gombosi, R. I., De Zeeuw, D. L., Combi, M. R., and Powell, K. G., *Science* **276**, 939 (1997).
18. Neugebauer, M., Goldstein, R., Goldstein, B. E., Fuselier, S. A., Balsiger, H., and Ip, W.-H., *Astrophys. J.* **372**, 291–300 (1991).
19. Fuselier, S. A., Shelley, E. G., Goldstein, B. E., Goldstein, R., Neugebauer, M., I, W.-H., Balsiger, H., and Rème, H., *Astrophys. J.* **379**, 734–740 (1991).
20. Brandt, J. C., *Introduction to the Solar Wind*, Freeman, San Francisco, (1997).
21. Bochsler, P., *Physica Scripta* **T18**, 55–60 (1987).

22. Northrop T. G., Lisse C. M., Mumma M. J., and Desch, M. D., *Icarus* **127**, 246–250 (1997).
23. Northrop T. G., *Icarus* **128**, 480–482 (1997).
24. Uchida M., Morikawa, M., Kubotani, H., Mouri, H., *Astrophys. J.* **498**, 863–870 (1998).
25. Brandt J., Lisse C., Yi Y., *Bull. Am. Astron. Soc.* **189**, 25.05 (1996).
26. Bingham R., Dawson J. M., Shapiro V. D., Mendis D. A., Kellet B. J., *Science* **275**, 49–51 (1997).
27. Shapiro V. D., Bingham, R. Dawson J. M., Dobe Z., and Mendis D. A., *J. Geophys. Res.* submitted.
28. Ip W. H., and Chow V. W., *Icarus* **130**, 217–221 (1997).
29. Krasnopolsky V. A., Mumma M. J., Abbott M., Flynn B. C., Meech K. J., Yeomans D. K., Feldman P. D., and Cosmovici C. B., *Science* **277**, 1488–1491 (1997).
30. Owens A., Parmar A. N., Oosterbroek T., Orr A., Antonelli L. A., Fiore F., Schulz R., Tozzi G. P., Maccarone M. C., and Piro L., *Astrophys. J. Lett.* **493**, L 67 (1998).
31. Schmitt, J. H. M. M. Snowden, S. L., Aschenbach, B., Hasinger, G., Pfeffermann, E., Predehl, P., Trümper, J., *Nature* **349**, 583–587 (1991).

Ultrafast Structural Studies on Biological Molecules by X-Rays

Janos Hajdu, Richard Neutze, Remco Wouts, and David van der Spoel

Department of Biochemistry, Uppsala University, Box 576, S-751 23 Uppsala, Sweden

Abstract. Dramatic changes can be expected within the next few years in science that depends on synchrotron radiation today. An analysis of available data on short-pulse, high-intensity x-ray and electron sources indicate possibilities for imminent developments with x-ray free-electron lasers (driven primarily by the high energy physics community), and with table-top femtosecond x-ray sources based on laser-induced plasmas and wakefield acceleration. New sources could generate femtosecond x-ray pulses with as much as 12 orders of magnitude increase in peak brilliance and power over third-generation synchrotron storage rings. Such developments would create revolutionary new research opportunities in condensed matter physics, biology and medicine. Some of the possibilities are discussed here.

FEMTOSECOND TIME RESOLUTION IN X-RAY DIFFRACTION EXPERIMENTS

Diffraction exposures are limited by the pulse length of available x-ray sources. Pulses of 50–500 picosecond duration can be produced by synchrotrons, and picosecond-subpicosecond pulses have been reported from laser-induced plasma sources. Crystallographic data with temporal resolutions of nanoseconds have been recorded (see, e.g., Refs. [1–5]) and interpretable, near-picosecond macromolecular x-ray diffraction data collected [6,7].

While picosecond exposures are of considerable interest, basic chemical steps such as the breaking of a bond or coherent molecular motions following photo-excitation are usually over within a few hundred femtoseconds [8,9]. Given the wealth of fundamental chemical and biological phenomena which can be studied through femtosecond photo-excitation of macromolecules and the definitive structural information which is accessible only through x-ray crystallography, it is natural to anticipate the merger of these two fields [10,11].

Two options are considered here for the synthesis of femtosecond spectroscopy and x-ray diffraction.

Femtosecond Time Resolution with Picosecond x-ray Pulses

The standard procedure in time-resolved x-ray crystallography is to record a series of diffraction snapshots at known times, following reaction initiation [12–15]. Structure factors are then calculated by treating each temporally distinct data set as an instantaneous 'photograph' of the average electron density from which a 'movie' of the process may be constructed through discrete Fourier transformation.

Any light-initiated reaction requires a finite time for the femtosecond laser pulse to propagate through the body of a crystal and excite molecules. Thus the time evolution of the reaction at separated lattice points is not synchronised but is out of phase by a position dependent quantity as the pulse sweeps through the body of the crystal like a Mexican wave in a foot ball stadium. Recognising this lack of global crystal synchronisation, topographic x-ray diffraction was proposed [10] to achieve femtosecond temporal resolution from reflection profiles with x-ray exposures of picosecond or longer duration. Consider a 0.3 mm long crystal (most crystals used in x-ray diffraction experiments are about this size) onto which falls a picosecond long x-ray pulse so that the whole body of the crystal is bathed in the x-ray beam. The diffracted beam will produce a topographic image of the crystal. When white x-rays are used in the pulse, the topogram will have reflections arranged in a Laue pattern. From high symmetry crystals, a single exposure like this would be sufficient to obtain about 80% or more of the unique reflection set and obtain a structure.

Synchronized with the ps x-ray pulse, a fs-long (i.e. 0.3 μm) laser excitation pulse is fired at the crystal so that the direction of the UV-VIS laser pulse is perpendicular to the direction of the x-ray pulse. With crystal dimensions given above, the time required for the UV-VIS laser pulse to pass through the 0.3 mm crystal will be about 1 ps. This is also the length of the x-ray pulse. The profile of each reflection recorded by the ps x-ray pulse in the topogram will have a certain time component along the direction of the femtosecond laser pulse. The analysis of the reflection profiles on the Laue photograph (topogram) along this direction may give a movie with sub-ps time resolution.

The case is more complicated than outlined here (for details see Ref. [10]) but this possibility is worth of further scrutiny as diffraction experiments on the picosecond-femtosecond time scale could open up a new era in structural biology and chemistry. A few points to consider:

1. Each slice of the reflection profiles (sliced perpendicular to the direction of the UV-VIS laser pulse) contains information from a mixture of "old" and "new" structures. The mixing is different at different positions. However, each reflection pulse while "in flight" contains pure and unperturbed temporal information on the crystal. An averaging of the structures takes place when the reflection hits the detector and the wave fronts reporting from the different time points pile up on top of each other. If coherent x-rays are used, the coherence of the beam could perhaps be used to tackle this problem.

2. Apart from averaging, there will be a speed difference between the UV and x-ray pulses due to the different refractive indices for the light and x-ray pulses in the crystal. The Heisenberg principle is lurking in the background, e.g. the fs UV-VIS laser pulse is not monochromatic anymore. This will result in some loss of temporal resolution.

3. Currently, the electronic "jitter" in high performance circuits is around 0.5-2 ps. This makes the timing of the arrival of the UV-VIS laser and x-ray pulses difficult but a detector mounted from the z-direction (x = x-ray, y = laser, z = fluorescence detector) could measure time differences between the two pulses by measuring scattered radiation (both laser light and x-ray) from the crystal.

Picosecond x-ray pulses are currently available, and this approach can be used to study femtosecond reaction dynamics at atomic resolution on crystals of both small- and macromolecules at time scales characteristic of transition state lifetimes.

Femtosecond X-ray Pulses

Consensus in the physics community shows that the future of accelerator based research is related to the construction of linear electron/positron accelerators. Such accelerator could be used to feed several x-ray lasers delivering femtosecond light pulses in the wavelength range down to 1.0 Å (see, e.g. Ref. [16]). The average spectral brilliance and the gain in peak brilliance of such free-electron lasers can be 10–15 orders of magnitude higher than those of 3rd generation synchrotron radiation sources. The radiation is polarised and exhibits a very high degree of transverse coherence.

Other exciting possibilities for creating femtosecond x-ray pulses are offered by (i) table-top x-ray lasers based on wakefield acceleration [17–23], (ii) collision-dominated plasma sources [24–31], (iii) Compton scattering of laser photons on electron bunches [32], and (iv) 90-degree Thomson scattering [33,34]. Recent developments show that attosecond pulses may be produced [35–40].

Much of recent attention has been centered on laser wakefield accelerators. When a pulse of short, high power laser light interacts with some inert gas, the laser strips atoms of their electrons, and creates a wave within the resulting plasma that propagates along the laser pulse at nearly the speed of light. Trailing behind the laser pulse is a so called "wakefield" (similar to the wake behind a fast moving ship). The wakefield captures electrons from the plasma gas and accelerates them to energies of many MeV in the direction of pulse propagation. The electron beam accelerated this way interacts with the surrounding plasma, producing a coherent beam of radiation in the "plasma undulator."

New Opportunities in Structural Sciences with Intense Femtosecond X-ray Pulses

The extreme short duration of individual pulses from new sources will allow for novel experiments. The high peak power at short wavelengths offers new possibilities to study highly nonlinear physical phenomena:

1. Short pulses at high brilliance: Snapshots of physical and chemical processes on the fs time scale.

2. Tunable wavelength: Spectroscopic studies on gases, liquids, solids and plasmas, and snapshots of atoms at excited states.

3. Time and space coherence: Microscopy, holography, diffraction experiments.

These sources may allow observation of coherent reaction dynamics in three-dimensions or the imaging of single molecules at atomic resolution. While these are truly exciting possibilities, there will be challenges to overcome in their pursuit. A numerical analysis of diffraction of subpicosecond x-ray pulses from physical crystals shows, for instance, that the three-dimensional nature of the diffracting crystal and the finite size of the x-ray pulse will lead to a considerable broadening of the diffracted femtosecond x-ray pulse [41]. At extreme peak powers, this phenomenon may appear in combination with highly non-linear optical effects.

Implications for Structural Biology

The study of protein molecules has witnessed a shifting from studies on individual proteins to macromolecular complexes, where there have been some notable successes. Many such target complexes are difficult to crystallise or produce only very small crystals. Complexes of membrane proteins are perhaps the most notorious in this respect. Today, the bottle neck in atomic resolution imaging is a fundamental need for crystals. This limits the scope of detailed structural analysis to complexes which can be tricked into forming stable single crystals. However, many biologically important targets cannot be crystallised, and in order to extend structural studies to these systems, we need to undertake a radical reassessment of our current concepts, and to initiate theoretical and experimental studies to explore new and untried frontiers at the physical limits of macromolecular imaging. For this end, a joint European research network has been set up in Uppsala with funding from the EU-Biotech programme [42]. Here we describe some aspects of these studies. A more comprehensive description of the results will be published elsewhere.

The Radiation Damage Barrier

Radiation damage prevents the structural determination of single biological molecules at atomic resolution in conventional x-ray or electron scattering experiments [43,44]. Although cooling can slow down chemical deterioration, it cannot eliminate the accumulation of defects in the sample within the time needed to complete conventional measurements. The main question to address is this: What are the possibilities of obtaining an interpretable diffraction image at atomic resolution using ultra short bursts of x-rays before information on the starting structure is lost and the sample is destroyed by the intense x-ray pulse?

It is necessary to determine theoretically the smallest sample size from which it may be physically possible to obtain useful structural information. This may be a single molecule (could we get rid of the crystal in crystallography?), a closed periodic cluster of molecules (e.g. an oligomeric protein, or a virus), or an open periodic structure (e.g. a nanocrystal or a microcrystal). As the potential benefits from studies in this field are substantial, a comprehensive and rigorous investigation of these ideas is necessary and timely.

Scattering with ultrashort x-ray pulses

Preliminary calculations based on conservative performance estimates and on optimal boundary conditions indicate that extremely intense fs/as photon pulses may allow the direct imaging of Fourier transforms of single macromolecules, viruses or larger structures (e.g. intact cells) without the need to amplify scattered radiation through Bragg reflections. Resolution under such conditions does not depend on sample quality as in conventional crystallography, but is a function of radiation intensity, pulse duration, wavelength, detector quantum efficiency, spatial resolution, and the speed of sample destruction (i.e. movement).

Design parameters for hard x-ray lasers give pulse durations of 50 fs with around 10^{12-15} x-ray photons per pulse. Such coherent radiation pulses could, in principle, be focused to the diffraction limit. Under such conditions, the electric field strength in the focal spot would rapidly create a high temperature plasma from matter [45]. We estimate that with a focal spot diameter of 0.05 micrometer, the number of ionisation events would be around one per atom within the duration of a 50 fs x-ray pulse. Initial molecular dynamics calculations indicate that if all charges are simultaneously created (the worst case scenario), then a protein molecule would still retain its original structure for about 8-10 fs (with a root mean square displacement of atoms less than 0.2 Å compared to the starting structure). Under less severe conditions, we estimate that the sample life could extend to 50 fs, i.e. the duration of the pulse. A small protein molecule (e.g. lysozyme) would scatter x-rays to about 30–50 Å resolution with $I_{\text{scattered}} > 3\,\sigma\,(I_{\text{scattered}})$ under these conditions. This suggests that averaging a few hundred images collected on separate molecules could improve the signal to noise ratio to extract structural data far beyond the initial resolution, i.e. to about 1–2 Å resolution. Structural data from most protein crystals do not extend to such high resolution. However, averaging techniques can only be used on "reproducible structures" and events.

Further calculations on the intensity distribution of scattered x-rays from a single protein molecule and from various clusters of this molecule, indicate that regular nanoclusters will very likely produce interpretable diffraction images to atomic resolutions.

Construction and Handling of Macromolecular Nano-Clusters

Studies have been initiated to prepare regular nano-clusters or nano-crystals of proteins using a variety of techniques: (i) recombinant phage antibody systems, (ii) fused multiple binding units for use as core clusters, (iii) crosslinking with bifunctional reagents, (iii) various templates as epitactic surfaces.

When working with nanoclusters or single molecules of proteins, standard techniques for sample manipulation are no longer appropriate. We propose to inject samples into a vacuum chamber at high velocity. The micro-droplets can be dried to a desired level before entering the measurement area. The latest separation technology is required for preserving non-covalent macromolecular assemblies involving nano-flow sample introduction at low temperatures. Initial studies show that native structures can be retained under these conditions in the gas phase. An exciting possibility is to select and store desired particles in ion trapping devices, then inject them into the beam.

As the potential benefits of ultrafast imaging are large, we consider that a comprehensive investigation of these ideas is both necessary and timely. Crystallographers are used to imagining things within the confines of crystals. Not a bad system, but imagine producing a femtosecond snapshot of a large membrane protein or a virus or perhaps even a cell at atomic resolution. Visualising protein folding, the workings of a ribosome, observing membrane channels in action, checking which repressors are bound and where on the DNA, looking at the progress of a viral infection in atomic details, or reading out the sequence of the genom of an individual. These are of course dreams at present but we have to check them out as they may offer a chance.

ACKNOWLEDGMENTS

We thank Edgar Weckert, Gyula Faigel, Sven Hovmöller, Jan Davidsson, Karin Markides, and David Barnage for discussions. This work is supported by the Swedish research council NFR and the EC-Biotech Programme (Brussels).

REFERENCES

1. Jamet, F., "Laue diffraction on exploding single crystal alumium with a 300 ns x-ray flash from a tungsten anode," *C. R. Acad. Sci. Paris* **B271**, 714–717 (1970).
2. Johnson, Q., Keeler, R. N., and Lyle, J. W., "X-ray diffraction experiments in nanosecond time intervals," *Nature* **213**, 1114–1115 (1967).
3. Johnson, Q., Mitchell, A. C., Keeler, R. N., and Evans, L., "X-ray diffraction during shock wave compression," *Phys. Rev. Lett.* **25**, 1099–1101 (1970).
4. Larson, B. C., White, C. W., Noggle, T. S., Barhorst, J. F., and Mills, D., "Time-resolved x-ray diffraction measurement of the temperature and temperature gradients in silicon during pulsed laser annealing," *Appl. Phys. Lett.* **42**, 282–283 (1983).
5. Wark, J. S., Woolsey, N. C., and Whitlock, R. R., "Novel measurements of high-dynamic crystal strength by picosecond x-ray diffraction," *Appl. Phys. Lett.* **61**, 651–653 (1992).
6. Srajer, V., Teng, T., Ursby, T., Pradervand, C., Ren, Z., Adachi, S., Schildkamp, W., Bourgeois, D., Wulff, M., Moffat, K., "Photolysis of the carbon-monoxide complex of myoglobin - nanosecond time-resolved crystallography," *Science* **274**, 1726–1729 (1996).
7. Wulff, M., Schotte, F., Naylor, G., Bourgeois, D., Moffat, K., and Mourou, G., "Time-resolved structures of macromolecules at the ESRF: Single-pulse Laue diffraction, stroboscopic data collection and femtosecond flash photolysis," *Nucl. Inst. Meth. Phys. Res.* A**398**, 69–84 (1997).
8. Vos, M. H., Jones, M. R., Hunter, C. N., Breton, J., and Martin, J. L., "Coherent nuclear-dynamics at room-temperature in bacterial reaction centers. *Proc. Natl. Acad. Sci. U.S.* **91**, 12701–12705 (1994).
9. Ludowise, P., Blackwell, M., and Chen, Y., "Femtosecond time-resolved mass and photoelectron spectroscopic study of OClO photodissociation. Coherent energy transfer in a stepwise reaction," *Chem. Phys. Lett.* **273**, 211–218 (1997).
10. Neutze, R., and Hajdu, J. "Femtosecond time resolution in x-ray diffraction experiments," *Proc. Natl. Acad. Sci. U.S.A.* **94**, 5651–5655 (1997).
11. Rischel, C., Rousse, A., Uschmann, I., Albouy, P. A., Geindre, J. P., Audebert, P., Gauthier, J. C., Forster, E., Martin, J. L., and Antonetti, A., "Femtosecond time-resolved x-ray diffraction from laser-heated organic films," *Nature* **390**, No.6659, 490–492 (1997).
12. Hajdu, J., Acharya, K. R., Stuart, D. I., McLaughlin, P. J., Barford, D., Klein, H. and Johnson, L. N., "Time-resolved structural studies on catalysis in the crystal with glycogen phosphorylase b," *Biochem. Soc. Trans.* **14**, 538–541 (1986).
13. Hajdu, J., Acharya, K. R., Stuart, D. I., McLaughlin, P. J., Barford, D., Oikonomakos, N. G., Klein, H. and Johnson, L. N., "Catalysis in the crystal: Synchrotron radiation studies with glycogen phosphorylase b," *EMBO J.*, **6**, 539–546 (1987).
14. Hajdu, J., Machin, P. A., Campbell, J. W., Greenhough, T. J., Clifton, I. J., Zurek, S., Gover, S., Johnson, L. N. and Elder, M., "Millisecond x-ray diffraction: First electron density map from Laue photographs of a protein crystal," *Nature* **329**, 178–181 (1987).
15. Hajdu, J. and Andersson, I., "Fast X-ray Crystallography and Time-Resolved Struc-

tures," *Ann. Rev. Biophys. Biomol. Struct.* **22**, 467–498 (1993).
16. Brinkmann, R., Materlik, G., Rossbach, J., Wagner, A. (editors), "Conceptual design of a 500 GeV e+e- linear collider with integrated x-ray laser facility," DESY 1997-048 and ECFA 1997-182 (1997).
17. Fraenkel, M., Zigler, A., Faenov, A. Y., Pikuz, T. A., "Generation of intense collimated monochromatic x-ray beam using femtosecond table-top laser," *Physica Scripta* **56**, 571–573 (1997).
18. Li, Y. L., Schillinger, H., Ziener, C., and Sauerbrey, R., "Reinvestigation of the Duguay soft x-ray laser: a new parameter space for high power femtosecond laser pumped systems," *Opt. Commun.* **144**, 118–124 (1997).
19. Nakajima, K., Kando, M., Kawakubo, T., Nakanishi, T., and Ogata, A., "A tabletop x-ray fel based on the laser wakefield accelerator-undulator system," *Nucl. Inst. Meth. Phys. Res.* **A375**, 593–596 (1996).
20. Ogata, A., Nakajima, K., Kozawa, T., and Yoshida, Y., "Femtosecond singlebunched linac for pulse-radiolysis based on laser wakefield acceleration," *IEEE Trans. Plasma Sci.* **24**, 453–459 (1996).
21. Spielmann, C., Burnett, N. H., Sartania, S., Koppitsch, R., Schnurer, M., Kan, C., Lenzner, M., Wobrauschek, P., and Krausz, F., "Generation of coherent x-rays in the water window using 5- femtosecond laser pulses," *Science* **278**, 661–664 (1997).
22. Ting, A., Moore, C. I., Krushelnick, K., Manka, C., Esarey, E., Sprangle, P., Hubbard, R., Burris, H. R., Fischer, R., Baine, M., "Plasma wakefield generation and electron acceleration in a self-modulated laser wakefield accelerator experiment," *Phys. Plasmas* **4**, 1889–1899.
23. Umstadter, D., "Terawatt lasers produce faster electron acceleration." *Laser Focus World* **32**, 101–107 (1996).
24. Eisenberger, P., Suckewer, S., "Subpicosecond x-ray pulses," *Science* **274**, 201–202 (1996).
25. Gauthier, J. C., Audebert, P., Bastiani, S., Geindre, J. P., "Recent progress in the production and use of x-ray radiation above kev by femtosecond plasma," *Ann. Physique* **22**, 61–68 (1997).
26. Gibbon, P., Förster, E., "Short pulse laser-plasma interactions," *Top. Rev. Plasma Phys. Contr. Fusion* **38**, 769–793 (1996).
27. Kalman, P., and Brabec, T., "Generation of coherent hard-x-ray radiation in crystalline solids by high-intensity femtosecond laser-pulses," *Phys. Rev.* **A52**, R21–R24 (1995).
28. Kalman, P., and Brabec, T., "Evolution of coherent hard-x-ray radiation generated in crystalline solids by high-intensity femtosecond laser-pulses," *Phys. Rev.* **A53**, 627–629 (1996).
29. Makeev, N. G., and Rumyantsev, V. G., "Feasible designs of a pulsed high-current plasma x-ray tube," *Instr. Exp. Tech.* **38**, No. 6 Pt. 2, 779–782 (1995).
30. Murnane, M. M., Kapteyn, H. C., Rosen, M. D., and Falcone, R. W. "Ultrafast x-ray pulses from laser-induced plasmas," *Science* **251**, 531–536 (1991).
31. Murnane, M. M., Kapteyn, H. C., Gordon, S. P., and Falcone, R. W. "Ultrashort x-ray pulses," *Appl. Phys.* **B58**, 261–266 (1994).
32. Zholents, A. A., and Zolotorev, M. S., "A proposal for the generation of ultra-short

x-ray pulses," *Nucl. Inst. Meth. Phys. Res. A***358**, 455–458 (1995).
33. Leemans, W. P., Schoenlein, R. W., Volfbeyn, P., Chin, A. H., Glover, T. E., Balling, P., Zolotorev, M., Kim, K. J., Chattopadhyay, S., and Shank, C.V., "X-ray based subpicosecond electron bunch characterization using 90-degree Thomson scattering," *Phys. Rev. Lett.* **77**, 4182–4185 (1996).
34. Schoenlein, R. W., Leemans, W. P., Chin, A. H., Volfbeyn, P., Glover, T. E., Balling, P., Zolotorev, M., Kim, K. J., Chattopadhyay, S., and Shank, C. V., "Femtosecond x-ray pulses at 0.4 angstrom generated by 90-degree Thomson scattering — a tool for probing the structural dynamics of materials," *Science* **274**, 236–238 (1996).
35. Christov, I. P., Murnane, M. M., Kapteyn, H. C., "Generation of single-cycle attosecond pulses in the vacuum ultraviolet," *Opt. Commun.* **148**, 75–78 (1998).
36. Christov, I. P., Murnane, M. M., Kapteyn, H. C., "Generation and propagation of attosecond x-ray pulses in gaseous media," *Phys. Rev. A***57**, R2285–R2288 (1998).
37. Csonka, P. L., Kroo, N., "Methods to generate femtosecond and attosecond electron and x- ray pulses," *Nucl. Inst. Meth. Phys. Res. A***376**, 283–290 (1996).
38. Molchanov, A. G., "Attosecond x-ray pulse generation by relativistic frequency conversion," *Inst. Phys. Conf. Ser.* **151**, 463–465 (1996).
39. Protopapas, M., Lappas, D. G., Keitel, C. H., and Knight, P. L., "Recollisions, bremsstrahlung, and attosecond pulses from intense laser fields," *Phys. Rev. A***53**, R2933–R2936 (1996).
40. Vartak, S. D., and Lawandy, N. M., "Breaking the femtosecond barrier—a method for generating attosecond pulses of electrons and photons," *Opt. Commun.* **120**, 184–188 (1995).
41. Tomov, I. V., Chen, P., and Rentzepis, P. M., "Pulse broadening in femtosecond x-ray diffraction," *J. Appl. Phys.* **83**, 5546–5548 (1998).
42. Hajdu, J., "New Frontiers in Structural Biology" in *Area 6, Structural Biology, EC Biotechnology Programme, 4th Call* (European Commission, Brussel) pp. 34–35 (1998).
43. Henderson, R., "Cryoprotection of protein crystals against radiation-damage in electron and x-ray-diffraction," *Proc. R. Soc., Series B-Biological Sciences* **241**, 6–8 (1990).
44. Henderson, R., "The potential and limitations of neutrons, electrons and x-rays for atomic-resolution microscopy of unstained biological molecules," *Quart. Rev. Biophys.* **28**, 171–193 (1995).
45. Doniach, S., "Studies of the structure of matter with photons from an x-ray free-electron laser," *J. Synchrotron Rad.* **3**, 260–267 (1996).

Atomic Electron Correlations in Intense Laser Fields

L. F. DiMauro*, B. Sheehy*, B. Walker*, P. A. Agostini[†]
and K. C. Kulander[‡]

*Brookhaven National Laboratory, Upton, NY 11973
[†]SPAM, Centre d'Etudes de Saclay, 91191 Gif Sur Yvette, France
[‡]TAMP, Lawrence Livermore National Laboratory, Livermore, CA 94551

Abstract. This talk examines two distinct cases in strong optical fields where electron correlation plays an important role in the dynamics. In the first example, strong coupling in a two-electron-like system is manifested as an intensity-dependent splitting in the ionized electron energy distribution. This two-electron phenomenon (dubbed continuum-continuum Autler-Townes effect) is analogous to a strongly coupled two-level, one-electron atom but raises some intriguing questions regarding the exact nature of electron-electron correlation. The second case examines the evidence for two-electron ionization in the strong-field tunneling limit. Although our ability to describe the one-electron dynamics has obtained a quantitative level of understanding, a description of the two (multiple) electron ionization remains unclear.

INTRODUCTION

Many-body effects form the basis of a problem which is fundamental and central to our understanding of physics. In atomic physics, electronic correlation has been shown to play a prominent role both in the atomic structure and dynamics [1,2]. The study of multielectron atoms in intense laser fields raises similar issues concerning the influence of correlation in multiphoton excitation and ionization. As early as the mid-70s, correlation was considered an important element in explaining the anomalous multiple charge state distributions observed in the multiphoton ionization of alkaline earth atoms [3]. Although this assignment in the end was in error, twenty years of ensuing intense field investigations have produced only a few unambiguous cases where correlation is relevant. The reader is referred to a recent comprehensive review of two electron atoms in intense fields by Lambropoulos et al. [4].

In this paper, we present two intense field experimental studies were the role of electron-electron correlation is important. The first scenario deals with the two-photon ionization of calcium [5], a two-electron like atom, with intense femtosecond light that is "resonant" with a core transition. The fundamental issue is what influence does a strongly driven core excitation have on the outgoing (ionized) photoelectron? In a second case, nonresonant ionization of helium by low energy (1.6 eV) photons results in the anomalous double ionization yield [6] similar in character to that first observed in alkaline earth atoms by Suran and Zapescohyni [3]. However, unlike previous double ionization studies, helium ionization occurs in the strong field limit where the dynamics are dominated by tunneling. Even though much of the ambiguity present in the interpretation of a multiphoton ionization experiment is absent in the tunneling regime, the mechanism responsible for strong-field double ionization remains unclear. One significant difference between the two cases considered here is the external field strength as compared to the field between the valence electron and the core. In the first case, the laser field strength is 10^{-3} a.u. (perturbative) while for helium the amplitude is approaching 1 a.u. (50 V/A) resulting in nonperturbative behavior. Thus in the language of Keldysh [7], calcium ionization occurs in the multiphoton limit ($\gamma > 1$) with strongly-coupled levels while helium tunnel ionizes ($\gamma < 1$). The Keldysh adiabaticity parameter, γ, is defined as the ratio of the laser frequency to the tunneling rate.

Strongly Coupled Two-Electron Atom: The Continuum-Continuum Autler-Townes Effect

Two bound states strongly coupled by an ac-field manifest an energy splitting which is due to the oscillation of population between the states in the presence of the driving field. This phenomenon, well known as the Autler-Townes doublet [8] when probed by a transition to a third level, or the Mollow triplet [9] when probed by resonance fluorescence, is usually *not* observed between a single bound state coupled to the continuum or between two coupled continua. In general, there is no population oscillation, as the breadth of the accessible phase space over which the coupling strength is distributed in the continuum makes excitation out of the initial state essentially irreversible. Thus, the initial state decays exponentially; saturation is reached without any splitting. For example, in multiphoton ionization, as the coupling between ionization continua is increased, only a broader distribution of the photoelectrons among those continua, separated by the photon energy, is achieved [above-threshold ionization (ATI)], with no splitting.

The situation is different for a two-electron atom where coupling between continua can produce a final state splitting if the driving laser field is resonant with some ionic core transition [10–13]. Physically, the reason for this is that the electron-electron interaction transfers the energy shift of the core electron to the outgoing electron. This has been dubbed "coherence transfer" by Ref. [10].

One simple way to understand this effect is to think of the final ionic state as split by the resonant (core)-interaction, thus the outgoing electron sees two asymptotic energy limits separated by the Rabi frequency $\Omega = \mu_\pm \mathcal{E}/\hbar$, where $\mu_\pm$ is the ionic dipole and $\mathcal{E}$ the electric field. To emphasize the fact that it is actually two continua

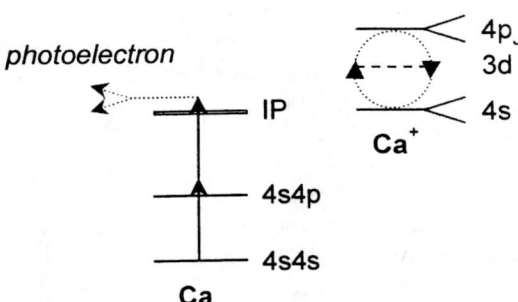

FIGURE 1. Simplified level diagram of calcium showing the 2-photon ionization of the neutral and the $4s \Leftrightarrow 4p_J$ coupled (split) core transition. The splitting of the photoelectron energy due to the continuum-continuum coupling is illustrated by the dotted line.

that are coupled, one can talk about continuum-continuum Autler-Townes splitting. Dynamically, the outer electron is being ionized and, at the same time, the core-electron is driven in a Rabi oscillation. Note that the splitting would be reflected in Rydberg states as well [11]. Actually, the time-evolution of a Rydberg wave-packet under strong coupling of the core-electron gives rise to very interesting effects as discussed by Hanson and Lambropoulos [15]. For the phenomenology of strong-field optical resonance in two-level systems, the reader is referred to the literature [16]. We just summarize the general behavior of the photoelectron energy spectrum "on" resonance. At low intensity, the spectrum would consist of a single energy peak, as the intensity increases the peak will be symmetrically split by an amount proportional to the square root of the intensity. One should also recall that "on" resonance, the states are actually a linear superposition of bare states, thus any labeling of the split components by bare state quantum numbers is arbitrary.

The experimental realization [5] of this phenomenon uses a two-photon ionization scheme, as illustrated in Fig. 1. The initial bound state is the calcium ground state $4s^2\ {}^1S_0$, and the first and second continua are the $|4s, \epsilon\rangle$ and $|4p_J, \epsilon\rangle$, respectively. Grobe and Eberly [10] showed that these are the minimum ingredients necessary to generate the effect. As a first approximation, the continuum-continuum coupling can be estimated by using the known bound $4s \rightarrow 4p$ transition strength in Ca$^+$. The dipole strength is approximately 1.5 atomic units (a.u.), which for a moderate field strength of 3×10^{-3} a.u. (intensity equal to 300 GW/cm^2) yields an easily observable Rabi splitting (Ω) of about 120 meV. The $4s \rightarrow 4p$ ionic transition frequency is approximately 25300 cm^{-1} ($\lambda \sim 395$ nm) and neutral calcium is ionized by absorbing two of the corresponding blue photons. Additional levels present in the calcium atom (Fig. 1) which are also coupled by the laser field complicate the minimal model described above. For instance, the $|4p, \epsilon\rangle \Leftrightarrow |5s, \epsilon\rangle$ and the two-photon $|4s, \epsilon\rangle \Leftrightarrow |5s, \epsilon\rangle$ couplings can be of the same order of magnitude as the $|4s, \epsilon\rangle \Leftrightarrow |4p, \epsilon\rangle$ coupling. The presence of additional peaks may be traced to the influence of the fine structure of the $4p$ ionic state [17]. Likewise, calculations [13] show that neutral resonances such as the $4s^2 \to 4s4p$ transition can also contribute.

The experiment uses a frequency-doubled, regeneratively amplified titanium sapphire laser which produces tunable (380-405 nm), 180 fs pulses. The pulse bandwidth (~ 15 meV) is less than twice the transform limit and the intensity fluctuations are $\leq 6\%$. Spectral measurements were made on the fundamental light with a monochromator and an optical multichannel analyzer calibrated with a krypton arc lamp. The spectral resolution was 0.5 nm. The calcium was produced in an 775 K atomic beam and background contamination was less than 0.01%. Various lenses with f-numbers ranging from 7 to 25 focused the light into the atomic beam. The laser's confocal length exceeded the atomic beam's cross-sectional length, ensuring a flat intensity distribution in the interaction volume. Electron energy analysis was performed with a time-of-flight spectrometer with 2π solid angle collection and an energy resolution of 30 meV.

Figure 2 shows the change in the photoelectron energy spectrum (PES) with increasing photon energy (bottom to top) for the low and high intensity limits. In the low intensity perturbative limit [Fig. 2(a)], the spectra reflect 2-photon excitation to an unperturbed $4s$ ion ground state. The electron emission is confined in a narrow peak (~ 30 meV width) centered at $(2E_\omega - IP)$, where IP is the neutral calcium ionization potential (6.11 eV). The high intensity spectra shown in

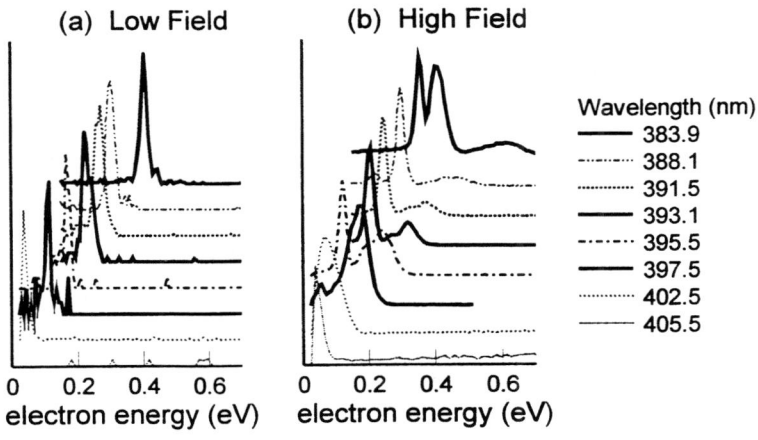

FIGURE 2. PES for calcium at several wavelengths at intensities of (a) 10 and (b) 300 GW/cm^2. The solid lines correspond to "on" resonance spectra.

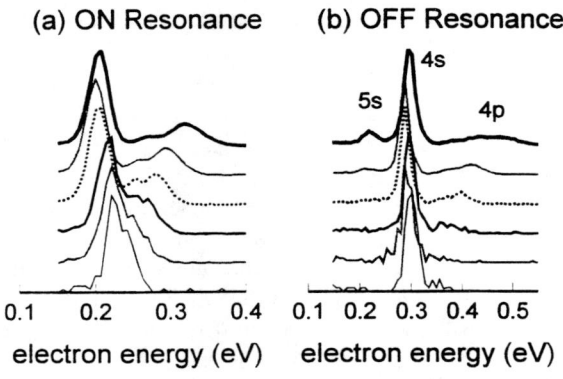

FIGURE 3. Electron spectra (a) "on resonance" ($\lambda = 393.5$ nm) and "off resonance" ($\lambda = 388.1$ nm) for different intensities labeled in fraction of the saturation intensity (300 GW/cm^2).

Fig. 2(b) are taken at the 2-photon saturation intensity (300 GW/cm^2) for neutral ionization. Examination of the (m/q) resolved total yields reveals that the fraction of Ca^{2+} present at this intensity is negligible ($< 10^{-3}$). Several new peaks appear for wavelengths shorter than the $4s \rightarrow 4p_{1/2}$ core resonance (397 nm) whose relative amplitudes evolve with the laser wavelength. The most prominent features appear at wavelengths near the 1-photon $4s \rightarrow 4p_{3/2}$ (393.5 nm) and 2-photon $4s \rightarrow 5s$ (383.4 nm) core transitions. The same features are reproduced in the ATI peaks e.g. (2+1)-photon ionization, with amplitudes which are a few percent of those of the main peak for wavelengths shorter than 400 nm.

Figure 3 shows the intensity dependence of the photoelectron energy spectra at constant wavelength. In Fig. 3(a) the laser is tuned "on resonance" with the ionic $4s_{1/2}$-$4p_{3/2}$ transition (393.5 nm) for intensities ranging from about 10^{10} to 3×10^{11} W/cm^2. The Rabi splitting for this intensity range varies from 4 to 120 meV. At the lowest intensity only one peak emerges at the expected energy for the two-photon ionization, with a small shoulder evident on the high energy side. As the intensity increases, the main feature is red-shifted while the shoulder develops into new structures on the high energy side becoming progressively blue-shifted. In fact, the blue shifted structure resolves into a clear doublet, and the relative amplitude of the two components depends on the intensity. The splitting at saturation is 120 meV and scales as the $\sqrt{I}$ ($\propto E$), as expected for an "on" resonance scenario. The laser is tuned "off resonance" in Fig. 3(b). Besides the trivial shift due to the change in photon energy, the intensity dependence of the spectrum is somewhat different: the main peak is basically unshifted, a weak component is increasingly blue-shifted and at the highest intensity, a new feature appears on the red side of the main peak.

The above experiment demonstrate a qualitative behavior as a function of the wavelength and intensity which is certainly consistent with the predictions of the continuum-continuum Autler-Townes model. Our analysis [17] and the work by several other groups [12,13,18] show that it is the complexity of the atomic structure which produces deviations beyond the minimal model [10]. All calculations rely on an "essential states" approximation and give excellent agreement with the above measurements. However, the experiment clearly illustrates the importance of correlation in a strongly-coupled two electron atom and the ability of an intense laser field to modify continuum structure.

Double Ionization in the Strong Field Limit

As formalized by Keldysh [7] in 1965, the character of ionization changes with increasing intensity. In weaker fields a bound electron will be promoted into the continuum by the simultaneous absorption of enough photons to increase its energy above its ionization potential. This is called multiphoton ionization (MPI). However, as the laser intensity increases, a completely different mode of escape becomes possible. At large distances from the nucleus the electrostatic attraction of the ion core can be overwhelmed by the laser's instantaneous electric field, producing a barrier through which the valence electron can tunnel. In this regime a quasi-static tunneling picture becomes appropriate: the laser field varies so slowly compared to the response time of the electron that the ionization rate becomes simply the cycle-average of the instantaneous dc-tunneling rate. In the language of Keldysh, tunneling ionization (TI) becomes dominant when the ratio of the frequency of the applied field to the tunneling rate becomes less than unity.

The Keldysh theory prediction of the evolution to TI in strong fields has been confirmed by various rigorous theoretical methods [19]. However, experimental access to the tunneling regime has been limited, hampering quantitative comparisons with various strong-field models. The reason for this is simply that for visible laser pulses, even as short as 50 fs, ionization depletes the ground state (saturation) before the atom can experience intensities where $\gamma < 1$. Consequently, the majority of experimental studies on neutral atoms exposed to intense, short pulse laser fields have been carried out in the MPI or mixed regime ($\gamma > 1$).

Recently a comprehensive understanding of the underlying dynamics of how a tunnel ionized electron leaves the atom has been achieved. The reader is referred to recent review paper for a more complete treatment [20]. This advance has been driven by significant progress in both experimental and theoretical capabilities. Experimentally, the advent of kilohertz repetition rate, high peak power lasers [21] has provided an essential tool necessary to span the entire intensity range of importance. At the same time numerical solutions of the time-dependent Schrödinger equation have provided accurate and informative views of the excited electron dynamics [22]. The culmination of these is an intuitive model of strong field rescattering [23,24] based on simple quasi-classical notions. Once an electron in a strong field has made the transition into the continuum from its initial bound state, its motion is dominated by its interaction with the external laser field. Approximately one-half of an optical cycle after the electron enters the continuum, the field can drive the electron back into the vicinity of the ion core where it can undergo elastic or inelastic scattering, or be recaptured into the initial ground state by emitting a high energy photon (high harmonic generation). The essential physics underlying the production of the observed high energy photons and electrons is contained in these (re)collision events.

In this section, we describe a series of systematic studies of the strong-field ionization of helium and neon atoms in the tunneling regime. It has been shown [6,25] that because of their large binding energies, these two atoms tunnel ionize ($\gamma \sim 0.5$) near saturation with femtosecond, titanium sapphire pulses. Thus, these atoms form a paradigm for our theoretical and experimental investigation of the subtle consequences produced by the rescattering of a tunneled wave packet with its parent ionic core. Furthermore, these experiments provide unambiguous evidence for double ionization in the tunneling limit, insights into the dynamics and stringent tests for theoretical models.

Let us begin by examining the experimental evidence for double ionization. Figure 4 shows the helium ion yield curves for 160 fsec, 780 nm excitation. Each data point (symbols) contains $\geq$ 60,000 laser shots. Results from five separate scans with three different spot sizes are plotted. It should be noted that the data spans twelve-orders of magnitude in counting range which is only possible due to the enhanced utility of high powered, kilohertz repetition rate lasers. The He$^+$ yield increases nonperturbatively up to a measured saturation intensity of 8×10^{14} W/cm^2. Beyond this point the yield increases as I$^{3/2}$ consistent with an expanding Gaussian focal volume, a purely geometric growth. However, the He^{2+} curve shows the characteristics of two rate kinetics; nonsequential (NS) two-electron production at low intensities ($1.5 - 8 \times 10^{14}$ W/cm^2), a saturated regime ($0.8 - 3 \times 10^{15}$ W/cm^2), sequential production above 3 PW/cm^2 ($He^+ \rightarrow He^{2+}$) and saturation at 8 PW/cm^2. It is this behavior in the He^{2+} yield which suggests the existence of correlated double ionization.

Further inspection of the data reveals some additional clues into the double ionization mechanism. First, Fig. 4 shows the He$^+$ yields calculated by numerically solving the time-dependent Schrödinger equation using a single-active electron (SAE) approximation [23] (solid line) and ac-tunneling (ADK) rates [26] (dashed line). SAE provides the total He ionization rate, including both the multiphoton and tunneling pathways. For the He$^+$ yield, both calculations agree with the data at high intensity but the ADK curve falls below the measured yield at low intensity. By contrast, the SAE results are in agreement over the full dynamic range of the experiment showing the multiphoton contribution becoming increasingly important below 0.5 PW/cm^2. For the He^{2+} yield, the SAE and ADK result in overlapping curves (solid line) and agree with the measured yield above 3 PW/cm^2. All the He^{2+} yield at low intensity is beyond any SAE approximation. Second, note that the He$^+$ and He^{2+}, labeled NS, yields follow each other over ten orders of magnitude in signal, saturating *simultaneously*. This behavior, verified by measurements [6], firmly establishes that the NS production is connected with the depletion of

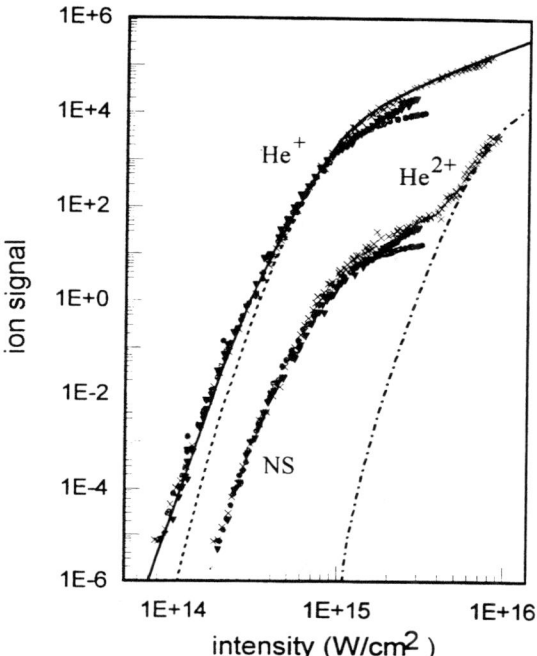

FIGURE 4. Measured He ion yields for linear polarized, 100 fsec, 780 nm light. Calculations are shown as solid (SAE) and dashed (ac-tunneling) lines. The dashed-dotted curve on right is the calculated sequential He^{2+} yield.

ground state neutral helium. Third, the efficient NS ionization is unlikely to be attributable to resonance effects because the helium doubly excited states are well above the first ionization threshold, by over 35 eV, so that they cannot be expected to be strongly excited by the optical field. Finally, the NS rate is found to have a much stronger dependence on the ellipticity of the laser field than the sequential process [27,28] and is essential extinguished with circular polarized light.

A sensitive measure of the NS dynamics is provided by plotting the intensity dependence of the He^{2+}/He^+ ratio, shown in Fig. 5. To ensure accuracy, the two ions were concurrently collected at a fixed intensity and averaged for at least 10^6 laser shots. Although the ion curves in Fig. 4 show a strong intensity dependence varying by 7-orders of magnitude between 0.15 and 5.0 PW/cm^2, the ratio exhibits a gentle slope of only $I^{1.3}$. The ratio is constant (0.0020[3]) from about 5 PW/cm^2 until the sequential production of He^{2+} becomes significant. The onset of sequential He^{2+} production is corroborated by the unambiguous appearance of a high energy tail in the PES beyond 3 PW/cm^2 [6]. Assuming that the NS rate is given by the pure ac-tunneling rate times a constant, which is defined by the measured ratio of He^{2+}/He^+ at saturation, the dotted NS yield curve in Fig. 4 results. Furthermore,

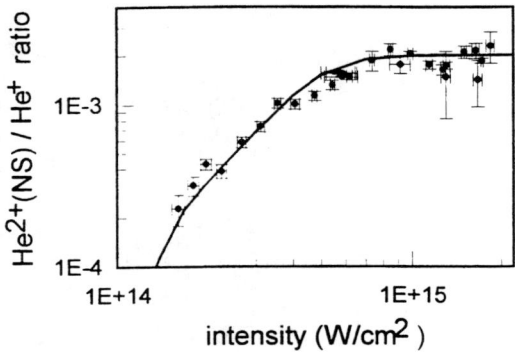

FIGURE 5. Intensity dependence of $He^{2+}(NS)/He^+$ ratio for 780 nm. Error bars indicate 1 standard deviation. Solid line is calculated; see text for details.

the ratio of this curve to the SAE He^+ yield curve produces the solid line in Fig. 5. The striking agreement with the data implies that tunneling is correlated with NS double ionization. We emphasize two differences between tunneling and MPI. First, in tunneling, electrons are emitted in bursts near the maxima in the oscillating electric field while the multiphoton excitation is constant throughout the optical cycle. Second, the multiphoton ionized electrons appear in the continuum near the nucleus whereas tunneling electrons originate at the outer turning point of the instantaneous potential barrier, $6\text{-}10a_o$ from the nucleus. These differences mean that the dynamics of the electrons, after reaching the continuum by these separate pathways, can be significantly different.

We have tested [25] strong-field rescattering using a complete quasi-classical model which incorporates recollision of a field-driven electron with realistic core potentials. We divide the optical cycle into a large number of equal time intervals. The model assumes that at each phase a tunnel ionized wave packet propagates in the combined fields of the laser and the ion core along the classical trajectories. The wave packet is a freely spreading Gaussian which is allowed to have only a single return to the core. We can then calculate the differential elastic scattering cross section for comparison to the measured photoelectron distributions. While the e-2e inelastic process, which leads to the production of double ionization, is calculated using a modified Lotz cross section [29] which accounts for both excitation and ionization. Spatial and temporal averaging is performed for comparison to the experimental measurements.

The calculation yields excellent agreement for the photoelectron energy and angular distributions for both helium and neon. Figure 6 shows the measured (solid line) total PES for helium at 0.8 PW/cm^2 compared to three different calculate curves. The dashed-dotted curve is the result calculated in the absence of rescattering and shows the necessity of the additional electron-core interaction for producing high energy electrons. The dashed curve incorporates rescattering with a realistic he-

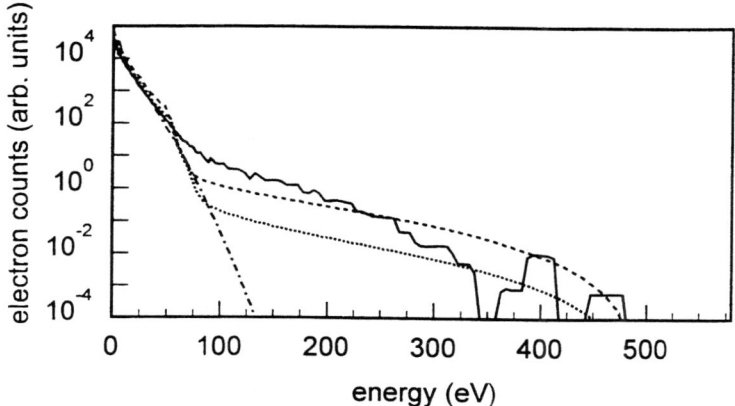

FIGURE 6. Total helium PE energy distribution for 0.78 μm excitation 8×10^{14} W/cm^2. The experimental and calculated distributions using the complete semi-classical theory presented here correspond to the solid and dashed lines, respectively. The dotted line results from pure Coulomb rescattering and the dashed-dotted is without rescattering. The Keldysh parameter, γ, equals 0.5.

lium core potential while the dotted curve is pure Coulomb. The importance of the short range physics is exemplified by the better agreement achieved with a realistic potential. Similar agreement is found at different intensities, as well as for neon ionization. Again, we find that the use of a realistic core potential is a necessity.

The complete quasiclassical calculation, described above, can be used to predict the double-to-single ionization ratio produced from e–2e inelastic rescattering. Figure 7 shows the measured (open circles) and calculated (solid line) ratio for helium and the measured (solid circles) and calculated (dashed line) ratio for neon. These results are computed using the same initial conditions and core potentials used to calculate the photoelectron spectrum of Fig. 6. A modified "field-free" e–2e Lotz cross-section [29] is used to account for double ionization contributions from both core excitation and direct ionization. In these strong fields, the returning electron needs only to excite one of the remaining electrons of the ion in order to produce the doubly charged ion. Any excited state will be immediately ionized when the oscillating field of the laser reaches its next maximum. Also, it has been shown [30] that the use of field-free cross sections is a reasonable approximation since the slowly varying electric field from the laser has a very small effect on the inelastic scattering processes. Clearly, the e–2e rescattering severely underestimates the absolute measured ratio, as well as the shape. The ratio of the experimental to calculated value at saturation is 47 for helium and 5 for neon. The lack of agreement suggests that more than inelastic rescattering is involved in the physics of the nonsequential ionization.

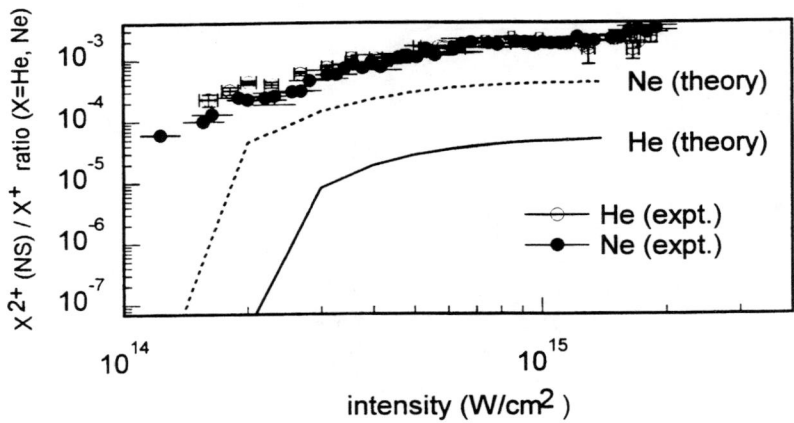

FIGURE 7. Compiled experimental (symbols) and calculated (lines) ratio of nonsequential double ionization to single ionization for helium and neon for 0.78 μm excitation. The helium and neon calculated curves are given by the solid and dashed lines, respectively.

The disagreement between the rescattering predictions and the experiment goes beyond underestimating the absolute value. As seen in Fig. 7, the experimental data shows a smooth decrease in the ratio with decreasing intensity, whereas the calculations show a sharp and abrupt cutoff. The origin of the cutoff is clear, as the intensity is lowered the electron's return energy decreases to the point that it can no longer free the second electron. Again, no such behavior is seen in the experiment. Additionally, it is difficult to rationalize in a rescattering picture why the double ionization ratios would be the same for helium and neon considering the order of magnitude difference in the e–2e cross sections. The good agreement found between the complete quasiclassical calculation and the experimental electron distributions demonstrated the important distinction produced by the atom's short-range potential. Obviously the calculated curves in Fig. 7 reflect the difference in the ionization cross sections, while the experiment does not.

The disagreement between the measured and calculated double ionization ratio could indicate deficiencies in our rescattering model. Certainly our assumptions of the wave packet spread and cross sections which yield accurate predictions for the electron distributions seen in Fig. 6 support the estimates used in our model. Two other relevant approximations are that the e–2e inelastic cross section is unaffected by the external field and that the wave packet has a maximum of one interaction with the core. The consequences of our first assumption was examined using a semiclassical model for helium [30]. It was found that the oscillating field has negligible effect on the collision-induced transition probabilities of the more tightly bound ion core states. This means that using field-free cross sections in model calculations for the intensities that ionize the first electron will yield reasonably accurate results.

A rescattering calculation by Brabec et al. [31] which treats both electrons classically, examines the influence of higher order rescattering. The main conclusion of this study was that, although there is a small probability for impact ionization of the bound electron during the first return of the free electron, inclusion of additional returns can significantly enhance the efficiency of double ionization by trajectories with low drift velocities. This leads to an overall factor of 30 increase in the total NS yield relative to that obtained by considering only the first return. Brabec et al. attribute this enhancement to the refocusing of the trajectories by the Coulomb field so that later returns produce much higher charge density near the nucleus. They conclude that this refocusing overwhelms the transverse expansion of the TI wave packet while it propagates (most of the time) in the region beyond the effective range of the ion core potential.

We can test the importance of refocusing for a real, quantum tunnel ionized (TI) wave packet using the SAE approximation to calculate the strength of the generated high harmonics as a function of time after the wave packet is created. Since the harmonics are produced by transitions back to the ground state, this is an ideal probe of the density distribution of rescattering electrons near the nucleus. We use a constant intensity pulse to produce a TI wave packet during the first half cycle. At this point, the time-dependent wave function is orthogonalized to the ground state and the subsequent evolution represents only that of the excited state component of the total wave function. As this TI wave packet is driven back and forth across the ion core, we can Fourier transform (FT) the dipole matrix element between the wave packet and the ground state. We can consider the spectra generated by different "returns" by restricting the time interval in the FT. We find that the emission rate during the first return is at least a factor of ten stronger than that from the next two cycles, with later returns falling by more orders of magnitude. We must conclude that the Coulomb focusing is not sufficient to explain the substantial enhancement Brabec et al. find in their trajectory calculations. A more likely explanation for their result is that when the TI electron first returns, it can transfer a small amount of energy to the bound electron, becoming trapped in a low-lying "doubly excited" state. These states, which cannot exist in the quantum system, are allowed classically because the density of states is continuous. The captured electron requires more collisions before it can re-escape. This will produce a very large enhancement of the NS ionization yield that would be completely absent in the real, quantum system.

A number of more elaborate two-electron quantum calculations [32–34] have been reported to explain the helium measurements described above. However, treatment of these results are beyond the scope of this paper and the reader is referenced to the appropriate papers. Each of these calculations give reasonable agreement with the helium double ionization for linear polarized light presented above. In fact, the conclusion of Ref. [34] is that rescattering mechanism dominates. However, we feel that considering the more global experimental evidence and some of the model approximations, no convincing theoretical demonstration of any explicit mechanism of strong field double ionization currently exists. Over twenty years has passed since the first observation of Suran and Zapesochnyi [3], yet the question of how electron correlation influences strong-field double ionization remains largely unanswered and remains a significant challenge for future investigations.

ACKNOWLEDGMENTS

This research was carried out in part at Brookhaven National Laboratory under contract No. DE-AC02-98CH10886 with the U.S. Department of Energy and supported by its Division of Chemical Sciences, Office of Basic Energy Sciences, and in part under the auspices of the U. S. Department of Energy at the Lawrence Livermore National Laboratory under contract No. W-7405-ENG-48. L. F. D. and P. A. acknowledge travel support from NATO under Contract No. SA.5-2-05(RG910678).

REFERENCES

1. Fano, U., and Rau, A. R. P., *Atomic Collisions and Spectra*, New York: Academic Press, 1986.
2. Chang, T. N., *Many Body Theory of Atomic Structure and Photoionization*, Singapore: World Scientific, 1993.
3. Suran, V. V., and Zapesochnyi, I. P., *Sov. Tech. Phys. Lett.* **1**, 420 (1975).
4. Lambropoulos, P., Maragakis, P., and Zhang, J., *Phys. Repts.* (1998) in press and references therein.
5. Walker, B., Kaluža, M., Sheehy, B., Agostini, P., and DiMauro, L. F., *Phys. Rev. Lett.* **75**, 633 (1995).
6. Walker, B., Sheehy, B., DiMauro, L. F., Agostini, P., Schafer, K. J., and Kulander, K. C., *Phys. Rev. Lett.* **73**, 1227 (1994).
7. Keldysh, L. V., *Zh. Eksp. Teor. Fiz.* **47**, 1945 (1964).
8. Autler, S. L., and Townes, C. H., *Phys. Rev.* **100**, 703 (1955).
9. Mollow, B. R., *Phys. Rev. A.* **5**, 2217 (1972).
10. Grobe, R., and Eberly, J. H., *Phys. Rev. A* **48**, 623 (1993).
11. Robicheaux, F., *Phys. Rev. A* **47**, 1391 (1993).
12. Grobe, R., and Haan, S. L., *J. Phys. B* **27**, L735 (1994).
13. Hanson, L. G., Zhang, J., and Lambropoulos, P., *Europhys. Lett.* **30**, 81 (1995).
14. Knight, P. L., *J. Phys. B* **11**, L511 (1978).
15. Hanson, L. G., and Lambropoulos, P., *Phys. Rev. Lett.* **74**, 5009 (1995).
16. Allen, L., and Eberly, J. H., *Optical Resonance and Two-level Atoms*, New York: John Wiley & Sons, 1975.
17. Walker, B., Sheehy, B., Kaluža, M., DiMauro, L. F., Trahin, M., and Agostini, P., in *Super Intense Laser Atom Physics IV: NATO ASI Series*, Dordrecht: Kluwer Publishing, 1996, pp. 295-304.
18. Haan, S. L., Bolt, M., Nymeyer, H., and Grobe, R., *Phys. Rev. A* **51**, 4640 (1995).
19. Kulander, K. C., and Schafer, K. J., *Multiphoton Processes*, Singapore: World Scientific, 1993, pp 391-409.

20. For a review, see DiMauro, L. F., and Agostini, P., in *Advances in Atomic, Molecular, and Optical Physics 35*, San Diego: Academic Press, 1995, pp 79-120.
21. Saeed, M., DiMauro, L. F., and Tornegard, S., *Laser Focus World* **27**, 57 (1991).
22. For a review of time-dependent methods, see Kulander, K. C., Schafer, K. J., and Krause, J. L., in *Atoms in Intense Radiation Fields*, New York: Academic Press, 1992, pp. 247-300
23. Schafer, K. J., Yang, B., DiMauro, L. F., and Kulander, K. C., *Phys. Rev. Lett.* **70**, 1599 (1993).
24. Corkum, P. B., *Phys. Rev. Lett.* **71**, 1994 (1993).
25. Walker, B., Sheehy, B., Kulander, K. C., and DiMauro, L. F., *Phys. Rev. Lett.* **77**, 5031 (1996).
26. Ammosov, M. V., Delone, N. B., and Krainov, V. P., *Soviet Phys. JEPT* **64**, 1191 (1986).
27. Walker, B., Mevel, E., Yang, B., Berger, P., Chambaret, J. P., Antonetti, A., DiMauro, L. F., and Agostini, P. A., *Phys. Rev. A* **48**, R894 (1993).
28. Dietrich, P., Burnett, N. H., Ivanov, M. V., and Corkum, P. B., *Phys. Rev. A* **50**, R3585 (1994).
29. Lotz, W., *Z. Phys.* **216**, 241 (1968).
30. Kulander, K. C., Cooper, J., and Schafer, K. J., *Phys. Rev. A* **A51**, 561 (1995).
31. Brabec, T., Ivanov, M. V., and Corkum, P. B., *Phys. Rev. A* **54**, R2551 (1996).
32. Becker, A., and Faisal, F. H. M., *J. Phys.B* **29**, L197 (1996).
33. Faisal, F. H. M., and Becker, A., *Laser Phys.* **7**, 684 (1997).
34. Watson, J. B., Sanpera, A., Lappas, D. G., Knight, P. L., and Burnett, K., *Phys. Rev. Lett.* **78**, 1884 (1997).

Recollisions and High-Harmonic Generation

P. L. Knight, A. Patel, M. Protopapas, N. J. Kylstra,
and D. G. Lappas

Optics Section, Blackett Laboratory, Imperial College London, London SW7 2BZ, U. K.

K. Burnett, A. Sanpera, S. Shaw, and J. Watson

Clarendon Laboratory, Oxford University, Parks Road, Oxford OX1 3PU, U. K.

Abstract. We describe the Recollision Model of high-harmonic generation, in which electrons escape from atomic Coulomb binding in intense laser fields by tunneling, are ponderomotively accelerated, and recollide on return to the parent ionic core. High harmonics are generated by Bremsstrahlung, modulated by the coherent periodic nature of the tunneling process. Harmonic plateaux involve the time-dependent dipole induced by the returning wave packets. We show how atomic wave packets are generated and manipulated by strong laser driving fields, and link the recollision to nonsequential ionization.

INTRODUCTION

When atoms are exposed to intense laser light with sufficient intensity to cause tunnel ionization, a rich variety of new physics emerges, with high-harmonic generation (HHG) being one of the most important [1]. In HHG, the strongly driven atom re-radiates energy at odd multiples (because of the symmetry of the laser excitation) of the laser frequency, right into the soft x-ray water window [2]. A "standard model" of HHG has emerged, the recollision model [3], which has two elements: the emission due to a recolliding, ponderomotively driven wave packet generated by tunneling, and the quasi-periodic emission of wave packets in successive cycles. An electron escapes from the suppressed Coulomb binding potential by tunneling, provided the laser intensity is of the order of a few percent of the atomic unit of intensity (3.51×10^{16} W cm^{-2}). Once it has escaped, the electron will oscillate under the influence of the laser electric field; this ponderomotive oscillation is essentially classical. Depending on the phase at which the tunneling electron is produced, this electron will have some probability of returning to the nucleus. If it does so, it experiences a rapid acceleration and emits high-frequency radiation. From a simple analysis of the possible classical trajectories of a free electron in the laser field, the maximum radiated energy (the cut-off) is given by

$$E_c = I_p + 3.17 U_p,\qquad(1)$$

where I_p is the binding energy of the initial state, and U_p is the ponderomotive energy, which for a laser frequency ω_L and amplitude E_L is given by $U_p = E_L^2/4\omega_L^2$. This classical recollision picture agrees with results from numerical integration of the time-dependent Schrödinger equation (TDSE) and with a two-step quantum model [4] known as the strong-field approximation. We have investigated the role of the repeated, periodic nature of the recollision, linking the harmonic spectrum to a periodically modulated Bremsstrahlung spectrum derived from a single encounter [5]. In a pulsed field where the laser amplitude rapidly varies with time, the phase shift within each ponderomotive swing changes as the pulsed field amplitude rises, leading to a blue shift of the emitted harmonics. Quantum interferences between the returning wave packet and the surviving undepleted population in the core turn out to be very important in determining the plateau structure of the harmonic spectrum.

Here we describe the basic theoretical methods used to describe recollisions and high-harmonic generation. We start with numerical methods to integrate the TDSE, firstly for linear laser polarization and secondly we describe the newly developed approach to the TDSE for elliptical polarizations. We show that the single active electron approximation (SAE) works well for the description of multiphoton dynamics for multielectron atoms for many cases, and then turn to a description of how it breaks down in the strong-field ionization of helium, where nonsequential ionization has been observed, indicating the importance of electron-electron dynamics [6]. Harmonic radiation is due to the dipole induced by the interference between the continuum wave packet and the component of the wave packet that remains in the ground state. We discuss the two mechanisms proposed to explain this nonsequential double-ionization, that of "shake-off" and of $e - 2e$ recollision. Fully correlated theoretical models of helium in strong fields are being developed, in one dimension [7] and in three dimensions [8], but are computationally intensive. We describe our simple extension of the Active Electron Approximation, which includes the time-dependent changes on the "inner" electron caused by the tunneling and recollision of the "outer" electron [9], which we have termed the Crapola model [10]. We describe recent developments of the Crapola model linking HHG to the nonsequential double-ionization [11] and how this may be manipulated using elliptic polarizations [5]

NUMERICAL APPROACHES TO INTENSE FIELD WAVE PACKET GENERATION

We are here only interested in the single-atom response: problems of multi-atom response, propagation and phase-matching are not covered here (see [13] for an account of the relationship between the single-atom and bulk responses). It is sometimes convenient to employ a simple 1-D model of the atom to explore basic phenomena (see e.g. [1] and references therein). We then integrate the TDSE numerically, using a Crank-Nicholson method. To avoid the singularity at the 1D core, we use the soft-core Rochester potential

$$V(x) = \frac{-Z}{(a_0^2 + x^2)^{1/2}} \qquad (2)$$

where Z is the nuclear charge and a_0 an appropriate smoothing parameter. In 3D we expand the atomic wavefunction in terms of spherical harmonics

$$\psi(\underline{r}) = \sum_{lm} \frac{1}{r} |lm\rangle \qquad (3)$$

If we assume the laser to be linearly polarized (which it is for HHG studies), the problem reduces to 2-D and may be solved efficiently, for example by means of a split operator method. For multielectron atoms, it often suffices to make the single active electron (SAE) approximation, in which the active electron is driven while the core is frozen [14].

If the laser driving field is linearly polarized, electron trajectories do indeed recollide with the parent ion. But what of the elliptic or circular polarizations? The above numerical schemes fail, as the basis expansion becomes prohibitively large. We have recently implemented a new method [15] for numerically solving the Schrödinger equation in *two* dimensions, which allows us to investigate atom-laser interactions with *arbitrary* laser polarizations. The 2-D Schrödinger equation we solve is

$$i\frac{\partial}{\partial t}\Psi(x,y;t) = [-\frac{1}{2}\left(\frac{\partial^2}{\partial x^2} + \frac{\partial^2}{\partial y^2}\right) - \frac{1}{\sqrt{a^2+x^2+y^2}}$$

$$+ (xE\sin\omega t + \epsilon y E\cos\omega t)f(t)]\Psi(x,y;t), \qquad (4)$$

where $\Psi(x,y;t)$ is the two-dimensional wave function, $f(t)$ the pulse envelope and ϵ the polarization ellipticity. We integrate this equation in the $p.A$ gauge by the split-step method in which the minimal coupling term $(p-A)^2$ is split in two parts. The part of the time-evolution operator containing momentum acts on the wavefunction in the Fourier space and the part containing the atomic potential acts in the configuration space. Each time step is accurate to order Δt^3. Typical runs on a workstation employ a 512×512 square spatial grid for the wavefunction

$\Psi(x, y)$. An absorber is used to remove any part of the wave packet reaching the boundaries, so that artificial reflections are avoided. The remaining norm gives information about the degree of ionization.

In Fig. 1 we show the structure of wave packets produced by driving fields of linear (a) and circular (b) polarization.

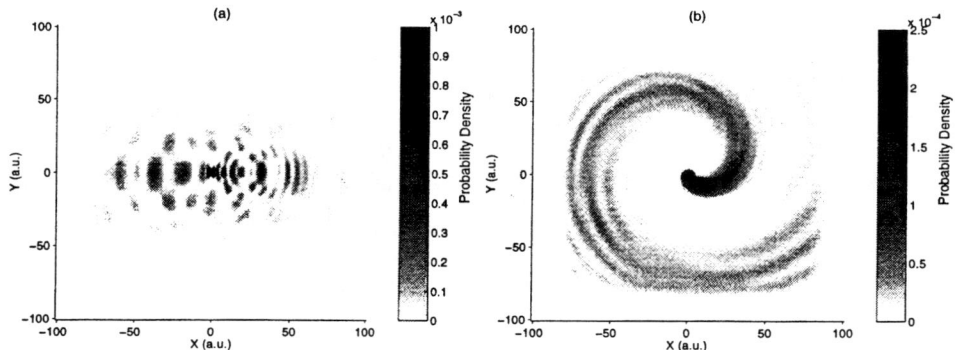

FIGURE 1. Snapshots of the probability density taken after 3.5 cycles for (a) linear polarization and (b) circular polarization with intensity 3.51×10^{14} W cm^{-2} and wavelength 526 nm. Distances (X,Y) are shown in atomic units (a.u.).

THE SINGLE ACTIVE ELECTRON APPROXIMATION AND ITS BREAKDOWN

The numerical calculation of the response of a correlated two-electron atom in an intense laser field is computationally demanding, while for three or more electrons it remains intractable. Fortunately, for most purposes it is unnecessary, as the independent-particle model, in which each electron moves in the mean field of the other electrons, works remarkably well. In this "single active electron" approximation (SAE) we assume the active electron (often called the outer electron) moves in the effective static potential representing the frozen core of all the other electrons, so that the external laser field acts only on the active electron [15]. Once the outer electron is removed by ionization, the laser field can then be assumed to drive an inner electron, which now acts as the new active electron. For the case of helium, we denote the independent-particle SAE wavefunctions for the outer and inner electrons by χ_2 and ϕ_1; they obey the approximate Schrödinger equations

$$i\frac{\partial \chi_2}{\partial t} = (-\frac{1}{2}\nabla_2^2 + V_{\text{eff}}(\mathbf{r}_2) + \mathbf{r}_2 \cdot \mathbf{E}\sin(\omega t))\chi_2(r_2), \quad (5)$$

$$i\frac{\partial \phi_1}{\partial t} = (-\frac{1}{2}\nabla_1^2 - \frac{2}{r_1} + \mathbf{r}_1 \cdot \mathbf{E}\sin(\omega t))\phi_1(r_1), \quad (6)$$

where E is the laser electric field and $V_{\text{eff}}(r_2)$ is the static frozen-core Hartree-Fock potential seen by the outer electron. This approach explicitly assumes sequential ionization.

The SAE has been used very effectively to describe a number of strong-field experiments. A notable example is the comparison of theory and experiment in the generation of high harmonics from rare gas atoms driven by 350fs KrF pulses of peak focused intensities around 10^{17} W cm^{-2} by Preston et al. [16], which demonstrated excellent agreement provided that the response of neutrals and daughter ions were described by the SAE.

Signs that the SAE was not sufficient to describe all observed strong-field phenomena started to emerge with the development of very high repetition rate pulsed lasers, which enabled ionization yields to be measured over a very wide range of intensities. Enhanced double-electron emission associated with nonsequential (NS) ionization was observed by a number of groups [6]; Walker et al. in particular observed a substantial "shoulder" of excess NS ionization from helium in the intensity range where the single-ionization dynamics begins to be well-described by tunneling.

At least two basic mechanisms have been proposed to explain the existence of this "shoulder." The first is the "shake-off" mechanism proposed by Walker et al. [6], based on the idea that the rapid ionization of the inner electron can leave the remaining electron in an excited (or a continuum) state of the helium ion. The second one proposed is a "recollision" mechanism by Corkum [17], which is based on the recollision model of high-harmonic generation, where the outer electron ionizes, then recollides with the core to eject a second electron. We here comment on the relative roles of each of these.

THE CRAPOLA MODEL

As we mentioned in the introduction, fully correlated two-electron models, both in 3-D [8] and in 1-D [7] are being developed in response to experimental developments, particularly the observation of nonsequential double ionization. But given the successes of the single active electron approximation, it is worthwhile to develop a model that builds on the SAE and adds corrections to describe the time-evolving electron-electron interaction. This we have done in the so-called "Crapola" model [9–11] where we have assumed (following Hartree-Fock solutions for the He ground state) that the two helium electrons can be labeled as an outer and an inner electron, and where the outer electron ionizes very rapidly whereas the inner one does so much more slowly. We therefore solve for the outer-electron dynamics using the single active electron approximation, but describe the inner electron dynamics using a time-dependent potential due to the nucleus plus the outer electron.

For simplicity, we describe the Crapola model first in 1-D. We write the two-electron wavefunction as a product

$$\Psi(x_1, x_2, t) = \psi_1(x_1, t)\psi_2(x_2, t), \tag{7}$$

where $\psi_1(x_1,t)$ describes the outer and $\psi_2(x_2,t)$ the inner electron. The outer electron obeys the SAE time-dependent Schrödinger equation

$$i\frac{\partial \psi_1(x,t)}{\partial t} = -\frac{1}{2}\frac{\partial^2 \psi_1(x_1,t)}{\partial x^2} + V_{\text{eff}}(x_1)\psi_1(x,t), \tag{8}$$

where V_{eff} describes the smoothed 1-D interaction and is chosen to reproduce the Helium ionization potential. The inner electron is described by a similar equation except that V_{eff} is supplemented by a time-dependent V_{eff}

$$V_{\text{eff}} = -\frac{2}{(a_2^2 + x_2^2)^{1/2}} + \int dx_1 \frac{\psi_1*(x_1,t)\psi_1(x_1,t)}{(2 + (x_1 - x_2)^2)^{1/2}}. \tag{9}$$

The first term here represents the interaction of the inner electron with the He^+ core (and a_2 is chosen to reproduce the He^+ ionization potential). The second term represents the time-dependent effect of the outer electron on the inner (and is smoothed in the usual manner of Rochester potentials (e.g. [1] and references therein)). Solving the equations for ψ_1 and ψ_2 leads to double ionization yields that clearly contain a shoulder compared with the SAE predictions.

Encouraged by the predictions of the simple 1-D Crapola model, we constructed a full 3-D approach without the need for the rather ad-hoc smoothing required in 1-D models. We have two equations of the form

$$i\frac{\partial \psi_n}{\partial t} = \left[-\frac{\nabla_n^2}{2} + V_n(\mathbf{r}_n, t) + V_{int}(\mathbf{r}_n, t)\right]\psi_n(\mathbf{r}_n, t), \tag{10}$$

where V_{int} is the electron-laser interaction and V_n is the atomic potential. For the outer electron, V_1 is a time-independent effective potential due to the nucleus and the frozen inner electron. For the inner electron, V_2 contains a constant term due to the nucleus plus a time-dependent term due to the outer electron

$$V_2(\mathbf{r}_n, t) = -\frac{2}{r_2} + \int d\tau \frac{|\psi_1(\mathbf{r}_1, t)|^2}{|\mathbf{r}_1 - \mathbf{r}_2|}. \tag{11}$$

We can expand the Coulomb interaction between electrons using

$$\frac{1}{|\mathbf{r}_1 - \mathbf{r}_2|} = \sum_l \frac{r_<^l}{r_>^{l+1}} P_l(\cos(\theta_{12})). \tag{12}$$

The $l = 0$ term represents a time-dependent shielding of the nucleus. The $l = 1$ term is the one that dominates the nonsequential ionization dynamics. We calculate the outer electron dynamics assuming the SAE, solving by a split operator method and using absorbing boundary conditions. At each time step, we calculate the effective mean field experienced by the inner electron due to the outer. Again we see clear evidence of the shoulder of excess double ionization, in the intensity range around 10^{15} W cm^{-2}. Once our theoretical results are spatially averaged over a

Gaussian laser-beam profile and multiplied by the density of atoms, we see, with no adjustable free parameters, a very close fit between our predictions and the results of Walker et al. [6].

The Crapola model allows us to investigate the relative importance of shake-off and recollision mechanisms. In particular, we can inhibit the return of the outer electron by placing an absorber very close to the core, so that the outer electron is ejected but is consumed by the artificial absorber and prevented from recolliding. With an absorber placed around 20 a.u. from the core (noting that at the relevant field strengths the ponderomotive swing α_0 lies between 30 and 150 a.u.), then the He^{2+} yield (the "shoulder") is decreased by an order of magnitude as shown in Fig. 2.

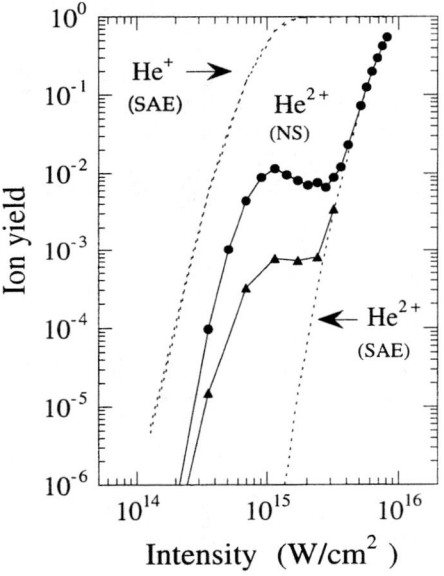

FIGURE 2. The single- and double-ionization yield calculated by the simple three-dimensional model (from Watson et al. 1997), and also the double-ionization yield calculated using the SAE approximation for the same laser parameters as those used by Walker et al. The full circles correspond to the full 3-D model whereas the full triangles correspond to the cases where the recolliding trajectories of the outer electron have been eliminated by an absorbing boundary.

Thus, for these experimental conditions (wavelengths, pulse-lengths and peak field intensities), recollision seems to be the dominant mechanism for nonsequential ionization. This may be surprising, as it is known that a simple approach to recollision-driven NS ionization fails: if the cross-sections for $e - 2e$ collisions are used together with the known energy distribution of the returning electrons, no substantial NS ionization is predicted. This is because the energies of the returning electrons seem to be too low to be effective in knocking out secondary electrons. One clue to understanding this puzzle is to note the importance of quantum interference between the returning wave packet and that part of the initial-state wave packet that had not tunneled out: such interferences are known to lie at the heart of the plateau structure of the high harmonics [12]. We have investigated the role of this interference within a simple model where we can manually suppress interferences: again we [11] see a huge enhancement of the low-energy $e - 2e$ process when interferences are included.

NON-SEQUENTIAL IONIZATION AND HIGH HARMONICS

There is, as a careful reader of the previous section of this paper have deduced, a close link between nonsequential ionization and high-harmonic generation. If we expand the Coulomb interaction term between the electrons in the Crapola model as in Eq. (12), we obtain a multipole series of terms driving the electron-electron interaction. In the previous section, we described how the dipole $l = 1$ term dominates the nonsequential ionization. We can write it as

$$V(l=1) = z_2 \int \frac{\cos(\theta_1)}{r_1^2} \psi_1*(\mathbf{r_1},t)\psi_1(\mathbf{r_1},t) d\mathbf{r_1} \tag{13}$$

which by Ehrenfest's theorem can be expressed as

$$V(l=1) = z_2 \left\langle \frac{\partial V_{coulomb}}{\partial z} \right\rangle = z_2 \langle \ddot{r}_1 \rangle \tag{14}$$

But this is precisely the term that drives the radiation spectrum of the outer electron. In other words, the strong dipole field from the outer-electron dynamics acts on the inner electron to generate nonsequential ionization, as well as generating a radiated field.

DRIVING ATOMS WITH ELLIPTICALLY POLARIZED LASER FIELDS

Finally, we bring together the two themes of this paper: arbitrary polarization fields driving atomic wave packets, and recollisions resulting in high-harmonic generation and nonsequential ionization. In a linearly polarized field, the emerging electron is driven along the polarization axis and must re-encounter the atomic core. But if we use elliptical polarization, there will be a transverse component to the electric field, so that the oscillating electron can "miss" the atomic core, reducing both high-harmonic generation and nonsequential double ionization. Dietrich *et al.* [18] have demonstrated experimentally how both high-harmonic generation and double-electron ionization depend strongly on laser ellipticity.

We have recently implemented a conjunction of our arbitrary polarization code and our Crapola time-dependent electron-electron interaction to investigate the role of ellipticity on high-harmonic generation and nonsequential ionization. We use a split operator method on the Edinburgh PCC CRAY-T3D parallel machine, and our preliminary results confirm the strong ellipticity dependence seen in experiments.

CONCLUSIONS

We have shown in this article how quite simple ideas (tunneling, ponderomotive acceleration, recollisions) are capable of describing a wealth of high-intensity phenomena in atomic physics. We have shown how sophisticated wave-packet computer codes contain within them these essential ideas, and describe accurately ionization, high-harmonic generation and nonsequential ionization.

ACKNOWLEDGMENTS

This work was supported in part by the U. K. Engineering and Physical Sciences Research Council and the European Union.

REFERENCES

1. See, e.g., Protopapas, M., Keitel, C. H., and Knight, P. L., *Rep. Prog. Phys.* **60** 389 (1997) and references therein.
2. Chang, Z., *et al.*, *Phys. Rev. Lett.* **79** 2967 (1997); Spielmann, Ch., *et al.*, *Science* **278**, 661 (1997); Schnürer, M., *et al.*, Phys. Rev. Lett. **80** 3236 (1998); see also Krausz, F., *et al.*, *Optics and Photonics News* **7**, 46 (1998).
3. Corkum, P. B., *Phy. Rev. Lett.* **71** 1994 (1993); Kulander, K. C., Schafer, K. J., and Krause, J. L., "Super-Intense Laser-Atom Physics (SILAP III), edited by B. Piraux, A. L'Huillier and K. Rzazewski, Vol. 316 of NATO Advanced Study Institute, Series B: Physics (Plenum Press, New York 1993), p. 95; Corkum, P. B., *et al.*, in Atomic Physics **14** (Proc. 14th International Conference on Atomic Physics), eds. Wineland, D. J., Wieman, C. E., and Smith, S. J., (American Institute of Physics, New York 1995), p. 405.
4. Lewenststein, M., Balcou, Ph., Ivanov, M. Yu., L'Huillier, A., and Corkum, P. B., *Phys. Rev. A* **49**, 2177 (1994).
5. Protopapas, M., Lappas, D. G., Keitel, C. H., and Knight, P. L., *Phys. Rev. A* **53** R2933 (1996); Watson, J. B., Sanpera, A., Burnett, K., and Knight, P. L., *Phys. Rev. A* **55**, 1224 (1997); Lappas, D. G., and Knight, P. L., *Comments At. Mol. Phys.* **33**, 237 (1997)
6. Fittinghoff, D. N. *et al.*, *Phys. Rev. Lett.* **69**, 2642 (1992); Walker, B., *et al.*, *Phys. Rev. Lett.* **73**, 1227 (1994); Larochelle, S., *et al. J. Phys. B* **31** 1201 (1997).

7. Lappas, D. G., Sanpera, A., Watson, J. B., Burnett, K., Knight, P. L., Grobe, R., and Eberly, J. H., *J. Phys. B* **29**, L619 (1996); Lappas, D. G., and van Leeuwen, R., *J. Phys. B* **31** L249 (1998).
8. Parker, J., Taylor, K. T., Clark, C. W., and Blodgett-Ford, S., J. Phys. B **29**, 133 (1996)
9. Watson, J. B., A. Sanpera, A., Lappas, D., Knight, P. L., and Burnett, K., *Phys. Rev. Lett.* **78**, 1884 (1997)
10. Burnett, K., Watson, J. B., Sanpera, A., and Knight, P. L., *Phil. Trans. R. Soc. Lond. A* **356** 317 (1998)
11. Sanpera, A., Watson, J. B., Shaw, S. E. J., Knight, P.L., Burnett, K., and Lewenstein, M., submitted to *J. Phys. B* (1998)
12. Protopapas, M., Lappas, D. G., and Knight, P. L., *Phys. Rev. Lett.* **79**, 4550 (1997); Lappas, D. G., and L'Huillier, A., to be published (1998); Watson, J. B., *et al.*, in preparation.
13. L'Huillier, A., *et al.*, in 'Atoms in Intense Laser Fields', ed. M. Gavrila (Academic, New York 1992) p. 139
14. Kulander, K. C., *Phys. Rev. A* **36**, 2726 (1987).
15. Protopapas, M., Lappas, D. G., and Knight, P. L., *Phys. Rev. Lett.* **79**, 4550 (1997); Patel, A., Protopapas, M., Lappas, D. G., and Knight, P. L. *Phys. Rev. A* (in press), (1998).
16. Preston, S. G. *et al. Phys. Rev. A* **35** R31 (1996).
17. Corkum, P. B., *Phys. Rev. Lett.* **71**, 1599 (1994).
18. Dietrich, P., et al., *Phys. Rev. A* **50**, R3585, (1994).

AUTHOR INDEX

A

Agostini, P. A., 386
Andersen, N., 237
Artoni, M., 42

B

Bartschat, K., 254
Bergquist, J. C., 29
Berkeland, D. J., 29
Bollen, G., 322
Bollinger, J. J., 87
Bosch, F., 344
Briegel, H., 170
Brown, H. N., 334
Brune, M., 209
Bunce, G., 334
Burnett, K., 74, 400

C

Cancio, P., 42
Carey, R. M., 334
Choi, S., 74
Cirac, J. I., 170
Crowe, A., 278
Cruz, F. C., 29
Cushman, P., 334

D

Danby, G. T., 334
Debevec, P. T., 334
Deng, H., 334
Deninger, W., 334
Dennerl, K., 361
Dhawan, S. K., 334
DiMauro, L. F., 386
Dodd, R. J., 74
Druzhinin, V. F., 334
Duong, L., 334
Dür, W., 170

E

Earle, W., 334
Efstathiadis, E., 334

F

Farley, F. J. M., 334
Fedotovich, G. V., 334
Flambaum, V. V., 14
Fujita, J., 223

G

Gerton, J. M., 58
Giron, S., 334
Giusfredi, G., 42
Gomer, V., 118
Gray, F., 334
Grimm, R., 100
Grosse Perdekamp, M., 334
Grossmann, A., 334

H

Haeberlen, U., 334
Hagley, E., 209
Hajdu, J., 377
Hare, M., 334
Haroche, S., 209
Hazen, E. S., 334
Heinzen, D. J., 132
Hertzog, D. W., 334
Huang, X.-P., 29, 87
Hughes, V. W., 334
Hulet, R. G., 58
Hutchinson, D. A. W., 74

I

Inguscio, M., 42
Itano, W. M., 29, 87
Iwasaki, M., 334

J

Jaecks, D. H., 296
Jelenković, B. M., 87
Julienne, P. S., 144
Jungmann, K., 334

K

Kawall, D., 334
Kawamura, M., 334
Khazin, B. I., 334
Kindem, J., 334
Kishimoto, T., 223
Knappe, S., 118
Knight, P. L., 400
Krienen, F., 334
Kronkvist, I., 334
Kulander, K. C., 386
Kylstra, N. J., 400

L

Lappas, D. G., 400
Larsen, R., 334
Lee, Y. Y., 334
Leggett, A. J., 154
Leo, P., 144
Liu, W., 334
Logashenko, I., 334

M

Macek, J. H., 234
Maître, X., 209
Manek, I., 100
McNabb, R., 334
Meng, W., 334
Meschede, D., 118
Mi, J.-L., 334
Miller, J. D., 29
Miller, J. P., 334
Minardi, F., 42
Mitake, S., 223
Mitchell, T. B., 87
Morgan, S. A., 74
Morinaga, M., 223

Morse, W. M., 334
Moslener, U., 100

N

Neutze, R., 377
Nogues, G., 209
Noordam, L. D., 197

O

Onderwater, C. J. G., 334
Orlov, Y., 334
Ovchinnikov, Yu. B., 100

P

Pai, C., 334
Patel, A., 400
Pavone, F. S., 42
Polly, C., 334
Pretz, J., 334
Prigl, R., 334
Protopapas, M., 400
Proukakis, N. P., 74

R

Rafac, R. J., 29
Raimond, J. M., 209
Redin, S. I., 334
Reiter, U., 118
Rind, O., 334
Roberts, B. L., 334
Robicheaux, F., 197
Rosenbusch, P., 100
Rusch, M., 74
Ryskulov, N., 334

S

Sackett, C. A., 58
Sanders, R., 334
Sanpera, A., 400
Schadwinkel, H., 118

Sedykh, S., 334
Semertzidis, Y. K., 334
Serednyakov, S., 334
Shaw, S., 400
Sheehy, B., 386
Shatunov, Yu. M., 334
Shimizu, F., 223
Sidorov, A. I., 100
Solodov, E., 334
Sossong, M., 334
Steinmetz, A., 334
Strauch, F., 118
Sulak, L. R., 334

T

Tan, J. N., 87
Tiesinga, E., 144
Timmermans, C., 334
Trofimov, A., 334

U

Ueberholz, B., 118
Urner, D., 334

V

van der Spoel, D., 377
Van Enk, S., 170
van Wijngaarden, W. A., 305

W

Walker, B., 386
Walther, H., 179
Warburton, D., 334
Wasik, G., 100
Watson, J., 400
Welling, M., 58
Wieman, C. E., 1
Wiese, L. M., 296
Williams, C. J., 144
Wineland, D. J., 29, 87
Winn, D., 334
Wouts, R., 377
Wunderlich, C., 209

X

Xu, Q., 334

Y

Yamamoto, A., 334
Yenen, O., 296
Young, B. C., 29

Z

Zielonkowski, M., 100
Zimmerman, D., 334
Zoller, P., 170
zu Putlitz, G., 334